# GLOBAL CHALLENGE

# Political Map of the World

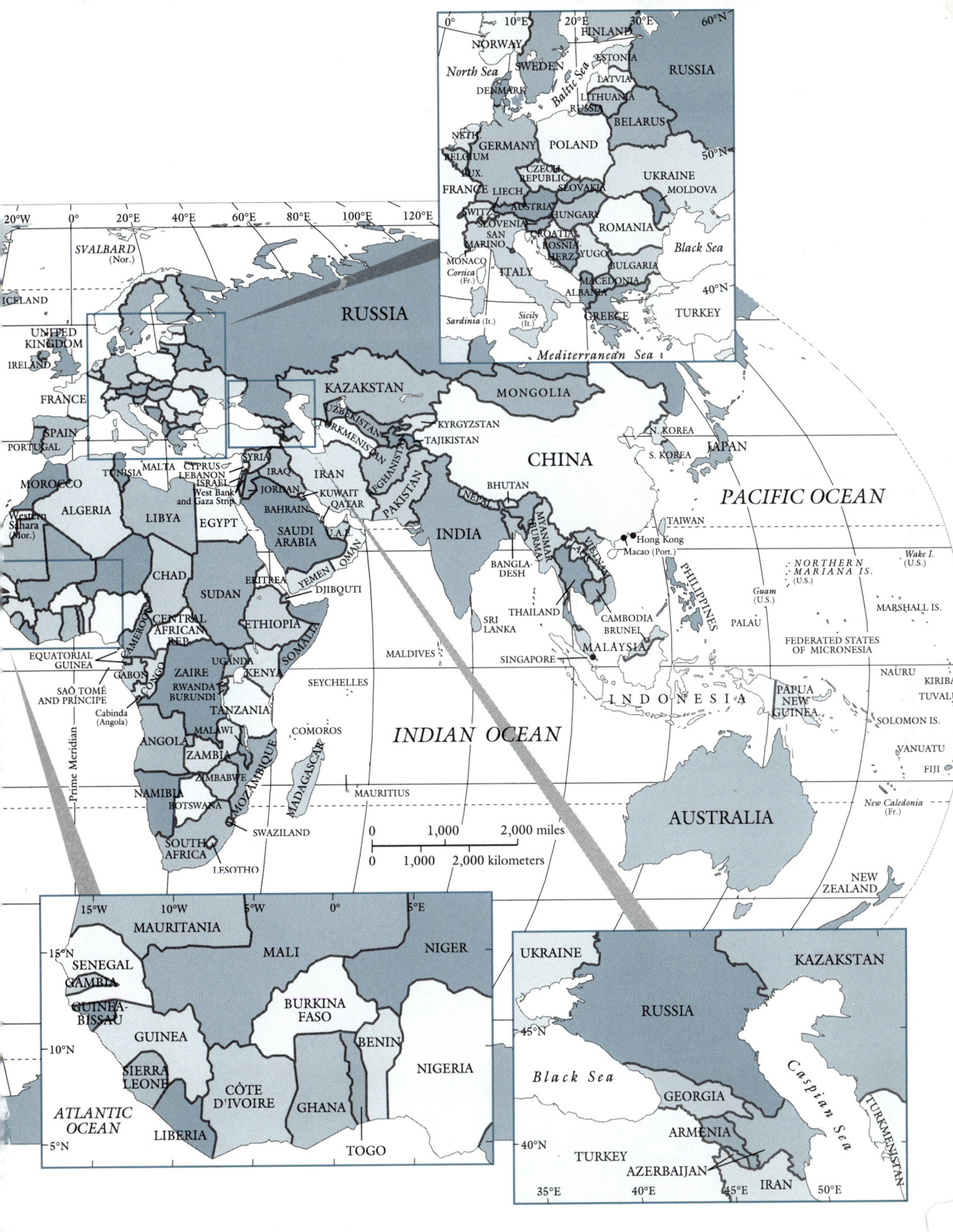

# GLOBAL CHALLENGE
## *Change and Continuity in World Politics*

**LLOYD JENSEN**
Temple University

**LYNN H. MILLER**
Temple University

**Harcourt Brace College Publishers**

Fort Worth   Philadelphia   San Diego   New York   Orlando   Austin   San Antonio
Toronto   Montreal   London   Sydney   Tokyo

|  |  |
|---|---|
| **Publisher** | Christopher P. Klein |
| **Senior Acquisitions Editor** | David Tatom |
| **Developmental Editors** | Christopher B. Nelson |
|  | Fritz Schanz |
| **Product Manager** | Steve Drummond |
| **Senior Project Editor** | Charles J. Dierker |
| **Production Manager** | Jane Tyndall Ponceti |
| **Art Director** | Brian Salisbury |

ISBN: 0-15-501961-9

Library of Congress Catalog Card Number: 96-75656

Copyright © 1997 by Harcourt Brace & Company

All rights reserved. No part of this publication may be reproduced or transmitted in any form or by any means, electronic or mechanical, including photocopy, recording or any information storage and retrieval system, without permission in writing from the publisher.

Requests for permission to make copies of any part of the work should be mailed to:
Permissions Department, Harcourt Brace & Company, 6277 Sea Harbor Drive, Orlando, Florida 32887-6777.

*Address for Editorial Correspondence:*
Harcourt Brace College Publishers, 301 Commerce Street, Suite 3700, Fort Worth, TX 76102.

*Address for Orders:*
Harcourt Brace & Company, 6277 Sea Harbor Drive, Orlando, FL 32887-6777. 1-800-782-4479, or 1-800-433-0001 (in Florida).

Harcourt Brace College Publishers may provide complimentary instructional aids and supplements or supplement packages to those adopters qualified under our adoption policy. Please contact your sales representative for more information. If as an adopter or potential user you receive supplements you do not need, please return them to your sales representative or send them to:

Attn: Returns Department
Troy Warehouse
465 South Lincoln Drive
Troy, MO 63379

Photo credits: Cover, Lenin statue-Patrick Piel/Gamma Liaison; p. 1, Wide World Photos; p. 27, Bettmann Archive; p. 81, AP/Wide World Photos; p. 94, Wide World Photos; p. 118, AP/Wide World Photos; p. 139, Reuter/Bettmann; p. 239, © P. Abril/Gamma Liaison; p. 261, AP/Wide World; p. 302, Reuters/Bettmann; p. 329, © Lauren Van De Stockt/Gamma Liaison; p. 361, AP/Wide World Photos

Printed in the United States of America

6 7 8 9 0 1 2 3 4 5    039    10 9 8 7 6 5 4 3 2 1

**To our mothers**

*In memory of
Edna Fackrell Jensen (1909–1995)
and
Janet Hellwarth Miller (1907–1996)*

# Preface

The challenge of global politics arises out of many factors. The very subject matter seems constantly to grow larger and more complex; new forces are empowering new actors to compete with states on the world stage as never before in the modern world; and the established truths of the past have been upset, while explanations to help guide us into the future apparently are in short supply.

We have tried to respond to each of these challenges while synthesizing what is known about the operation of the contemporary global system. Our goal in writing *Global Challenge: Change and Continuity in World Politics* was to provide the student with a thorough introduction to the theoretical premises, historical setting, structure and process, and most important contemporary issues of world politics. A consistent theme is the extent to which the interstate system itself is being challenged by the changes we explore.

We also chose to write a text that clearly intertwined empirical knowledge about the subject matter of world politics with the normative choices that confront decision makers and private citizens. Nearly every chapter contains two feature boxes—one setting out a number of global changes relevant to the text, the other raising normative dilemmas—with questions for the reader meant to stimulate an awareness of the consequences of values and the choices made in their name.

The book is organized into five sections. Part I ("Perspectives on World Politics") begins, in Chapter 1, by laying out an analytic framework and introducing themes that recur throughout the text. These include levels of analysis and the various theoretical paradigms that grow out of two basic world perspectives: realism and idealism. Chapters 2 and 3 treat the history of the modern nation-state system and the ways in which it is challenged today.

Part II ("Explaining Foreign Policy") turns to the most traditional, state-centric subfield of the subject. Chapter 4 examines the impact of ideas and beliefs on foreign-policy goals, whereas Chapter 5 provides analyses of foreign-policy decision making and the difficulties it presents.

Part III ("Conflict and the Search for Security") approaches the most fundamental problem of societies interacting within a nation-state system, that of group security. Chapter 6 examines capabilities and power; Chapter 7 considers the role of military force, including strategies of deterrence and arms control; Chapter 8 is concerned with the various explanations for violence; and Chapter 9 explores novel developments in the "traditional" subjects of diplomacy and negotiation as a way of commenting on changes in the global system.

Part IV ("The World Political Economy") devotes four chapters to this burgeoning subfield of world politics. Chapter 10 provides an overview of the development of a world economy, including regional and global trading agreements and the international monetary system. The focus then returns to the state level of analysis in Chapter 11, which examines economic statecraft and private economic activity. Chapter 12 is concerned with the problems of economic development in the context of North-South politics, and Chapter 13 then assesses resource and environmental challenges produced by economic development.

Finally, Part V ("The Construction of a Global Society") treats the diverse responses to the growing web of social interconnections that are developing across state lines. Chapter 14 examines the large and contradictory trends of interdependence, integration, and separatism; Chapter 15 explores many of the normative issues raised in the study of international law; and Chapter 16 considers why human rights issues have found their way onto the agenda of world politics. Chapter 17 concludes by examining the extent to which humanity's organizing tendencies are creating a structural framework of peace for global society.

We are happy to acknowledge our debt to the many who have assisted in bringing this book to fruition. At Harcourt Brace, both Fritz Schanz and Christopher Nelson have been intelligent shepherds of our manuscript, and Charles J. Dierker an understanding editor. We are deeply appreciative of the efforts of numerous individuals who read and commented on all or parts of the manuscript, and whose suggestions are often reflected in the current text. Among them are David L. Bartlett, Vanderbilt University; William DeMars, University of Notre Dame; Ralph B. A. DiMuccio, University of Florida; Daniel Druckman, National Academy of Sciences; Larry Elowitz, Georgia College; Richard H. Foster, Idaho State University; John R. Freeman, University of Minnesota; Kathryn Bryk Friedman, State University of New York at Buffalo; Larry George, California State University, Long Beach; Martin Goldstein, Widener University; Steven W. Hook, University of Missouri-Columbia; Philip L. Kelly, Emporia State University; Carmen Barker Lemay, Ball State University; Joseph Lepgold, Georgetown University; Roy Licklider, Rutgers University; Karen Litfin, University of Washington; Fred R. Mabbutt, Rancho Santiago College; Susan McMillan, Penn State University; James Meernick, University of North Texas; Karen Mingst, University of Kentucky; Susan Northcutt, University of South Florida; John W. Outland, University of Richmond; Won K. Paik, Central Michigan University; Arthur W. Rohloff Jr., Texas Tech University; Donald Rothchild, University of California-Davis; Lars Skålness, University of Oregon; Barry H. Steiner, California State University, Long Beach; Kenneth Thompson, University of Virginia; James Wirtz, Naval Postgraduate School; Frank C. Zagare, State University of New York at Buffalo; and I. William Zartman, Johns Hopkins University.

It is a truism to note that teachers are taught by their students. Several generations of our students have challenged us and taught us how to formulate what we understand about world politics in ways that are instructive to them. For their inquiring minds, we are grateful. Three of those students—Kathryn Lambert, Mark Mills, and Jessica White—were able research assistants. We deeply appreciate, finally, the understanding and support we have received from our faculty colleagues at Temple University.

# Table of Contents

Preface   ix
List of Figures, Maps, and Tables   xxi
About the Authors   xxiii

## PART I
## Perspectives on World Politics   1

### CHAPTER 1
### Understanding World Politics   3

Defining World Politics   3
    Terminology: From International to World Politics   5
    National and International Politics: Different yet Alike   6
The Subfields of World Politics   6
    Foreign Policy   7
    Security Studies   7
    Political Economy   7
    International Organization   7
Three Levels of Analysis   8
    Explaining World War II: The Individual   8
    Explaining World War II: The State   9
    Explaining World War II: The World Political System   10
Contending Theories   14
    Two Perspectives on World Politics   14
        Political Realism   14
        Political Idealism   16
    Contemporary Paradigms and Approaches   18
        Realism and Power Politics   18
        Neorealism   19
        Globalism   20
        Pluralism or Neoliberalism   21
        Feminism   21
    How Perspectives Affect Understanding   22
Analysis and Practice: Policy for Whom?   23
Conclusion   24

### CHAPTER 2
### Westphalia and the Nation-state System   25

The Legacy of Rome   26
The Foundations of the Nation-state System   30
The Idea of the Sovereign Equality of States   33
The Utility of the Concept of Equality   34

The Reality of Unequal State Power  36
    The Evolution of Nation-states  37
        European Imperialism and the Double Standard  38
        Small Powers and Great Powers  39
The Twilight of a European International System  41
    The Explosion of Sovereign Statehood  45
    The Problem of Nationalism  45
Conclusion  47

## CHAPTER 3
## Contemporary Challenges to the Nation-State System  48

Nationalism versus Imperialism in the Twentieth Century  48
Westphalia Becomes a Global System  50
    The Troubled Triumph of the Nation-state  56
        The Problem of Tiny States  59
        The Problem of Disintegrating States  61
The Growth of Nonstate Actors in World Politics  63
    Intergovernmental Organizations  63
    Nongovernmental Organizations  66
    Multinational Corporations  67
The Shrinking Planet  70
    The Interdependence of Security  70
    The Communications Revolution  73
    Economic Dependence and Interdependence  74
    The Assault on the Biosphere  76
Conclusion  79

## PART II
## Explaining Foreign Policy  81

## CHAPTER 4
## Objectives, Beliefs, and Foreign Policy  83

Changing Westphalian Objectives  83
    National Interests and Foreign Policy  84
    Normative Values in Foreign Policy  86
The Foreign Policy Role of Beliefs  86
Competing Belief Systems  90
    Marxist-Leninist Beliefs  90
        Soviet Marxism  90
            Gorbachev and the Collapse of Communism  92
            China and Marxist-Leninist-Maoist Thought  93
    Liberal-Democratic Belief Systems  95
        America's Isolationism and Interventionism  95
        America's Moralism  96
    Religious Belief Systems  98
        Islamic Beliefs  98
        The Positive Functions of Religion  101
    The Impact of Belief Systems  102

The Future of National Belief Systems    104
    Nationalism and Multiculturalism    105
    Morality and Foreign Policy    106
    Idealism, Realism, and Moral Foreign Policy    106
Conclusion    109

## CHAPTER 5
# Foreign Policy Decision Making    111

The Rational Actor Model    111
The Individual and Foreign Policy Choice    112
    Personality Traits and Foreign Policy    113
        Aberrant Personalities    115
        Compatibility of Personalities    116
    Motivation of Decision Makers    116
    Perception and Foreign Policy Choice    119
        Historical Experiences    119
        Personal Expectations    121
        Cognitive Consistency    121
        The Problem of Misperception    123
Domestic Structures and Foreign Policy Choice    124
    Democratic versus Authoritarian Regimes    124
    Bureaucracies and Foreign Policies    127
        Bureaucratic Trends    128
        Lack of Creativity and Decisiveness    129
        Bureaucratic Fragmentation of Policy    129
        How Important Are Bureaucracies in Decision Making?    131
    The Role of Public Opinion    133
        Decisional Latitude    134
        Manipulation of Public Opinion    134
Decision Making in a World of Complex Interdependence    136
Conclusion    137

# PART III
# Conflict and the Search for Security    139

## CHAPTER 6
# Capabilities and Power    141

Defining Power    141
    The Meaning of Power    141
    The Elusiveness of Power    143
Tangible Sources of Power    145
    Geopolitical Sources of Power    145
        Geopolitical Theories    146
    Population    148
    Economic Wealth    149
        Measuring Wealth    150
    Military Capabilities    151

Delivery Systems    151
Difficulties in Comparing Military Strength    153
Intangible Sources of Power    154
Leadership and Decision-Making Capability    155
Diplomatic and Strategic Skills    156
Intelligence Capabilities    156
Communication Capabilities    157
National Morale    158
Miscellaneous Intangibles    158
Measuring Power    159
Conclusion    160

## CHAPTER 7
# The Role of Military Force    161

Balance of Power    162
Balance of Power Prerequisites    162
Military Deterrence    163
Communicating Credible Threats    164
Deterrent Strategies    166
Does Military Deterrence Provide Security?    168
Destabilizing Factors in Deterrence    169
Stabilizing Deterrence    171
Continuing Threats to National Security    171
Nuclear Proliferation    172
Conventional Arms Spread    173
International Terrorism    175
History of Terrorism    175
Misconceptions about Terrorism    176
Dealing with Terrorists    177
Arms Control and Disarmament    179
The Relationship between Arms Control and Deterrence    179
The History of Arms Control    180
Arms Control after the Cold War    183
Conclusion    186

## CHAPTER 8
# Explaining Violence    187

History of Global Violence    187
Historical Trends    187
Civil Violence    190
Analysis of Violent Conflict    191
Systemic Explanations    192
Polarity and War    192
Bipolarity and War    192
Multipolarity and War    193
Arms Races    195
National and Societal Explanations    197
Nationalism    198

Positive and Negative Aspects of Nationalism     198
Democracy and War     200
  Why Are Democracies More Peaceful?     201
Domestic Instability     201
The Military-Industrial Complex     203
Individual Explanations     206
  Human Aggression     206
    Violence as an Innate Response     206
    The Frustration-Aggression Hypothesis     206
    Violence as Learned Behavior     207
  Misperception and Miscalculation     208
    Reasons for Misperception     208
Conclusion     210

## CHAPTER 9
# Diplomacy and Negotiation     212
History of Diplomacy     212
  Pre-Westphalian Diplomacy     212
  Early Westphalian and Modern Diplomacy     213
    Strains in Modern Diplomacy     214
    Diplomatic Representation     215
    Functions of Diplomats     216
The Negotiating Structure     217
  Summit Negotiations     218
    Dangers of Summitry     220
  Open and Closed Diplomacy     221
    Problems of Open Diplomacy     222
  Bilateral and Multilateral Negotiation     223
    Negative Aspects of Large-Group Negotiations     224
    Positive Aspects of Large-Group Negotiations     224
    Challenges to the Westphalian Negotiating Structure     225
The Negotiating Process     226
  Nonagreement Objectives of Negotiation     227
  Setting the Agenda     227
  The Role of Concessions     228
  National Negotiating Styles     232
  The Negotiating Style of Women     234
  Negotiating an End to Violence     236
Conclusion     236

# PART IV
# The World Political Economy     239

## CHAPTER 10
# The Development of a World Economy     241
Perspectives on International Political Economy     241
  Mercantilism     242

Economic Liberalism     243
    Hegemonic Stability Theory     243
Marxism     244
The Rise of the Trading State     244
The Debate over Free Trade     245
    Arguments for Free Trade     246
    The Case for Protectionism     248
    Restraints upon Trade     249
Creating a Liberal Trading Environment     251
    GATT and the World Trade Organization     251
    Enhancing Regional Trade     253
        The European Common Market     253
        The European Free Trade Association     255
        The North American Free Trade Agreement     255
    The Future of Regional Trade Agreements     256
    Conflict over Bilateral Trade     257
The International Monetary System     259
    Currency Valuation     260
    The Balance of Payments Problem     261
    Currency Values and the Flow of Trade and Investment     262
Consolidating Capitalism     263
Conclusion     264

## CHAPTER 11

## Economic Statecraft and Private Economic Activity     266

Economic Statecraft     266
    Economic Imperialism     267
        Capitalism and Imperialism     267
    Foreign Aid as an Instrument of Policy     268
    Economic Sanctions     270
        Types of Economic Sanctions     270
        Effectiveness of Sanctions     272
The Role of the Private Sector     275
    Market Forces versus the State     275
    The Multinational Corporation     276
The Impact of Economic Interdependence     278
Conclusion     281

## CHAPTER 12

## North-South Economic Relations     282

The North-South Gap     283
    Modernization Theory     284
    Dependency Theory     285
The Dominance of the Liberal Economic Order     287
    Financing Development: The World Bank Group     287
        The World Bank     288
        The International Finance Corporation     288

        The International Development Association   288
        The International Monetary Fund   288
    Official Development Assistance   289
        The Changing Picture of Development Aid   290
    Foreign Direct Investment   291
IGOs in the North-South Debate   293
    United Nations Development Program   293
    UNCTAD and the NIEO   295
        The Failure of the NIEO   296
Newly Industrialized Countries   297
Communism's Collapse and Economic Development   300
    Economic Development under Communism   300
        "Reform" of Communist Systems in the 1980s   300
    The Failure of Authoritarian Economic Planning   301
    Economic Liberalism Uncontested?   302
A 1990s Assessment   303
Conclusion   305

## CHAPTER 13
Resource and Environmental Challenges   307

Scarcity versus Development Demands   308
    Oil Consumption and Oil Depletion   308
    Economic Growth versus Nature's Limits   309
    The Pressure of Population Growth   311
Development's Threat to the Planet Earth   315
    Global Warming   315
        Burning Fossil Fuels   315
        Deforestation   316
        Global Action   317
    Destruction of the Ozone Layer   319
        Ending CFC Production   320
The Challenge to the Global Commons   322
    Laissez-Faire and Freedom of the Seas   322
        1982 Law of the Sea Treaty   323
    Antarctica as a Global Commons   324
How to Manage the Environment?   325
Conclusion   326

# PART V
## The Construction of a Global Society   329

## CHAPTER 14
Interdependence, Transnational Integration, and Separatism   331

Interdependence, Integration, and Amalgamation   332
The European Union   333
    Proposed Amalgamation after World War I   333

The Functional Approach after World War II 334
    The European Coal and Steel Community 334
    The European Economic Community 335
The Institutions of the European Union 335
    The Treaty of Maastricht 337
The EU: Assessment and Projections 338
Other Trends in Economic Cooperation 340
    MNCs in the Global Village 341
The Disintegration of Multinational States 345
    Separatism within the Former Soviet Empire 346
        Imposed Unity and Its Aftermath 346
        The Emergence of Pluralism 346
        The Example of Western Economic Progress 347
    The Result of Separatism 348
        Constraints against Intervention in Separatist Conflicts 348
        The Lessons of Bosnia 349
Who's in Charge of the World? 350
    From Geopolitics to Infopolitics? 351
        Europe's New Economic Motors 351
        Leading the Infopolitical Way 353
Conclusion 353

# CHAPTER 15
# International Law and Global Order 355

Domestic versus International Law 356
    The Hierarchy of Domestic Law 356
    The Lack of Hierarchy in International Law 356
        The Absence of World Government 357
        Three Purposes of International Law 357
The Rights and Duties of States 358
    The Ground Rules for States 358
    Codified and Customary Law 359
        Reciprocity 359
The Impact of State Power on Global Order 360
    The Special Roles of Great Powers 361
        Great Power Intervention 362
        Why Freedom of the Seas? 363
International Law and Foreign Policy 363
    International Law's Utility in Foreign Policy 364
        When the State's Interest and the Law Coincide 365
        Law as an Instrument of Communication 365
        Law's Use in Conflict Management 365
        Law's Use in Mobilizing Opinion 366
    The Conflict between Foreign Policy and International Law 367
        When Governments Ignore Legal Rules 367
        When Governments Violate Legal Rules 367
        Retaliation and Reprisal 368

Judicial Institutions    369
      Arbitration    369
      The World Court    370
  Conclusion    372

# CHAPTER 16
# Human Rights    374

  Bringing Justice to Human Beings    374
      Creating International Standards for Human Rights    375
          The Universal Declaration of Human Rights    375
          International Covenants on Civil and Political Rights and on Economic, Social and Cultural Rights    376
          Other U.N. Treaties on Human Rights    376
      The U.N. Framework for Enforcing Human Rights    377
      Regional Efforts to Advance Human Rights    380
  A Human Rights Consensus?    382
      The Human Rights Challenge to States    385
      Political Conflict and the Challenge to Human Rights    386
  International Criminal Law    389
      Nuremberg and Tokyo Trials    389
          Crimes against the Peace    389
          War Crimes    390
          Crimes against Humanity    391
      War Crimes Tribunals in the 1990s    392
      An International Criminal Court?    393
  Conclusion    394

# CHAPTER 17
# The World's Structural Framework for Peace    396

  International Institutions for a Shrinking Planet    396
      IGOs: Supporting Sovereignty    397
      IGOs: Challenging Sovereignty    397
  Collective Security and the League of Nations    398
      The Requirements for Collective Security    399
      The Failure of Collective Security under the League    400
          The Ethiopia Case    400
          Successes of the League of Nations    401
  The U.N. Framework for Peace    401
      Alternative Approaches to Peace    402
      Collective Enforcement under the Security Council    403
      Bipolarity in the Cold War    405
          The Korean Conflict    405
              The Uniqueness of the Korean Operation    406
      The Evolution of U.N. Peacekeeping    406
          The Purposes of Peacekeeping    408
              No Act of Aggression is Determined    408
              Use of Interpositionary Force    408

        No Settlement Is Enforced    408
        Operations Are Voluntary    408
        Administration by the Secretary-General    409
    Peacekeeping during the Cold War    409
    Peacekeeping after the Cold War    410
    The United Nations, Collective Security, and the Gulf War    413
        The Imposition of Sanctions    413
        The Armed Attack against Iraq    413
        Lessons about Collective Security    413
    Improving the United Nations' Ability to Keep the Peace    414
Securing Peace through Regional Frameworks    415
    Europe: The Cold War and After    415
        The North Atlantic Treaty Organization    416
        The Warsaw Pact    417
        The Cold War Peace Structure    417
        European Security Structures after the Cold War    418
    Other Regional Structures for Peace    419
        The Organization of American States    419
        The Middle East, Africa, and Asia    420
The Construction of International Regimes    421
Conclusion    422

Glossary    425

Author Index    437

Subject Index    441

# List of Figures, Maps, and Tables

## FIGURES

2.1  Growth in the Number of Generally Recognized Nation-states, 1800–1995   46
3.1  World Population Growth, 1750–2100   77
6.1  Nuclear Warheads   152
7.1  International Terrorist Incidents, 1975–1994   177
10.1  World Export Volume, 1720–1992   246
11.1  Largest Suppliers and Users of Net Capital Flows, 1989–1993   276
12.1  Inflows of Foreign Direct Investment, 1988–1993   292
12.2  The United Nations System: Economic Development Agencies and other Organizations   294
13.1  World Population Growth   312
13.2  Carbon Emissions from Fossil Fuel Burning by Economic Region, 1950–1993   316
13.3  Global Average Temperature, 1881–1994   317
13.4  Tropical Deforestation, Fading Biodiversity   318
16.1  Percentage of Countries with Democratic Systems, 1985–1986 and 1995–1996   384
16.2  Percentage of Free, Partly Free, and Not-Free Countries, 1985–1986 and 1995–1996   384
17.1  United Nations System   404
17.2  The United Nations at Age 50   407
17.3  United Nations: Referrals and Success Rates   410

## MAPS

2.1  Europe and the Mideast, 1500   30
2.2  Central Europe after the Peace of Westphalia, 1648   31
2.3  The Changing Contour of Germany   43
3.1  The British and French Empires in 1900   51
3.2  German Mastery of Europe, 1942   52
3.3  Japanese Expansion, 1931–1942   55
3.4  British India in 1900, and India, Pakistan, and Bangladesh Today   57
3.5  The Ethnic Patchwork of Nigeria   59
3.6  Nationalities in Former Soviet Republics   62
4.1  Distribution of Muslims throughout the World   100
13.1  Projected World Population in 2030, by Region   313
14.1  European Union Member Countries   337
14.2  Europe's New Economic Motors   352
17.1  U.N. Peacekeeping Operations, Past and Present   412

## TABLES

- 1.1  Theoretical Approaches   19
- 3.1  Independence Dates of New States from 1945 to the Present   53
- 3.2  Size of Annual Product of Selected Countries and Multinational Corporations   69
- 3.3  World Population, Total and Annual Addition, 1950–1994   78
- 4.1  Comparative Beliefs   101
- 6.1  Great Powers in History   144
- 6.2  Gross Domestic Product, 1994 and 2020   150
- 7.1  Major Arms-Control Agreements   181
- 7.2  U.S. and Russian Nuclear Warheads   185
- 8.1  Battle Deaths in War, 1815–1994   189
- 8.2  Armed Conflicts by Type and Region, 1945–1995   190
- 8.3  Minorities at Risk of Conflict   199
- 10.1  The Balance of Payments Ledger   262
- 12.1  Relative Percentage Shares of World Manufacturing, 1750–1900   283
- 12.2  Selected Official Development-Assistance Contributors, 1993   290
- 12.3  Geographical Distribution of ODA, 1982–1983 and 1992–1993   291
- 13.1  World Agricultural Energy Use and Grain Production, 1950–1985   309
- 13.2  World Population, Economic Output, and Fossil Fuel Consumption, 1990–1995   310
- 14.1  Regional IGOs Designed for Economic Cooperation   342
- 14.2  The Disintegration of the Soviet Bloc   347
- 16.1  Major Human-Rights Treaties, with Ratifications   378

# About the Authors

Lloyd Jensen received his Ph.D. from the University of Michigan. He has taught at the University of Illinois, Northwestern University, and the University of Kentucky, where he was director of the Patterson School of Diplomacy and International Commerce. Since 1978 he has been a full professor of political science at Temple University. He has also been a professorial lecturer at the School of Advanced International Studies at Johns Hopkins University. His research and publications have focused on comparative foreign policy, bargaining behavior, and national security policy.

Lynn H. Miller received his Ph.D. from Princeton University. He taught at the University of California at Los Angeles and has been a full professor of political science at Temple University since 1984. He has been a visiting lecturer at the University of Pennsylvania and chaired Temple's Political Science Department from 1990 to 1995. His research and writing have concentrated on problems of international order, international law, and organization.

# PART I

# Perspectives on World Politics

Anyone with even minimal awareness of the remarkable events that are occurring in world politics at the end of the twentieth century knows that political change is in the air. That makes the time we live in today an unusually exciting period. Yet, we are told that traditionally in China the wish, "May you live in exciting times!" was offered not as a blessing, but as a curse. While not all agree that exciting times are inevitably more troublesome than times of placidity, the Chinese expression does remind us that a period of fundamental political change challenges our minds and our imaginations in unprecedented ways. Our curse may come if we fail to even try to respond effectively to such challenges.

Effective responses to political challenges are possible only when we can make sense of the seemingly infinite number of events that continually shape world politics. Making sense of things means being able to explain them; only then do arguably sensible responses follow. Chapter 1 will acquaint the reader with some of the most important ways in which analysts have sought to understand—and thus explain and react to—the myriad phenomena of world politics.

Sensible understanding of current world politics also requires us to know something of the historical nature and evolution of political life on the planet. Therefore, Chapters 2 and 3 provide the reader with bird's-eye views of those historical trends most responsible for the global political dynamic of today. There are two principal reasons why a historical understanding is important.

First, by examining world history we can find patterns of behavior and recurrent problems that are relevant to issues confronting us today. Each of us is situated in history and is more or less constrained by what history has made possible. We may be unable to communicate with those who speak only Amharic or Tagalog; yet those linguistic barriers are deeply rooted in history, and help define the separation of communities where those languages are spoken into the nation-states we know today as Ethiopia and the Philippines.

Second, to have a grasp of the basic historical changes that have occurred in world politics also helps us understand trends and, therefore, the potential for changes in the way global problems will be treated today and in the future. The historical record is a story of change—or more precisely, of developmental change, since novel possibilities for human life unfold as the result of past events. The life span for humans has been gradually extended as a result of modern knowledge regarding sanitation and health care; slavery and indentured servitude could be abolished once machines were invented to take the place of much human labor. Both developments have revolutionary consequences for the ways in which our social and political lives are organized, separating us in distinctive ways from the political worlds of the Aztecs or of the Mogul conquerors of India.

So have world politics revealed some constant features over long periods, even while they have developed in novel ways—sometimes quickly, as with the ending of the cold war; sometimes over the course of centuries, as with the gradual development of nationalism into a powerful force across the globe. ∎

*Overleaf* • **A dawn ecumenical service at the Earth Summit, Rio de Janeiro, June 1992.**

# CHAPTER 1

# Understanding World Politics

The last years of the twentieth century will surely be remembered as a time of stunning change in world politics. Among the most dramatic events have been those that swept across much of the vast landmass of Eurasia, bringing pro-democracy demonstrations in Tiananmen Square in Beijing; the dismantling of the Berlin Wall; a bloody war in the Persian Gulf; and a failed coup in Moscow. This last was followed by the rapid dismantling of the Soviet Union, the world's largest remaining empire that had enclosed nearly half the area of Eurasia. It has been mainly the youngest generation that has played the critical role in these and other events that have shaken the world. As the inheritors of these sweeping changes, they and their contemporaries have had to make sense of the implications of these changes in their own lives. This has become all the more important as global developments appear to have a greater and more complex impact upon the lives of more individuals today than at any time in the past.

But to speak of significant world political change in our time clearly does not mean that everything about our subject has been transformed. Indeed, the short-term result of the collapse of Soviet Communism plunged some regions, particularly parts of the Balkans, into the kinds of ethnic conflicts that most resembled those of the start of the twentieth century. One of our central aims is to help our readers assess the nature and extent of the changes that characterize world politics today, so they may differentiate what is likely to remain constant from that which is undergoing unprecedented change in a time of rapid flux.

## DEFINING WORLD POLITICS

It is no longer true, if it ever was, that either "states" or "governments" are the sole actors to be reckoned with in world politics. In the broadest sense, our subject is *political relationships and interactions within a global social system*. More specifically, *the study of world politics focuses on events that are played out beyond the political arena of a particular state*. A number of actors in addition to governments and states may have an impact. Apart from the highest-level governmental officials, these are rarely individuals operating on their own. However, occasionally someone without any official standing whatsoever can influence world politics. One Gavrilo Princip fired a shot in August 1914 that killed the heir to the throne of Austro-Hungary, thus setting in motion the events that we know as World War I.

Various kinds of groups are more typically important. Frequently included among other **nonstate actors** are private associations—sometimes organized as political parties (for example, social democratic parties from a number of countries), sometimes as pressure groups (Oxfam, Greenpeace), and sometimes as economic enterprises (Mitsubishi, Exxon, Shell).

But **associations of governments** themselves may act as a unit in international politics. They may be joined in alliances (the North Atlantic Treaty Organization), in trade or development organizations (the European Union, the Association of South-East Asian Nations), or in specific or general-purpose organizations uniting most of the governments of the world (the International Labor Organization, the United Nations).

Cooperation and conflict both are the stuff of world politics, which encompasses a continuum of behaviors. At one pole stands the organized violence of warfare, with a variety of cooperative actions across state boundaries at the other pole. Like all politics, world politics is about the way human beings, usually organized in groups, strive to advance certain collective values in opposition to those of others. When the effort is undertaken in concert with like-minded groups (parties, corporations, governments, and so on), cooperation is the result; when it opposes the goals of others, the outcome is conflict.

**Conflict** itself takes many forms. It may be present in an exchange of notes between foreign ministries, in the raising of tariffs on goods imported by one country from another, in the effort of a government to stem a flood of refugees from entering its country, in protests over the treatment of ethnic minorities by a foreign government, in disputes over who controls scarce water or other resources, and much more. Occasionally, conflicts over which policies and interests shall be advanced lead to the form of violence we know as warfare.

**War** remains the greatest problem of world politics because of its impact on human life. Yet, international war, defined as the organized killing of citizens of two or more states in the pursuit of conflicting governmental policies, has long been regarded as an integral, if tragic, force in world politics in the sense that it is a last resort of foreign policy. According to a famous nineteenth century Prussian strategist, "War is the continuation of policy by other means."[1]

**Cooperation,** too, occurs in many ways. States cooperate for their security interests when they form alliances. They engage in economic cooperation when they

---

[1] Carl von Clausewitz, *On War*, ed. and trans. Michael Howard and Peter Paret (Princeton: Princeton University Press, 1976), 87.

facilitate trade and the flow of money. A growing number of international organizations are generally meant to provide vehicles for intergovernmental cooperation. But so are many nongovernmental efforts to bring together groups of people from different societies to advance particular shared political interests—from opposing nuclear testing to preventing the extinction of elephants.

In world politics, as in all politics, cooperation and conflict among human beings are seldom very far apart. Friends, lovers, and allies all may quarrel even though their fundamental or underlying behavior is cooperative and mutually supportive. Conversely, even enemies may cooperate during war, such as when they agree to treat each other's prisoners humanely or refrain from using some weapon available to them (poison gas was an example in World War II).

## TERMINOLOGY: FROM INTERNATIONAL TO WORLD POLITICS

Until recently, most books on this subject referred to **international politics.** For the past several centuries, our subject matter essentially involved the interactions of nation-states. Today, however, that label seems too restrictive. Instead, **world politics,** which focuses upon the contemporary global situation, seems more appropriate because of growing evidence that nation-states no longer securely hold the world stage as the only actors capable of affecting the political drama.

*International* literally means *between nations.* Yet states (traditionally, our principal "actors") may or may not always be nations. A **nation** is actually a cultural term, referring to a large group of people who regard themselves as constituting a distinct community. A **state,** in contrast, connotes a particular political unit with a **government** capable of commanding those who live within it. To call a state **sovereign** means that its government recognizes no legal authority higher than itself, although it presumably acknowledges that the many other sovereign states in the world likewise are not bound by a higher law.[2] While it would clearly be more accurate for most purposes to speak of *interstate* rather than *international* politics, we all accept the convention that has made the latter term more common.

Frequently, it is difficult to know whether a reference is literally to a whole "state" in the sense just defined, or merely to the government that speaks and acts on its behalf. Further, much of the writing in this field has tended to equate the interests of a state's government with those of all (or perhaps a majority) of those under its rule. This no doubt reflects our tendency to suppose, unthinkingly, that governments do act on behalf of their societies. But even in states whose governments are comparatively representative, there is dissent from foreign policy, whereas in others the loyalty of citizens may be maintained by fear. The point is that students of world politics need also to consider whether the interests of certain governments are identical to the interests of most (rarely all) of the people in whose name they usually profess to act.

---

[2] For a discussion of how the concept of sovereignty developed in the state system of Europe, see John H. Herz, *International Politics in the Atomic Age* (New York: Columbia University Press, 1962), 49–61.

## NATIONAL AND INTERNATIONAL POLITICS: DIFFERENT YET ALIKE

At the end of the twentieth century, it is easy to see some very characteristic differences between domestic or national and trans- or international politics. The life of every nation-state is shaped by a government that typically can maintain law and order through control of a police force and other governmental institutions. In contrast, world politics take place where no police force exists because no central government is present to rule the earth, where today more than a hundred and eighty separate governments of nation-states compete and cooperate with one another. A well-run state is characterized by an ability to provide civil and political order, in contrast to the seeming anarchy of the global arena that still provides the breeding ground for war.

The following two generalizations appear to be contradictory because they emphasize different truths about our political life. The first, as just described, points out that there are *fundamental differences* between interstate relations on the one hand and domestic or national political life on the other. These differences are expressed in different political and social systems at national and international levels—different structures, values, and rules of behavior whose implications must be explored to understand the global political realm.

The second generalization is that there is an even more *fundamental unity* of all political and social behavior. That is, all of the achievements of domestic tranquility, justice, and social order that have been accomplished by the most "civilized" of states are the products of human intelligence and ingenuity. All presumably can be advanced among human societies (as some already have been) that wish to live together in greater peace and justice. That basic distinction between domestic and international political life turns out to be a difference only of degree. Violence and threats of anarchy never entirely disappear from any human community, and even the relatively more anarchic realm beyond the state is by no means in a constant state of war.

When we emphasize the basic difference in our expectations about domestic and international political behavior, we tend to focus on a description of humanity's experience on earth that was generally, but by no means universally, accurate throughout most of recorded history.[3] When we stress those qualities that give all political life a fundamental unity, we are likely to note possibilities for solving international political problems—that is, those that bring nation-states into conflict—by applying techniques of conflict resolution that humankind has learned in the interaction of smaller social groups.

## THE SUBFIELDS OF WORLD POLITICS

The scope of our field, always broad, has been widening as a number of actors in addition to states take their place as important players on the world stage. There

---

[3] Medieval Europeans, from about the fifth through the fourteenth centuries, would not have been able to make so clear a distinction between *domestic* and *international* politics, for there was little effective government at any level. See Marc Bloch, *Feudal Society*, trans. L. A. Manyon (Chicago: University of Chicago Press, 1961).

is so much information about so many things relevant to world politics that no one can be expected to know—or even try to know—all of it. Therefore, some conventional ways of dividing up the field have developed to facilitate mastering at least some parts of it. This book provides overviews of the most familiar of these: foreign policy, security studies, political economy, and international organization.

## FOREIGN POLICY

This is the most familiar and conventional subfield of world politics. It examines the policies of a specific state or states toward other nation-states. Foreign policy analysis typically is concerned with how a state's government seeks to advance the values and interests of its society, as it perceives them, in a world where others have different goals. It examines the phenomena of world politics in terms of their impact on the state under scrutiny. Foreign policy analysis may seek to understand how domestic societal influences shape the government's foreign policy, just as it may explore how the actions of external groups, organizations, or governments have an impact on the policy of a particular state. This state is the "lens" that brings the rest of the world into focus.

## SECURITY STUDIES

Some students of world politics focus on questions of security in the international system. Usually, the emphasis is upon military security—and, often, of particular states or groups of states—since states typically have sought to enhance their security through what they considered to be appropriate levels of military preparedness. Few would agree that the security of particular states—not to mention world society as a whole—is attainable solely through military power, however, so this subfield includes nonmilitary aspects of security as well. Nonetheless, in this subfield the study of war—and the threat and use of force—is the lens through which world politics is viewed.

## POLITICAL ECONOMY

Economic issues have always had a large effect on political interests and, therefore, the interactions of states. Today, the quest for wealth is increasingly global in its dimensions. It has created actors and enterprises that compete with, as well as influence, governments. Our global economic activity, in turn, has produced new issues for the world's political agenda, including the depletion of nonrenewable resources, the environmental costs of economic development, and new social problems connected with the widening gap between the world's wealthy and its poor. As a result, the study of the international political economy has probably never been so powerful a prism through which to analyze world politics as it is today.

## INTERNATIONAL ORGANIZATION

Our world is characterized by the growth of all sorts of transnational organizational ties between particular groups and governments. Much of this growth is the result of our ever-greater ease of communication and association in the modern world. Most of it has been viewed as necessary to make predictable, organized interactions possible where the conditions of modern life require them. Through this international organization lens, the focus of attention is upon the ways in which the global society governs itself in the absence of world government. This subfield examines the workings of such international institutions as the Council of Europe and the International Monetary Fund. But it also may consider how organization impacts upon issues—and vice versa—to determine the effect for world politics.

# THREE LEVELS OF ANALYSIS

A sensible adage says that what you see depends upon where you stand or sit. When referring to world politics, this adage helps us understand why we have no single, grand theory that attempts to explain everything in this field. That is because our explanations are related to our perspectives on the world, which are meaningful only from particular vantage points.

In studying world politics, three different positions provide different "views" of the world and, by the same token, three places, or levels, of analysis. In abstract terms, these are referred to as the *micro-, mid-,* and *macro-analytic* levels. More concretely, these levels often correspond to the individual, the state, and the world system.[4] As we shall see, other units enter into the analysis of world politics today, but these are the most basic and familiar.

Each of these levels of analysis offers a useful "place to stand" from which to examine global politics; but each asks unique questions as a result of these different perspectives. In response, our answers to what kinds of things cause international events will be different. It is those differing answers that frequently prevent honest students of world politics from feeling that they have arrived at some complete, final, and unchallengeable explanation for why such-and-such has happened in the world. Let us consider how we might try to explain the set of events we know as World War II.

## EXPLAINING WORLD WAR II: THE INDIVIDUAL

If asked to identify one person with whom they associate World War II, most Americans would probably name Adolf Hitler. Certainly, the policies of this authoritarian leader of Germany's Third Reich produced the conquests of much of central Europe that, in turn, brought a military response from a coalition of Hitler's enemies and the resulting military engagements of the war. So, from the individual level of analysis, one might interpret the war by explaining Hitler's behavior: how a man of his personality managed to win the support, and even the adulation, of the masses who were mobilized to carry out the conquests of the Nazis. Those interested in these relationships would want to explore what is known, first, about personality for-

---

[4]Two important studies that explore the implications of world politics levels of analysis are J. David Singer, "The Level-of-Analysis Problem in International Relations," in *The International System,* ed. Klaus Knorr and Sidney Verba (Princeton: Princeton University Press, 1969), 77–92; and Kenneth Waltz, *Man, the State, and War* (New York: Columbia University Press, 1954).

mation as a way of explaining Hitler's own militarism, anti-Semitism, and racism and, second, about human psychology more generally to try to understand how this man was able to achieve and maintain the loyalty of a large population.

This example already suggests two paths one might take when the individual is used as the unit of analysis. One leads toward the motivation of specific leaders who have played important roles in particular international events. The other is directed toward identifying what has made those leaders succeed with their followers—the source of their power. Clearly, one could study a number of leaders of a number of countries—Allied as well as Axis, democratic as well as authoritarian—who had important roles in the unfolding drama of World War II. One might want to compare the similarities and differences of these various leaders' appeals to mobilize their populations for war. It might also be useful to try to explain how the similarities and differences in the policies of relevant political leaders reflected the values and goals of their own cultures.

We may seek answers to a great number of questions relating to the behavior of individuals that should shed light on the reasons why World War II unfolded as it did. A sample might include the following: To what extent did Hitler adapt Mussolini's doctrine of Fascism to his own purposes, or, to reverse the question, how was Mussolini influenced by Hitler's outlook and policies toward his neighbors? What prepared Winston Churchill to become an effective wartime leader of Britain, whereas his predecessor, Neville Chamberlain, was not? Why was Joseph Stalin so ill-prepared for the Nazi attack on the Soviet Union? What prompted the militarist leaders of Japan to attack Pearl Harbor without apparent provocation? What was there in Franklin Roosevelt's personality that made him a successful wartime leader? Why did Harry Truman decide to drop atomic bombs on Hiroshima and Nagasaki, and how did that decision affect his relationship with the Japanese and Soviet governments at the war's conclusion?

Questions of this kind are the stuff of political biography, with its focus on the lives of individuals, and of the discipline of psychology. So-called *great man* approaches to history arise from the individual level of analysis, which should also remind us that some political leaders (such as Hitler or Stalin) no doubt will get more of this treatment than do others (e.g., Chamberlain or Philippe Pétain, the marshal of France). But an *analysis from the individual level need not be restricted to the particular names of well-known leaders. It also includes efforts to understand political behavior based upon assumptions about the nature of all human beings.* Political philosophers from Confucius and Plato to Marx and Gramsci have held clearly identifiable views of human nature from which they have formulated their ideas about what is and what ought to be the nature of humanity's political life. Similarly, the much more modern discipline of human psychology has informed the analysis of social behavior generally, and of human interactions as they relate to world politics more specifically.

## EXPLAINING WORLD WAR II: THE STATE

We may also attempt to understand World War II by considering those particular states that became important participants in the war. Much presumably can be learned about the 1939–1945 conflict by examining aspects of relevant state behavior. We might start with Germany and the impact on that society when the Nazis came to power. Or, we might ask how British disarmament after World War I or America's return to isolationism in the same period affected the foreign policies of those particular states, perhaps helping to set the stage for the next war.

As these examples suggest, the focus of our attention at this level may be upon the **foreign policies** of states. More precisely, the effort here is to understand how those policies are the products of factors that arise from within particular states. We may concentrate on how the domestic political process influences decisions made in the name of the state. However, this perspective also allows examination of such **societal** phenomena as a nation's culture and traditions or its level of social, educational, and economic development. *The fundamental assumption underlying research using the state as the unit of analysis is that the actions of states (governments) are crucial to understanding what causes particular international events.*

The following kinds of questions might emerge by examining World War II with the state as the unit of analysis: How did Germany's political system allow Adolf Hitler to become its chancellor? Why was the Nazi leader's push to rearm Germany in the 1930s so successful? What factors in Japanese political culture during that decade allowed a military government to rule that country? What domestic political factors prevented President Roosevelt from substantially aiding the European democracies after they had declared war on Nazi Germany in 1939? What main features in the Soviet defense of its territory prevented the Nazi conquest of the Soviet Union? To what extent did the resistance movements in Nazi-occupied countries assist in the ultimate liberation of those societies by the Allies? One problem for analysis at this level is determining whether to equate the *state* with the *government* that speaks for it in foreign policy. Clearly, these are not the same thing, although it is fundamental to the modern world that it is the governments of states—rather than private groups of citizens or the societies as a whole—that officially interact with one another. That is why much of the foreign policy literature often uses the two terms almost interchangeably.

This usage points out an additonal basic problem in much foreign policy analysis: state reification. To reify something (in this case a state) is to treat an abstraction as if it were a concrete or material entity—often, a human being. Thus, when we say that, in the period leading up to World War II, "Britain sought to appease Hitler" or "Japan set out to conquer China," we can assume that the reference is really to the governmental policies of Britain and Japan. But when we say that "America embraced isolationism" or that "France felt secure behind the Maginot Line," it is less clear whether we are assigning those feelings and actions simply to the governments of the United States and France or to all or most of their citizens. The reification is apparent when we note that no abstraction—"America" or "France"—is capable of "embracing" an idea or even "feeling" a human emotion.

# EXPLAINING WORLD WAR II: THE WORLD POLITICAL SYSTEM

Finally, one may address international politics from a vantage point "above" the globe, so to speak, in order to concentrate on the interrelationships of relevant units seen as parts of a complex system. From this perspective, such details as the

"Come on in. I'll treat you right. I used to know your Daddy."
C.D. Batchelor. *Daily News*, New York, 1936.

personalities of individual leaders, or even the natural resources available in the territory of a particular state, will fade in comparison to the physical relationships between all or groups of states, their connections through alliances or trade arrangements, and other continuing features of state interactions. Among these are assumptions that states are sovereign, that is, that they interact in a system of relative anarchy; that some are more influential than others; and that all are more or less inhibited from intervening in the affairs of others.

However, we are not limited to viewing states as the only units acting in the world political system; commercial enterprises, interest groups, political parties, and international organizations are a few of the additional actors one may notice when looking at world politics from this whole point of view. At this level, we might seek to understand World War II by examining factors that seemed to affect much of the whole political system of the world in the years before 1939. For example, the impact of the Great Depression might account for the rise of economic nationalism in many of the most highly developed states at the time. The Depression might account, too, for a turn to highly authoritarian leadership in those societies most adversely affected by the economic downturn, or perhaps for the appeal of untried ideologies (Fascism, Communism) in certain societies whose old ways of life seemed threatened. Alternatively, a system-wide analysis might note that the greatest *revisionist* powers of the interwar period—that is, those most willing to engage in aggressive foreign policies to achieve their goals—were also those with little or no overseas empires comparable to those of the Western powers that would soon become their adversaries in the war.

The following questions relate to a world political system understanding of World War II: To what extent did the failure of the League of Nations sanctions against Italy in the 1930s embolden the Axis powers to launch the aggressions that produced a second world war? How did the rise of several great-power governments with extreme demands of their neighbors prevent the "normal" reestablishment of a balance of power before World War II? To what extent were the Axis governments' attempts to conquer other territories impelled by the desire of their industrial sectors to acquire new markets? Did the refusal of major Western democracies to intervene in the affairs of others facilitate the outbreak of the war? How did the major belligerents in World War II most effectively employ whatever material resources they commanded to try to achieve victory? Where was naval power used most effectively in the conduct of the war, and what was its impact on the overall outcome?

It is clear that most of these questions do not relate to every state on earth in the World War II period. Analysis at this level may see certain units or actors as more relevant to a particular aspect than others. Just as this perspective shuts out some of the detail visible at closer range, it also permits one to focus on some of the actors in the overall system while ignoring others. *The critical assumption usually underlying the decision to approach the world from this level of analysis is that relevant parts of the system are interrelated in understandable ways.* For example, if your purpose is to examine the place of economic development assistance in contemporary international politics, you will certainly want to know which governments are the major donors and which are the major recipients of aid; but you will also need to learn about the role of commercial banks, about the international organizational structure that administers assistance, and about the impact of such aid on development.

This level of analysis also permits us to look at some specific part of the world whose units, or political actors, are interrelated in a **subsystem.** During the cold war period, students of world politics examined the "Communist international system" or the "Soviet bloc," for the obvious reason that certain states whose governments looked to the Soviet Union for leadership were interconnected in ways that made that bloc an international (but not *the* international) system. During World War II, one might have studied either the Allied or the Axis subsystems—or, perhaps, the belligerent subsystem as a whole—as a way of looking at the interrelationships of all the adversaries in the war.

Finally, we should consider that a systems-level perspective may be growing in importance as our planet figuratively shrinks according to the nature of modern life. This development makes more of the issues than ever before susceptible to a holistic analysis that considers a great many more actors than the states alone. Concern over the prospect of global warming is a clear example. If, in fact, the temperature of the earth is heating up dangerously, the fault does not lie solely with this or that government, and cannot even be remedied by an agreement among all the governments on earth. Rather, global warming is the result of myriad factors, which include rapid population growth (mainly through improved health care), the spread of industrialization, and an increase in the burning of fossil fuels—particularly in the en-

## Explaining the End of the Cold War: Which Level of Analysis?

The following quotations from six different experts each purport to explain why the cold war ended during the last several years of the 1980s. Each uses one of the three levels of analysis just discussed. As you read, identify each comment's level of analysis.

1. "In just less than seven years, Mikhail Gorbachev transformed the world. He turned his own country upside down. . . . He tossed away the Soviet empire in Eastern Europe with no more than a fare-thee-well. He ended the cold war that had dominated world politics and consumed the wealth of nations for nearly half a century" (Robert G. Kaiser, 1992).
2. "No other industrialized state [than the Soviet Union] in the world for so long spent so much of its national wealth on armaments and military forces. Soviet militarism, in harness with communism, destroyed the Soviet economy and thus hastened the self-destruction of the Soviet empire" (Fred Charles Iklé, 1991–1992).
3. "The metamorphosis in the U.S.-Soviet relationship was the result of two interconnected factors: a formal recognition by the Soviet Union that to tackle its extraordinary economic difficulties it had to seek a permanent settlement with the capitalist world, and a growing recognition in Washington that to keep the world stable while it addressed its own economic problems (some the result of Reagan's policies), a deal with the Soviet Union would be highly desirable" (Michael Cox, 1990).
4. "Russia did not lose the Cold War. The Communists did. . . . A democratic Russia deserves credit for delivering the knockout blow to communism in its motherland" (Richard M. Nixon, 1993).
5. "The acute phase of the fall of communism started outside of the Soviet Union and then spread to the Union itself. By 1987, Gorbachev made it clear that he would not interfere with internal experiments in Soviet bloc countries. As it turned out, this was a vast blunder. . . . If Poland could become independent, why not Lithuania and Georgia? Once communism fell in Eastern Europe, the alternative in the Soviet Union became civil war or dissolution. The collapse of the Soviet Union might well be called the revenge of the colonies" (Daniel Klenbort, 1993).
6. "[The end of the Cold War was possible] primarily because of one man—Mikhail Gorbachev. The transformations we are dealing with now would not have begun were it not for him. His place in history is secure" (James A. Baker III, 1991).

Quotations 1 and 6 both use the individual level of analysis; quotations 2 and 4 look at the cold war's end from the perspective of the Soviet state; and quotations 3 and 5 offer explanations by examining aspects of the international system. Note that of the two state-level explanations, quotation 2 blames Soviet government policies, while quotation 4 regards groups within the country—communists and democrats—as responsible. And of the two systems-level explanations, quotation 3 treats the U.S.-Soviet relationship as a system, while the viewpoint in quotation 5 is based on a Soviet bloc or international communist system.

SOURCE: Charles W. Kegley, Jr., "How Did the Cold War Die: Principles for an Autopsy," *Mershon International Studies Review* XXXVIII, suppl. 1 (April 1994), 22–24.

gines of motor vehicles. The causes of global warming must be understood and dealt with across the planet at a host of levels, many of which, as in the above examples, did not traditionally fall within the province of international politics. Yet, such complex phenomena and interrelationships will increasingly affect life in the global village of the coming century.

# Contending Theories

We have noted that it is impossible for anyone to grasp the full reality of global political life and thereby achieve unchallengeable or perfect knowledge about world politics. That is also true for other aspects of our lives; we seek clarity and meaning through what must always be limited and frequently confusing perceptions of the universe. Therefore, human beings create mental constructs to help them arrange the chaotic data of nature so that they can try to understand and assign meaning to their lives. Such constructs may take the form of religious dogmas, which may try to explain the nature and purpose of the whole cosmos, or scientific theories, which focus only on particular dimensions—evolution theory for biology, quantum mechanics for physics, supply and demand theory for economics, harmony for music, and so on.

These mental constructs may be the primary method of cognition, since making sense of things is essential to the learning process. Some of our constructs and theories advance us toward the truth more clearly than do others. But because we frequently cannot be sure of the "truth" in the absence of a testable hypothesis or theory, it is often preferable to judge theories by their usefulness in leading us toward available (or desirable) "truths" than to argue that the theories are themselves true or false. For example, the ancient theory that the sun, moon, and planets all revolved around the earth did not interfere with the development of human beings for thousands of years; in fact, it sufficed during all the centuries when astronomy was limited to what could be seen with the naked eye. But once Copernicus and Galileo, through the invention of the telescope, proved the theory false and concluded that the earth and other planets revolved around the sun, many advances in human knowledge of the universe became possible that could not have been made on the basis of the outmoded theory.

Similarly, in trying to understand human political life on the planet, our explanatory theories must adjust to whatever changes in political capabilities and ideas have rendered old explanations obsolete. One of our ongoing concerns throughout this book is to explore the extent to which traditional ways of looking at international politics remain useful in a world where human capabilities still are changing at an often dizzying rate, and may be altering much of the very content of global political life in novel ways as well.

## TWO PERSPECTIVES ON WORLD POLITICS

It should be helpful to start by considering what have been the two most basic perceptual constructs—or worldviews—for the study of world politics for centuries. Out of these come the more precise theories that contemporary students of world politics have developed to guide their understanding.

**POLITICAL REALISM** At a general level, we know what it means to be "realistic." It assumes taking the facts of one's life as they are and facing them squarely, without wishful thinking about what ought to—but presumably cannot—be. That is the basis for the realist outlook in world politics as well, although it has more specific directives and implications than the everyday sense of the word.

First, typically, *the realist emphasizes the differences between intra- and international political life*.[5] She or he argues that the comparative anarchy of world politics heightens the importance of such material factors as a nation-state's raw power and minimizes the role of moral or legal authority in resolving conflicts among states. Within the state, police officers, judges, courts, and prisons dispense justice; between states, power must be balanced and wars sometimes fought in the effort for each to try to assure a state's own self-interest, since it can hardly be bothered with questions of the greater good of others.

Second, this emphasis typically leads realists to *assume that a double standard is acceptable for decision makers, depending upon whether they are acting at home or abroad*. Such behavior as duplicity in dealing with others, the use of force to achieve goals, or the making of alliances with one's recent foes all may be perfectly sensible, even "moral" for a leader when dealing with foreign governments, in a way that they would not be appropriate for the same leader acting within the nation's domestic political life.

Third, *realists usually view this difference between intra- and international life as largely unchanging over time*. When Thucydides, in the fifth century B.C., analyzed the uses of military capability and diplomatic prowess among the Greek city-states in the course of the Peloponnesian War, he described behavior in a tiny interstate system that is almost exactly duplicated by the realist analysis of the modern global international system.[6]

Fourth, as implied above, *realism is inherently a preservative outlook* in the sense that it tends to regard constancy in the nature of international politics throughout human history as a more powerful and commanding force than the changes in the ways human communities have interacted with one another. Realists draw from the analysis of other realists who wrote in distant times and places—Sun Tzu and Machiavelli[7] are prime examples—out of their conviction that interstate political life is fundamentally the same today as in Greece in the fifth century B.C., or the Italian peninsula in the fifteenth century A.D., when Machiavelli was born.

Fifth, the "real" in realism also refers to a *philosophic outlook that places primacy in the physical or material nature of the universe*, meaning that which is knowable

---

[5] The classic realist writers, not coincidentally, have lived in times and places of considerable interstate conflict—a marked contrast to the comparative stability of the political unit to which they owed allegiance. One of the earliest recorded realist writings, that of Sun Tzu in the fourth century B.C., came out of the "Warring States" period of Chinese history, when the earlier stability of the Chinese state system had broken down. At roughly the same time in ancient Greece, Thucydides' account of the Peloponnesian War was written in the aftermath of a system-wide conflict of the Greek city-states that would permanently weaken the Greek system. And in the early sixteenth century, Niccolo Machiavelli, a Florentine diplomat and the founder of modern realism, was immersed in the warring culture of the Italian city-states.

[6] Thucydides, *History of the Peloponnesian War,* trans. Rex Warner (Harmondsworth, England, and New York: Penguin Books, 1982).

[7] See Ralph D. Sawyer, *Sun Tzu: Art of War* (Boulder, Colo.: Westview Press, 1994). For a discussion of Machiavelli's realist thought, see the introduction by Max Lerner in Niccolo Machiavelli, *The Prince and the Discourses* (New York: Modern Library, 1950), xxv–xxvi.

through the use of the senses. Thus, a political realist is likely to place great store in the tangible or measurable attributes of a state's capability, including its physical size, location, natural resources, productivity, and so on. Since it is these "real" factors that are thought to be the principal influences on a state's behavior (or more precisely, that of its government), it generally follows in realist analysis that a large and populous state, rich in resources, and with an educated and energetic citizenry will likely become a major power in world politics (e.g., the United States). On the other hand, a small, landlocked state may choose neutrality in foreign policy to avoid being drawn into conflict between larger neighbors (Austria since World War II is a good example).

**POLITICAL IDEALISM** Another outlook on the world of politics typically competes with realism and is virtually its mirror image; that is, the characteristics of idealism are more or less the opposite of realism.[8] This term also has a more specific meaning in the analysis of international politics than its everyday usage.

First, political *idealists note the fundamental unity of all political and social behavior from the smallest local or group setting, through the nation, to the world beyond,* rather than emphasizing the differences between domestic and international political life. An idealist may argue that, just as most large political communities today have learned to function in relative peace, so should it be possible to develop more peaceful habits when it comes to international relationships. As one result, the idealist may wish to replicate at the interstate level some of the institutions and procedures available to governments for resolving conflict peaceably.

Second, it follows that *idealists insist upon a single or universal moral standard to judge the behavior of individuals,* whether private citizens or the foreign policy officials of governments. If cheating, deception, and the use of force are regarded as unacceptable in private life, then so should such behavior be ruled out of the conduct of international affairs. The "should" in that sentence is important, since no idealist would argue that these things always *are* ruled out of the conduct of foreign policy; rather, they *ought* to be a matter of consistency in moral conduct. It is often said that idealists are concerned with moral values (or what ought to be) in the conduct of world politics, while realists are not; yet it is more accurate to say that the former hold up the higher single standard of conduct for foreign policy decision makers, whereas realists permit a double standard that demands less of those acting on behalf of the state than is required for individuals in their private lives.

Third, often *idealists emphasize what they see as evolutionary or organic change in world history,* viewing the problems and the behavior of the various political units active today as more or less different from those at earlier stages in history. They might argue that, just as the state has evolved in complex ways from its prehistoric origins, so are interstate relationships changing and evolving in ways that can make the future of world politics look quite different from much of the past.

---

[8]For a classic idealist analysis of world politics, first published in 1795, see Immanuel Kant, *Perpetual Peace: A Philosophical Essay* (Indianapolis: Bobbs-Merrill, 1957). Those trained in international law generally are inclined to an idealist view of world politics, as is evident in Chapter 15 of this book. See, for example, the work of the "father of international law," Hugo Grotius (1583–1645), *The Rights of War and Peace,* trans. A. C. Campbell (New York: M. Walter Dunne, 1901).

Fourth, for all of the reasons implied above, *idealism is inherently progressive in what it expects of international political life.* If human social and political interaction is constantly evolving, then the behavioral patterns of the past will not necessarily be those of the present, nor will those of the present inevitably predict those of the future. To speak of the progressive development of human history literally means only that it evolves and unfolds; it need not equate "progress" with humanity's "betterment" in moral or physical terms, which is the usual way we define that concept. Yet it is certainly the case that an idealist's outlook typically does suppose that the future can be better than the present or the past. The reason for this becomes clearer when we consider the following, most important characteristic of idealism.

Fifth, idealism is tied to an ancient philosophical outlook. Underlying the apparent reality of the material world is *an ideal or perfect realm* that *is only imperfectly manifested in the tangible world accessible to us through our senses.* Human beings can

---

## Realist or Idealist Views of World Politics?

Identify which of the following quotations from a variety of past and present commentators on world politics reflect essentially realist versus idealist views of world politics?

1. "Consequently, the art of using troops is this: when ten to the enemy's one, surround him. . . . When five times his strength, attack him. . . . If double his strength, divide him. . . . If equally matched you may engage him. . . . If weaker numerically, be capable of withdrawing. . . . And if on all respects unequal, be capable of eluding him, for a small force is but booty for one more powerful." (Sun Tzu, *The Art of War,* fourth century B.C.)

2. "All nations are . . . under a strict obligation to cultivate justice towards each other, to observe it scrupulously, and carefully to abstain from every thing that may violate it. Each ought to render to the others what belongs to them, to respect their rights, and to leave them in the peaceable enjoyment of them." (Emmerich de Vattel, *The Law of Nations,* 1758)

3. "It was always clear that a lasting peace (in Vietnam) could come about only if neither side sought to achieve everything it had wanted; indeed, that stability depended on the relative satisfaction, and therefore on the relative dissatisfaction, of all the parties concerned." (Secretary of State Henry Kissinger, at a January 24, 1973, news conference)

4. "An evident principle runs through the whole program I have outlined. It is the principle of justice to all peoples and nationalities, and their right to live on equal terms of liberty and safety with one another, whether they be strong or weak. Unless this principle be made its foundation, no part of the structure of international justice can stand." (Woodrow Wilson, "The Fourteen Points" Address to Congress, January 8, 1918)

5. "And you are to understand that a Prince . . . cannot observe all those rules of conduct in respect whereof men are accounted good, being often forced, in order to preserve his Princedom, to act in opposition to good faith, charity, humanity, and religion. . . . [H]e ought not to quit good courses if he can help it, but should know how to follow evil courses if he must." (Niccolo Machiavelli, *The Prince,* 1532)

6. "I, personally, would wait, if need be, for ages rather than seek to attain the freedom of my country through bloody means. I feel in the innermost recesses of my heart . . . that the world is sick unto death of blood-spilling. The world is seeking a way out, and I flatter myself with the belief that perhaps it will be the privilege of the ancient land of India to show the way out to the hungering world." (Mohandas K. Gandhi, radio broadcast to the American people, September 13, 1931)

grasp this concept as an idea (an "ideal" is the "idea" of perfection), that transcends the material world and, thus, is not empirically verifiable. The ideal cannot be attained in life as we know it, though it may guide us in our moral conduct as the perfection toward which we strive. Among world religions, the Judeo-Christian and Islamic traditions have been particularly influenced by idealist notions that a transcendent reality exists beyond this earth, but it is hard to imagine any body of ethical doctrine that does not assume some concept of the ideal to guide human behavior.

## CONTEMPORARY PARADIGMS AND APPROACHES

When the principles, presuppositions, and expectations about human behavior are fully articulated, providing one's rationale for studying the phenomena of world politics in a certain way, we may refer to that intellectual product as a **theoretical approach** or **paradigm**. Table 1.1 and the following discussion highlight the most dominant paradigms today concerning the analysis of world politics.

**REALISM AND POWER POLITICS** The power politics approach was the dominant paradigm for analyzing world politics for most of the half century following World War II. Based on the realist tradition, its most influential text in the United States was written by a German-born American political scientist, Hans J. Morgenthau. In his *Politics among Nations,* first published in 1948, Morgenthau argued that "the main signpost that helps political realism to find its way through the landscape of international politics is the concept of interest defined in terms of power."[9] Interest, for Morgenthau, meant the interest of the state, for his approach was state-centric in two respects. First, it viewed the world more often from the mid- than from the macro-analytic level and, second, it regarded states as the fundamentally important actors in a world in which they sought to maximize their power.

As the latter phrase suggests, from Morgenthau's vantage point, the world system was characterized by the "ever temporary balancing of (state) interests and the ever precarious settlement of conflicts."[10] That meant that for Morgenthau and other influential realist writers of the 1940s and 1950s, the role of the professional diplomat was all important as the "manager" of those precarious accommodations that could maintain a state's interests in an uneasy world.[11] Understanding the corresponding need for their counterparts in other countries to protect their interests as well, diplomats should seek a **balance of power** to accommodate everyone's interests by allowing no single state to dominate the system.

---

[9]Hans J. Morgenthau and Kenneth W. Thompson, *Politics among Nations,* 6th ed. (New York: Alfred A. Knopf, 1985), 5.

[10]Ibid., 3.

[11]The most influential scholar-diplomat of this period was George F. Kennan, a United States Foreign Service officer whose many writings on world politics criticized what he saw as the "legalism and moralism" of much U.S. foreign policy. See Kennan's *American Diplomacy: 1900–1950* (Chicago: University of Chicago Press, 1951). Kennan was the author of the "containment" policy toward the Soviet Union in the early days of the cold war. That policy was outlined in an article he published anonymously (as "X"), "The Sources of Soviet Conduct," *Foreign Affairs* XXV (July 1947): 566–582. It called for skillful, nonideological diplomacy to counter Soviet efforts to extend their influence.

**TABLE 1.1** Theoretical Approaches

| | DOMINANT LEVEL OF ANALYSIS | DOMINANT THEORETICAL ASSUMPTIONS | IMPORTANT ACTORS IN WORLD POLITICS |
|---|---|---|---|
| Power politics | State | Anarchy at global level; states seek power and influence | State (in a world of states) |
| Neorealism | State and world system | Anarchy at global level; states are constrained by global structure | State (in a world of states and other subnational, transnational actors) |
| Pluralism | Individual, substate, trans-state groups | Groups compete within and across states; transnational groups and structures influential; checks and balances explain outcomes | Substate groups; groups within governments; trans-state groups |
| Neoliberalism | Substate groups, nonstate actors | Groups compete within and across states; economic interdependence; growing governance beyond state | Substate groups; groups within governments, trans-state groups |
| Globalism: dependency | World system | Economic realtionships are important; dependency explains politics | Dominant economic agents (MNCs, economics ministers) |
| world order | World system | World is a single social system; law and organization increase world order | Governments; IGOs; NGOs; treaty regimes |
| Feminism | Individual; groups | State is a male construct; foreign policy is dominated by men; feminism releases alternate values | More than half the world's human population (women) |

**NEOREALISM** Younger scholars in the power politics tradition have refined and modified this approach to the point that they are now often referred to as *neorealists*. Neorealism, like realism, sees the anarchy of the international environment as quite durable, just as it also assumes the primacy of the state in world politics. It modifies the realist viewpoint, however, in several respects.

First, neorealists tend to grant more importance than their predecessors did to the structure of the global system and how it influences the behavior of states. For example, Kenneth Waltz views the bipolar structure of the cold war period as inherently more stable than one in which power is more widely distributed among a number of states.[12] It is clear, in his view, that responsibility for managing power lay with the two bloc leaders, who knew they must pay more attention to each other's moves than to other actors. Second, neorealists recognize that the contemporary world has given rise to organized forums in which states frequently interact (international organizations) and to binding or semi-binding agreements on a host of matters (international regimes). They view these developments as having significantly altered the ways in which the game of international politics is played, although states remain the fundamental decision makers, for no central authority yet overrules state sovereignty. Finally, neorealists are likely to insist that what states seek is their own security in the world, rather than the acquisition of power for its own sake.[13]

Although these varying paradigms out of the realist tradition have dominated the study of world politics for the past half century, the idealist tradition has influenced important theoretical approaches, too. Characteristically, the main focus here is at the macro-analytic or global level. Note the differences in the dominant level of analysis for the following three theoretical approaches. What characteristics do they share suggesting that each is influenced by idealism?

**GLOBALISM** Globalism is a name given to one or more neo-idealist approaches. They share the view of the people of the world as constituting a single, immense social system whose component parts are increasingly interdependent. One must understand the forces at work throughout the world system, of which the actions of states are only a part. Much analysis based on this approach asserts the dominance in the contemporary era of capitalism in world politics.[14] Among the assumptions that may follow are, first, the perennial dependence of the members of some societies upon economic decisions made in others; second, the sense that state sovereignty is being challenged as never before; and third, the need for long-term historical comparison to understand the dynamics of today's world, since both capitalism and the state system are comparatively modern inventions.[15]

Another school of globalism explores the kinds of policy choices available throughout global society for overcoming anarchy by strengthening the rule of law

---

[12]Kenneth N. Waltz, "The Stability of a Bipolar World," *Daedalus* 93 (Summer), pp. 881–909. Waltz expresses his fullest theoretical statement in *Theory of International Politics* (Reading, Mass.: Addison-Wesley, 1979).

[13]For additional neorealist writings, see Hedley Bull, *The Anarchical Society* (New York: Columbia University Press, 1977); Robert Gilpin, *War and Change in World Politics* (New York: Cambridge University Press, 1981); and Kenneth A. Oye, ed., *Cooperation under Anarchy* (Princeton University Press, 1986).

[14]This approach is greatly influenced by Marxist analysis, which—while never very influential in the United States—has been an important paradigm for analyzing world politics in much of the world throughout most of the twentieth century. See Richard Lowenthal, *Marxism in the Modern World* (Stanford, Calif.: Stanford University Press, 1965).

[15]For analysis based on this approach, see Albert Bergesen, ed., *Crises in the World-System* (Beverly Hills, Calif.: Sage Publications, 1983); Gabriel Kolko, *The Politics of War* (New York: Vintage Books, 1968); and Immanuel Wallerstein, *Geopolitics and Geoculture: Essays on the Changing World-System* (New York: Cambridge University Press, 1991).

and order in the world.[16] A concern with policy choices requires an effort to determine what kinds of actions can strengthen humane conditions and practices across the globe. Rather than viewing the well-being of the state as the final arbiter of political action, these globalists attempt to determine what is best (or at least, better) for human development more generally. Often, they prescribe preferred policies based upon that analysis.

**PLURALISM OR NEOLIBERALISM[17]** Pluralism combines elements of both realist and idealist worldviews. It emphasizes the importance in world politics of actors in addition to states.[18] It is mainly distinguished by an insistence that the "state" concept is an abstraction that masks the multiplicity of conflicting forces within it. Similarly, to speak of a state's "foreign policy" is really to refer to a number of decisions made competitively by a number of actors—bureaucrats, interest groups, and elected officials. Similarly, a variety of actors operate across national borders, which are, according to pluralists, comparatively soft and permeable when it comes to giving and receiving influences. It follows, then, that the agenda of world politics goes well beyond the traditional military-security dimension, but includes such issues as pollution, energy use, individual rights, and the appropriate uses of the oceans.[19]

**FEMINISM** Feminism criticizes much of the traditional thinking about world politics on the grounds that it has been biased in favor of males. The experience of men too often is unthinkingly regarded as that of the whole species, even though women are socialized to have somewhat different roles and goals than those of men. An obvious part of the critique is that men have been the dominant decision makers for many centuries—as kings, prime ministers, diplomats, and soldiers. Nearly as obvious are the masculine attitudes and behaviors that have long been at the heart of this subject—conflict, the quest for power over others, militarism, and warfare. A feminist approach tends to see the state itself as a masculine construct. Historically, women have made their influence felt in the family and the community, where they advance

---

[16]Examples of works using this approach are Daniel Deudney, *Whole Earth Security: Towards a Geopolitics of Peace*, Worldwatch paper no. 55 (Washington, D.C.: Worldwatch Institute, July, 1983); Richard Falk, *The Promise of World Order* (Philadelphia: Temple University Press, 1987); and Warren Wagar, *Building the City of Man: Outlines of a World Civilization* (San Francisco: W. H. Freeman, 1971).

[17]In addition to Table 1.1, for some distinctions between pluralism and neoliberalism see Charles W. Kegley Jr., *Controversies in International Relations Theory: Realism and the Neoliberal Challenge* (New York: St. Martin's Press, 1995).

[18]Students of democracy often give the name *liberalism* to pluralism, at least in domestic polities. In this sense, liberal democratic systems are those that expect various groups to compete for the right to rule, and rulership is always contingent upon protecting the rights of minorities so that they may continue to compete for primacy. Clearly, liberalism (or pluralism) has borrowed from both the realist and the idealist traditions, though our use of *liberal* associates it more with the latter than the former worldview. To confuse matters further, *liberalism* and *neoliberalism* in the international political economy refer to the doctrines underlying contemporary world capitalism.

[19]Pluralist analysis includes Robert Axelrod, ed., *Structure of Decision: The Cognitive Maps of Political Elites* (Princeton, N.J.: Princeton University Press, 1976); Robert O. Keohane and Joseph S. Nye Jr., *Power and Interdependence: World Politics in Transition* (Boston: Little, Brown, 1977); and James N. Rosenau, *Turbulence in World Politics: A Theory of Change and Continuity* (Princeton, N.J.: Princeton University Press, 1990).

Gable/*Globe and Mail,* Toronto, Canada/Cartoonists & Writers Syndicate.

values that frequently have not been shared by the state. These include such welfare matters as protection for children, health care, and peaceful solutions to conflicts.[20]

## HOW PERSPECTIVES AFFECT UNDERSTANDING

Although this does not exhaust the list of recent approaches and paradigms relevant to the study of world politics, most others are variants on the approaches examined above. It is our contention that all of us develop worldviews that fit somewhere within the realist-idealist framework. Mature adults understand both the imperfections of the material world and their individual capacity, nonetheless, to strive toward an ideal.[21] At the same time, most nonexperts in the study of world

---

[20]Feminist critiques of the study of world politics include Cynthia Enloe, *Bananas, Beaches and Bases: Making Feminist Sense of International Politics* (Berkeley: University of California Press, 1989); Christine Sylvester, *Feminist Theory and International Relations in a Postmodern Era* (Cambridge: Cambridge University Press, 1992); and J. Ann Tickner, *Gender in International Relations: Feminist Perspectives on Achieving Global Security* (New York: Columbia University Press, 1992).

[21]In the view of an important analyst of the two traditions in the study of international politics, realism and idealism reflect *analysis* and *purpose* in the science of international politics. Thus, even though they are opposites or antitheses, "sound political thought and sound political life will be found only where both have their place." E. H. Carr, *The Twenty-Years Crisis: 1919–1939* (London: Macmillan and Co., 1952), 10.

politics do not approach the world with mental constructs as rigorous as those we describe as paradigms.

It is clear to social scientists that the answers regarding complex phenomena are always a function of the questions themselves. And the questions one frames tend, consciously or unconsciously, to reflect one's worldview. So, a realist will tend to ask questions that relate to a state's power, its national interests, the quality of its diplomacy in defending those interests against the competing ones of other states, and so on. In contrast, one with a more idealist perspective may want to know whether international legal standards are effective in certain areas (that is, legal norms correspond to legal ideals), how alternatives to the use of force can be furthered in the affairs of states and the like.

## Analysis and Practice: Policy for Whom?

We have seen that perspectives of world politics vary with each level of analysis. Our perspective, combined with our approach, also largely determines our conclusions. One ever-present issue in the study of world politics is the matter of distinguishing objective analysis of a situation from policy preferences and goals. All work in the social sciences is built upon the premise that intelligent analysis is objective and should precede policy recommendations—or, conversely, that sound policy follows accurate analysis of one's options, objectively considered.

But note that in international politics (in the traditional, literal meaning of the term), most of those individuals engaged in making and recommending policy are, in fact, responsible for the foreign policy of a particular country. That is, they are meant to represent the interests of one state, rather than those of many or all societies, or of other social groups. Their view of the world is principally, therefore, from the state level of analysis. While these policy-makers no doubt are trained to understand their world from both individual and international systemic levels of analysis, the fact remains that their recommendations are meant to further a particular state's foreign policy goals, which may or may not correspond to the goals of you and me as citizens. Pluralist analysis contributes to this goal clarification, since it can reveal the interests of nonstate actors as these impact upon global policy. So do globalist and feminist approaches, which may help clarify the extent to which important individual and group values conflict and coincide with policies pursued by governments, corporations, and the like. Our purpose is to help the student understand the political system of the planet Earth in order to develop sound global policy preferences.

In each chapter, beginning with the next, the reader will find two different features that focus on different aspects of the world political system. The first, *Global Changes,* suggests a few of the ways in which our objective analysis of world politics has had to—or now must—take account of new developments in the lives of human societies. The second feature, *Normative Dilemmas,* presents alternative choices to the reader about policy issues. Many of these represent the kinds of dilemmas statespersons face as they determine what action they ought to take; some represent choices for informed citizens that impact upon the world we live in. All of them should be considered with this question in mind: Policy for whom? For myself and my associates (however they are defined) even at the expense of other groups? To satisfy some larger society (however that is defined) without doing too much injury to the values of another? For the benefit of generations yet to come, even if that means a denial of some of what I may want for myself?

# Conclusion

We have examined some of the complexities of studying world politics in this chapter, complexities that may seem particularly formidable during this time of rapid change. The very language we use to discuss world politics often does not represent literally the complexity, or the evolution, of the social and political phenomena it is meant to describe. That is because world politics is rich, complex, and continually evolving.

We considered micro-, mid-, and macro-levels of analysis as they apply to the study of world politics—where the individual, the state, or the international system dominate what we see, depending upon the place or level from which we choose to examine the world. We noted the differences in the kinds of questions we would ask (and, therefore, answers we might receive), depending upon the level of analysis we chose in trying to understand the causes of World War II and the end of the cold war.

It is feasible to regard national and international political life as fundamentally quite different from each other, or as fundamentally alike. The matter of which conception we choose relates to our perception of the world. Our perceptions are built upon mental constructs that help us order the chaotic information we receive through our senses so that we can arrange that information in meaningful ways.

The most general of these orientations constitutes one's worldview. We considered the assumptions of realism and idealism. Realism emphasizes the contrast between domestic and international life, sees this difference as unchanging over time, assumes that a double standard of behavior is acceptable in statecraft, is inherently preservative, and stems from a materialist conception of the universe. Its opposite is political idealism, which tends to assume the fundamental unity of all political and social behavior, notices evolutionary change in political phenomena through history, assumes a single standard of behavior for decision-makers, is inherently developmental, and is informed by philosophic idealism. From these two worldviews a number of analytic approaches or paradigms have grown. Realism, neorealism, pluralism, globalism, and feminism have developed as theoretical approaches or paradigms in response to late twentieth-century developments across the globe.

Finally, we noted the basic importance of arriving at policy choices as the result of objective, unbiased analysis of the issues. We noted, further, that most of the commanding policy choices made in today's world are made on behalf of particular nation-states, by governmental officials whose interests may or may not be identical to those of their larger societies, to segments of them, or to larger interests or groups of humankind.

# CHAPTER 2

# Westphalia and the Nation-state System

In August 1990, the army of Iraq swept across the border of its tiny neighbor, Kuwait, and conquered it. Within weeks, the Iraqi government had formally annexed the territory, designating Kuwait as a new province of Iraq. From start to finish, Iraq's action seemed consistent with the ancient story of world politics. On innumerable occasions throughout history, power has been successfully wielded or lost by human communities, with consequent changes in their status in relation to others. Some of the issues still at the heart of global politics—especially, war and peace—have been central from the beginning. Saddam Hussein's conquest of Kuwait in 1990 resembled the invasion of Mesopotamia (today's Iraq) by the Egyptian Pharaoh Thothmes II in 1458 B.C., as well as countless other conquests in the intervening 3,448 years. But it may be that historians of the future will view the 1990 action in different terms because of the great changes in the conditions of human life and attitudes that have developed since that distant time. Such changes include our current ability to communicate instantaneously across the globe, the far greater destructiveness of our weapons, and our economic interconnectedness. These or other factors could produce outcomes very different from the consequences of similar actions in the ancient world. Saddam's invasion ultimately failed because of the collective action by many of the world's political actors. Therefore, it may be perceived very differently by historians from Thothmes' successful effort to exact tribute from the rulers of ancient Mesopotamia.

Regardless of how the Kuwaiti case comes to be regarded, the current era may prove to be a rare transition point in world politics, comparable perhaps to the birth of the interstate system in Western Europe some three and one-half centuries ago. Much evidence suggests that the current period may indicate a real departure from the way human societies have interacted for the past several centuries. In many respects, we seem to be developing novel kinds of interrelationships and instrumentalities for interaction on a global scale. Such a transition would indicate that the nature and shape of world politics are in some sense less predictable than before, when we could assume substantial continuity between the expectations of past, present, and future.

Yet, even should this prove to be such an epochal transformation, global political change is not limitless. Even radical departures must consider past events. Nor can we be certain that what today appear to be historic shifts will seem so in another decade or another century. We often exaggerate the importance of a dramatic event while overlooking what is less visible below the surface. For these reasons, it is important to be open to the possibilities for unprecedented change while understanding our history.

Clearly, not every aspect of history—not even the history of intergroup conflict—carries equal weight in the study of modern global politics. An analysis of world politics must extract from the historical record, drawing from the vast storehouse of human experience to provide generalizations, suggest hypotheses, and develop theories about the nature and prospects for world political life today and in the near future.

This chapter provides a historical overview of the principal factors that have created the setting in which world politics are played today. As seen throughout human history, there have been many places on the globe where comparatively large societies have interacted in characteristic ways over time. In Egypt, China, India, Mexico, Peru, West Africa, and Europe, ancient peoples were sometimes linked in large imperial systems—and sometimes disconnected during times of political fragmentation. But by the early modern period, the Europeans had developed technologies—first, gunpowder, then mechanized means of travel and communication—that allowed them increasingly to subdue non-European peoples. As they did so, from about the fifteenth to the beginning of the twentieth century, they carried with them their ideas and practices of intergroup politics.

Today, even though the long period of European dominance has come to an end, the European political arrangement for interstate relations remains largely in place—albeit with modifications as it has been extended across the globe. Everyone alive today—including those who are non-Western or professedly anti-Western—lives within a formal world system that was largely the invention of Europeans several centuries ago. The creation of this system represented a significant departure from earlier conceptions of political order.

# THE LEGACY OF ROME

For about one thousand years after the fall of Rome (in the fifth century A.D.), political life in Europe was characterized by the generally shared *ideal* of European unity in the midst of a great deal of *real* fragmentation and disunity.[1]

That ideal was expressed in a number of ways. What kept it alive above all was the Europeans' spiritual unity in the Western (or Roman Catholic) Church.

---

[1] See the Chapter 1 discussion of the relation of *realism* and *idealism* to contemporary perspectives on world politics.

Commodore Perry in Tokyo, opening Japan to the West in 1854.

Christianity had become the official religion of the Roman Empire early in the fourth century, at which time the barbarian hordes that sacked and destroyed the imperial capital either had already been Christianized or soon adopted the new faith. After the collapse of Rome as a political empire, the Church continued to govern the spiritual life of Rome's former subjects through a hierarchical organizational structure similar to that which had governed the empire. The concept of the unity of Christendom replaced the unity of the Roman world. The Pope became the ultimate spiritual authority of a region that now lacked an effective political or secular authority. The ideal of European unity was also maintained by the shared cultural legacy of Rome—including a common language (Latin) in which the educated elite could communicate, a well-developed legal system, and a dim awareness that the political unity of the past constituted a golden age by comparison to the strife and fragmentation of the present.[2]

---

[2] For a more detailed discussion of medieval European political organization, see Walter Ullmann, *Principles of Government and Politics in the Middle Ages* (New York: Barnes and Noble, 1966).

In reality, Rome's conquerors were divided among themselves. After destroying the empire, they offered no alternative political system with which to replace it. Without an ability to maintain centralized control over people and territories, political authority was fragmented. Those who ruled usually had to do so indirectly through complex bargains made with their vassals, who themselves commanded important resources of their own. The basic units of political life typically were extremely small: the lands surrounding a fortified castle, a walled town, or a tiny duchy. The succession of authority frequently was unsettled, which required more deal making, the mortgaging or sale of territories, and frequent conflicts among those who would seize and hold a throne. Since huge territories came to be regarded as the personal domain of those who had inherited or otherwise acquired them (often through conquest), the loyalties of whole populations theoretically could shift with a nobleman's marriage or inheritance.

Out of this conjunction of real political disunion and Christendom's ideal of unity was born the political theory of medieval Europe. It assumed the God-given hierarchy of human social and political life: Some were born to be milkmaids or shepherds, others were given land and subjects to rule, while still others received divine callings to the priesthood. Each sphere of life, secular and religious, was structured by authority and obedience—from the family to the king's domains, from the parish church to the archdiocese. To live in submission to that structure was to follow God's will as the only way to ensure social order.

Occasionally, political fragmentation was overcome by the unity of substantial regions of Europe. Charlemagne became the emperor of much of Europe in the year 800 as the result of his inheritance, skill in battle, ruthlessness toward adversaries, and support for a beleaguered pope. For a brief time under his rule, the ideal of Christendom's unity under an emperor of all Europe was nearly a political reality as well. But after Charlemagne's death in 814, the empire was divided once again. A century and a half later, central Europe was again reunited under reign of the Holy Roman Empire. Even though the empire lasted longer than eight hundred years, this effort to maintain the ancient Roman ideal throughout Christendom never managed to exercise its authority over more than a fraction of the continent.

The weakness of the Holy Roman Empire stemmed partly from the fact that, starting in the late Middle Ages, several large kingdoms were consolidated in what had once been distant outposts of Rome—where the ancient ideal had perhaps never received strong support. Within these lands, kings gradually asserted their centralized control, thereby reducing their need to strike bargains with their own vassals. Particularly successful were the monarchs of France, England, and Portugal, who established control over states still recognized today. As they did so, the prospect of a truly united Europe grew less real, for these monarchies developed as a new kind of unit. These territorially based nation-states would become the basic actors in the politics of Europe.[3]

---

[3]See Gianfranco Poggi, *The Development of the Modern State* (Stanford, Calif.: Stanford University Press, 1978); and Charles Tilly, ed., *The Formation of National States in Western Europe* (Princeton: Princeton University Press, 1975).

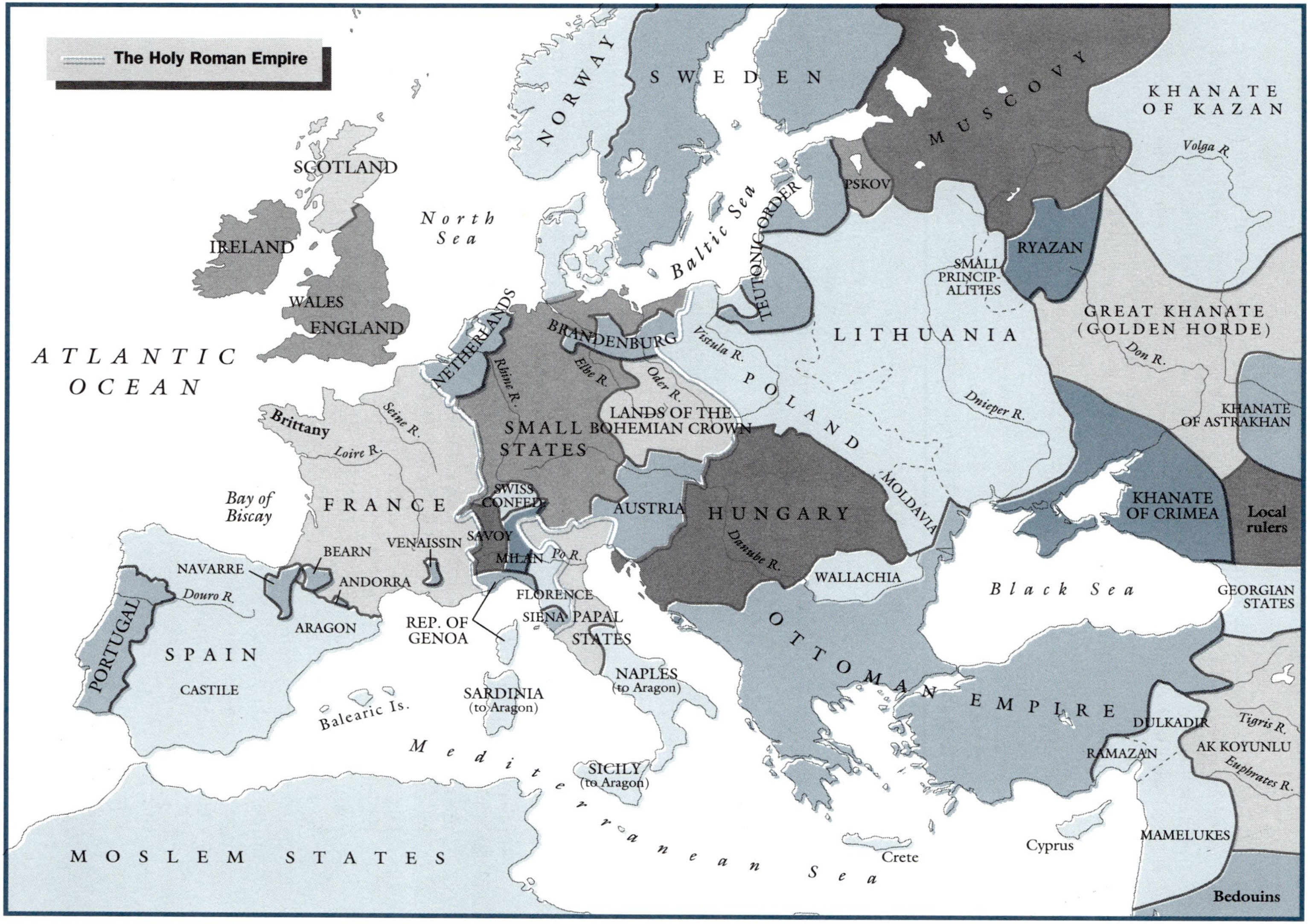

**MAP 2.1** Europe and the Mideast, 1500

Then, early in the sixteenth century, the spiritual unity of Christendom was torn apart by a wave of conflicts resulting from the Protestant Reformation. Wherever Protestantism gained adherents, the authority of the Bishop of Rome was denied, and with it, the claim that the Catholic Church was the sole authoritative voice of Christendom. The Western Church became politically divided for the first time. These divisions resembled and helped to justify the real separations that had long existed among Europe's secular authorities. After nearly one thousand years, the stage at last was set to abandon the myth that Europe maintained a Roman legacy of political and spiritual unity.

# THE FOUNDATIONS OF THE NATION-STATE SYSTEM

The last of the terrible religious wars, which had wracked Europe since the Reformation, began in 1618 and devastated much of northern Europe before it finally ground to a halt three decades later. The Thirty Years' War had a revolutionary impact on the European political system, for it effectively ended the ideal of a unified Europe. Specifically, it marked a profound change to the continent in both real (or material) and ideological respects.

At the material level, the war caused the lingering death of the Holy Roman Empire, which had so long embodied the medieval concept of political order. While turning the empire into a hollow shell from which it would never recover, the war made casualties of at least one-third—and perhaps as many as one-half—of the inhabitants of northern Europe. Huge numbers of peasants, in particular, were the inadvertent victims of marauding armies, who brought death from famine and disease in the wake of their widespread destruction. New independent actors—the Netherlands and Switzerland—emerged from the crippled empire, as did a number of small German states.

The course of the war revealed how religion had ceased to be the dominant issue among the various political actors of Europe and was beginning to be replaced by a new "secular religion": nationalism. What began as a struggle by Protestant nobles of Bohemia to dethrone the Catholic King Ferdinand of Austria soon became far more complex, with new belligerents entering the fray from dynastic and territorial, as well as religious, motives. Late in the war, Catholic France, Protestant Sweden, and a number of German Protestant princes were aligned against Catholic Austria. National interests were thus emerging as more important than those uniting unrelated peoples in a common religious faith. When at last the conflict was over, an exhausted Europe tacitly agreed that coercion and bloodshed had failed, and clearly would not produce unity again.

The peace treaty that ended this conflict addressed many matters specific to that time and place. But what is of interest today is that the **Peace of Westphalia** was the first expression of a new conception of how the independent political units of Europe ought to interact in the future. This conception was built upon analysis of their real capabilities in the first half of the seventeenth century and is still fundamental to the way in which world politics are organized and conducted. It advanced the notion that states (which at the time meant only some European states)

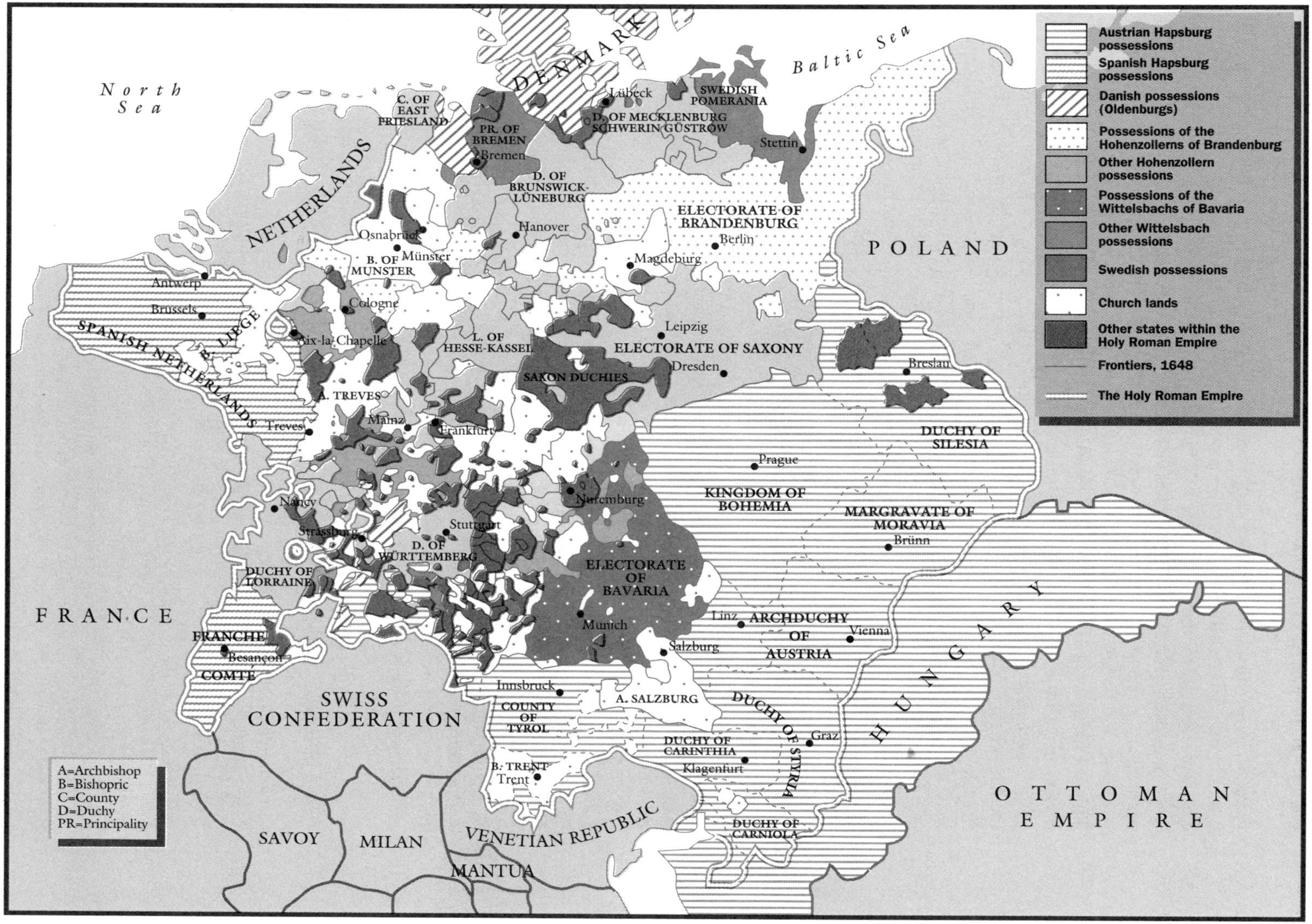

MAP 2.2 Central Europe after the Peace of Westphalia, 1648

were sovereign and independent and, therefore, legally equal to one another in their interactions. This would henceforth become the basic principle of international theory, replacing the medieval concept of a hierarchy of authority embracing all Europe.[4]

Stated differently, the capabilities of many European actors had evolved to the point that, by 1648, it seemed necessary to agree to a new idea or organizing principle that might help reduce the potential destructiveness available to governments. This meant recognizing the rough equality or parity of those who commanded the loyalty of various populations and establishing new rules for their interaction.

**Sovereignty** is a term that was beginning to be used by the end of the sixteenth century to describe the absolute authority of the monarch within his (only occasionally, her) realm, limited only by the laws of God and nature.[5] This concept may seem thoroughly outmoded today, since there is almost nowhere in the world in which absolute monarchy still exists. But it gave birth to the idea of the absolute authority of the state—regardless of how the state was ruled—within its own territorial jurisdiction. Since Westphalia, the nation-state—whether it is governed by kings, dictatorships, or through such representative bodies as congresses and parliaments—has been sovereign. The concept of sovereignty formed the basis for the definition of our subject: International relations refer to the relations among these sovereign units. And, because sovereignty means absolute authority within the state's realm, the rules of the game are qualitatively different in the relations among states than they are in the relations among individuals and groups.[6]

The sovereign **state,** by definition, differs from all other human groups in the following respects:

1. It is based on territory, which belongs exclusively to one state and not another;
2. Its authority is all-inclusive within that territorial base;
3. It possesses a monopoly over the instruments of force within its jurisdiction, which includes denial of that right to others;
4. Its membership is compulsory—that is, it requires (almost always) exclusive citizenship; and
5. Its sovereignty or supreme authority translates into its independence under international law.[7]

The state represents only half of the term, *nation-state,* which is used to describe the new sovereign actors of European politics after Westphalia. While the state

---

[4]An interesting analysis of Westphalia's influence over the next three centuries can be found in Leo Gross, "The Peace of Westphalia, 1648–1948," *The American Journal of International Law,* XLII, no. 1 (January 1948): 20–41. See also Cornelius J. Murphy, "The Grotian Vision of World Order," *The American Journal of International Law,* 76, no. 3 (July 1982): 477-498.

[5]Some years before the outbreak of the Thirty Years' War, Jean Bodin developed the idea of sovereignty in his *Six Books of the Republic.* For a summary of Bodin's views, see George H. Sabine, *A History of Political Theory,* 4th ed., rev. Thomas Landon Thorson (Hinsdale, Ill.: Dryden Press, 1973), chapter 21.

[6]Alan James, *Sovereign Statehood* (London: Allen and Unwin, 1986).

[7]This definition does not distinguish between a **state** and its **government,** which, of course, are not identical. As the official actor on behalf of the state, the government is implicitly referred to by the second, third, and fifth listed characteristics.

is a legal entity, the **nation** is a cultural and anthropological concept. It refers to that large social group whose members view one another as one people, usually because they are tied by a common linguistic, religious, ethnic, and cultural heritage. In many times and places, including premodern Europe, national groups or their equivalents have not formed independent states. Often the members of one nation have ruled over others (as in such multinational empires as the Holy Roman [962–1806 A.D.], the Ottoman [1300–1923 A.D.], or the Cambodian [the ninth to the fifteenth centuries A.D.]). For nearly 2,000 years, rulers from the Han Chinese held together a government and civilization as large as that of ancient Rome, making the Han the dominant ethnic group in the Chinese "nation." Or the members of one national or ethnic group have been divided into more than one state (as has been the case with German-speaking peoples in central Europe; Armenians split among Turkey, Russia, and Persia in the Caucasus; the Masai people of modern Kenya, Tanzania, and Uganda; or the Arab "nation," long divided into a number of states).

Not until more than a century after the Westphalia settlement did nationalism become a potent political force in European and then in world politics. But, even at the time of Westphalia, several of the most important new sovereign actors already were virtual nation-states (including France, England, Portugal, Spain, Sweden, and the Netherlands). Today, nationalism has been so linked to independent statehood that we may assume that connection has been "ordained by nature." Yet, a number of anomalies still remind us that reality does not always match this ideal image.

First are the divided nations, such as Italy until 1861; Germany from 1945 until 1990; Korea from 1945 to the present; and Vietnam from 1952 to 1975. Then there are the nations that have been absorbed by others for periods in which they have endured as states-in-waiting, such as Palestine and a number of other national groups at present; Poland through 125 years of partition from the eighteenth to the early twentieth centuries; and Tibet under Chinese rule today. Finally, there are those states whose separate "nationalities" are more the accident of colonial history, such as sub-Saharan Africa, where the unity of tribal groups was often split by the exigencies of European colonial rule. Similarly, a bit earlier in the Western Hemisphere, European colonization created English-, Spanish-, Portuguese-, and French-speaking "nations"[8] on the basis of rather arbitrary patterns of settlement, generally coloring those new nations with the culture of indigenous peoples while largely absorbing and overwhelming them. (Consider Aztec Mexico, Inca Peru, or Iroquois New York.)

## THE IDEA OF THE SOVEREIGN EQUALITY OF STATES

If states are sovereign and independent of any higher authority, it follows logically that they can only interact as equals under **international law.** All states must be assumed to have equal rights and duties when they interact; if they do not, some will be subordinate to others and so, by definition, not sovereign. This concept is

---

[8]The North American French "nation" of Quebec is a province of Canada rather than a nation-state, although Quebecois separatism continues to be a powerful force in Canadian politics.

almost exactly the same as constitution-based legal systems in which all citizens are considered equal under the law. These legal systems do not say that all are equal in power, in the circumstances of their birth, or in any of the other ways in which physical or material attributes are measured. Rather, they insist that the society's system of justice is built upon the premise that all citizens have identical rights and duties and enjoy equal protection of the law.

Thus the notion of sovereign equality deliberately ignores real physical and material differences of states. It creates instead an ordering arrangement that grants those defined as sovereigns with equal rights, duties, and protection of international law. It must have been clear in 1648, as it is today, that in fact the various units defined as sovereign states often had very unequal material capabilities. Why, then, was a new legal ideal accepted that so distorted the real material capability of states?

The chief explanation is that those new sovereigns to whom the Westphalia treaties applied possessed material capabilities that at least made them equal to one another in this important respect: Each could protect and defend its sovereign piece of territory in a way that earlier authorities could not. Soldiers now fought with gunpowder, which revolutionized warfare by permitting large armies to secure comparatively impenetrable borders. Such armies were made possible by the monarch's increased ability to raise conscripted armies from the populace; the armies, in turn, supported the increased autonomy of the monarch, who no longer had to rely upon his feudal barons and their private armies. Armies could now defend fixed borders more effectively than in the past, which meant that the sovereign in whose name they acted could also exercise more effective internal control over that territory.

This new capability was obviously limited by the similar ability of other, neighboring monarchs to do the same. The most painful lesson of the Thirty Years' War had been that even when a more powerful monarch might overrun the territory of a weaker enemy, the invader could not sustain his control over time (the rival still possessed an army equipped with guns and gunpowder) to make it worth the huge cost in lives and treasure. At Westphalia, it seemed preferable to try to strengthen the notion that states had fixed boundaries that could no longer be breached by the religious or dynastic claims of external actors. That might help prevent the kind of warfare Western Europe had just endured by encouraging these new sovereigns to turn more of their attention inward, to developing their own states.[9]

# THE UTILITY OF THE CONCEPT OF EQUALITY

Thus, the idea of the equality of states was in 1648, and still is today, a legal fiction. But it was then, and in important respects remains, a useful fiction for several reasons.

First, the idea of the equality of states was a less destructive adaptation of reality than the idea of a single, centralized world government. The effort to unify

---

[9]For a discussion of how the notion of equal sovereignty produced rules about what a state could and could not do, see David J. Gerber, "Beyond Balancing: International Law Restraints on the Reach of National Law," *Yale Journal of International Law* 185 (1985): 194–195.

Christendom through armed force had been responsible for much of Europe's warfare for centuries, including the terrible destruction of the Thirty Years' War. In the past, conquest and submission by the sword often had succeeded in producing order throughout a substantial population. But in 1648, the signatories at Westphalia agreed that separate sovereign governments that might coexist in peace were preferable to the pretense of unity that had produced resistance, widespread killing, and material exhaustion. The price for the political unity of Europe in 1648—and presumably for all humankind today—was unacceptably high if it could not be achieved even after decades of the fruitless destruction and loss of human life. To recognize that sovereigns could control their realms effectively was simply to recognize a new reality and avoid bloodshed in unrealistic causes.

Second, the idea of the equality of sovereign states embodied a principle of pluralism and autonomy that was, and evidently still is, politically and socially desirable. Most human attachments are fairly localized, which makes a community's material situation—rather than its theoretical tie to groups with which it may never directly interact—its primary concern. By 1648, material capabilities in Europe could be increasingly organized in response to human needs and desires by these large (but still, in a sense, localized) new units called nation-states. The development of larger and more complex economic systems was made possible by centralized power in the hands of the monarch. Goods could be traded securely throughout the state; a central treasury could regulate the supply of money more effectively than in the past; and taxes could be collected on a regular basis. Economic life in general could expand beyond the village marketplace to become more or less coterminous with the state. That, in turn, would lay the foundation for the initiation (in England about a century later) of the Industrial Revolution.

Third, the concept of the sovereign equality of states permitted "rules of the game" of international politics to be created that were simple and clear, allowing the easy addition of new sovereign players as political life evolved across the globe. A new actor that wished to be recognized as sovereign essentially had to meet a threefold test. It needed to demonstrate that it was ruled by a *government* over a fixed *territory* capable of commanding the loyalty of that territory's *population*. Once other sovereigns were satisfied that the new claimant met that test, they, acting individually as sovereigns, could extend recognition to the new actor so that they might interact with each other in accordance with the established rules. Admitting new sovereigns in this way would generally be much less costly than to try to keep them forever excluded by subduing them. The struggle for dominance and hierarchical control had been ruinous to seventeenth-century Europe; to allow equal participation hopefully would discourage the desire to try to dominate others. When it did not, other sovereigns were encouraged by the rules of the game to take whatever measures they could, including the use of force, to thwart the would-be dominator.

These three explanations for the idea of the sovereign equality of nation-states should suggest its continuing importance for interstate relations today.[10] Because it remains so basic, we shall explore its implications more fully throughout the rest of

---

[10] For a discussion of sovereignty's relationship to international law, see Mark W. Janis, *An Introduction to International Law,* 2d ed. (Boston: Little, Brown, 1993), 151–162.

> **Global Changes**
>
> # The Changing World System: 1648–1919
>
> At the time of the Thirty Years' War (1618–1648), Europe was divided into several hundred political units. When World War I broke out in 1914, Europe contained some 25 sovereign states.
>
> - How might the Westphalian idea of sovereignty help account for this reduction in the number of political "actors" in Europe over roughly 250 years?
>
> In 1875, over one-half of the world's people and two-thirds of its land mass was ruled by the governments of seven European states: Austria, Great Britain, France, Netherlands, Russia, Portugal, and Spain.
>
> - Did nationalism bring an end to these empires first in Europe or outside Europe? Why?
>
> Nearly a dozen new states were born out of the former empires of Austria and Turkey after their defeat in World War I.
>
> - What ideology was most responsible for ending these empires? Have their successors proved viable as nation-states? Why or why not?

this book. Now we need to consider the chief respects in which all the complex material capabilities of political actors diverged from—even struggled against—the idea that they had equal rights and duties within the kind of international framework that was largely constructed at Westphalia in 1648.

# THE REALITY OF UNEQUAL STATE POWER

Although by 1648 a number of European rulers could govern their various societies more or less equally well, they were typically far from equal according to most material standards of measurement. Some ruled over much larger populations than others; some were rich while others were poor; some had interests halfway around the globe while others were restricted to small lands in Europe. As a result, they did not behave identically in their interactions. Even though certain basic rules of the game had been established at Westphalia, enormous leeway was left to the players to develop different strategies based upon their particular needs and situations in the world. Indeed, the nature of world politics has been evolving ever since, largely as the result of these kinds of variations in the power and capability of states, and because their very sovereignty allows them to determine—and redetermine—the rules as they play.

The countless variations in the material resources of nation-states no doubt explain why no two of them have ever behaved in exactly the same way in the real world. To understand all the differences, one would have to be intimately familiar with the historical record of the hundreds of sovereign units that have played an independent political role in the modern world. That is unlikely, even for the highly trained expert (although it suggests why professional practitioners of foreign policy may need to be expert in the affairs of particular states). Fortunately, however, some generalizations do help explain the major kinds of differences and similarities in the behavior of states over the more than three-hundred-year history of the Westphalian international system.

We shall consider three such factors, each of which runs counter to the abstract principle of sovereign equality invented at Westphalia. As we have emphasized, that principle does not consider the real capabilities of the participants, resulting in an inevitable tension in their relations with one another.

## THE EVOLUTION OF NATION-STATES

The first factor conflicted with the principle of the sovereign equality of nation-states from the beginning: Those who signed the Peace of Westphalia did not everywhere create nation-states—not even across Europe—with the stroke of a pen in 1648. It would be more accurate to say that this treaty assumed as its basic model the sovereign actor that would increasingly resemble the nation-state we know today. But the expression of medieval political theory that still endured as the Holy Roman Empire remained intact until 1806 (when France's Napoleon forced the last elected emperor to renounce the imperial title).

The evolutionary process was not completed for more than two and one-half centuries. Many of the north German states began to form loose confederations among themselves by the end of the eighteenth century. But not until the largest of them, Prussia, fought several wars in the 1860s and 1870s—first with Denmark, then Austria, and finally France—did the Prussian statesman Bismarck create a unified German Empire in 1871. Although the southern reaches of the Holy Roman Empire had not extended very far down the Italian peninsula since about the twelfth century, Italy, too, remained fragmented among independent republics, kingdoms, imperial provinces, and papal states until a united nation-state finally was achieved in 1866 (not until 1870 was Rome included). Finally, the Austrian Empire endured as a multinational throwback to the Middle Ages right up to the end of World War I in 1918. Ruled by the Hapsburg family since 1282, its varied lands and peoples had been acquired through inheritance, deft political marriages, and success in war. At its end, the Hapsburg empire ruled over such disparate nationalities as the politically dominant Austrians (themselves the southern branch of German-speaking peoples), Hungarians, Croats, Serbs, Slovenians, Czechs, Slovaks, Poles, Romanians, Greeks, Italians, Ruthenians, and others.

Early in the twentieth century, nationalism finally made sovereign states of almost all the nationality groups in Europe.[11] But, as we shall see, it did not solve all the problems of ethnic demands or of intra-European politics generally. It did bring to a close Westphalia's first great chapter, in which the international system created in 1648 had been limited largely to Europe and European outposts.

---

[11] For more on nationalism and its impact on contemporary world politics, see Paul Brass, *Ethnicity and Nationalism* (Newbury Park, Calif.: Sage, 1992); Naomi Chazan, ed., *Irredentism and International Politics* (Boulder, Colo.: Lynn Rienner, 1990); Eric J. Hobsbaum, *Nations and Nationalism Since 1780* (Cambridge: Cambridge University Press, 1990); and William Pfaff, *The Wrath of Nations: Civilizations and the Furies of Nationalism* (New York: Simon and Schuster, 1993).

# EUROPEAN IMPERIALISM AND THE DOUBLE STANDARD

A second factor has also discouraged the realization of sovereign equality. For some three centuries, European actors generally refused to apply the principle of sovereign equality to those with whom they interacted outside Europe. Therefore, when non-Europeans, including Asian or African political communities, could be subdued, they were viewed as legitimate targets for imperial conquest. This assumption had a somewhat different effect in the New World, where colonies were populated by Europeans themselves, presumably (so their rulers must have supposed in 1648) to remain permanent overseas branches of the sovereign state in Europe—the "mother country." These differences between European imperialism in the Old World of Africa and Asia and in the New World of America would produce different results, bringing sovereignty to much of the New World a century or more before most of Africa and Asia.

When the British added India to their empire in the eighteenth century, or when they and other European powers carved up Africa a century later, they did so for material gain: more human power, natural resources, trading privileges, or strategic advantage. They had no intention of treating these communities like sovereign equals; they made them colonial holdings meant to serve their own polities in Europe. The fact that these distant peoples were not a part of the European inheritance no doubt made it easy to justify *not* including them within the European-made international system. These imperial holdings were perceived as something like pieces of the sovereign's territory that happened to be located far from home, with indigenous populations that might be enslaved (in the case of Africans) or otherwise made to work to enrich the Europeans.[12] This fiction perpetuated notions of racial and cultural superiority on the part of Europeans, and, as long as it could be maintained, kept the control of world politics effectively in European hands.

The situation was different for European settlements in North and South America in the sixteenth and seventeenth centuries, and in Australia and New Zealand a bit later. There the governments were established for transplanted Europeans, who emigrated in substantial numbers. They tended to overwhelm indigenous peoples, pushing them aside, absorbing them, or destroying them altogether. Those Europeanized colonies became replicas of their parent societies; as they thrived, the logic of sovereign equality became more and more appealing to those who lived there. After more than one hundred and fifty years of British rule in North America, residents of her thirteen colonies began to insist that it was unjust for them to be taxed without representation in Parliament. In effect, they said that they did not deserve to be second-class citizens of Britain, simply because they happened to live far from the capital. And when their grievances remained unmet, they took matters into their own hands, obtained their independence, and made themselves into a sovereign equal to the very state of which they had recently been a part.

---

[12]From the late nineteenth century to the present, the most perceptive explanations for European imperialism have viewed it as driven by economic more than security motives. See, for example, John A. Hobson, *Imperialism: A Study* (Ann Arbor: University of Michigan Press, 1965); and V. I. Lenin, *Imperialism: The Highest Stage of Capitalism* (New York: International Publishers, 1967). Note that economic explanations of imperialism remain powerful for those persuaded that *neo-imperialism* and *economic dependency* continue in the contemporary world, after formal colonialism itself has largely ended.

English colonists along North America's east coast were first to achieve sovereign independence, but they were soon followed by the descendants of Spanish and Portuguese settlers in South America. Canada, Australia, and New Zealand achieved independence without having to resort to revolution. In every case, the imperial centers in Europe resisted losing these valuable territories, although once the North American cause triumphed, similar causes were more difficult to thwart elsewhere. But after their independence, these new states were readily admitted into the international system. That was proof, in a way, of the soundness of Westphalia's organizing principle. Once a people could demonstrate their sovereignty by taking effectve charge of their government and territory, other sovereigns would do well to accept them into the club, so that all might then share in the burdens—as well as the rights—of membership. Eventually, these examples also would make it hard to resist similar claims to sovereignty by the non-European peoples of Africa and Asia. Yet not until the middle of the twentieth century would most of these people be granted the right to participate as sovereign equals with the states of the European world.

## SMALL POWERS AND GREAT POWERS

The third material factor that has always impinged on the Westphalian ideal of sovereign equality is clear in the historical record. Small powers tend to behave differently than great powers, and the characteristic behavior of each has remained similar throughout the modern period.[13] Great powers typically take a more active role in foreign policy than small powers, as can be seen in a variety of ways. Usually it has meant that powerful states have maintained more and larger diplomatic missions in other countries; frequently they have raised larger armies or navies; sometimes they are willing to employ their armed forces on behalf of some nation-state in addition to their own. Great powers have entered into more international agreements, have fought more wars, and generally have made their influence felt more widely than have their smaller brethren. Conversely, small powers typically have devoted most of their limited military capability to policing their own borders; they have usually not expressed a strong interest in distant political events, but have been more inward-looking. Often they have sought to protect themselves through neutrality in the quarrels of others; sometimes they have looked for protection from a great power—typically the rival of a great power that threatens them.[14]

This is not to say that a state's status as a great or small power is unchanging over time. The state's power, which really means its capability to act as it wishes in the world, is a relative or relational matter. That is, its relevant material factors can

---

[13]It is instructive that we call them *powers* in this context rather than *states* or *sovereigns,* since power, which equals capability, is never distributed equally among individuals.

[14]A great many writers have investigated characteristic differences in the behavior of small and large powers. A sample includes Maurice A. East, "Size and Foreign Policy Behavior," *World Politics* XXV (1973), 556–576; Robert L. Rothstein, *The Weak in the World of the Strong* (New York: Columbia University Press, 1977); and Max Singer and Aaron Wildavsky, *The Real World Order: Zones of Peace/Zones of Turmoil* (Chatham, N.J.: Chatham House, 1993). A related body of research examines the factors that produce change in the rankings of state actors over time. See, for example, Paul Kennedy, *The Rise and Fall of the Great Powers: Economic Change and Military Conflict from 1500 to 2000* (London: Unwin Hyman, 1988); and Richard Rosecrance, *The Rise of the Trading State: Commerce and Conquest in the Modern World* (New York: Basic Books, 1985).

# A Profile of Switzerland

The fact that Switzerland's place has changed very little over three centuries is quite unique, even among small states. Its stability owes much to the comparative isolation of its mountain territory in the very heart of Europe. Sparsely populated from early times, the region was nominally included within the Holy Roman Empire after 1033. Yet real political control remained highly localized. During the next two centuries, several communities high in the Alps increasingly opposed growing encroachments on their rights by the more powerful feudal houses. In 1291, three of these communities, or *cantons*, formed a defensive league against the house of Hapsburg. During the next century, they were joined by five neighboring cantons and, through their success on the battlefield, maintained their autonomy from Austria. They added further territories to their confederation throughout the fifteenth century. In 1499, the Emperor Maximilian I granted Switzerland virtual independence.

What began as a league of peoples who spoke a German dialect had by the sixteenth century grown to include both French- and Italian-speaking populations. Protestantism swept across much of the country during the Reformation, but was then fiercely resisted in a number of the cantons. A 1531 treaty halted the spread of Protestantism in Catholic areas, but with its population now divided by language and religion, the creation of a Swiss nation-state seemed thwarted. Still, even during the Thirty Years' War, Switzerland remained neutral. Swiss cantons prospered while the areas all about them suffered miserably. With formal recognition of its sovereign independence at Westphalia, the stage was set for the further evolution of Switzerland as a nation. In the absence of linguistic or religious unity to bind it together, Swiss statehood was built upon the common interest its varied people shared in maintaining themselves as an island of prosperity within a frequently war-torn continent.

Switzerland has generally succeeded in remaining free from foreign intervention and from entanglements in the foreign policy concerns of others. (Switzerland is currently the only significant state that has refused to join the United Nations on grounds that membership might undermine its determination not to take sides in world events!)* This freedom has been reinforced by certain unchanging realities, including Switzerland's small population (little more than six million today), its location in something of a natural fortress, and its lack of valuable natural resources. It is also surrounded by much larger states that have often been in conflict with one another.

In some respects, Switzerland qualifies as the world's oldest democracy. Ironically, however, Switzerland was the last country in Europe to extend the right to vote to women. That only occurred as the result of a (male-only) referendum in 1971!

1. Is it accurate to describe Switzerland as a nation-state today? Why or why not?
2. How has the Swiss practice of armed neutrality helped its society to maintain peace?
3. How has Switzerland managed to prosper despite its lack of resources? To what extent has its livelihood supported peace with its neighbors, and vice versa?

---

*Switzerland is officially an "observer" at the United Nations along with Monaco, Vatican City, and the Palestine Liberation Organization. Nonetheless, Switzerland hosts the European U.N. headquarters in Geneva.

---

only be roughly measured in comparison to those of some other state *at a particular time*. Today, both Portugal and the Netherlands are considered small powers, and that is borne out by the kinds of foreign policies they conduct. But Portugal in the fifteenth century, and the Netherlands in the sixteenth century, were decidedly great powers. Why? Each then was among the most highly developed societies of its era. Both produced a proportionately large number of citizens sufficiently skilled and wealthy to take the lead in navigation and exploration, money and banking, and the pursuit of new knowledge. During that time, each state was able to exercise an influence in the world—which included establishing overseas empires—relatively greater than their absolute size would indicate.

These generalizations only begin to explain the behavior of both great and small powers. Nor do they attempt to address the behavior of states that cannot be clearly categorized. One further point should be made about these typical differences. Nation-states—like other human communities—seek to use whatever capabilities they possess to maintain or improve upon their way of life. When they interact with others, they are likely to have defined interests that they seek to defend or advance through the material forces they can command. These may include, in addition to military power, the natural resources they can claim and the economic, cultural, and intellectual capabilities of their populations. Differences in these material forces help explain the varied interests and actions of states.

# THE TWILIGHT OF A EUROPEAN INTERNATIONAL SYSTEM

This overview of the development of the nation-state system so far has focused upon Europe—for the simple reason that European states developed the ground rules at a time when they were beginning to command the world stage. As European influence was extended outside Europe through commerce, colonialism, and imperialism, the Westphalian arrangement gradually expanded to embrace the whole world.[15] Westphalia had to be able to include new political actors as they met the test of sovereign statehood and showed a willingness to play by the European rules of the game.

First came the system's expansion within Europe itself. We have glimpsed examples of this in the way modern Germany was gradually formed from a cluster of small political units dating from the Middle Ages, and in tracing how the Austrian empire of the Hapsburgs clung to the medieval justification for its existence before disintegrating into a whole new group of nation-states at the close of World War I. Meanwhile, the Westphalian order was extended eastward across the continent to include nation-states that, though Christian, had been part of the Orthodox rather than the Roman Catholic tradition and thereby were largely excluded from the medieval political order of Western Europe.[16] The prime example was Russia, which viewed itself as the keeper of Christian Orthodoxy after most of the Byzantine em-

---

[15]See William F. McNeill, *The Rise of the West* (Chicago: University of Chicago Press, 1963) for a historical account of Europe's development to the point of global dominance in the modern period. Chapter 3 provides a more complete discussion of Westphalia's impact on the non-European world.

[16]The Roman Empire was permanently divided into East (Byzantine) and West (Roman) components after the year 395. That marked the decline of Rome's political influence and the ascendancy of Byzantium. The last Roman Emperor was deposed by the Goths in 476, effectively ending the Western Empire. Meanwhile, Christianity was developing differently under Byzantium and Rome. In the East, it passed directly from being a persecuted sect to the imperial control instituted by the Emperor Constantine. In the West, the Church remained far freer of political control because of the weakness of political authorities. As a result, the two branches of Christianity gradually evolved different cultures and practices of worship. In the ninth century, their separation was virtually legalized in the Great Schism, after which it is appropriate to recognize the Eastern Orthodox and Roman Catholic Churches.

# A Profile of Germany

What we know today as Germany represents a flat and largely indefensible region of east central Europe. For many centuries, instead of German unity, periodic invasions by Slavs, Magyars, and the Norse produced a localization of security and economic life. Germany remained a patchwork of tiny principalities and free cities throughout the Middle Ages and well into the modern era. The Thirty Years' War ravaged the German lands, which lost up to half their populations. The Peace of Westphalia established a loose confederation of the German states under the nominal rule of the Holy Roman emperor.

By the seventeenth century, one of the easternmost of these states, Prussia, began to emerge as more powerful than the others. A century later, under Frederick the Great (1712–1786), Prussia emerged as one of the great powers of Europe. Yet Prussia's growth did little to unify the German states as a whole. Napoleon briefly rearranged the map of Europe, but the Congress of Vienna in 1815 then restored it after Napoleon's defeat. The foreign minister of Austria, Prince Metternich, played a dominant role in discouraging a unified Germany. Even though Austria was a German-speaking state, Metternich did not want it to unite with the other German states; his interest was in perpetuating the polyglot Hapsburg empire that had descended from the Middle Ages. Nor did Metternich wish to see the other German states unite into a powerful federation that would challenge Austria. Instead, he succeeded in persuading the Congress to create a new confederation of north German states, largely controlled by Austria.

Not for another half century would German unity finally be achieved—without Austria. Then the lead was taken by Austria's rival, Prussia, which by 1862 became the engine of a German state. After a brief war with Austria in 1866, Prussia excluded its defeated rival from the reconfigured North German Confederation. While Austria was at last silenced regarding German unity, the French government grew alarmed at Prussia's heightened influence. The resulting Franco-Prussian War of 1870–1871 brought the complete defeat of France, its government's collapse, and the triumphant proclamation of the German Empire. The Prussian king became Kaiser (emperor) Wilhelm I of the new state.

Yet, German unity had been achieved at the cost of France's humiliation, which included annexation of the French provinces of Alsace and Lorraine, and even the proclamation of the German Reich in Paris. The resulting enmity between France and Germany fostered the two worst conflicts of the twentieth century. The harsh settlement that France exacted from the defeated Germany to end World War I in 1919 seemed a revenge for what France had suffered nearly fifty years before. That settlement, the weakness of the new republican government that replaced the monarchy of Kaiser Wilhem II, and, a decade later, the economic upheaval produced by the Great Depression all combined to set the stage for the leadership of Adolf Hitler and the second world war within twenty-five years.

Once again, a massive coalition brought the defeat of Germany. But when the victorious allies of World War II quickly split apart at the start of the cold war, their "temporary" zones for occupying Germany were hardened into a long-term division of the German state. Within five years, the French, British, and U.S. occupation zones in the western two-thirds of the country had joined to form the Federal Republic of Germany, while the Soviet zone in the east (shorn of most of Prussia, which went to Poland and the USSR) took the name of the German Democratic Republic. For the next four decades, the frontier that slashed through Germany became the front line known as the Iron Curtain; there rival military forces were built into the greatest armed confrontation in history. The cold war came to an end when the allies of World War II agreed upon the terms for German reunification in 1989. But once the two Germanies were reunited, the huge disparities in the wealth of the former eastern and western sectors led to serious problems of adjustment.

1. What historical reasons explain why Austria, a German-speaking state, is not a part of modern Germany?
2. What major factors contributed to the unified Germany's very late arrival (for Europe) as a nation-state?
3. Although Germany has generally acted as a great power since its creation in 1871, it did not do so from 1945 to 1990. Why not?
4. What role is a reunited Germany playing in the post–cold war era?

pire of the Orthodox faith fell to Islam, beginning in the eleventh century. But Russia did not play a major role in the politics of western Europe until Tsar Peter I (the Great) began the sweeping Westernization of his country in 1698. He was eager to adopt Western standards in all things, including statecraft. From his time onward, Russia was clearly a great power participant in the system devised at Westphalia.

Next, a non-Christian power, Ottoman Turkey, was brought into the European system. After the fall of Constantinople (today's Istanbul) to the Turks in 1453, the Ottoman presence in Europe could not be ignored. During the next century, Turkish rule expanded into central Europe, threatening to capture the Hapsburg capital of Vienna in 1683. A powerful, medieval empire at a time when nation-states were becoming ascendant in the West, Turkey also posed a difficult ideological problem to the Westphalian core: as an Islamic state, it did not share the political traditions of Christendom. Much of the Ottoman Empire was in Asia Minor, which added to the perception by Western states that Turkey was not truly a European player. Nonetheless, Turkey gradually came to behave in its relations with others much like other European powers, losing chunks of its empire to European rivals in the process. In 1856, the European great powers, meeting as a concert in Paris, declared that Turkey was entitled to full status in the European system.

Also in the nineteenth century, the Westphalian system was enlarged to include new actors far from Europe—most notably, the United States of America and Japan. The first was, in fact, European in everything but its physical location. A newly independent United States assured its acceptance into the system by showing that it would adhere to established principles of conduct, although it played an insignificant role in world affairs for almost the next century. Also in the New World, newly independent colonies of Spain and Portugal took their places as sovereign states. Isolated Japan was forcibly opened to Western trade after 1854 and made to accept unequal treaties and the rapid intrusion of Western influences into Japanese life. Japan soon responded by adopting Western political ideas and adapting them to Japanese culture. By the end of the century, Japan was showing promise of beating the Europeans at their own game: Nationalism and imperialism had been unleashed, as had enormous economic and military growth. Japan was on the way to becoming a great power in Westphalian terms.

By early in this century, the Westphalian state system had shown how readily it could be extended to embrace the non-European world.[17] That expansion would continue after World War I, although not until the post–World War II period would it explode to include the entire planet. Meanwhile, within Europe itself, the logic of Westphalia was at last almost complete. There had been approximately two hundred, more or less separate—but not always sovereign—political entities scattered about Europe at the time of the Thirty Years' War. By 1900, these had been consolidated (the costs of which had included many wars in the intervening two and one-half centuries) into some twenty-five recognizable nation-states. The main exceptions were

---

[17]The fact that the system of sovereign states could be extended across the globe does not mean that it thereby "solved" the question of what kinds of social units should interact in the world by forcing all human communities into a single mold. Nor did it make European-style nation-states of every political unit that achieved sovereignty in the twentieth century. Chapter 3 explores these points and their implications for world politics.

**MAP 2.3** The Changing Contour of Germany

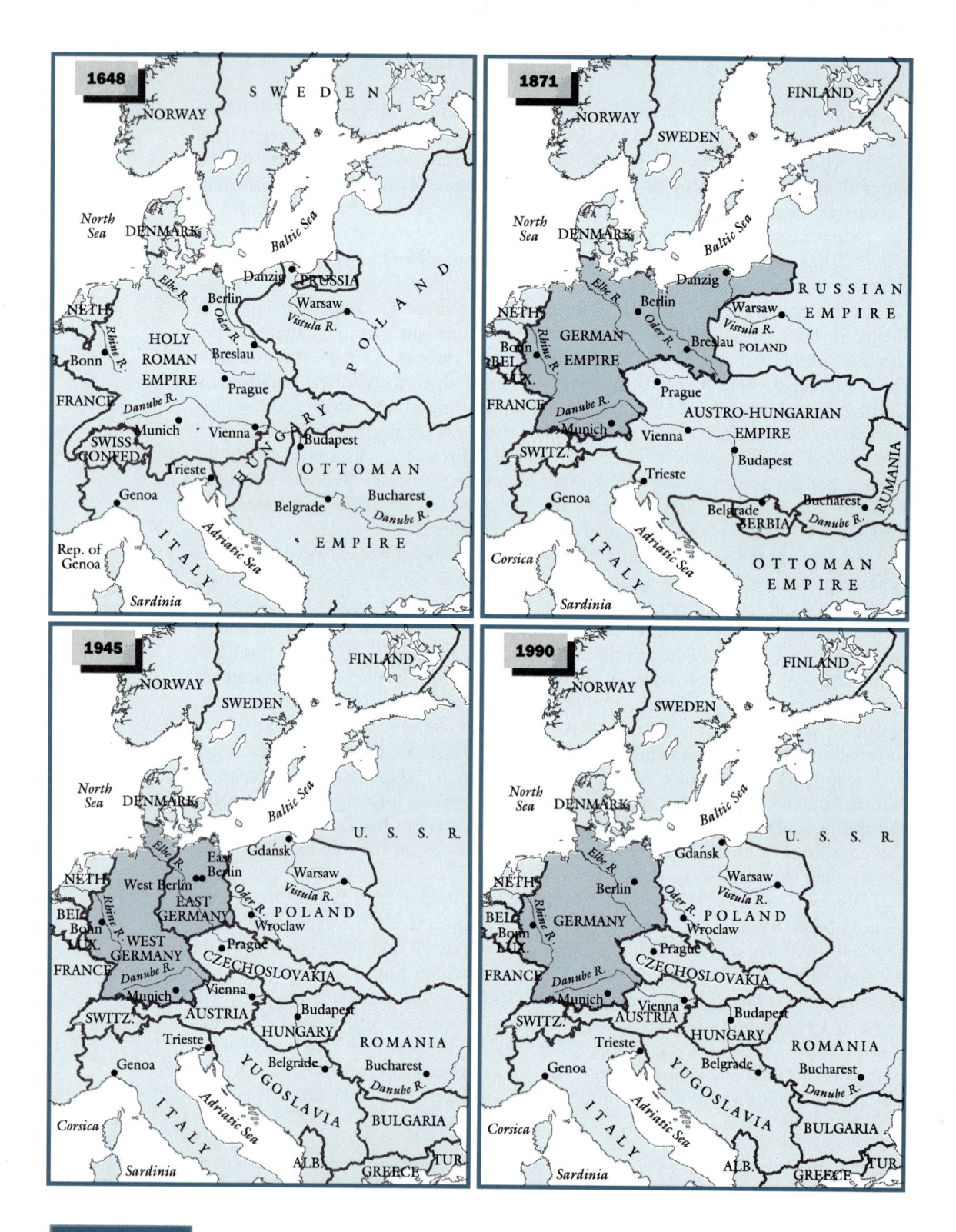

still the outmoded Austrian and Ottoman empires, which would cling to life until their defeat in World War I. Then a number of new nation-states would be created from their ashes, with two implications—the exponential growth of sovereign states and nationalism—signaling future world trends.

## THE EXPLOSION OF SOVEREIGN STATEHOOD

The era of state consolidation came to an end following World War I. There would never again be as few as twenty-five states in Europe; the Ottoman and Austrian empires alone were replaced by more than a dozen sovereign units. As overseas empires of Europeans began to achieve independence, the number of sovereign participants in the international system grew exponentially, from some 30 in 1900, to about 70 at mid-century, to more than 180 by 1990. By the last decade of the twentieth century, this process of continued state-creation seemed to be returning once again to Europe—its birthplace.

In the process, the original meaning of sovereignty sometimes seemed to have been lost—or at least diluted. When invented, sovereignty described the political reality of a number of large (for their day) European units that had demonstrated their near self-sufficiency as human societies whose governments could provide a measure of security against those who threatened them from outside. But by the late twentieth century, some sovereign state polities were far too small to stand on their feet in those terms (an extreme example is the Republic of Nauru in the South Pacific, whose population of less than 10,000 lives on an island about one-tenth the size of Washington, D.C.!). Conversely, even the largest sovereign states today are neither as self-sufficient nor are their people as secure from the prospect of terrorist violence as were their much smaller predecessors in 1648 (see Figure 2.1).

## THE PROBLEM OF NATIONALISM

Another implication of modern world trends has been the driving force of nationalism behind statehood in the twentieth century. Just after seeing nationalism make sovereign states out of empires across the non-European world, we have witnessed its return to a number of the "old" states of Europe—particularly the former Soviet Union, the United Kingdom of Great Britain and Northern Ireland, and Spain, whose minority nationality populations had long been quiescent. In the 1990s, murderous interethnic conflicts in the Balkans, Rwanda, and elsewhere shook humanity into seeing that nationalism had by no means been tamed in our time.[18] These and other bloody conflicts are also our inheritance from Westphalia's view that sovereignty ought to follow for whatever group defines itself as a nation.

Still, nationalism is an elusive force. A sense of nationality is the product, usually over many centuries, of a shared way of life. But it is virtually impossible to disconnect the development of nationality (and the nationalism that may follow) from

---

[18]We use the term *nationalism* in its broadest sense to mean the self-definition of people as a community distinct from others. Thus it may apply to the desire for independence or autonomy on the part of societies that have never yet formed nation-states, such as the Kurds of the Middle East, the Tatars of central Russia, or the Ibos of eastern Nigeria.

**FIGURE 2.1** Growth in the Number of Generally Recognized Nation-states, 1800–1995

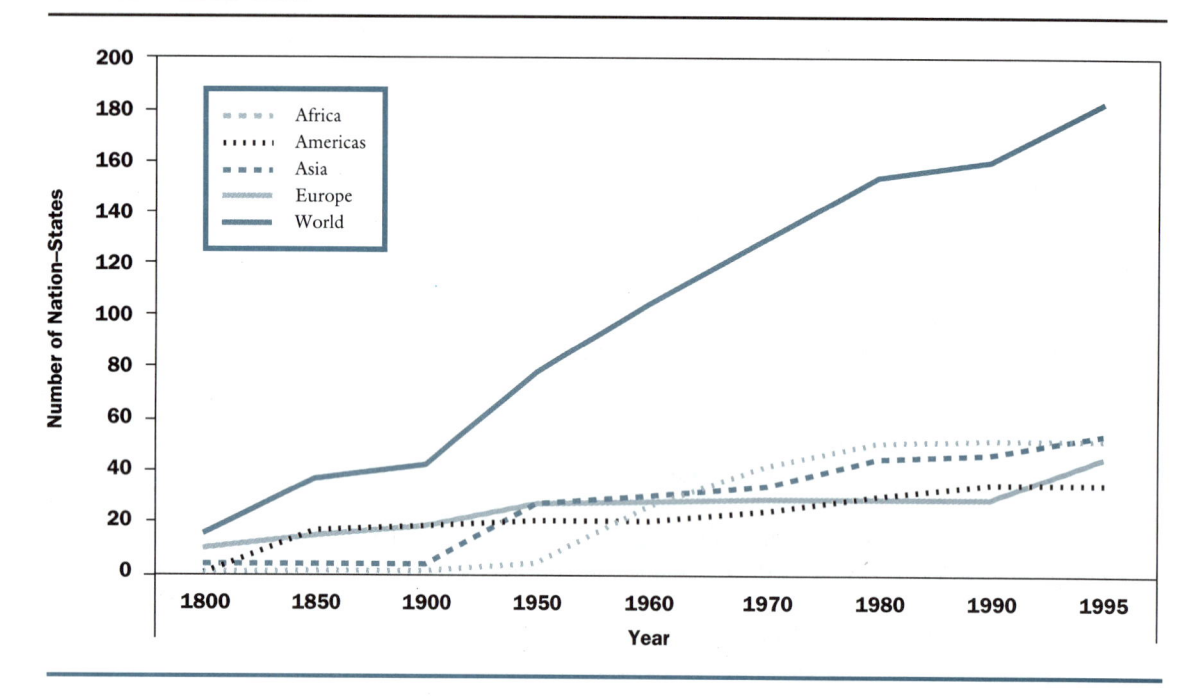

the ways in which people have been politically organized over long stretches of history. Durable political organization begets social cohesion that in turn shapes national identity.[19]

That is clearly true in China, a nation first centralized by conquest more than 2,000 years ago. It was then dominated by a dynasty (the Han, 206 B.C. to 220 A.D.) so effective and long-lasting that the culture it advanced and spread is what we recognize today as distinctively Chinese.[20] It is just as true of many much more recent "nations," as well. In the case of, say, Mexico, nationality is built upon the cultures of both the ancient Meso-American civilizations that were conquered by the Spanish in the sixteenth century, and the Europeans, who intermarried with native Americans over the next three centuries of Spanish rule. Political events and consequent traditions that have developed in a way unique to that country since Mexico's independence in 1821 further developed its national identity.

[19] For a classic study of European nationalism, see Carlton J. H. Hayes, *The Historical Evolution of Modern Nationalism* (New York: Macmillan, 1931). See also Charles Tilly, ed., *The Formation of National States in Western Europe* (Princeton, N.J.: Princeton University Press, 1975). For a recent anthology of the writings on nationalism up to the 1990s, see Anthony Smith and John Hutchinson, eds., *Nationalism* (Oxford: Oxford University Press, 1995).

[20] Not until China was absorbed into the Westphalian system was it appropriate to speak of China as a "nation." In the words of William Pfaff, "to be Chinese was to belong to a civilization which was presumed to be universal, or, if not universal, to have only barbarians beyond it" *(The Wrath of Nations: Civilizations and the Furies of Nationalism* [New York: Simon and Schuster, 1993], 17).

But it is clearly not the case that minority groups within every strong state have been assimilated within the larger nation. Much of the post–cold war bloodshed in the world has testified to the durability of intercommunal hatreds even after many decades in which multinational states have been ruled by a united sovereign. A less obvious observation about current social behavior also deserves mention. Today, multicultural influences are having an unprecedented impact upon many societies. The ability of people in the modern world to move swiftly and easily has opened floodgates of foreign immigrants and temporary residents in many countries. There, the lifestyles of the newcomers frequently conflict with those of the older residents, sometimes creating new political and social problems in the form of resistance to new influences. Yet, the sheer magnitude of this multicultural mixing today suggests that it may also be shaping many nationhoods in ways we can scarcely imagine. As a result, the future of nationality and nationhood may look very different from that of the past.

## Conclusion

The modern nation-state system was largely born with the Treaty of Westphalia of 1648. Although there have been other interstate systems in other times and places—including ancient China, South Asia, India, and Africa—the Westphalian system, which was based upon independent nation-states, came to dominate the modern world. That process began in Europe, then moved to the Americas, Asia, and Africa through European imperialism and subsequent decolonization.

The Thirty Years' War that ended at Westphalia underscored the fact that security in Europe could no longer be provided by the unity of Christendom, which had been split by the Reformation into Protestant and Catholic communities. Westphalia gave secular leaders sovereignty or control over their respective territories, and they were no longer made to answer to a higher authority.

Entry into the new Westphalian order required three basic ingredients: territory, population, and a government. Ideally, the territory of the state would be based upon a commonly recognized national identity, although that has often not been the case during the Westphalian era. Many states have been multinational, while other nationality groups have been divided among two or more states. Many nationalities—including the Kurds, Palestinians, and Tibetans, to name but a few—are aggrieved by their lack of state representation.

The ideal of the Westphalian system was one of sovereign equality, although this has never been achieved in real terms because states' individual capabilities vary greatly. As a result, large and small states usually behave quite differently in world politics. Nor was the ideal of sovereign equality adhered to by the Europeans when they carved up Africa and Asia into their own empires. Only in the latter half of the twentieth century did these areas take their place in the Westphalian system. As they did so, these "new" states (new only in terms of their participation in the Westphalian order, since many of their cultures date back to ancient times), with their different values and experiences, presented new challenges to the nation-state system at the end of the twentieth century.

# CHAPTER 3

# Contemporary Challenges to the Nation-state System

In this chapter, we shall examine the major developments of the twentieth century—some of which have been dramatic, some nearly invisible—that appear to be challenging the ability of the Westphalian nation-state system to maintain itself by satisfying fundamental needs of the human community.

That last phrase is crucial: Nation-states originated as social and political units that seemed best able to meet basic psychological, economic, and security needs and aspirations of one small segment of humankind in western Europe. Gradually, more such independent states developed throughout the world, even though many of the newer arrivals bore scant resemblance to the originals. The triumph of the Westphalian principle would face its greatest test: To what extent do sovereign states still satisfy humanity's needs at the end of the twentieth century? Is the nation-state system being fundamentally altered by any of the novel global developments and trends of our time?

## NATIONALISM VERSUS IMPERIALISM IN THE TWENTIETH CENTURY

Nationalism has both served and disrupted the authority of sovereign states. With ever increasing force since 1648, nationalism has been the source legitimizing sovereign statehood.[1] The enlargement of the Westphalian system and the justification for ending imperialism—first within Europe and, later, throughout the world—have been founded largely on the modern ideal that nations should be sovereign. But the very power of that idea continues to make it a two-edged sword, for it can still justify the breakup of long-established states (in particular, those multinational empires created before nationalism became the justification for statehood) in the name

---

[1] See John Breuilly, *Nationalism and the State*, 2d ed. (Chicago: University of Chicago Press, 1994).

of minority national groups living within them. Since established governments almost never freely relinquish their control over others, including other nationalities, the resulting conflicts have filled the pages of modern world history while producing much of its violence.

Nationalism—the force typically linking the cultural group (the nation) to the independent state—is the principal ideology opposing **imperialism,** which means the centralized rule of many diverse peoples. From today's vantage point, it seems that nationalism has in fact won the contest throughout the world. The imperialism that was such a predominant fact of life at the start of this century has almost everywhere given way to triumphant nationalism. The most recent major example is within the sphere of the former Soviet Union, which succeeded the Russian Empire in 1917. This sweeping end to imperialism—at least in the sense of one nation's ruling over peoples outside it—is one of the most important political trends of this century.[2]

World War I was the most devastating conflict for many peoples since the Thirty Years' War. Triggered by an act of nationalist violence within the Austrian Empire, "the war nobody wanted" quickly spiraled into a general European conflict. Each new contestant entered out of a sense of duty to maintain alliance commitments to others already threatened, thereby redressing some dangerous imbalance, real or anticipated, in the configuration of power on the continent.[3] The United States became a belligerent only late in the war—in part because its merchant fleet was suffering at the hands of German U-boat (submarine) attacks intended for the Allies. Once the United States entered, the scale eventually was tipped in favor of the Allied cause. Germany, Austria, and Ottoman Turkey were defeated, and the latter two were destroyed as empires, thanks in part to assistance from internal forces in the final stages of the defeat.

The diplomats who went to Paris in 1919 to write the peace treaties ending World War I had the formidable task of creating new nation-states out of the Austrian and Ottoman Empires. They did not fully succeed, largely because of the complex way in which various nationality groups were spread across the lands of the former empires.[4] In particular, German-speaking people and closely related Slavic groups were widely scattered across central and eastern Europe. The so-called *pan-nationalist* movements by these groups in the nineteenth century helped to undermine the Austrian and Ottoman Empires. While most of the southern Slavs were united in the new country of Yugoslavia *(yugo* means *south)* after 1919, defeated Germany had its territory and population reduced. In addition, sizeable German communities that had been under Hapsburg rule now were assigned to new, non-German nation-states. As a result, crises would erupt in much of central and southeastern Europe in the years to come.

---

[2]Imperialism is sometimes (mis)joined to nationalism when used as the justification of one "nation" to colonize other peoples—often the product of feelings of national superiority rather than sovereign equality with others. No doubt the most extreme example was the Nazi-led effort to absorb virtually all of Europe into a German Empire, in which non-"Aryan" peoples were to either serve as slave laborers or be exterminated.

[3]For analysis of the origins of World War I, see Barbara Tuchman, *The Guns of August* (New York: Dell, 1962).

[4]See American political scientist James T. Shotwell's first-hand account of the difficulties faced by the peace conference: *At the Paris Peace Conference* (New York: Macmillan, 1937).

The 1920s and 1930s have frequently been characterized as a period of struggle between forces supporting the self-determination of the infant nation-states of central and eastern Europe and those—led by Nazi Germany and Fascist Italy—that sought to impose their own imperial domination of those regions. Yet, the leading victor states in World War I were themselves the greatest imperial powers on earth. What distinguished Great Britain and France in 1918 from the Austrians and Turks as imperial powers was their colonial holdings outside the continent of Europe, where the rules of a Europe-centered Westphalian order were not yet enforced. At the end of World War I, Great Britain's empire encompassed approximately one-quarter of the world's territory and, with 500 million people, one-fourth of the world's population. France's empire was about one-third that size, but nonetheless included more than 100 million people.

In contrast, the total size of the territories ruled by the three great powers that would form the Axis in World War II—Germany, Italy, and Japan—covered a total area little more than one-twelfth the size of the territory ruled by Britain and France. Yet these three states had no imperial holdings whatever in 1919 and total populations that were about one-third larger than the total home populations of Britain and France. The global struggle that ended in World War II was in important respects the doomed effort of states without colonies to enhance their great power status by acquiring empires like those of their major European rivals. Hitler tried to conquer much of Europe to build a German Empire (using pan-German nationalism as an initial excuse); Mussolini brought Ethiopia under Italian rule; and the Japanese warlords sought an empire in East Asia.

World War II produced even more far-reaching effects than the halt of these new imperialist challenges. The war began, after Hitler's attack on Poland in September 1939, as an Allied effort to repulse the Third Reich's aggressions in Europe; then, after the attack on Pearl Harbor late in 1941, the war challenged Japanese imperialism in Asia. After a much greater loss of blood and treasure than even that of World War I, the Axis powers were finally defeated in their efforts to gain empires of their own.

But the imperial holdings of the victorious Allied powers soon became untenable as well. The war had unleashed national liberation movements throughout much of Asia and Africa; Britain and France had been reduced to second-rank powers by the very war they had won; and the ideologies of the two new "superpowers" (the United States and the USSR) had strongly anti-imperialist strains, whatever the realities of their histories and foreign policies. Once the anticolonial tide began to sweep across the empires of the British and the French, the much smaller colonial powers—Belgium, the Netherlands, Portugal, and Spain—would also find the process difficult to thwart.[5]

# WESTPHALIA BECOMES A GLOBAL SYSTEM

Decolonization of the empires of the Europeans began almost immediately after the end of World War II in 1945 and was largely completed within the next

---

[5]For an interpretation of how nationalism was shaped as decolonization brought an end to the European empires, see Frank Furedi, *Colonial Wars and the Politics of Third World Nationalism* (New York: St. Martin's Press, 1994).

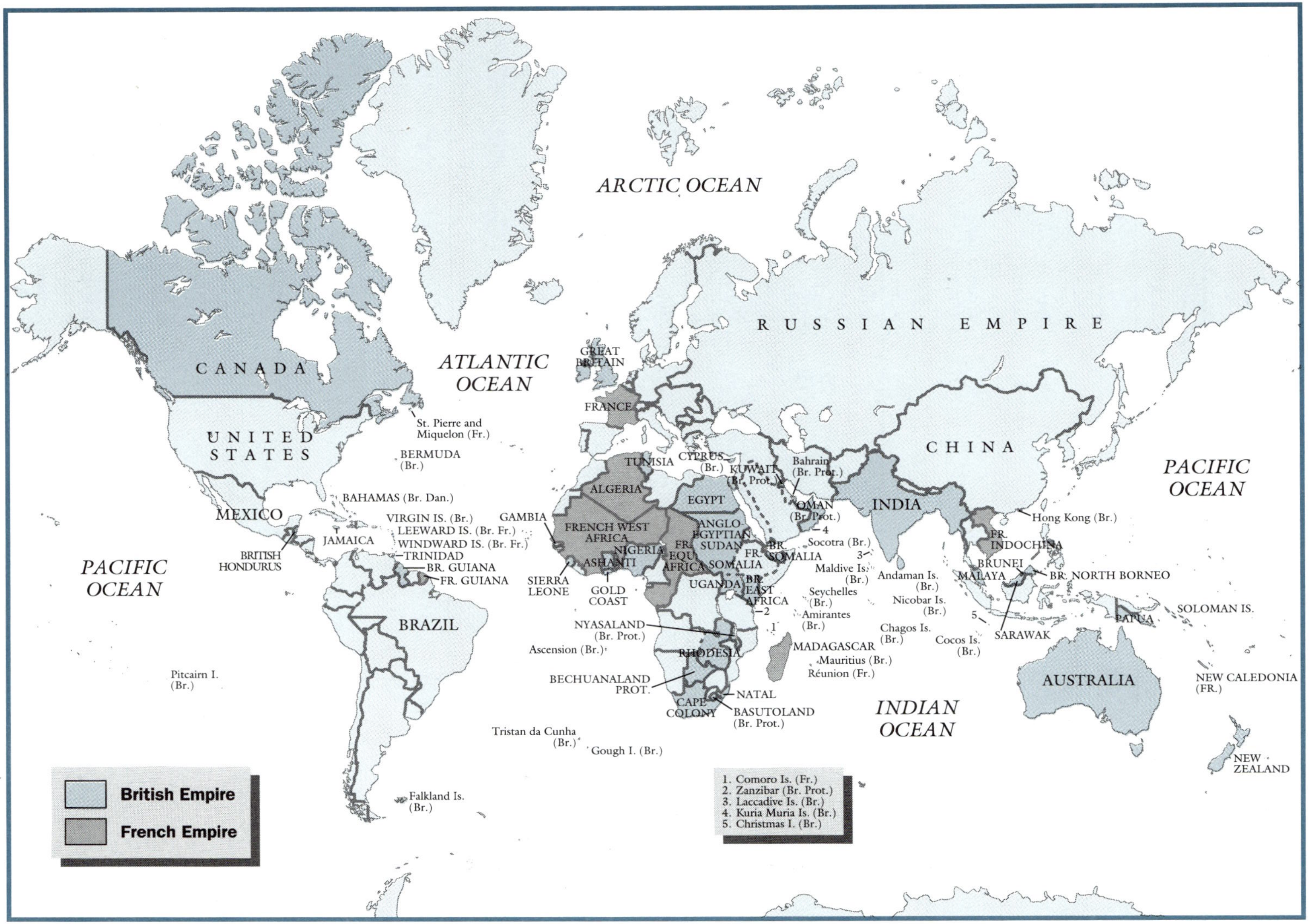

MAP 3.1 The British and French Empires in 1900

### TABLE 3.1  Independence Dates of New States from 1945 to the Present

| YEAR OF INDEPENDENCE | STATE | FORMER COLONIZER | YEAR OF INDEPENDENCE | STATE | FORMER COLONIZER |
|---|---|---|---|---|---|
| 1945 | Indonesia | P, H | 1961 (cont.) | Tanzania | P, G, E |
| | Vietnam | F | 1962 | Algeria | O, F |
| 1946 | Jordan | O, E | | Burundi | G, B |
| | Philippines | S, US | | Jamaica | S, E |
| | Syria | O, F | | Rwanda | G, B |
| 1947 | India | E, F, H, P | | Trinidad and Tobago | S, E, F |
| | Pakistan | E | | Uganda | E |
| 1948 | Burma | E | 1963 | Kenya | P, E |
| | Israel | O | 1964 | Malawi | E |
| | North Korea | J | | Malta | E |
| | South Korea | J | | Zambia | E |
| | Sri Lanka | P, H, E | 1965 | Gambia | E |
| 1949 | Bhutan | IN | | Maldives | P, H, E |
| | Cambodia | F | | Singapore | E |
| | Laos | F | 1966 | Barbardos | E |
| 1951 | Libya | O, I | | Botswana | E |
| | Oman | P, E | | Guyana | E, S |
| 1956 | Morocco | O, F, S | | Lesotho | E |
| | Sudan | E | 1968 | Equatorial Guinea | S |
| | Tunisia | O, F | | Mauritius | H, F, G |
| 1957 | Ghana | E | | Nauru | G, E |
| | Malaysia | P, H, E | | Swaziland | E |
| 1958 | Guinea | F | 1970 | Fiji | E |
| 1960 | Benin | F | | Tonga | E |
| | Burkina Faso | F | 1971 | Bahrain | O, E |
| | Cameroon | E, F, G | | Bangladesh | E |
| | Central African Republic | F | | Qatar | O, E |
| | Chad | F | | United Arab Emirates | E |
| | Congo | F | 1973 | Bahamas | E |
| | Cyprus | O, E | 1974 | Grenada | F, E |
| | Gabon | F | | Guinea-Bissau | P |
| | Ivory Coast | P, F | 1975 | Angola | P |
| | Madagascar | E, F | | Cape Verde | P |
| | Mali | F | | Comoros | F |
| | Mauritania | F | | Mozambique | P |
| | Niger | F | | Papua New Guinea | P, G, H, E |
| | Nigeria | E | | Sao Tome and Principe | P |
| | Senegal | F | | Surinam | H |
| | Somalia | E, F, I | | | |
| | Togo | F, G | | | |
| | Zaire | B | | | |
| 1961 | Kuwait | O, E | | | |
| | Sierra Leone | E | | | |

## TABLE 3.1 *continued*

| YEAR OF INDEPENDENCE | STATE | FORMER COLONIZER | YEAR OF INDEPENDENCE | STATE | FORMER COLONIZER |
|---|---|---|---|---|---|
| 1976 | Seychelles | F | 1991 | Armenia | O, R |
| 1977 | Djibouti | F | | Azerbaijan | R |
| 1978 | Dominica | E | | Belarus | R |
| | Solomon Islands | E | | Croatia | A, O |
| | Tuvalu | E | | Estonia | R |
| 1979 | Kiribati | E | | Georgia | R |
| | Saint Lucia | E, F | | Kazakhstan | R |
| | Saint Vincent and the Grenadines | E, F | | Kyrgyzstan | R |
| | | | | Latvia | R |
| | | | | Lithuania | R |
| | | | | Macedonia | O |
| 1980 | Vanuatu | E/F Condominium | | Moldova | O, R |
| | | | | Oman | P, E |
| | Zimbabwe | E | | Slovenia | A |
| 1981 | Antigua and Barbuda | E | | Tajikistan | R |
| | | | | Turkmenistan | R |
| | Belize | E | | Ukraine | R |
| | | | | Uzbekistan | R |
| 1983 | Saint Kitts and Nevis | E | 1992 | Bosnia and Herzegovina | O |
| 1984 | Brunei | E | | Serbia and Montenegro | O, A |
| 1986 | Federated States of Micronesia | US | 1993 | Czech Republic | A |
| | Marshall Islands | US | | Slovakia | A |
| 1990 | Namibia | G, SA | | | |
| | Yemen | O | | | |

KEY: **A** Austrian (Hapsburg) Empire, **B** Belgium, **E** England (Great Britain), **F** France, **G** Germany, **H** Holland (Netherlands), **I** Italy, **IN** India, **J** Japan, **O** Ottoman Turkey, **P** Portugal, **R** Russia, **S** Spain, **SA** South Africa, **US** United States

twenty-five years.[6] That was a remarkably brief period in which to effect the complete transformation of the international system—at least in its formal or ideal terms—with a more than threefold increase in the number of its sovereign participants. That legal transformation did not instantly transform the capabilities of these new sovereign actors, as we shall see. But giving independent voices to those who had never before participated in the Westphalian order was to open up—in that sense, democratize—the system in a way that made it truly global for the first time. It thereby largely eliminated the double standard that had for so many centuries justified the coexistence of imperialism with the equality of sovereign states (see Table 3.1).

---

[6]By the 1970s, only a comparative handful of non-European territories were still ruled by European (or Western) states, and these generally had populations of less than 100,000.

**MAP 3.2** German Mastery of Europe, 1942

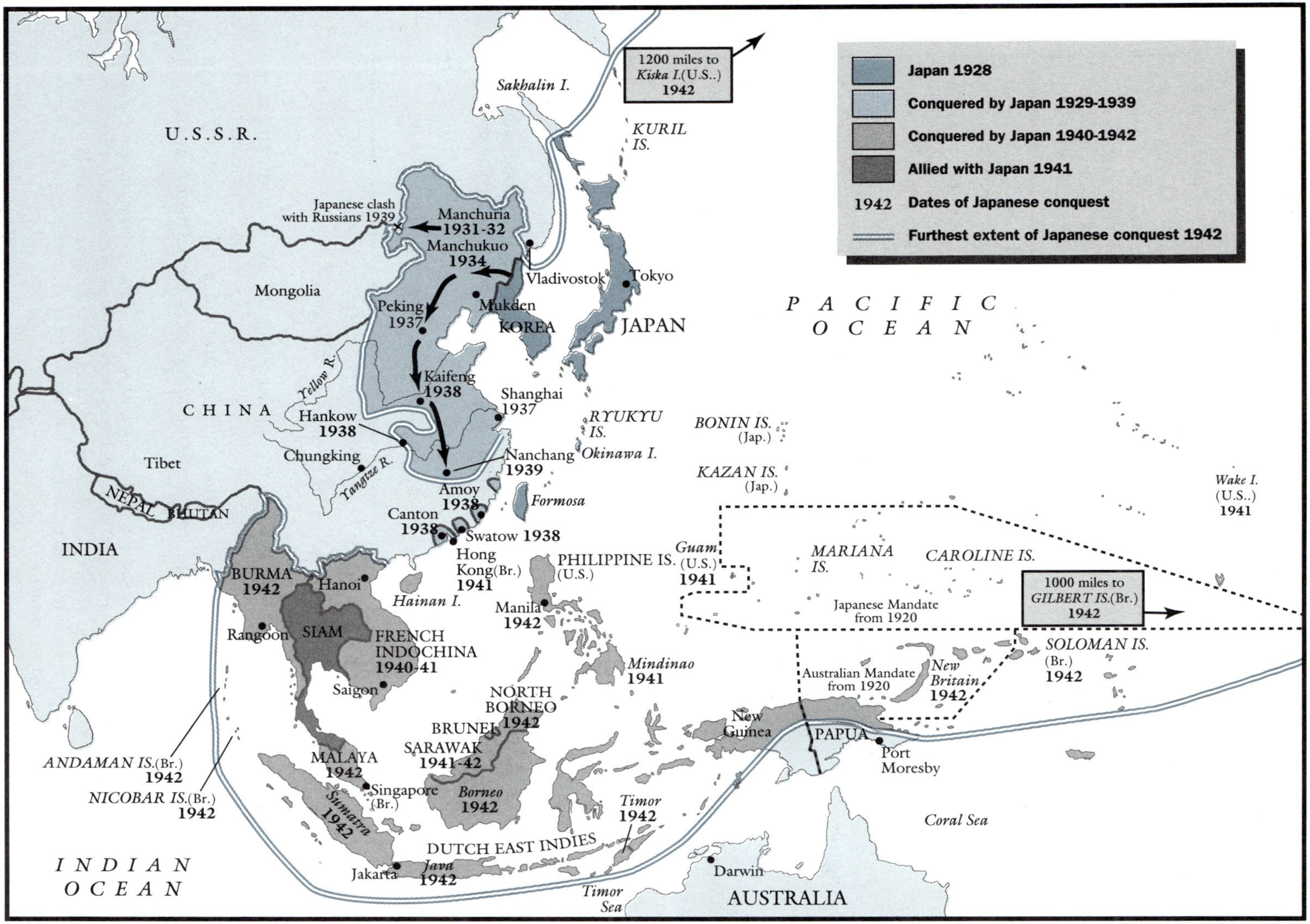

**MAP 3.3** Japanese Expansion, 1931–1942

## A Profile of India

India is the home of one of the earliest civilizations on earth, originating in the Indus Valley more than 5,000 years ago. As its culture developed and spread, parts of the Indian subcontinent were periodically unified by conquest, but political fragmentation was the norm over many centuries. In the sixteenth century, the Mogul dynasty, based in Afghanistan, conquered most of the subcontinent. A large minority of India's population was converted to Islam, the conqueror's religion.

European trade with India began with the Portuguese explorer, Vasco da Gama, in 1498. As Mogul power declined, Britain became dominant; its army defeated a French-Mogul force in 1757, which then led to further British conquests. By 1849, Britain's rule was established across nearly all of India. Early in the twentieth century, however, divisions between the two dominant religious communities—Hindu and Muslim—were becoming political divisions as well, with Muslim leaders insisting upon a separate, independent state.

The independence from Britain in 1947 therefore prompted a formal partition between the republics of India and Pakistan. Although India was largely Hindu, it was created as a secular democracy without an established religion. In contrast, Islam was made the state religion of Pakistan. Partition produced widespread flights of people during the communal rioting that occurred as members of one religious group found themselves in alien territory. Nor was that to be the end of national divisions in former British India. Pakistan itself, originally separated into two sectors by more than one thousand miles of Indian territory, could not sustain its own unity when its crowded eastern region rebelled against the superior political and economic status given to West Pakistan. In 1971, East Pakistan achieved its own independence—with help from India—as the result of civil war, to form the new nation-state of Bangladesh.

India's governmental institutions are adaptations of the British system, and English remains the common language uniting educated people in a country embracing speakers of more than 1,600 different languages and dialects. The country contains a rich ethnic mix as well, reflecting migrations from throughout Eurasia dating from ancient times. A democratic political system has withstood the early dominance of one party, a state of emergency enacted by Prime Minister Indira Gandhi in the 1970s and, later, the assassinations of two prime ministers—Ms. Gandhi in 1984 and her son, Rajiv, in 1991.

1. What principal factors account for British India's split into three separate nation-states?
2. In what respects do these factors explain the phenomenon of nationalism?
3. Does India's huge population (second only to China) qualify it as a great power? Why or why not?
4. What aspects of India's experience under British rule have shaped its political culture today?

## THE TROUBLED TRIUMPH OF THE NATION-STATE

The process of decolonization seemed to represent the obvious triumph of nationalism over imperialism; yet it may be more accurate to say that it marked the triumph of the nation-state as the basis for sovereign and independent participation in global politics.[7] Westphalia had long since made the rules for participation clear. In

---

[7] Neorealism (see Chapter 1) emerged as an analytic approach in the 1970s, modifying the power politics that had dominated the study of this field in the immediate postwar period. With its greater emphasis on the structure of the international system, neorealism reflected in part this huge expansion in the number of states in the world, along with the fact that states now tended to be grouped in relation to the bipolar structure of the cold war period.

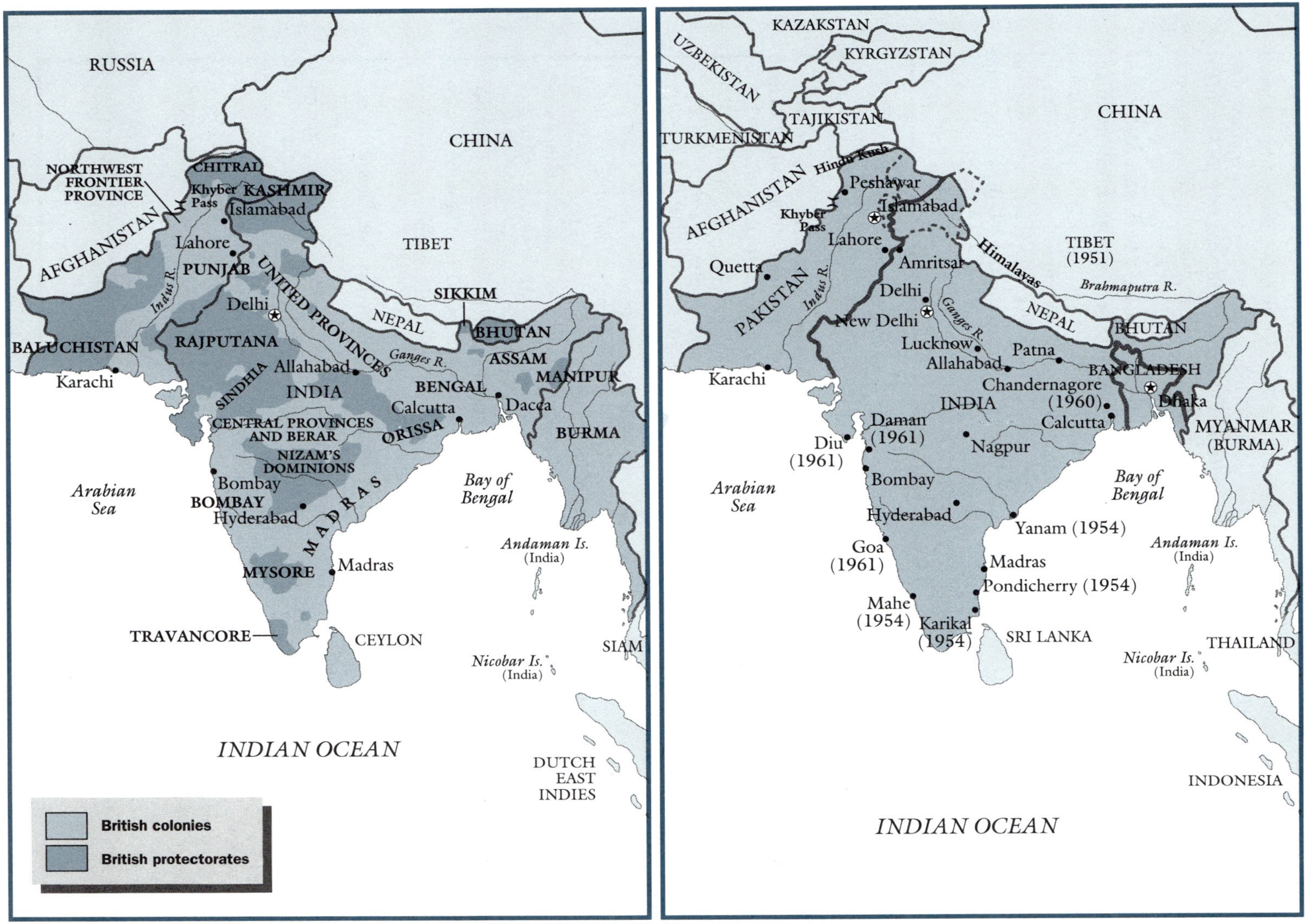

MAP 3.4  British India in 1900 (left), and India, Pakistan, and Bangladesh Today (right)

the post–World War II era, non-European societies throughout the world achieved their independence by adapting their political life as necessary to the nation-state model. The creation of nation-states outside Europe typically has required combining various traditional social groups whose culture and political organization were different from those of nations in the Western sense.

Most of the non-European states in today's world have been shaped by their experience with European colonialism, which was imposed on a great variety of ancient cultures. One result has been fabricated nation-states, most typically in much of Africa, some lacking the necessary social cohesion for political stability. That, in turn, has continued to produce conflicts that find their way onto the agenda of world politics. The civil wars of Nigeria in the 1960s, Zimbabwe in the 1970s, and Rwanda

---

# A Profile of Nigeria

Nigeria, with its 128 million inhabitants, is by far the most populous country in Africa. The earliest culture in this area developed more than two thousand years ago. Over the next thirteen centuries, many ethnic groups migrated into the region. In the north, Hausa and Fulani peoples dominated; in the west, Yoruba and Benin kingdoms arose; and in the east, the Ibo people were in the majority.

Contact with Europeans began with the arrival of the Portuguese in the fifteenth century. There followed more than two centuries of European trade in African slaves, who typically were acquired from the interior. While that trade in humans flourished, no European power attempted to control the region. But in 1807, Britain abolished the slave trade and later stationed a naval squadron off the coast to intercept the slave ships of other countries. British merchant ships then began to trade in palm oil and other local products, gradually creating a deeper British interest in the region. In 1861, Britain annexed the bustling port city of Lagos, and twenty-five years later made a colony of the rest of Nigeria. Not until 1914 were the two territories combined.

Nigeria achieved its independence in 1960. Although representative institutions were established, the history of independent Nigeria has been marked by frequent military coups against elected officials, followed by the suspension of democratic institutions. As a result, no democratically elected government has lasted very long. This history reflects underlying conflicts that result from overlapping ethnic, religious, and regional divisions within Nigeria. Diverse tribal groups speak more than 250 languages. Islam is the dominant religion in the North; the West has a mix of Christians and Muslims; and the East is predominantly Christian. At least 20 percent of the population still adheres to other traditional religions.

In 1967, civil war broke out after an Ibo-dominated military regime was in turn overthrown by northerners. Hundreds of thousands of Ibos who had settled in the north fled to their ancestral homeland in the east following a massacre. Soon, the government of the Eastern Region declared itself the independent Republic of Biafra. Not until January 1970 did the Nigerian army defeat the breakaway government, forcibly reuniting the country.

1. What accounts for the fact that Nigeria has experienced civil war in modern times while Switzerland has not, even though both countries are divided along ethnic and religious lines?
2. How could the doctrine of nationalism have been used to support the effort of the Ibos of Biafra to create an independent state for themselves?
3. In what respects did the existence of the state system encourage the Nigerian government to defeat Biafra's attempt to achieve its independence?

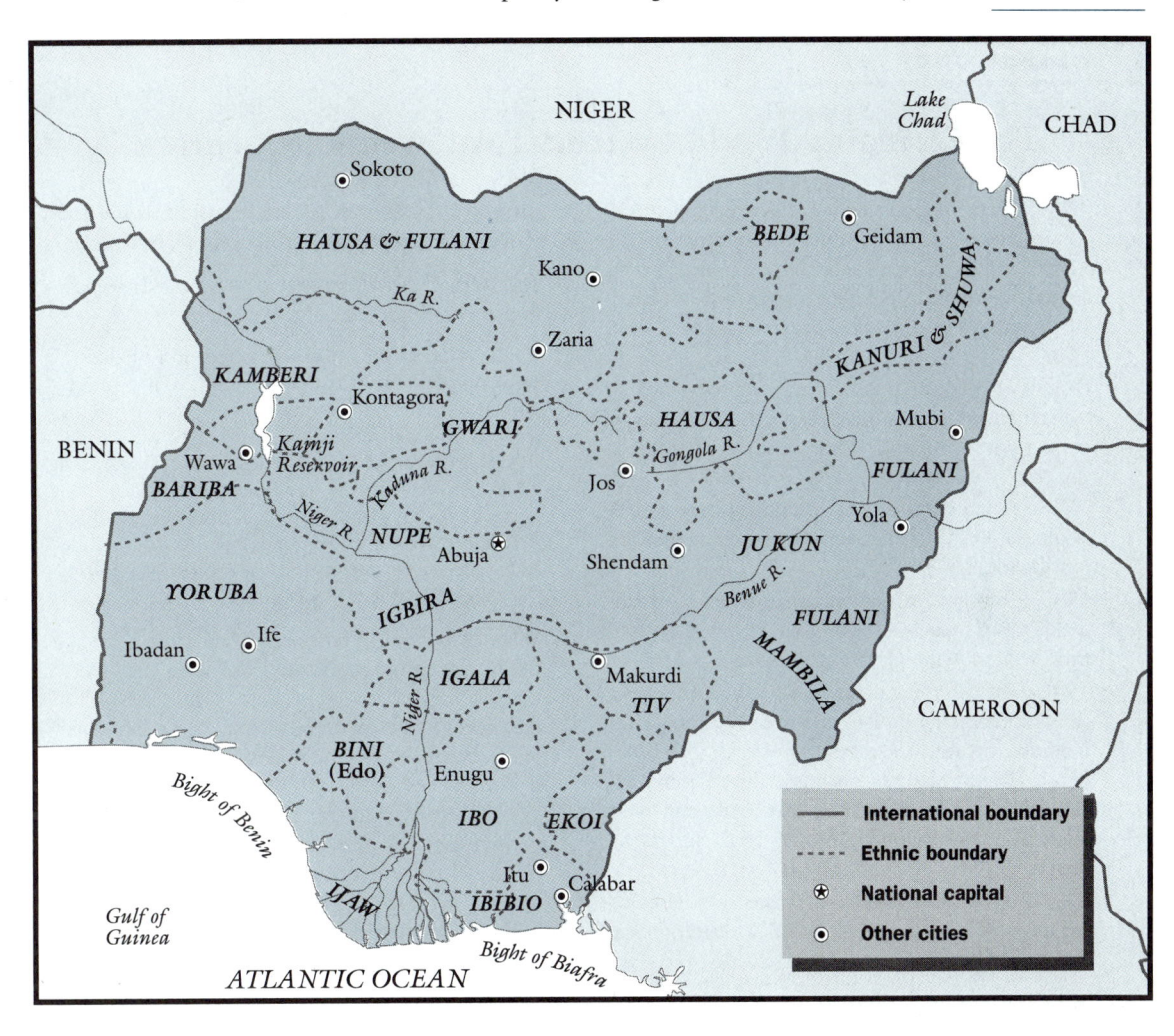

**MAP 3.5** The Ethnic Patchwork of Nigeria

in the 1990s all were the products of conflicts between distinct ethnic communities within each of these states.

**THE PROBLEM OF TINY STATES** In addition to fabricated nation-states, decolonization has sometimes produced states too small to have been thought viable fifty or one hundred years ago. The 1960s saw the independence first of a number of *ministates* with populations of less than one million: Cyprus in the eastern Mediterranean, Gambia in West Africa, Fiji in the South Pacific, and Guyana in South America. Then came the *microstates,* which include several island nations in the Caribbean and two in the Indian Ocean and the South Pacific with less than 100,000 people. One of the smallest "states" in the world, San Marino, with fewer than 24,000

> **Global Changes**
>
> ## The Changing World System: The Twentieth Century
>
> - 1918: The empire of Great Britain contained 550 million people and covered one-quarter of the earth's land surface. The empire of France contained 140 million people and covered one-sixth of the earth's continents. Texas.
>
>   1995: Great Britain, divested of its empire, contained 58 million people in a territory slightly smaller than Oregon. France, shorn of its empire, contained 58 million people in a territory about the size of Texas.
>
> - 1918: Intergovernmental organizations worldwide: about 50. Nongovernmental organizations worldwide: about 300.
>
>   1995: Intergovernmental organizations worldwide: more than 300. Nongovernmental organizations worldwide: several thousand.
>
> - 1920: The representatives of 42 countries met in the first Assembly of the League of Nations. They included two from Africa, six from Asia, and seventeen from the Americas (the United States refused to join). In addition to Australia and New Zealand, the remaining fifteen were European states.
>
>   1995: At the fiftieth meeting of the U.N. General Assembly, 185 members were in attendance. Forty-nine were from Africa, forty-nine from Asia, thirty-five from the Americas, eight from the South Pacific, and forty-three were from Europe.
>
> In 1990, Giesbert suggested the hazards of predicting technological change:
>
> > In the thirties, the American president Franklin D. Roosevelt commissioned his administration to undertake a vast study of the coming technologies. When the study was published it made a very big impression. Indeed, it was enthralling. There was just one problem: it had not predicted the coming of television, nor that of plastic, or jet planes, or organ transplants, or laser beams, not even of ballpoint pens! (In Alexander King and Bertrand Schneider, *The First Global Revolution* [New York: Pantheon Books, 1991], p. xxiv.)
>
> 1. Were Britain and France relatively more influential actors in the world in 1918 than they were in 1995?
> 2. Are intergovernmental organizations likely to play a more important role in world politics today than they did in 1918? Why or why not?
> 3. If, as Giesbert suggests, it is hazardous to try to predict technological change, which technological developments of the past ten or twenty years are now having an effect on world politics? Can you predict their likely impact over the next ten or twenty years?

people, became a U.N. member in 1992.[8] Clearly states of such sizes, while legally "sovereign," can scarcely be expected to staff embassies around the world, engage in major foreign policy ventures, or do much to thwart invasions from abroad—even though these have traditionally seemed essential to membership in the international system. Nor are these tiny states considered "nations" according to the older European definition.

In sum, Westphalia's impact upon the non-European world has been, at best, a mixed blessing. While it has encompassed, and eventually may have encouraged, the change from imperialism to nationalism, it has induced a proliferation of independent states. By granting independence on the basis of the political structures established by the colonial powers, and by allowing or even fostering the expression

---

[8]The tiniest Caribbean states include Antigua (64,000), Dominica (87,000), Grenada (84,000), and St. Kitts (40,000); the smallest Pacific states are Seychelles (70,000), and Nauru (10,000).

of nationalism, the Westphalian system has ensured continuing conflict, both within and among many "new" nations of the non-European world. By forcing the conduct of world politics into a single structure, the Westphalian order too frequently has ignored the realities of social life, thereby encouraging conflicts of a certain type.

**THE PROBLEM OF DISINTEGRATING STATES** Nationalism is still an extremely potent force today. It has recently exploded within several of the older states of Europe that had long seemed to be models of nation-statehood. Most striking at the end of the twentieth century is the disintegration of the Soviet Union, a successor to the Russian Empire whose earlier imperialism differed from Portugal or France in that Russia's rulers gradually conquered contiguous populations—first southeastward toward the Ural Mountains and the Caspian Sea, then in eastern Europe and the Baltic region, and finally clear across the Siberian reaches of Asia to the Pacific. That history, combined with Moscow's highly centralized rule, has meant that anyone looking at a map during most of this century might easily have supposed that the enormous landmass labeled *USSR* in fact comprised not just a single state, but a single nation as well.

Yet, with the coming of revolutionary reforms introduced into the Soviet Union after 1985, restive national groups under Soviet rule began to make their voices heard—with increasingly violent results.[9] By the end of 1991, the Soviet Union was disbanded and a Commonwealth of Independent States was created in the effort to provide a much looser link among eleven of the fifteen former Soviet republics. Meanwhile, sporadic violence had broken out between Armenians and Azerbaijanis in the Caucasus region, while there and elsewhere, Russian soldiers and restive Georgians, Uzbeks, Chechens, and others clashed.

The still potent idea of self-determination for nations also affected other states that had been created from the ashes of the Hapsburg and Ottoman Empires, including Czechoslovakia and Yugoslavia. Czechoslovakia split in 1992 into two states representing their Czech and Slovak populations, and Yugoslavia splintered into five new republics, with warfare the result.

Nor have other "older" states been immune from these kinds of nationality problems. In Chapter 2 we noted the similar problems stemming from British rule in Northern Ireland and Spanish control over the Basque population in Spain. One might add many more to this list: Turks in Bulgaria, Hungarians in Romania, and French-speaking Canadians in largely English-speaking Canada, to name just a few.[10]

In the worst cases, conflicting social groups have actually produced the virtual collapse of a few states, making it impossible for any group to govern them effectively. Lebanon qualified as a "failed" state from the mid-1970s through most of the 1980s, when near anarchy prevailed in most of the country. In the 1990s, Somalia, too, was torn apart as the result of fighting among the leaders of various clans. The

---

[9]For analysis of the last years of the Soviet Union and the emergence of new states from the Soviet empire, see Rachel Walker, *Six Years That Shook the World: Perestroika—The Impossible Project* (Manchester: Manchester University Press, 1993).

[10]Scholars are beginning to recognize the need to reconceptualize the position of the state in the world. See, for example, Joseph A. Camilleri, Anthony P. Jarvis, and Albert J. Paolini, eds., *The State in Transition: Reimagining Political Space* (Boulder, Colo.: Lynne Rienner Publishers, 1995).

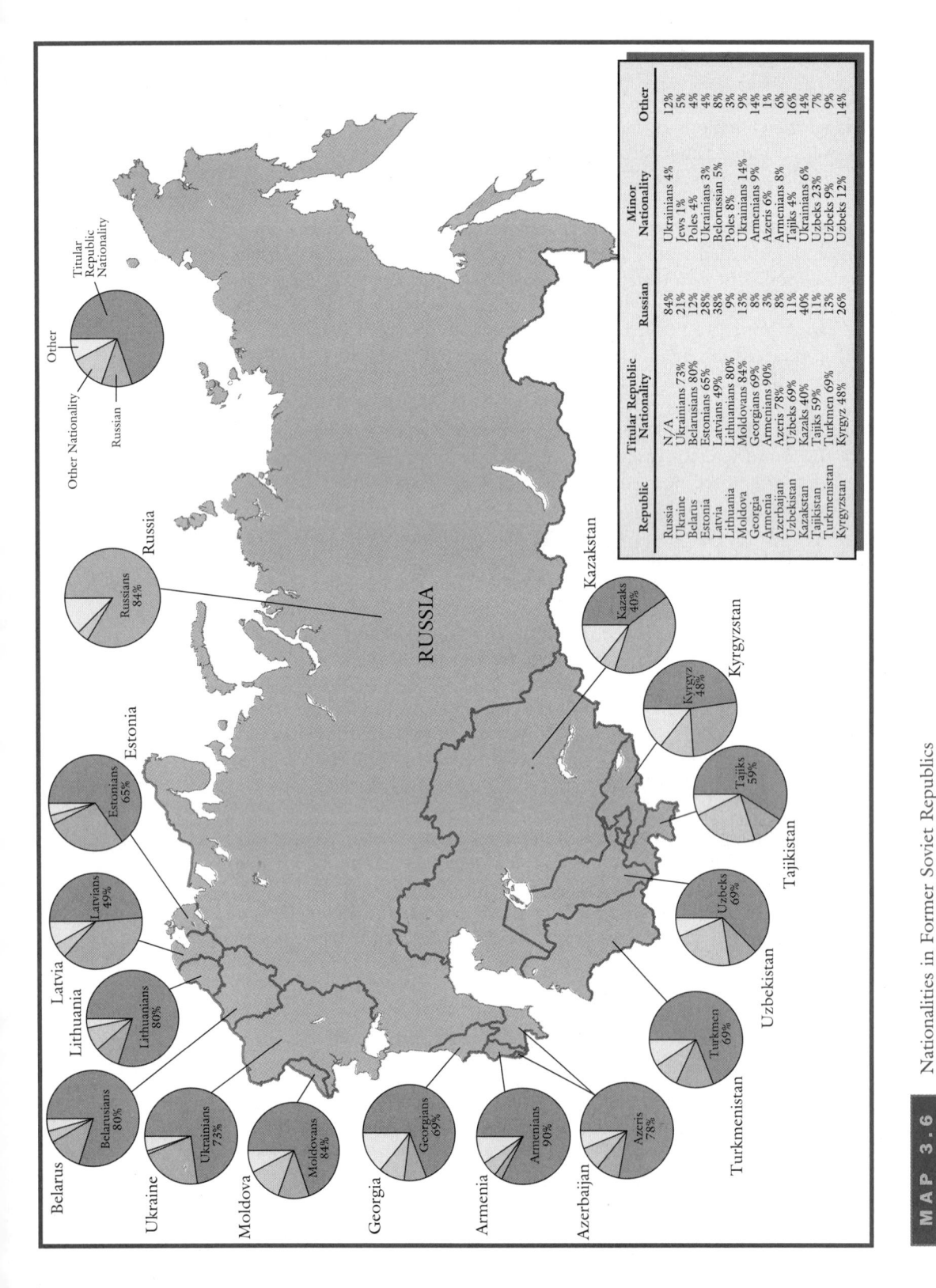

**MAP 3.6** Nationalities in Former Soviet Republics

Percentages are based on 1979 data.

| Republic | Titular Republic Nationality | Russian | Minor Nationality | Other |
|---|---|---|---|---|
| Russia | N/A | 84 | Ukrainians 4% | 12% |
| Ukraine | Ukrainians 73% | 21 | Jews 1% | 5% |
| Belarus | Belarusians 80% | 12 | Poles 4% | 4% |
| Estonia | Estonians 65% | 28 | Ukrainians 3% | 4% |
| Latvia | Latvians 49% | 38 | Belorussian 5% | 8% |
| Lithuania | Lithuanians 80% | 9 | Poles 8% | 3% |
| Moldova | Moldovans 84% | 13 | Ukrainians 14% | 9% |
| Georgia | Georgians 69% | 8 | Armenians 9% | 14% |
| Armenia | Armenians 90% | 3 | Azeris 6% | 1% |
| Azerbaijan | Azeris 78% | 8 | Armenians 8% | 6% |
| Uzbekistan | Uzbeks 69% | 11 | Tajiks 4% | 16% |
| Kazakstan | Kazaks 40% | 40 | Ukrainians 6% | 14% |
| Tajikistan | Tajiks 59% | 11 | Uzbeks 23% | 7% |
| Turkmenistan | Turkmen 69% | 13 | Uzbeks 9% | 9% |
| Kyrgyzstan | Kyrgyz 48% | 26 | Uzbeks 12% | 14% |

Westphalian order is of little help in such cases, since it is built upon the assumption that interacting states truly have sovereign capability at home.[11]

# THE GROWTH OF NONSTATE ACTORS IN WORLD POLITICS

Also in the twentieth century, a number of important participants in world politics have appeared that are neither sovereign states nor otherwise accounted for in Westphalian theory. The rise in importance of such nonstate actors as the United Nations and the European Community, Amnesty International and the Red Cross, or the Mitsubishi Corporation and Standard Oil means that the governments of sovereign states no longer exclusively dominate the field of international decision making. It also means that global society increasingly is being shaped by agents other than the governments that have the formal authority to speak and act on behalf of national communities. Some of these newly important actors and agents seem actively to help governments do their jobs, while others either oppose or ignore their sovereign authority. Among the most important of these nonstate actors in the world today are intergovernmental organizations, nongovernmental organizations, and multinational corporations.

## INTERGOVERNMENTAL ORGANIZATIONS

As their name suggests, **intergovernmental organizations (IGOs)** are standing bodies created by treaty among several or many sovereign governments to assist them in advancing certain kinds of policies. Familiar examples today include the United Nations (UN), the Organization of American States (OAS), and the Association of Southeast Asian Nations (ASEAN). Although such arrangements existed before the modern period—for example, the city-states of ancient Greece periodically conducted diplomatic relations through organized leagues—not until late in the nineteenth century did IGOs make a significant appearance within the Westphalian world. And only in the twentieth century did their number and importance increase exponentially. About 50 such organizations existed when World War I broke out in 1914; at the start of World War II in 1939, they numbered about 80. After World War II, they multiplied much more quickly, so that more than 300 were in existence by the 1990s.

Beginning in the 1870s with such little-known organizations as the Universal Postal Union (UPU) and the International Telegraph Union—today called the International Telecommunications Union (ITU)—these first modern IGOs were created by governments to assist them in carrying out important but noncontroversial tasks of benefit to people who increasingly interacted across state borders. These first two IGOs simply assured regular procedures for international communication, whether by post office or telegraph. The hundreds of comparable IGOs in existence today try to address the needs of different societies for dependable modes of inter-

---
[11] See I. W. Zartman, ed., *Collapsed States* (Boulder, Colo.: Lynne Rienner Publishers, 1995).

action when they engage in activities that need some regulation for the common good. An example would be commercial navigation on the Danube, which flows through several states.

At the close of World War I, the Allied diplomats who wrote the peace treaties at Paris also concluded a novel kind of treaty, the Covenant of the **League of Nations,** which created a new kind of IGO. Far more ambitious in intent than the IGOs that had appeared before, the League's central purpose was to transcend the failed effort of a handful of great powers to maintain the peace in Europe by uniting them with as many other states as would join them in a massive, collective commitment to keep world peace in the future.

After twenty years of some success in resolving international conflicts, the League of Nations failed to stop the aggressions that led to World War II, and there it died.[12] Yet it had succeeded in other respects that would be important for our time. It had proved to be a nearly indispensable tool, first, by providing a place where most governments of the world could interact on a regular basis and usefully address all sorts of foreign policy issues. The League had also shown that there were many issues in the modern world, particularly those that were not immediately peace-threatening, where progress in achieving human needs was possible. These included improving distribution of the world's food supply, ending the practice of forced labor, and helping to provide better health care around the world. Therefore, many were convinced as World War II ended that a new intergovernmental "umbrella" organization was necessary—one that would include virtually all the governments of the world and would pursue a host of measures, from trying to provide collective security to overseeing economic development, and more. The United Nations was the result.

The **United Nations** itself would soon begin to spawn a host of specialized agencies and other intergovernmental bodies of its own. But other IGOs were born after World War II that linked groups of states in myriad ways, creating whole networks or webs of interdependence among them.[13] Some, like the North Atlantic Treaty Organization (NATO) are in fact alliance arrangements, although they differ from alliances of an earlier era because these had permanent organizational structures, staffed with bureaucrats responsible for facilitating long-term military planning among member governments. Others, such as the Organization of African Unity (OAU), are multipurpose organizations meant to facilitate cooperation and conflict resolution among states of a particular region that nonetheless have diverse problems and interests. Still others, like the **European Union (EU),** are so ambitious in their plans to integrate the separate economies of members that they may lead to an unprecedented unity beyond their member states. As they succeed, they produce an analytical puzzle, for at some point, they should cease to be truly *intergovernmental* organizations and begin to act as *supranationally* integrated communities, with cen-

---

[12]For an account of the League, see F. P. Walters, *A History of the League of Nations* (Oxford: Oxford University Press, 1964).

[13]The terms *networks* and *webs* of interdependence have long been used by scholars to describe the result of this growth of NGOs. See, for example, Ernst Haas, *Tangle of Hopes* (Englewood Cliffs, N.J.: Prentice-Hall, 1969); and Harold K. Jacobson, *Networks of Interdependence,* 2d ed. (New York: Alfred A. Knopf, 1994).

tral authorities assuming such governmental functions as levying taxes and punishing those who ignore its rules.

This last kind of evolution, which is neither evident nor possibly intended in the work of most IGOs, nonetheless reminds us why we discuss the growth of intergovernmental organizations in a chapter on current challenges to the nation-state system. That may seem puzzling at first, for we have said that IGOs are the creation of sovereign governments to assist them in various ways in the conduct of their foreign policy. If that is so, how can they "challenge" the same state system they were designed to serve?

They do so for several reasons. First, the very existence of IGOs as standing arrangements ensures that no member's foreign policy is any longer just a series of one-to-one interactions with other states. That is, what used to be a bilateral relationship (state A asks B for something; B refuses; A insists more strongly; B gives A some of what it wants) has been changed into an ongoing, multilateral form of interaction (now when A asks B for something, states C through Z, all with interests in the outcome, are present to side with A or B, or to suggest alternative outcomes, etc.). We know that our individual interactions with another person are likely to take on different colorations when the two of us become part of a larger social group, perhaps especially when the larger group operates as a club or social organization. First, when a bilateral relationship is subsumed within a larger group interaction, with a new set of rules, procedures, and goals to guide its meetings, the social dynamics are likely to change, and individual members may find their goals and outlooks changing, too. Although the increase in IGOs clearly does not revolutionize Westphalian assumptions, it may be subtly changing the perceptions of governments about their fellow sovereigns while at the same time altering the ways in which they conduct their affairs.[14]

Second, while most IGOs may be largely instruments in the hands of member governments, a number of them were created to deal with problems that seem to be beyond the reach of the individual nation-state to solve. The work of the **Universal Postal Union (UPU)** is a simple example. It assures that a letter mailed from the United States to a person in Bolivia or Thailand will be delivered at the other end. That comes from the commitment of the Bolivian and Thai governments to honor U.S. postage, and from that of the U.S. government to do the same when letters from either country are addressed to persons in the United States. This creates a kind of supranational governmental authority, however modest in scope, uniting the United States, Bolivia, and Thailand (as well as all other members of the UPU) that seems to be demanded by our social interconnectedness. The need for this kind of governmental authority beyond the level of state government appears to be growing in pace with the rapid "shrinking" of our planet.

The sort of government beyond the nation-state exemplified in the work of the UPU may seem too trivial really to challenge the state's sovereign authority. Perhaps it is. Yet the ambitious plan to integrate fully the economic lives of many of the states of Europe through the EU is simply a larger and more complex example of the same phenomenon. It may be seen as the result of the nation-state's increasing inadequacy

---

[14]For an analysis of the impact of IGOs upon the state, see Paul Taylor, *International Organizations in the Modern World* (New York: Pinter, 1993).

to serve all the needs—perhaps first and foremost in economic life—demanded by the level of our social and political development. Will an integrated Europe simply copy, on a larger scale, the sovereign units typical of Westphalia, or will it evolve as a novel kind of social and political creature—something truly new under the sun?[15]

## NONGOVERNMENTAL ORGANIZATIONS

As their name indicates, **nongovernmental organizations (NGOs)** are distinguished from IGOs by the fact that they are not the creations of governments, but rather are the product of private groups of individuals whose interests or activities cross nation-state lines. NGOs began to appear in the nineteenth century and by World War I more than 300 existed; at the close of World War II, their number had more than doubled. Today there are thousands of NGOs—but their numbers fluctuate so much that no accurate count is possible. They are, if anything, even more diverse than IGOs.

One type of NGO unites the practitioners of particular professions internationally to facilitate exchanges of ideas among them in various ways, and to advance and protect their own individual and professional interests in general. Medical doctors, agronomists, musicologists, chemists—these and hundreds of others pursuing particular careers (or occasionally, hobbies) are joined in their own international associations. Such purely *professional NGOs* probably have only a slight impact on the work of sovereign states, but what they do express is the obvious fact that a great many professional interests and goals are quite irrelevant to one's nation-statehood; thanks to today's speed of communication, like-minded people can now be brought together for their common satisfaction as never before. At a minimum, this development is no doubt working to break down the simplistic—and frequently prejudiced—stereotyping of nationalities that stems from the lack of contact with members of other nationality groups.

Second are those NGOs that clearly do seek to influence state behavior in some way that is allegedly for the public good. These *public-interest NGOs* typically have a political agenda to advance, usually one that has encouraged them to organize like-minded individuals from various countries to lobby governments to change particular policies. Some of these have become familiar to informed publics throughout the world and are clearly having an impact on the way we expect to advance human aspirations. Two examples are *Amnesty International* and *International Physicians for the Prevention of Nuclear War,* both of which have been recipients of the Nobel Peace Prize. Amnesty International investigates charges of inhumane treatment of the political opponents of governments, publicizes abuses, and so directs both private and governmental pressure against oppressive regimes to release prisoners and otherwise improve their human rights behavior. International Physicians for the Prevention of Nuclear War was the brainchild of two doctors—one from the United States and the other from what was then the Soviet Union—concerned that governments seemed unwilling to make plain the unmanageable health costs that would

---

[15]For a clear discussion of the evolution of the European Union, see Desmond Dinan, *Ever Closer Union? An Introduction to the European Community* (Boulder, Colo.: Lynne Rienner Publishers, 1994).

be inflicted upon societies that survived a nuclear war. As an antinuclear pressure group, their organization's strength lies in the medical expertise of many of its members and in the commitment to saving lives that is central to the medical profession.

A third cluster of transnational bodies does not fit easily into either of the above categories. One such organization is the Red Cross. It resembles an IGO since it is the product of a governmental agreement, but it also functions as an association of the various national societies that are its members. Moreover, its main neutral body that assists prisoners in time of war—the International Committee of the Red Cross—is composed of a number of private citizens of Switzerland. There are also organizations established to convey and maintain a body of religious teaching throughout the world—the Roman Catholic Church is probably the most familiar example. While they are somewhat analogous to modern NGOs, they nonetheless exist to espouse a universalistic doctrine that, whatever its effect on politics, is largely meant to transcend the political concerns of this world.

Some public-interest NGOs are beginning to take on more direct roles in the global decision-making process. For example, NGOs with expertise (and policy goals) in environmental issues now often are participants at international conferences, where they may offer advice to government officials and may even be allowed to participate in the intergovernmental negotiations themselves. The result is a blurring of the once clear distinction between those who do and do not speak for sovereign governments in world politics.[16]

## MULTINATIONAL CORPORATIONS

A third kind of nonstate actor has emerged in recent decades to challenge the traditional ability of nation-states to regulate economic activity. **Multinational corporations (MNCs)** are private companies whose operations are so extensive across a number of countries that they are in a sense beyond the reach of any single national government.

At first glance, MNCs appear to be simply the bigger offspring of the already huge corporations of an earlier period. Yet, whereas such companies used to make their products within the confines of a single state, now their activities have been internationalized. That is, the components essential to their enterprises are produced in more than one country. Usually that means that their capital, management, and technology requirements are met in rich and economically advanced countries, where these things are in greatest supply, while their labor and raw materials come from those poorer countries that have the most abundant (and therefore cheapest) supply of these things. As a result, it is no longer possible to say that the goods produced were made in a particular country. Toyota cars, for example, although "Japanese" in the sense that management decisions are made at corporate headquarters in Japan, may be the product of mining operations in India (iron ore) and Zimbabwe (chromium), the manufacture of component parts in Korea, assembly at a plant in Mexico, and marketing plans from the United States.

---

[16]For more on this subject, see Thomas G. Weiss, ed., *NGOs, the United Nations, and Global Governance* (Boulder, Colo.: Lynne Rienner Publishers, 1996).

The sheer economic power of many MNCs is today comparable to that of many sovereign states. By 1994, the annual product of Mitsubishi was larger than that of all but the twenty-one richest nation-states, and ten of the forty largest economic entities in the world were corporations rather than states (see Table 3.2). These facts alone make clear that MNCs increasingly "compete" with nation-states as engines of the kinds of policies and actions that radically affect people's lives.

The growth of MNCs has several implications for the nation-state system today.[17] One is that, in some cases, the economic activity of an MNC in a small, poor, and underdeveloped state may place crucial decisions about the country's economic policy in the hands of a foreign board of directors whose first concern must always be its company's profits. To the extent that occurs, the very economic sovereignty of the host state must be questioned, whatever economic benefits it may receive from MNC activity.

A second implication is that the global interests of MNCs may make their corporate decisions impervious to the foreign policies of their home states. That is particularly likely when the corporation's desire to maximize profits runs counter to a foreign policy directive. Such was the case when a subsidiary of Gulf Oil (a U.S.-based corporation), after hesitating only briefly, turned over several hundred million dollars to the winning side in the 1976 Angolan civil war even though the U.S. government opposed and refused to recognize the new regime. Similarly, U.S. MNCs were charged with failure to provide needed petroleum supplies to the United States during the 1973–1974 oil boycott, preferring to sell their products more profitably elsewhere.[18]

Some observers see a third implication in MNCs, one that also views them as challenging state sovereignty, but in a way that may be more benign for the peaceful development of humankind.[19] Here MNCs are regarded as a positive force for social integration. In cutting across state lines they are seen to encourage the creation of a peaceful global environment, which is essential to the conduct of their own enterprises. Some argue that MNCs are effective in the sharing of technology, training, and the development of managerial skills across state boundaries, which contributes to an increasingly cosmopolitan workforce around the world. That, in turn, may be producing individuals with a self-interest in promoting peaceful solutions to intergovernmental disputes, for if such disputes ended in war, the economic lives of those employed by the MNC could be upset or destroyed.

These and other implications will be explored more fully in later chapters. Here we note that the growth of multinational corporations suggests that the most advanced economic life of the planet today has burst the bounds of the state in unprecedented ways. It may be that the development of MNCs, along with that of IGOs and NGOs, is an indication that important aspects of our political, social, and economic life can no longer be contained within the Westphalian framework.

---

[17] For a further exploration of these implications, see Edward M. Graham, *Global Firms and National Governments* (Washington, D.C.: Institute for International Economics, 1995).

[18] Lloyd Jensen, *Explaining Foreign Policy* (Englewood Cliffs, N.J.: Prentice-Hall, 1982), 165–166.

[19] For a discussion of the liberal (economic) perspective, see Robert Gilpin, *The Political Economy of International Relations* (Princeton: Princeton University Press, 1987).

**TABLE 3.2** Size of Annual Product of Selected Countries and Multinational Corporations

| | |
|---|---|
| 1. United States | 6,648,013 |
| 2. Japan | 4,590,971 |
| 3. Germany | 2,045,991 |
| 4. France | 1,330,381 |
| 5. Italy | 1,024,634 |
| 6. United Kingdom | 1,017,306 |
| 7. Brazil | 639,562 |
| 8. Canada | 542,954 |
| 9. China | 508,180 |
| 10. Spain | 482,841 |
| 11. South Korea | 376,505 |
| 12. Mexico | 375,476 |
| 13. Australia | 331,990 |
| 14. Netherlands | 329,768 |
| 15. India | 291,054 |
| 16. Argentina | 281,922 |
| 17. Russia | 274,929 |
| 18. Switzerland | 260,352 |
| 19. Belgium | 227,550 |
| 20. Austria | 196,546 |
| 21. Sweden | 196,441 |
| 22. *Mitsubishi* | 175,836 |
| 23. Indonesia | 174,638 |
| 24. *Mitsui* | 171,491 |
| 25. *Itochu* | 167,825 |
| 26. *Sumitomo* | 162,476 |
| 27. *General Motors* | 154,951 |
| 28. *Marubeni* | 150,187 |
| 29. Denmark | 146,075 |
| 30. Thailand | 143,209 |
| 31. Hong Kong | 131,881 |
| 32. Turkey | 131,014 |
| 33. *Ford Motors* | 128,439 |
| 34. South Africa | 124,726 |
| 35. Norway | 109,568 |
| 36. *Exxon* | 101,459 |
| 37. *Nissho Iwai* | 100,876 |
| 38. Finland | 97,961 |
| 39. Poland | 95,406 |
| 40. *Royal Dutch Shell* | 94,881 |

Figures are in millions of U.S. dollars and represent, for countries, their gross domestic product and, for MNCs, their revenues. All figures are for 1994.)

**SOURCE:** World Bank and *Fortune,* August 7, 1995, E1.

# The Shrinking Planet

Finally, we need to consider the impact of the *shrinking planet* on the traditional behavior of the international system. At the time of the Peace of Westphalia, the ability of humans to travel and communicate with one another across long distances had not improved since their ancestors thousands of years earlier. A ship sailing in 1648 from the Ottoman port of Tripoli in the province of Syria for the North African port of Tunis would have required as many days for the journey, and been subjected to the same hazards of nature, as its counterpart in the seventh century B.C. setting out from Byblos in Phoenician Syria for Carthage. On land, travel and communication remained little faster than the pace at which a human being or beast of burden could walk. Today, people can fly from the capitals of the comparatively recent nation-states, Syria and Tunisia, in about three hours, or communicate between Damascus and Tunis in an instant. With such new and growing capabilities as these, it is little wonder that our world today is referred to as a global village.

This shrinking of our habitat challenges the long-standing organization of our political life in unprecedented ways.[20] Since we will explore a number of those challenges in following chapters, here we shall only suggest several of the important consequences of our growing interdependence for world political life.

**Interdependence** means a *mutual* dependence, suggesting a fairly symmetrical relationship among those it binds together. **Dependency,** in contrast, implies unequal or asymmetrical relationships among units. Thus, a small child is typically dependent upon his or her parents for protection, nurturing, and growth; the parents themselves may function interdependently in the sense that both may contribute equally (but not identically) to their common welfare and to that of their dependent children.

## THE INTERDEPENDENCE OF SECURITY

The fundamental purpose of states—or more precisely, their governments—has always been to provide physical security to the societies they rule. That has meant that the lives of a few might have to be sacrificed in war to protect the security of the many. But increasingly today the security of some, or even all, civilian members of many populations seems placed in jeopardy by the novel developments of our century. These have produced both governmental and private forms of terrorism unknown to history, including what might be called the contagion of much contemporary violence.

Ever since the nation-state emerged in Western Europe, the basic test of its claim to sovereignty has been its government's protective capabilities. Such a police power goes hand in hand with government; where it fails, government typically falls. On the world stage, a government's police power has had to be effective against for-

---

[20]The shrinking planet also explains why neoliberal and globalist paradigms for the study of world politics now challenge the power politics tradition. The latter assumes "impermeable" territorial states as central to its analysis, whereas neoliberal and globalist approaches seek to understand a great many connections among social groups that cut across state boundaries.

eign adversaries for the sovereign state to remain viable. True, sovereign governments in the past have often had to contend with adversaries that were not other sovereigns. Revolutionary groups, secessionist movements, and lone assassins all threatened the viability of governments in nineteenth century France, the United States, and Russia, to name three examples. Still, because of the conditions of our modern social life, it may be that the nonstate enemies of states are more threatening today than at any time since the end of the Thirty Years' War.

The first of these conditions is found in the level of our technological development. In basic terms, our technology gives us power, especially power over our physical environment, as when we leap across continents in an airplane or press a button directing a computer to locate a book for us in a distant library. But because that kind of technology-derived power is so widely available to human beings everywhere, it can far more readily be employed by disaffected groups against "the system" today than in the past. The airplane may be hijacked by armed terrorists frustrated that the national group with which they identify does not have an independent homeland. A computer operator may wreak havoc in a government's communications network in a time of world crisis, conceivably paralyzing its ability to carry out a military operation.[21]

The very power of modern technology is, therefore, a fragile thing. The greater its ability to command the physical universe on our behalf, the more complex, intricate, and vulnerable technology is to disruption, either accidentally or in the deliberate effort to advance a cause at the expense of some government. One of the characteristic features of much recent terroristic activity, in contrast with that of an earlier day, is that now it is frequently directed at presumably innocent civilians. That may be partly because civilians are easier to target than are prominent officials, but it also seems to reflect an assumption that every individual member of a nation-state whose policies one opposes is responsible in some way for that government's behavior. Whatever the rationale, targeting civilians is a dramatic way to remind the public that the government is unable to fulfill its responsibility to provide security to its own citizens. As a result, fear is aroused throughout innocent populations, while other disaffected individuals may be inspired to attempt similar acts on behalf of their own causes. Thus begins the contagion of violence.

By the late 1980s, citizens of the United States were made aware of the difficulty its government faced in controlling the traffic in drugs, most of which were being produced in South America. While the drug trade was not the work of "terrorists" in the usual definition, such trade has been called *narco-terrorism*, in part because of the problems it poses to the Westphalian assumption that a state can effectively police its territory. International drug rings have been able to penetrate the security apparatus of nation-states with impunity—albeit for a cause that was entirely profit-based, self-serving, and anti-social, unlike the causes of political groups that have more often been labeled as terroristic. As drug cartels took on the trappings of

---

[21] A suggestion of how such an event might occur was the release of the "Michelangelo virus" in 1992, which threatened to disrupt computer operations throughout the world. The actual event was far less disruptive than many had feared. Even so, many record-keeping operations were destroyed in institutions from Europe to North America to South Africa.

> **Normative Dilemmas**
>
> ## The Impact of European Colonialism
>
> Colonialism's benefit to Europe:
>
> > [The territory in the Americas, Africa, and Asia colonized by Europe] was an empty land five to six times the size of western Europe, a land whose resources had not been exploited. When this great area was made available to the crowded and impoverished people of the (European) Metropolis, they swarmed out like bees to suck up the nectar of wealth, much of which they brought home to the mother hive. (Walter P. Webb, *The Great Frontier* [Austin: University of Texas Press, 1964], p. 13.)
>
> Colonialism's harm to non-Europeans:
>
> > [W]hen the native hears a speech about Western culture he pulls out his knife—or at least he makes sure it is within reach. The violence with which the supremacy of white values is affirmed and the aggressiveness which has permeated the victory of these values over the ways of life and of thought of the native mean that, in revenge, the native laughs in mockery when Western values are mentioned in front of him. (Frantz Fanon, *The Wretched of the Earth* [New York: Grove Press, 1963], p. 43.)
>
> How accurate was it for the author of the first quotation above to assert that the lands outside Europe colonized by Europeans were empty? What are that author's normative assumptions about the benefits of colonialism to Europeans? To their colonial subjects? Does his comment imply that subject peoples were enriched in equal measure with their colonial masters?
>
> Does the author of the second quotation suggest that European imperialists applied a double standard of values in their actions toward (a) themselves and (b) their colonized peoples? Do you agree? Why should a non-European subject to European colonialism "pull out his knife" when he hears a speech about Western values?
>
> ## The Impact of the Communications Revolution
>
> > Information is the oxygen of the modern age. Breezes of electronic beams flow through the Iron Curtain as if it were lace. Trying to control the flow of information is a hopeless, desperate cause. The Goliath of totalitarian control will rapidly be brought down by the David of the microchip. (President Ronald Reagan's Churchill Lecture to the English Speaking Union, Guildhall, London, June 13, 1989.)
>
> > The Western monopoly on the distribution of news—whereby even stories written about one Third World country for distribution in others are reported and transmitted by international news agencies based in New York, London and Paris—amounts to neocolonialism and cultural domination. (Mort Rosenblum, "Reporting from the Third World," *Foreign Affairs* [July 1977]: p. 816.)
>
> Does President Reagan's comment reveal that he has any reservations about the value of information reaching people throughout the world? Why not? Do you share Reagan's view that access to information is the "David" that helped slay the "Goliath" of Communism?
>
> Does Rosenblum suggest that all news media reflect the cultural biases of those who write and present the news? Is that fair? How different would be your view of the relative importance of major events in the news if you were (a) an executive responsible for marketing for the Sony Corporation; (b) a Buddhist priest held in prison for criticizing the military government of Myanmar (Burma); or (c) a young woman herding llamas in the Andes?

"sovereign" entities, they removed or murdered scores of government officials who stood in their way, effectively controlling a thriving industry that states were pledged to eradicate.[22]

---

[22]For example, in September 1989, the U.S. Defense Department reported that just to cut the flow of illegal drugs into the United States by half, more than one-third of the Navy's fleet would be needed, as well as more radar planes than the Air Force possessed, and nearly one hundred Army battalions—all of which would mean sacrificing many long-standing military missions to carry out such an interdiction role. *Philadelphia Inquirer,* Sept. 13, 1989, A1.

# THE COMMUNICATIONS REVOLUTION

The most revolutionary impact of the shrinking of our planet may result from modern developments in communications. The fact that we are now able to communicate with one another almost instantaneously across the globe is rapidly altering our very conception of human community. And communities have always been the building blocks of all political life and organization. Yet today such social clusterings no longer are dependent upon the happenstance of people living together in the same physical space, whether tribal lands, city-states, or nations.

> In the past, personal relationships relied on meeting at a cafe, signing a contract together, shaking hands, or interacting in the village square. With the advent of telephones, the fax machine, international publications, and computers, personal and professional relationships can be maintained irrespective of time and place. Today we are all members of international "non-place" communities.[23]

The Westphalian international order is built upon the premise that human communities (in this case, nation-states) *are* entirely confined to one geographical place. The revolution in communications that is "dissolving" place as fundamental to community building may also be undermining the presumption that states are the ultimate arbiters of how the largest, but still physically confined, social groups shall interact. That undermining has not yet been so severe that nation-states are everywhere in question. But our unprecedented ability to forge relationships that are not place-bound is certain to have far-reaching effects on human affairs and, thus, the interstate system.

The huge explosion in the speed of communication and the amount of information being passed around the globe is also producing what may seem to be a contradictory trend: the concentration of the mass media in a comparative handful of giant corporations. It is likely that, by the dawn of the twenty-first century, "five to ten corporate giants will control most of the world's important newspapers, magazines, books, broadcast stations, movies, recordings and videocassettes."[24] This development means that the opinion and ideas expressed through the mass media are being homogenized as never before in human history.

To the extent that concentration of information media encourages agreement upon common values and goals for human beings throughout the world, one may see a positive result for our political life on earth. It is a development that could reduce the divergent goals, and therefore the conflicts, that traditionally have led to large-scale violence between organized communities. But centralization of the information media is potentially dangerous, too. Any limitation on the number and variety of ideas to which we have access is a blow to our continuing ability to evolve as thinking and problem-solving animals. The worst-case scenarios in science fiction depict a future humanity regulated like robots in the hands of elites with monopolistic control of information and technology.

---

[23]Howard H. Frederick, *Global Communication and International Relations* (Belmont, Calif.: Wadsworth, 1993), 7.
[24]Ben Bagdikian, "The Lords of the Global Village," *The Nation*, June 12, 1989, 805.

Dan Wasserman © 1995, *Boston Globe*. Reprinted by permission.

# ECONOMIC DEPENDENCE AND INTERDEPENDENCE

Our ability to conquer distance—particularly in highly advanced societies—has allowed us to develop ever larger and more complex economic relationships as well as social ones. A strong impulse behind the creation of sovereign nation-states in much of Europe more than three centuries ago was the then-new capability of the emerging sovereign to assist in the creation of truly nationwide economic systems. The ability to collect taxes from a large population enriched the sovereign government, which then was able to redirect some of that wealth toward the construction of an infrastructure—including roads, a merchant fleet, educational opportunities, and the like. This encouraged economic growth and the development of economies that were truly national in scope. Today, however, few states are able to "contain" all of the activities of the most powerful economic actors. The rise of multinational corporations is the most obvious sign of how economic life is bursting the boundaries of individual states—and state control. In general, the more highly developed the state, the more dependent it is likely to be on resources, a labor force, and consumers (and, less frequently, investment capital and technology) originating in other states.[25] The European Union is the prime example of an evolving economic system on a transnational scale, for its members have deliberately sought to integrate their economies into a single economy embracing them all. Yet elsewhere, highly developed societies are also forging numerous economic links beyond their states as more and more economic activity takes on global dimensions.

Poorer societies, meanwhile, frequently remain dependent upon economic decisions made in the banking and financial centers of highly developed countries.[26] For them, the issue is not so much their interdependence as it is their inability to develop their own economies independently of forces and actors beyond their control. Such dependence can take a variety of forms, although two perhaps are the most basic.

The first is that experienced by the producer of primary products, such as rubber or coffee. A Malaysian rubber plantation may be owned directly by an MNC based in France or the United Kingdom or, if locally owned, it may be entirely dependent upon the demand for latex from a processing industry in the North. In either case, since the production of latex is one of Malaysia's principal means for earning foreign exchange, it is to that extent dependent for its economic well-being on forces beyond its borders. Among those forces is the fluctuating price of rubber in the world market that will result from changes in supply and demand.

A second form of dependence grew serious for many governments beginning in the 1980s when they were unable to repay debts owed to northern banks. The borrowing had been intended to finance development, particularly of the oil industry in such countries as Mexico and Venezuela that lacked the capacity for producing, refining, and marketing their large oil reserves. When oil prices dropped in the 1980s, they faced having to institute austerity measures at home in an effort to repay their loans. Such measures increased the economic hardship of their own people and no doubt caused them to resent that hardship all the more when it was perceived as the means to repay far more wealthy foreign investors.

Both dependence and interdependence reflect the growth of an economic marketplace that is becoming global in scope. If that trend continues—and there is every evidence that it will—our economic interdependencies will grow as well. It is wise to note that the subsistence economies of premodern agricultural societies were, *because* they were at primitive levels, far more self-sufficient and impervious to external economic forces than are any of the developing or highly developed economies of our time. The oil shocks induced by OPEC in the 1970s are quite literally unimaginable in pre-Columbian North America, in West Africa during the empire of Ghana,

---

[25] See Felix Rohatyn, "Restoring American Independence," *New York Review of Books*, February 18, 1988, 8–10; and Edward M. Graham, *Global Firms and National Governments* (Washington, D.C.: Institute for International Economics, 1995).

[26] A number of scholars have sought to explain persistent problems of economic underdevelopment in much of the world in terms of "dependency" theory. See, for example, Heraldo Munoz, ed., *From Dependency to Development: Strategies to Overcome Underdevelopment and Inequality* (Boulder, Colo.: Westview Press, 1981). See also our discussions in Chapter 1 and Chapter 12 of the globalist perspective that focuses on dependency under capitalist economics.

or anyplace else in the world before the coming of the Industrial Revolution; preindustrial economic life had only the slightest requirement for oil but generally could continue undisturbed as the result of decisions made locally.

## THE ASSAULT ON THE BIOSPHERE

Environmental issues received almost no attention in world politics prior to the late 1960s. Before then, problems of pollution and environmental damage were viewed almost wholly as matters of national concern. They only began to be included in the world's political agenda when the crowding of the planet and the increase in industrialization began to make it clear that a much larger and more coordinated effort was needed to protect our natural habitat. Environmental issues still took a back seat to more traditional issues of world politics until late in the 1980s. Then Chernobyl, the greenhouse effect, the hole in the ozone layer, the Exxon Valdez oil spill, acid rain, and the loss of tropical rain forests all made headlines. They described threats to the safety and well-being of humanity that could not be effectively treated by even the most enlightened policies of one or a few national governments. Indeed, these kinds of issues will require international attention for years to come.

Two interlocking trends largely account for the unprecedented current threat to the biosphere.[27] The first is the explosion in recent decades in the human population; the second is the even greater rate of increase in the amount of energy human beings consume, which is rapidly depleting the earth's nonrenewable resources while filling the biosphere with toxic waste. These trends are related because the industrialization of traditional societies has always been accompanied by sufficient improvements in health care to assure the survival of infants—where before many would not have lived to bear children of their own. Birth rates have overtaken death rates, and explosive population growth has been the result (see Figure 3.1).

A few figures tell the story. When the Roman Empire fell in the middle of the fifth century A.D., there may have been 400 million human beings scattered across the planet. Their numbers slowly increased during the next thousand years (perhaps having doubled by the time Christopher Columbus was born), then reaching one billion individuals by 1600. Within another two centuries, the Industrial Revolution began in Western Europe; its effects have been extended to nearly every quarter of the globe today. As it spread, population growth took off exponentially. Whereas previously it had taken one thousand years for the world population to double, the doubling from one to two billion living human beings occurred in some three hundred years, from about 1600 to 1900. The third billion arrived in fifty years, and the fourth in just thirty more years; thus, the latest doubling of the world's population occurred in only about eighty years, from 1900 to 1980.

Clearly, such rocket-like growth cannot be sustained by the resources of a finite planet, for human beings need food, land, and water in abundance. Standing-room-only conditions are absurdly unimaginable, even though a projection from recent trends might suggest that as a possibility in the not too distant future.

---

[27]The term *biosphere* was coined in 1926 by the Soviet physicist V. I. Vernadsky to describe the earth's evolution to the point that it was capable of sustaining life.

**FIGURE 3.1** World Population Growth, 1750–2100

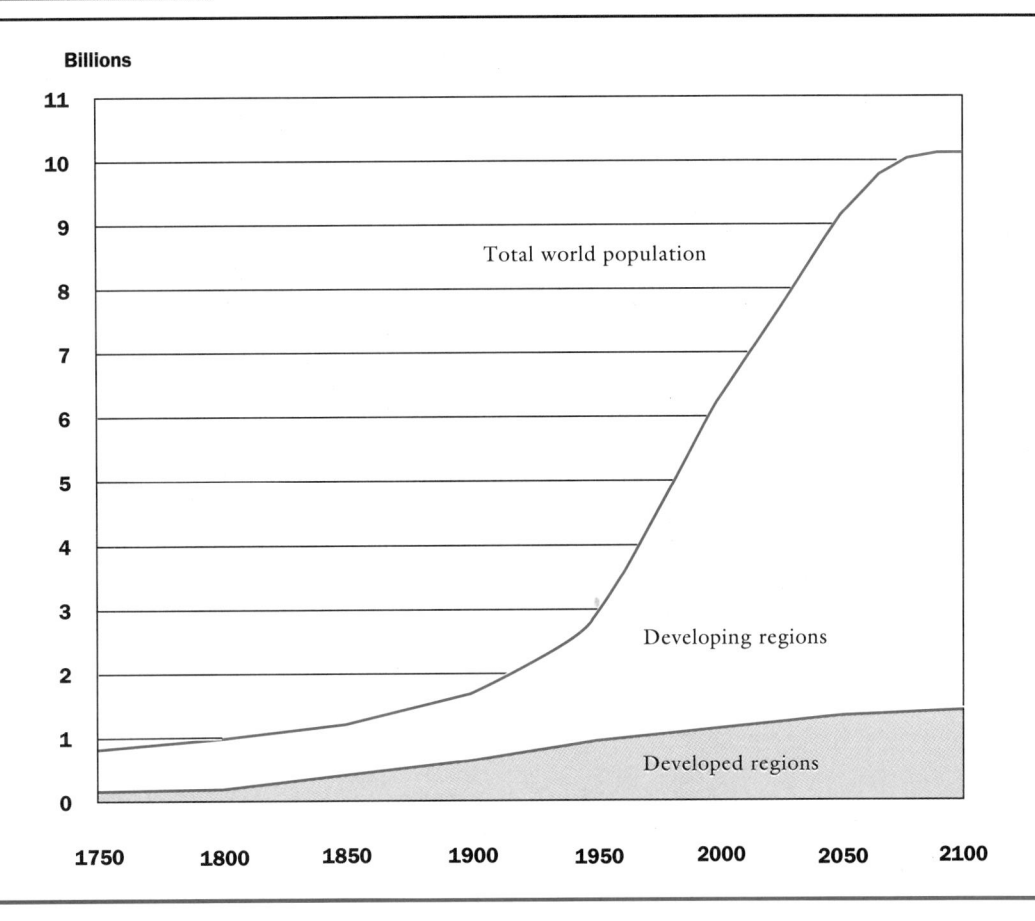

*[Graph showing world population in billions from 1750 to 2100, with curves for Total world population, Developing regions, and Developed regions. Total world population rises steeply from around 1950, approaching about 10 billion by 2100. Developing regions follow a similar curve. Developed regions grow slowly to about 1.5 billion.]*

In fact, we may already have reached the sustainable limits to our growth, for our current rates of energy consumption may be threatening the biosphere with collapse. The conditions of modern economic life demand the uses of energy on a scale unimagined by our ancestors. Each human being alive in 1970 consumed on average more than *ten times* the amount of energy than did each person living in 1900. Yet the earth's population had nearly doubled within that period, so that the earth's resources were being used up at a rate far in excess of the population growth rate alone.

But to speak of energy consumption can be misleading. While that accurately describes what we are doing to the planet's nonrenewable resources—particularly when we burn such fossil fuels as oil, natural gas, and coal—the resource is actually transformed in the process of providing us with energy so that its harmful by-products are returned to the biosphere as hazardous waste and pollution. We witnessed unprecedented industrial growth in the United States during the first half of the twentieth century, but saw Lake Erie and many other bodies of fresh water pol-

**TABLE 3.3** World Population, Total and Annual Addition, 1950–94

| YEAR | POPULATION (BILLION) | ANNUAL ADDITION (MILLION) |
| --- | --- | --- |
| 1950 | 2.555 | 37 |
| 1955 | 2.779 | 53 |
| 1960 | 3.038 | 41 |
| 1961 | 3.079 | 56 |
| 1962 | 3.135 | 70 |
| 1963 | 3.204 | 71 |
| 1964 | 3.276 | 69 |
| 1965 | 3.345 | 70 |
| 1966 | 3.414 | 69 |
| 1967 | 3.484 | 71 |
| 1968 | 3.555 | 74 |
| 1969 | 3.629 | 75 |
| 1970 | 3.704 | 78 |
| 1971 | 3.782 | 77 |
| 1972 | 3.859 | 77 |
| 1973 | 3.936 | 76 |
| 1974 | 4.012 | 74 |
| 1975 | 4.086 | 73 |
| 1976 | 4.159 | 73 |
| 1977 | 4.231 | 73 |
| 1978 | 4.304 | 76 |
| 1979 | 4.380 | 77 |
| 1980 | 4.457 | 77 |
| 1981 | 4.533 | 81 |
| 1982 | 4.614 | 81 |
| 1983 | 4.695 | 80 |
| 1984 | 4.775 | 81 |
| 1985 | 4.856 | 83 |
| 1986 | 4.941 | 87 |
| 1987 | 5.029 | 88 |
| 1988 | 5.117 | 88 |
| 1989 | 5.205 | 90 |
| 1990 | 5.295 | 86 |
| 1991 | 5.381 | 88 |
| 1992 | 5.469 | 88 |
| 1993 | 5.556 | 88 |
| 1994 (prel) | 5.644 | 88 |

**SOURCES:** U.S. Bureau of the Census, Center for International Research, private communication. February 6, 1995. Lester R. Brown, Nicholas Lenssen, Hal Kane, *Vital Signs 1995* (N.Y.: Norton, 1995), p. 95. Copyright by Worldwatch Institute.

luted to the point that they required massive clean-up efforts in the century's second half. We unleashed the atom's energy, but face catastrophe whenever human error releases radioactive fallout from a nuclear power plant. We have acquired more powerful machines than have ever served humans before, but may in producing them have induced a global warming that could alter climates and raise sea levels to the point of destroying every seaport on the planet.

This is only a part of the story. More people engage in more activities that threaten the health of the biosphere. We move ever larger tankers over the seas to transport the oil that drives our engines, but watch helplessly as plants and animals are destroyed when a tanker runs aground. We clear valuable land of its timber as the demand for both land and timber rises, although the destruction of our forests also destroys a principal source of the world's oxygen supply. We use quantities of chlorofluorocarbons (CFCs) to satisfy the ever rising demand for refrigerants and aerosol sprays, but now know that these chemical compounds are rapidly destroying the ozone layer that protects us from the lethal rays of the sun.

Since all living creatures, human beings included, depend for their very survival upon the health of the biosphere, our need to undo threats of these kinds—threats that have been almost entirely of our own making—is fast becoming a matter of the utmost urgency. Here our actions may determine, not just the fate of this or that nation-state, not even merely the survival of *Homo sapiens*, but that of every living species on the earth.

## CONCLUSION

A number of serious challenges confront the nation-state system as we approach a new century. Among these is nationalism, whose impact still seems more threatening to particular, non-national states than it does to the state system itself. In this century, nationalism has become the principal justification for creating independent states throughout the world, ending the empires of the European powers and inducing the creation of new nations (as well as new states) out of their ashes. Thus, while undermining what had been imperial states, nationalism has made nation-states the nearly universal, central actors in world politics.

As we look at ethnic settlement across the globe, we must conclude that the potential for serious conflict between dissident national minorities and the states in which they live remains nearly limitless. Few states are completely immune for the simple reason that most still incorporate national or ethnic minorities. Wherever these minorities become sufficiently unhappy with their treatment at the hands of the majority to claim a right to sovereign independence, the ethos of the Westphalian nation-state system will tend to support their claim, although it can do little to prevent the civil wars that secession may produce.

We have examined the growth of nonstate actors in world politics—in particular, the great increase in the number and roles of intergovernmental organizations, nongovernmental organizations, and multinational corporations. Of the former, the European Union is the most dramatic example of a new political actor that is actually challenging the Westphalian system's basis in the nation-state. NGOs and MNCs, meanwhile, are facilitating new social and economic associations that cut across state lines. Their growth suggests that human interaction is increasingly less confined within single states, and that these new transnational associations therefore challenge states in unprecedented ways.

The interdependence of peoples proceeds at a dizzying pace. Modern technology has made the security of most human beings ever more vulnerable. This vulnerability lies both in the ability of private groups to injure innocent civilians and in the potential of governments to commit mass murder—including warfare, the form of murder traditionally acceptable to world politics.

The communications revolution is allowing the formation of human relationships that are not nearly as limited by geographic proximity as was the case throughout history. As the amount of available information has exploded, its dispersal also is falling into the hands of ever fewer media giants. In our economic interactions, we are increasingly interconnected to markets and other forces beyond our borders. That condition is frequently one of mutual or interdependence for richer societies today, and of greater dependence on the rich for members of poorer societies. Finally, at the environmental level, we increasingly must recognize our global interdependence on a common biosphere, which humanity, through its burgeoning growth and economic activity, now threatens far more seriously than has ever been the case before in human history.

What all of these issues suggest most dramatically is that Westphalian arrangements for organizing humankind's political life look remarkably antiquated for at least some of the kinds of decisions and actions that seem required today. The very logic of Westphalia was, and is, that the human communities we call nation-states may develop and thrive in substantial isolation from one another; when they must interact, they may do so effectively without any higher authority to goad them into common action, or any centralized police power to assure compliance with the desires of substantial numbers of others. Yet the political, social, economic, and environmental issues we confront at the end of the twentieth century often seem to demand that we rethink how the state system has traditionally tried to provide security for large communities, and that we act in greater unity than has ever been necessary in the past to treat the common dangers and opportunities that we face as a species.

PART II

# Explaining Foreign Policy

Although current challenges confronting the global system may require that we rethink the institutional structure of the Westphalian state system, the fact is that the nation-state remains an important actor in the world system today—as the realist argues. The modern state, after all, retains dominant control over the instruments of violence and by and large determines how global resources are divided. It is therefore essential that we try to understand the various roles played by the state and why states behave as they do in the world arena.

To find answers to these questions, we shall first examine the foreign policy objectives of the various state actors and how certain belief systems appear to have shaped those objectives. Now that the cold war is over, are we likely to see the end of ideological conflict, as some have asserted? Or should we expect merely the reassertion of other ideologies, as illustrated by the rise of Islam, or the possible conflict between cultures, such as those represented in the North and the South?

What complicates whether foreign policy decision making can be rational is the very process in which those decisions are made. Particularly troubling at times in this regard are deficiencies in the perceptual capabilities of governmental decision makers, bureaucratic and governmental infighting, and restrictions placed upon rational choice by a host of selfish domestic interests and public opinion.

At the same time that we remain interested in the *why* and *what* questions that lead us into an explanation of empirical reality, we need to remind ourselves of an important purpose for raising such questions. Ultimately, we human beings are concerned with the normative answers revealed by our research. In other words, we think that the purpose of policy-related inquiry is to help us determine what ought to be done to create a more peaceful and prosperous world in which human dignity can prevail. ■

*Overleaf* • Governmental leaders of the seven most powerful industrialized states (G–7) meet in an economic summit at Naples, Italy, in July 1994.

CHAPTER 4

# Objectives, Beliefs, and Foreign Policy

If one expects to understand the myriad of foreign policies which have been adopted over the course of the Westphalian system, an obvious starting point would be that of the objectives and goals developed by the participants in the system. These are, of course, influenced by the **beliefs** that people hold regarding the world in terms of how it operates, whether it is changeable, or whether an individual can affect his or her own destiny. Beliefs involve convictions of what is right and wrong and thus have a strong normative component—that is, a sense of what ought to be. Since such beliefs have changed over time, the objectives of the participants in the Westphalian system have been far from static.

## CHANGING WESTPHALIAN OBJECTIVES

With the establishment of the Westphalian system in Europe, the primary objective of the emergent nation-state became that of providing security. By the seventeenth century, new military technologies had made smaller polities an anachronism, for they could no longer provide for the basic security needs of people. Gunpowder and battering rams rendered the walls that had defended fortified cities vulnerable to external attack, while much larger territories increasingly could be made secure by centralized governments with their newly acquired ability to raise conscript armies. Moreover, the decimation resulting from the religious wars of the sixteenth and seventeenth centuries had convinced many Europeans that the medieval conception of a political order built upon a united Christendom had been destroyed beyond repair.

Gradually, other state objectives tended to supplement the basic one of security. For example, economic well-being grew in importance as economic life developed, requiring more complex trade and monetary relationships among people throughout the state—which, for the first time, was becoming an integrated economic unit. During the seventeenth and eighteenth centuries, governments embraced **mercantilism**—a form of economic nationalism—by establishing restrictions on exports and imports, foreign investment, and other economic transactions, all in an effort to protect domestic economic interests.

As material needs for security and economic well-being increasingly were satisfied, still other objectives—such as propagating a government's ideological values—came to play a more important role in foreign policy. Religious groups have often sought to proselytize others—by the sword if necessary—as in the rapid spread of Islam in the seventh and eighth centuries, Christianity's crusades of the eleventh through the thirteenth centuries, and the wars of a divided Christendom in Europe in the early modern period. The successful effort to spread the values of European civilization occurred during the West's ascendancy, from about the eighteenth to the twentieth centuries. For the British, the French, and other European powers of that period, "civilizing" (which meant Europeanizing) colonial peoples became the "white man's burden." For the United States in the nineteenth century, extending control over a continent became the **manifest destiny** of American policymakers. In the twentieth century, violence has often been used in an effort to spread new ideologies, such as Fascism and Marxism-Leninism, which took on an almost messianic flavor. These twentieth-century ideologies in turn have been defeated, no longer providing the ideological engine for various foreign policies that they once did. This has led some to argue that the world has come to the end of ideological conflict, which has been replaced by the triumph of liberal-democratic values throughout the world.

## NATIONAL INTERESTS AND FOREIGN POLICY

As people increasingly identified with the power and glory of the state during the course of the Westphalian system, national prestige came to be a much sought-after value. Projects to send men and women to the moon and to plan flights to Mars have been undertaken largely from the desire to enhance the prestige of the state. They are in that respect today's equivalent of Sir Francis Drake's circumnavigation of the globe for England, or of Louis XIV's construction of a vast royal palace at Versailles. Similarly, the objective of some states to achieve a nuclear weapons capability has been motivated at least as much by the desire for prestige as it has from concern about national security.

Those ascribing to liberal democratic values increasingly have added to their foreign policy agendas the issue of how other governments treat the human rights of their own citizens. Idealists such as Presidents Woodrow Wilson and Jimmy Carter made democratic rights central to their foreign policies. However, when they were persuaded that human rights goals conflicted with certain of the more traditional or core interests of the United States, they sometimes allowed those realist interests to override their idealist goals. Thus, whereas President Wilson was preoccupied with national self-determination in the European peace settlement at the end of World War I, he seemed little concerned about it in the colonial areas of the world. Similarly, President Carter could find only three countries where human-rights violations were serious enough to require him to withhold aid in keeping with congressional legislation—and these three were all in Latin America and little threatened by communism. U.S. national security interests, however, were deemed too important to deny aid to such dictators as Ferdinand Marcos of the Philippines and Park Chung Hee of South Korea.

While it is often argued that bedrock **national interests**—maintaining the state's security, providing for economic and social welfare of its people, and so on—are extremely durable, the policies required to achieve those interests clearly must and do change in accordance with the times. For evidence, one need only reflect on the changing foreign policy issues of nations during the half century since the end of World War II. One major postwar agenda item was that of anticolonialism, but with the independence of virtually every non-European polity on the globe by the 1970s, that issue has nearly disappeared. The early postwar period also saw the rise of the cold war, which became in most respects the central feature of international politics for the next forty years; yet today, the cold war, too, has been relegated to the history books.

The early 1970s saw yet another abrupt change in agenda as the oil boycott and the energy crisis brought a radical shift in global power relationships. That shift in turn subsided somewhat with the oil glut of the 1980s. Also prominent, with the rise of sovereign voices from the South, was the developing world's attempt to build a new international economic order to replace the economic structure that many less developed countries viewed as biased toward the rich states of the North. More recently, environmental issues have taken precedence as people worry more about the impact of the depletion of the ozone layer and global warming trends, along with other potential environmental hazards. And with the fading of the cold war, suddenly the threat of nuclear annihilation no longer seems to be the overriding danger to humankind that it appeared to be to many only a few years earlier. Comparable changes no doubt will continue as states shift their foreign policy objectives to cope with an ever changing global environment.

To an appreciable degree, the goals that states pursue will depend upon which groups dominate the decisional structure of the state. For those states ruled by dynasties (as many used to be), continued political control by a family becomes the primary objective. Today, dynastic traditions remain largely confined to several Muslim states of the Middle East, although early in the Westphalian period, European dynasties were supported by the doctrine of the divine right of kings. Those systems dominated by strong economic interest groups with considerable political clout, not unexpectedly, will seek policies that advance the economic interests of those groups. Some small and poor states appear to have their domestic and foreign policy objectives shaped by the powerful economic interests of investors, often based abroad, who are typically more interested in profits for themselves than they are in the welfare of their host countries. The so-called banana republics of Central America have long been regarded as a case in point.

Those groups that speak on behalf of states actually have a hierarchy of goals. Some writers have sought to distinguish among such goals. First are those that can be viewed as of core importance, such as the survival of the state and the protection of its territory. Then come the more middle-range objectives, such as expanding trade relationships, enhancing the state's prestige, or perhaps even engaging in efforts to extend the state through economic or military imperialism. Finally, there are long-term goals, which envision such dreams as the communization or perhaps the democratization of the world.

## NORMATIVE VALUES IN FOREIGN POLICY

As we explore the ways in which a state's foreign policy goals are influenced by beliefs, we should understand what is implied in the very concept of a goal, either for an individual or for a society. By definition, a *goal* assumes some hoped-for departure from the empirical reality of one's situation. To have a goal is to propose that something *should* or *ought* to happen. That, in turn, assumes that the goal setter aspires to some standard or ideal outcome. The standards that guide human behavior are **norms,** and goal-oriented thought and action is therefore considered **normative** behavior. In other words, all foreign policy goals are driven by normative assumptions. Decision makers seek to change the reality of their state's situation according to their standard of what they prefer (what ought to be) for their state.

For a variety of reasons, it is unlikely that the foreign policymakers of different states will be guided by identical norms or standards. The foreign policymakers of a small, poor state will have a different set of preferred outcomes for their nation than those of their more powerful neighbors. Both groups may be able to articulate those differences in terms of their national interests. Yet the differing values, and therefore normative views, of individuals responsible for a state's foreign policy probably will mean that what is perceived as a national interest is itself subject to dispute. Some decision makers within a state may place a higher value than their colleagues do in seeking security by accommodating their neighbors, rather than preparing to oppose them militarily, and so on.

Clearly, an attempt to understand the normative values that decision makers hold is critical to an analysis of their goals, including the risks they may think they are taking in trying to achieve those goals.

> Values determine the choice of goals, influence the alternatives which are seen as 'thinkable' (assassination and nuclear bombing, for example, were ruled out [by the Bush administration] in the Persian Gulf War, despite some reasonable normative arguments in favor of the first and undoubted capability to carry out the second), and the degree of risk which can be tolerated.[1]

What, then, are the influences that shape our values?

## THE FOREIGN POLICY ROLE OF BELIEFS

As these considerations suggest, foreign policy objectives do not spring spontaneously from the minds of decision makers. In addition to being influenced by specific perceived needs, these objectives are a product of the past experiences of a na-

---

[1] Roy Licklider, *Policy Analysis and Argument* (New Brunswick, N.J.: Rutgers University, June 1993, mimeographed), 5.

## Belief and Value Conflict in the Bosnian Crisis

The Bosnian crisis, which began in 1992, pitted three Yugoslavian groups against one another: Muslims, Serbs, and Croats. The crisis was driven by contrasting beliefs and values involving conflict over the following:

### Goals: Something that should or ought to happen.

The Bosnian government, predominantly Muslim but represented by Croats and Serbs as well, pursued the goal of a unified Bosnia based upon the borders recognized by other states when the latter extended diplomatic recognition to Bosnia-Herzegovina in 1992. Yet many Bosnian Croats and Serbs hoped to carve Bosnia into ethnic divisions that would be linked with Croatia and Serbia, respectively.

### National or Ethnic Belief Systems: Hypotheses and theories that a group believes to be true; they may be derived from historical tradition, culture, or ideology.

Neither the unification of the southern Slavs in the state of Yugoslavia after World War I nor the area's unification under Tito's brand of communism sufficed to build a new and viable national belief system. Franjo Tudjman of Croatia and Slobodan Milosevic of Serbia took advantage of the fragile unity in Bosnia by calling attention to real and imagined past injustices over the centuries. This led to the creation of separate Serb and Croatian republics in other regions of what had been Yugoslavia, along with demands that all territories, including those in Bosnia, be partitioned to reflect these new nationalisms.

### Values: Beliefs about desirable behavior, objects, and situations along a continuum of relative importance.

Although there were a number of shared values and considerable intermarriage among Bosnians from each of the three groups, different experiences and religions have led many of them to see themselves as distinct.

### Ideology: A coherent body of ideas concerning the means for maintaining or changing the socioeconomic political order.

Several ideologies have gained adherents and influence thought as it relates to events in Bosnia. These include religious ideologies such as Islam, which provides a very complete guide to life in the Koran, and Christianity, which in turn reveals some of its divisions in the Roman Catholic Croatians and the Orthodox Christian Serbs. Because of their earlier influence, both fascism and Marxism remain relevant, albeit marginal. The region is also likely to be affected by growing global influence of liberal-democratic beliefs. These different ideas, of course, will continue to have a divisive effect upon the peoples of the region.

tion and the political beliefs and ideologies that have come to be accepted over the years. Whatever the source of a particular group's belief system, those beliefs influence the formulation and objectives of foreign policy in a variety of ways. First, a society's belief system affects what is seen and viewed as significant in the international system. It provides blinders by which certain international events might be denied or reinterpreted to be made compatible with the prevailing belief system. Some writers have likened the belief system to a *prism* through which decision makers view reality. Just as the light that passes through a prism is refracted and distorted, so, too, are new events colored differently as they pass through the prism of ideology and past experiences.

Belief systems also place certain *constraints* on the range of foreign-policy options. A group of decision makers, even in the most authoritarian regime, will find it difficult and perhaps politically suicidal to venture far from what is generally conceived as compatible with the belief system of their constituents. A democratic leadership may feel constrained by a population that is likely to disapprove of any action by its government that smacks of undemocratic or unfair manipulation of another state's political system. States with competing ideologies are likely to rule out alignments with each other as viable options. If a pact is made across ideological lines, it often tends to involve weak commitments, as in the nonaggression pact between the Communist Soviet Union and Nazi Germany in 1939, or the temporary alliance made by the Western democracies with the Soviet Union for the purpose of conducting the fight against Hitler during World War II.

National belief systems help provide *continuity in foreign policy*. The more cohesive a society's belief system, the more stable that nation's foreign policy will be. States with long historical traditions tend to have greater continuity in their foreign policy, making for greater predictability. On the other hand, newer states without such traditions of continuity do not have to overcome long-established ways of thinking or behaving in order to change their policies, which permits them considerable flexibility.

Belief systems provide a means for *rationalizing foreign-policy choices*—choices that are often made on the basis of interpretations of national security interests but are sold to the public on the basis of certain shared values. Throughout the cold war, Soviet decision makers were quick to note that any choice they made in foreign policy was guided by Marxist principles, and the United States tended to rationalize its choices in terms of "protecting the free world" or "making the world safe for democracy."

National beliefs may be used in *propaganda* as states seek to rationalize and justify their political positions. Efforts are made to convince others of the correctness of one's own views as ideologies struggle to gain people's attention through the use of such media as television, radio, newspapers, and magazines.

Belief systems can also serve as a device for *enhancing national unity* among those who subscribe to a given view. As such, beliefs are important factors in the development of nationalism and maintenance of a separate national identity. In newly independent states where national identity is not well established because of the lack of a continuous national history, leaders have sought to glorify past eras as a way of increasing national support. Abdel Nasser of Egypt could appeal to the era of the Pharaohs, while India's Jawaharlal Nehru drew from references to ancient Hindu empires. But a similar appeal to traditional culture has often served the nationalistic goals of the leaders of the older European states as well. Benito Mussolini publicized the success of the Roman Empire, and Charles de Gaulle looked toward the past glory of France in his effort to establish a sense of national pride.

In addition to enhancing the unity of a given country, common belief systems may *facilitate integration among nations*. The success that Europe has enjoyed in terms of international political integration during the postwar period (see Chapter 14) is a case in point. Shared beliefs in liberal democracy and other European values have made it easier for those states to give up certain sovereign powers to the

**Rendezvous**

David Low, *London Evening Standard,* 1939.

supranational authorities involved in the European Union. The acceptance of democratic values and free market economies have also become criteria for membership in the **North Atlantic Treaty Organization (NATO)** as a number of former Soviet bloc nations seek to join that institution.

Yet, common beliefs and values have not always served as a cohesive or integrative force across state lines. One study that coded various nations in terms of their common ideologies over the period from 1815 to 1939 found that such similarity did not predict the stability of alliances.[2] The evidence that ideological values do not cement relations is also shown in the deep schisms that long existed between such communist giants as the Soviet Union and the People's Republic of China. Indeed, China became a thorn in the side of the Soviets precisely because it espoused a communist belief system and hence threatened the Soviet leadership of the communist world. That those with similar belief systems can be the bitterest of enemies is also seen in the war between Iran and Iraq (1982–1989). The common bond of Islam in those two countries failed to minimize their mutual antipathy. Given the fact that the leadership of Iran and Iraq subscribed to different schools of Islamic thought, religion in this case was primarily a source of conflict, rather than unity.

Finally, belief systems may provide people with a *hope for the future.* Consequently, emphasis upon certain belief systems or ideologies can generate a willingness to sacrifice for the nation or the ideology. Thousands of underarmed and

---

[2]John D. Sullivan, "International Alliances," in *International Systems,* ed. Michael Haas (New York: Chandler, 1974), 99–122.

undertrained Iranian youths laid down their lives in the war against Iraq in the 1980s confident that, in the best tradition of Islamic thought, their deaths on the battlefield would be compensated by eternal rewards in paradise.

# COMPETING BELIEF SYSTEMS

Perhaps the larger impact of belief systems on foreign policy can be seen more vividly by examining the role that various belief systems transcending the nation-state have played in specific foreign policies. For this purpose we shall examine Marxist-Leninist, liberal-democratic, and different religious belief systems.[3] Then we shall consider how universally applicable ethical and normative principles accommodate these transnational, but frequently competing, ideologies.

## MARXIST-LENINIST BELIEFS

Governments espousing Marxism-Leninism fell from power across much of Eurasia by the start of the 1990s. But for decades before that, Marxist-Leninist ideology influenced the behavior of a number of states, including many of their foreign-policy objectives. That is still arguably the case for China and one or two small countries, such as Cuba and Vietnam. Central to Marxist thought is the idea of **economic determinism,** which holds that economic relationships are the basic determinant of social behavior—the substructure upon which all else is built—with politics a mere part of the superstructure. Such a premise leads to a tendency to interpret the foreign-policy behavior of the capitalist states in terms of the economic interests of dominant capitalist actors within them. Marxists therefore have given much attention to the economic role of multinational corporations, economic imperialism, and notions of the military-industrial complex in trying to explain and understand the foreign-policy behavior of the West.

Karl Marx essentially rejected the durability of the Westphalian state system, which he saw as the world political superstructure produced by capitalism. The state system was ill-suited, in Marx's view, for the international dictatorship of the working class that he expected would follow the overthrow of the capitalist system throughout the industrialized countries. Marx argued that only two important classes existed in the industrialized world of his day: the **proletariat** (workers) and the **bourgeoisie** (capitalists), which were locked in conflict with each other.

**SOVIET MARXISM** More than half a century after Marx's death, his influence on the Soviet leadership may partly explain why that government was slow to identify with third-world leaders as their countries became independent after World War II. The preindustrialized colonial empires of the West contained little in the way of a Marxist-style working class, and it was with such workers that Marx had exhorted his fol-

---

[3]For further reading, see Mustafa Rejai, *Political Ideologies: A Comparative Approach* (Armonk, N.Y.: M. E. Sharpe, 1991); Bernard Susser, *Political Ideology in the Modern World* (Boston: Allyn and Bacon, 1995); and Vernon Van Dyke, *Ideology and Political Choice* (Chatham, N.J.: Chatham Press, 1995).

lowers to unite. The anticolonial movement that swept the Southern Hemisphere at first must have looked largely irrelevant to the foreign-policy plans of the Kremlin. Soviet appreciation of the anticolonial movement did not fully develop until after 1956, when Nikita Khrushchev proclaimed a new doctrine arguing that Marxist and anticolonial societies were united in a "camp of peace" against the warmongering of the capitalist states.

Marx's belief that the state was only an anachronism that would eventually wither away after a successful proletarian revolution may have influenced Vladimir Lenin's initial decision to assume that a Soviet foreign minister was unnecessary. But the continuing realities of the nation-state system soon became apparent, leading the new Soviet government not only to create a foreign ministry, but also to establish diplomatic relations with capitalist states. For purposes of this discussion, the most critical addition Lenin made to Marx's thinking was therefore to adapt revolutionary communism to the continuance of the nation-state system.

Lenin's successor, Joseph Stalin, retreated even further into statist orthodoxy when he argued that the Soviet Union's encirclement by hostile capitalist states required it to maintain the kind of military capability that would permit defense of the new Soviet nation. And when his country was invaded by the Nazis during World War II, Stalin rallied his people in what he termed the "Great Patriotic War." Meanwhile, Stalin maintained totalitarian control of the instruments of the state, thereby transforming Marx's original cosmopolitanism into the most sinister excuse for maintaining a highly impervious and dictatorial state under Soviet rule.

The **historical determinism** rooted in Marxist ideology influenced the way the Soviet Union reacted externally. The belief that all people must follow the historical sequence from feudalism to capitalism, socialism, and then finally communism influenced predictions of where socialist revolutions might arise. Soon after the Bolsheviks came to power, they prepared policies anticipating that successful revolutions would occur first in Western Europe, where the bourgeois revolution had long since replaced **feudalism**.[4] Marxist theory hardly prepared Soviet leaders to expect a successful communist revolution in a China that was still largely feudal. Perhaps this partly explains why Stalin initially supported the nationalist Chinese leader, Chiang Kai-shek, instead of the Communist Chinese, although it is no doubt also the case that Stalin in this instance was behaving like the traditional nation-state leader in seeking an ally in nationalist China against the growing threat of fascism elsewhere in the world.

The belief in permanent conflict between classes that is inherent in Marxist thought may have led many to the assumption during the cold war period that little global stability was possible as long as contending classes remained in the international system. Such a belief may also have suggested to some Marxists that the communist world's very survival would require the destruction of capitalism.

The effect of the Marxist belief in the inevitable triumph of communism is somewhat more difficult to assess in terms of its possible impact on foreign policy.

---

[4]Feudalism is an economic and social system based upon hierarchical social relationships. Rulers—ostensibly of all the land—are connected to their vassals, who in turn are lords to those who live on the land, including, at the bottom of the social pyramid, the near-slave class of serfs.

It might seem that such a belief would make a state's leaders take greater risks on the assumption that even highly adventurous policies were destined by history to prevail. But one could conclude that such a leadership might behave cautiously, believing that it could obtain ultimate victory without expending much energy. It appears to have been the more optimistic interpretation of the inevitable triumph of communism that led Khrushchev to declare to the West, "We will bury you!" Certainly, little evidence followed that he intended to achieve such a goal through an adventurous foreign policy. Rather he was simply expressing the opinion that communism would outlast capitalism, not that the Communists would destroy the West.[5]

**GORBACHEV AND THE COLLAPSE OF COMMUNISM** Despite the fact that the Soviet Union had digressed a considerable degree from the tenets of Marxism, as Soviet leader after leader responded to the constraints of the international system just like any other national leader, one could hardly have anticipated the radical departures that Mikhail Gorbachev would initiate after he assumed leadership in the Soviet Union in 1985. As of late 1986, Gorbachev still adhered to Leninist dogma that emphasized the imperialistic and militaristic nature of capitalism. But, a year later, he suggested that capitalism could free itself of its tendency to exploit the developing world.[6] During his last years in power, Gorbachev went a step further by openly abandoning the claim that communism was the historically inevitable goal of Soviet society.[7]

To try to improve a badly deteriorated Soviet economic system, Gorbachev introduced certain free-market incentives **(perestroika),** undertook a number of unilateral initiatives involving extensive reductions of armaments, and declared an end to the cold war. While Leninism was held up as the presumed model for much economic reform through the first five years of Gorbachev's rule, nothing in Lenin's teaching could account for the much greater openness **(glasnost),** dissent, and even political competition that Gorbachev encouraged. He justified his radical departures in foreign policy on grounds that nuclear weapons, underdevelopment, and environmental destruction were now the common threats to all of humankind, requiring the joint action of East and West. While one could argue that this view was in keeping with Marx's own sense that social interests transcend the nation-state, it was scarcely like any Marxist doctrine known in history, and clearly shared much with the views of many non-Marxist thinkers as well.

Generally peaceful revolutions swept across virtually all of Eastern Europe in 1989 (only that in Romania was bloody). In one country after another where a communist party had held a monopoly of political power for some forty years, communists were forced to reform themselves into what appeared to be more like West European–style political parties and to plan free elections in which they would have to compete for power with noncommunist parties. Such a transformation provides

---

[5]Otto Klineberg, *The Human Dimension in International Relations* (New York: Holt, Rinehart and Winston, 1965), 153.

[6]Jeff Checkel, "Ideas, Institutions, and the Gorbachev Foreign Policy Revolution," *World Politics* 45 (October 1993), 287.

[7]Alfred B. Evans Jr., *Soviet-Marxism-Leninism: The Decline of Ideology* (Westport, Conn.: Praeger, 1993), 213.

the ultimate reminder of Marxism's adaptability, for the social democratic parties of the West have always claimed Marx as their ancestor, too.

The democratic processes unleashed by Gorbachev led ultimately to the collapse of the Soviet Union and to Gorbachev's loss of power to Boris Yeltsin in 1992. Leaders of most of the constituent Soviet republics late in 1991 declared that the Soviet Union no longer existed and that its place would be taken by a Commonwealth of Independent States. Real power, however, remains in the hands of the fifteen former and now totally independent republics. As each of these republics has moved toward a freer market economy, the symbols of Marxism and Leninism have likewise been systematically dismantled.[8]

**CHINA AND MARXIST-LENINIST-MAOIST THOUGHT** Meanwhile, Marxism-Leninism has taken nearly unrecognizable turns in the People's Republic of China, despite the fact that its leaders still claim to be communists. The belief system that shapes Chinese foreign policy is a product not only of communist ideology but also of distinctly Chinese historical and cultural forces. Above all, it reflects the legacy of Mao Zedong, who served as chairman of the Chinese Communist party for some forty-one years prior to his death in 1976.

That China fell to Mao and the Communists in 1949 was by no means attributable solely to the persuasiveness of Marxist-Leninist slogans. The promise of agrarian reform, as well as the ineptness and corruption of the regime of Chiang Kaishek, contributed to the success of the Communist Revolution. It has been suggested that Mao actually came to power "by waging a nationalistic struggle. He was a nationalist before he was a communist, and China [was] always . . . more important to him than world revolution."[9]

A number of Mao's views conflicted directly with Marxist thought. Whereas Marx argued that the structural qualities and inherent contradictions in feudal and capitalistic systems led to human injustice, Mao tended to emphasize the undesirable moral behavior of individuals within the class structure. Certainly the process by which Mao achieved his successful revolution in China violated Marxist theory, for China had not progressed beyond the feudal period, and considerable reliance had to be placed upon the peasants rather than the workers to achieve a successful revolution. Mao's belief that it would be possible to move directly from a semifeudal status to socialism without traversing the capitalist stage violated basic Marxism.

Mao's successors diverged even further from Marxism when they began to modernize China's economic system, borrowing practices from capitalistic states in an effort to stimulate economic productivity. Although the opening to the West began under Mao Zedong with the visit of President Richard Nixon in 1972, Mao's successor, Deng Xiaoping, went much further in opening China to foreign investment. Full diplomatic relations were established with the United States in 1979.[10]

---

[8]Peter Collins, *Ideology After the Fall of Communism* (New York: Boyars/Bowerdean, 1993); and Neal Robinson, *Ideology and the Collapse of the Soviet System* (Brookfield, Vt.: E. Elgar, 1995).

[9]Robert G. Wesson, *Why Marxism?* (New York: Basic Books, 1976), 107.

[10]For further reading see Yan Sun, *The Chinese Reassessment of Socialism, 1976–1992* (Princeton: Princeton University Press, 1995); and X. L. Ding, *The Decline of Communism in China* (New York: Cambridge University Press, 1994).

Whereas Deng and his colleagues were successful in modernizing China's economy, they were equally determined not to liberalize its political system. The repressiveness of the regime was probably less a function of Marxist-Leninist thought, which calls for a dictatorship of the proletariat until such a time as the communist revolution is a success, than it was the desire of a few elderly leaders to remain in control. When they quashed the student-led democratic movement in Tiananmen Square in 1989, they were also responding to a centuries-old tradition of authoritarian rule.

By and large, it is the deeply rooted culture of China that best explains modern Chinese foreign policy rather than its recent flirtation with Marxism-Leninism-Maoism, an ideology unlikely to outlive the current geriatric set of Chinese leadership. The more rooted the images, beliefs, and conceptions of a people based upon centuries of experience, the more difficult it becomes to eradicate them even if they no longer fit new realities. In other words, "even when China was weak and divided,

**A lone protestor confronts Chinese tanks at Tiananmen Square in Beijing, June, 1989.**

its leaders clung to traditional Sinocentric views of their country as the center of the world and a dominant power."[11]

A number of other political leaders, primarily in third world countries, declared themselves to be Marxist, but in many cases they never fully understood Marxist thought. What appealed to them was the revolutionary tone of Marxism and its accusation of exploitation by those groups in dominant positions in the capitalist world. With the collapse of Marxism in Eastern Europe, a similar trend followed in third world countries as a number of Marxist leaders declared themselves in favor of liberal democracy. According to the 1989 *Yearbook on International Communist Affairs*, there were twenty-five ruling communist parties in the world; three years later that number stood at only four: China, Cuba, North Korea, and Vietnam.

## LIBERAL-DEMOCRATIC BELIEF SYSTEMS

The liberal-democratic approach to foreign policy owes much to the laissez-faire notions of Adam Smith in the economic sphere and John Locke in the political arena.[12] Both Smith and Locke insisted that limited governments served society best, essentially by freeing up the energies of individuals, who are of paramount importance, to act with the least restraint necessary to public order. In Lockean terms, the essential role of government was to do little more than protect the life and liberty of the citizenry, whereas Smith advocated eliminating most governmental regulations of economic life. In the late twentieth century, the United States has been the chief exponent of a liberal-democratic belief system and thus merits our attention as we attempt to understand this particular set of beliefs.

**AMERICA'S ISOLATIONISM AND INTERVENTIONISM**  For the United States, the result of such liberal-democratic beliefs translated into policies built on moralism, or a sense of its higher moral conduct than that of other states. That moralism was first expressed as isolationism, but, as U.S. power grew, it was made to justify increasing interventionism in the affairs of others.

The American tradition of **isolationism** involved several aspects, including a sense of spiritual and philosophical separation from Europe's often autocratic monarchies, economic self-sufficiency, and the assurance of military security for the Western Hemisphere.[13] A variety of fortuitous circumstances made it possible for the United States to pursue a policy of isolationism throughout much of its history. Its two oceans were natural barriers against conquest; Britain, the dominant sea power of the nineteenth century, provided considerable protection by opposing the efforts of any other European power to gain territory at the expense of the United States. The fact that the United States soon became largely self-sufficient economically also per-

---

[11] Yaacov Y. I. Vertzberger, *The World in Their Minds: Information Processing, Cognition, and Perception in Foreign Policy Decisionmaking* (Stanford, Calif.: Stanford University Press, 1990), 265.

[12] Although both of these writers were British, one should not forget the important contributions to liberal-democratic thought provided by such continental writers as Charles-Louis de Secondat Montesquieu and Jean-Jacques Rousseau.

[13] Cecil V. Crabb Jr., *Policy Makers and Critics*, 2d ed. (New York: Praeger, 1986), 7–14.

mitted it the luxury of an isolationist foreign policy. Finally, the United States was not forced into world involvement, for as a new, fledgling state it did not threaten other states in the international system so long as it remained aloof from the quarrels of Europe.

After a century and a half in which isolationism remained the most prominent doctrine in American foreign policy, and after U.S. power had grown enormously, the sense of self-righteousness that this heritage encouraged came to be expressed increasingly through **interventionism.** After the American Civil War, this took the form of strong missionary movements, particularly in China. Such private and "humanitarian" interventionism was also reflected at the turn of the century in the U.S. government's Open Door policy toward China. The United States insisted that it be allowed to share in the lucrative trade relationship with that country—rationalized as opening China to the benefits of Western civilization—along with the European powers.

But the tradition of U.S. interventionism has had its most emphatic expression in Latin America. The **Monroe Doctrine** of 1823 had already communicated to major European states that they were not welcome in the Western Hemisphere. At the behest of various domestic interest groups such as the old United Fruit Company, the United States government took an increasingly active role in Latin American affairs. A peak was reached toward the end of the nineteenth century with American involvement in the Spanish-American War. Intervention in domestic Latin affairs was accelerated during the administration of William Howard Taft (1909–1913) with the strategy of so-called **dollar diplomacy,** in which it was asserted that the United States had a right to intervene in the domestic affairs of Latin American states in order to advance the economic interests of individual American companies. During President Wilson's administration (1913–1921) it was argued that the United States was morally obliged to aid in the development of representative governments in the hemisphere, for which American force might even be used as a last resort. Indeed, American troops were sent by President William Taft to Nicaragua in 1912 (where they stayed until 1925), and to the Dominican Republic and Haiti in 1915 by President Wilson (troops were not withdrawn from Haiti until 1934). Subsequent military interventions by the United States in Latin America, including covert actions, occurred again in Nicaragua from 1926–1933, in Guatemala in 1954, in Cuba in 1961, in the Dominican Republic in 1965, Grenada in 1983, Nicaragua once more throughout the 1980s, Panama in 1989–1990, and Haiti in 1994. The specter of Soviet intervention became the primary rationale in a majority of these cases.

**AMERICA'S MORALISM** By the early twentieth century, the ideology of liberal democracy also produced in American foreign policy a tendency to rely upon legal and moralistic arguments to rationalize its foreign-policy choices. President Wilson sought to justify United States involvement in World War I both to himself and to the world by saying that this would be a war to end all wars and would make the world safe for democracy. After the war, Wilson was the principal architect of the League of Nations—which his own country, in a retreat toward isolationism, refused to join. The League in some respects epitomized the legalistic-moralist and idealist approaches of liberal democracy, for it made treaty obligations the hoped-for foundation of peace and branded the resort to war as unacceptable in the foreign policy of any member state.

With the rejection of the League, the United States returned to isolationism, only to swing back to interventionism after World War II as the nation assumed the role of the world's police officer. Two such interventions, those in Korea and Vietnam, resulted in prolonged and full-scale wars. Critics such as Senator J. William Fulbright were led to inquire whether a new tradition was being built as the United States moved from isolationism to arrogance, involving itself in issues and problems throughout the world that were of minor concern to itself.[14] Fulbright challenged the United States' efforts to nominate itself as the Lord's agent on earth and saw the United States marching to disaster as the Athenians had against Syracuse and Napoleon and Hitler had against Russia. In his view, the solution was not to respond to dogmatism with dogmatism, and certainly not to imitate the perceived interventionism of the Soviet Union, but to exercise much greater restraint in the application of American power.

On the face of it, it may seem ironic that **liberalism,** with its effort to restrain governments in their domestic actions, could have been used to justify strong interventionist actions by the government of the United States abroad. Perhaps part of the American tradition of liberal interventionism can be attributed to the fact that the United States, in contrast to most European states, has not suffered the long history of devastating wars on its territory. Such experiences naturally lead many Europeans to be somewhat more pessimistic about their ability to control events, whereas Americans retain a passion for control, which may reflect their greater self-righteousness. The fact that the United States, unlike most European states, has not experienced a social revolution has been cited by Harvard professor Louis Hartz to help explain why the United States has not been adequately sensitive to democratic socialism in Europe and to revolutionary movements in Asia, and why it has sometimes adopted excessively paranoid views about the threat of communism. Hartz also sees moral absolutism, resulting from its different social history, as inspiring Americans either to withdraw from "alien things" or to seek to transform them.[15] Dramatic examples in the twentieth century of U.S. efforts to shape events were the American interventions in World War I and II, Korea, Vietnam, and the Persian Gulf. It is noteworthy that each of these actions was followed immediately by the desire on the part of many Americans to withdraw from world involvements.[16]

In recent years (and in contrast to the former Soviet Union), the United States has seen its basic belief system of liberal democracy spread throughout the world as state after state has held multiparty elections, many for the first time. With only three

---

[14]William J. Fulbright, *The Arrogance of Power* (New York: Random House, 1966).

[15]Louis Hartz, *The Liberal Tradition in America* (New York: Harcourt Brace Jovanovich, 1955), 286, 306.

[16]For further reading see Stanley Hoffmann, "The Crisis of Liberal Internationalism," *Foreign Policy* 98 (spring 1995), 159–177; Michael H. Hunt, *Ideology and U.S. Foreign Policy* (New Haven, Conn.: Yale University Press, 1987); Miroslav Nincic, *Democracy and Foreign Policy: The Fallacy of Political Realism* (New York: Columbia University Press, 1992); and Jerel A. Rosati, *The Carter Administration's Quest for Global Community: Beliefs and Their Impact Upon Behavior* (Columbia: University of South Carolina Press, 1987).

such regimes in 1790 (the United States, Switzerland, and France), that number increased to twenty-five in 1919, then dropped back to thirteen as a result of German aggression during World War II. But by the end of 1995, Freedom House placed the number of liberal democracies at an all-time high of 117 countries.[17]

# RELIGIOUS BELIEF SYSTEMS

The Treaty of Westphalia of 1648, marking the birth of the nation-state system, also brought an end to the century-long conflict between Catholics and Protestants that had culminated in the Thirty Years' War. This was not the first nor the last time that religious conflict would erupt as a political force in intergroup relations. Although Christianity spread across the Roman Empire without much bloodshed, from the eleventh through the thirteenth centuries, the Christianized Europeans clashed with Islam in an effort to regain the Holy Lands during the Crusades. Islam repeatedly sought to convert by the sword as it moved from the Middle East into North Africa, the Balkans, and much of Spain. That, with the tactic of Islamic leaders to obtain converts by using governmental power to reward the faithful and to punish the infidel, has brought Islam repeatedly into conflict with peoples of other faiths up to the present time. A perusal of ongoing conflicts, such as those between Islamic and Christian factions in Bosnia, between Protestants and Catholics in Northern Ireland, and between Hindus and Buddhists in Sri Lanka, serves to remind us that religion remains an important component of conflict between peoples.

It is likely that religion will have its greatest impact upon foreign policy when political power is assumed by the religious leadership itself, as in the case of Iran under the Ayatollah Khomeini and his Revolutionary Council. However, such secular leaders as General Mohammad Zia ul-Haq, who was president of Pakistan from 1976 until his death in 1988, and Libyan leader Mu`ammar Muhammad al-Gadhafi have attempted to establish Islamic republics based on the teachings of the Koran. Gadhafi used considerable amounts of Libyan oil revenues to support Islamic causes throughout the world. He claims to have been the first to have helped the Ayatollah Khomeini in his efforts to gain power in Iran and reportedly aided Pakistan in its quest for an "Islamic" nuclear bomb during Zia's tenure.

**ISLAMIC BELIEFS** Islam, like Christianity, is an activist or proselytizing religion, in contrast to Buddhism or Hinduism, which do not place a priority on converting nonbelievers. Also like Christianity, Islam has been willing to use military force to spread the faith, beginning with the Prophet Muhammad in the seventh century. Although using less militant instruments of policy today, Islam is the world's fastest growing major religion, claiming almost one billion adherents throughout the world.

Closely related to the notion of using the sword to obtain converts is the concept of holy war, or **jihad,** which has been particularly salient in the thinking of the dominant Islamic sect in Iran—the Shiites. Many Islamic leaders have called for holy wars against infidels and sometimes even against other Islamic states, as Khomeini

---

[17] Adrian Karatnycky, "Democracy and Despotism: Bipolarism Renewed," *Freedom Review,* 27, January/February 1996, 1.

did in his conflict with Iraq during the 1980s. The Islamic belief in the power of **martyrdom** has aided leaders who may ask others to sacrifice for the cause, as when the Ayatollah Khomeini induced many thousands of poorly armed Iranian youth to give up their lives in the holy war against Iraq. This in itself provides an important source of power, since morale is so vital in determining who is likely to prevail. A willingness to sacrifice more lives than your adversary can often compensate for a deficiency in number of troops.

When acceptance of martyrdom is combined with yet another characteristic of Islamic thought, respect for authority, the power of the leadership is strengthened even more. A particularly strong form of militant Islam can develop when the religious and political power of the state is concentrated in one man, as was the case with the Ayatollah Khomeini. This was reflected in the constitution drawn up for Iran in late 1979, in which ultimate power was given to the clergy, led by the **ayatollah,** who were to "safeguard against any deviations by various government organizations from their true Islamic functions and obligations."

Theocratic leaders, like the Ayatollah Khomeini, often believe that God is with them and that they have complete understanding and knowledge of the truth. With a highly emotional view of right and wrong, they are hardly likely to bargain in the usual sense of the term. The 444-day standoff over the issue of the American hostages in Iran and the decade-long, bitter war with Iraq in the 1980s illustrate the difficulties the true believer often has in making concessions.

Although the hope of Muslim fundamentalists is for unity among all Islamic peoples, Islam, like Christianity, has been subjected to many schisms and has had to compete with nationalism. The major split within the Islamic world has been between Shiite and Sunni, the latter of which represents the overwhelming majority of Muslims. The failure of some Islamic states to support the Ayatollah Khomeini was partially attributable to the fact that Iranians are primarily members of the minority **Shiite** sect, whereas most of the other Muslim states are **Sunni.** Anwar Sadat of Egypt was disturbed by the impact of Khomeini on the image of Islam; he even referred to the Iranian leader as a lunatic.[18] Sadat, like Sunni Muslims generally, came from the more tolerant religious culture that was part of the heritage of the Ottoman Empire. But his public opposition to Islamic fanaticism helps explain why Sadat was subsequently to be felled by the bullet of a fundamentalist Muslim.

Despite various efforts to establish Islamic unity, splits and violent conflicts have been a recurrent phenomenon among Islamic states, as in the 1980–1988 war between Iraq and Iran, Libya's border war with Egypt in 1973, and Syria and Egypt's attack on Jordan in 1970.[19] On one occasion, Syria even intervened in the Lebanese Civil War on the side of Christian forces rather than in support of its Islamic brethren. In 1990, both Islamic and Arab unity were thrown asunder, first with the Iraqi invasion of Kuwait by Saddam Hussein and, subsequently, by the adamant opposition to the invasion by Egypt and Saudi Arabia, among other Islamic states. Saudi Arabia,

---

[18]Christopher S. Wren, "Cairo Said to Worry about Islam's Image," *New York Times,* November 12, 1979.

[19]For further reading see Mir Zohair Husain, *Global Islamic Politics* (New York: Harper Collins, 1995); and John L. Espito, *The Islamic Threat: Myth or Reality?* (New York: Oxford University Press, 1995).

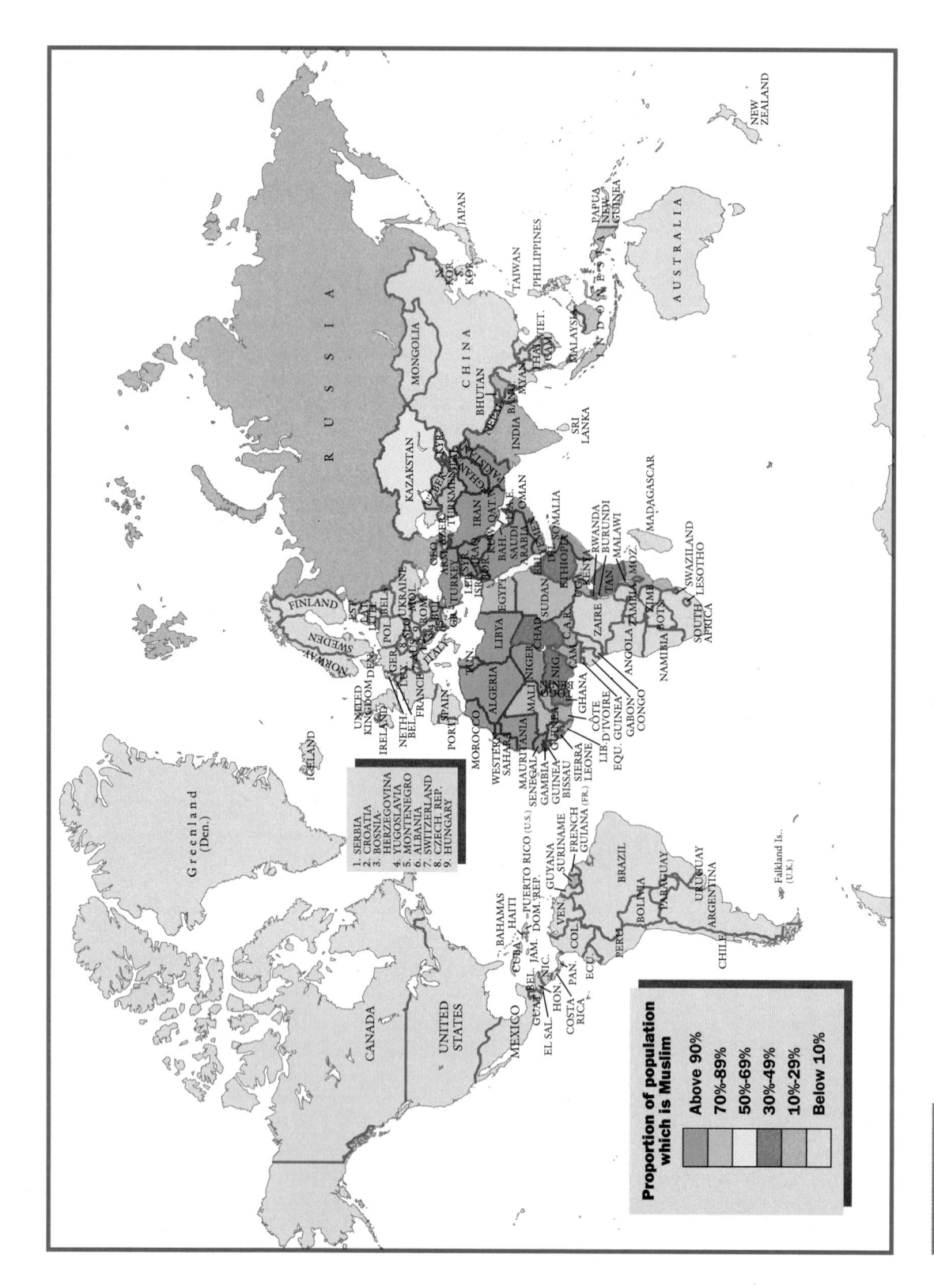

MAP 4.1 Distribution of Muslims throughout the World

despite great sensitivity about foreign cultural influence, even went so far as to invite several hundred thousand U.S. troops to position themselves on Saudi territory from which to launch an attack against Saddam. Such cases as these suggest that when religious values conflict with secular ones, leaders of nation-states generally opt for those policies supportive of their secular ambitions (see Table 4.1).

**THE POSITIVE FUNCTIONS OF RELIGION** The recitation of certain religious beliefs by political leaders does not necessarily make one religion more militant and less compromising than another; all religions have been used for both good and evil. Mohammadan texts, for example, assert that "Jews, Christians, and Sabians are in-

**TABLE 4.1** Comparative Beliefs

| BELIEF | MARXISM | DEMOCRATIC-LIBERAL | ISLAM |
|---|---|---|---|
| Human nature | Malleable; determined by class | Inherent capacity for liberty; rational | Predetermination by will of God, but moral striving also urged |
| Primary actors | Bourgeoisie and proletariat | Individuals and pluralistic groups | Faithful and infidel |
| Political authority | Communist Party to lead the revolution | Democratically elected authority | Fusion of temporal and spiritual authority under a caliph |
| Economic system | State directed until workers seize means of production | Laissez-faire; private ownership | Usury but not wealth is forbidden; the prophet began as a merchant |
| State system | Role in ending feudalism and creating capitalism | Provides security and furthers state interests | Westphalian state system expected to end |
| Conflict | Inevitable between classes; to end with creation of communist system | Although conflict may be inevitable, try to resolve by peaceful means | Inevitable as long as infidels remain |
| Future of government | Withers away when communism is achieved | Will continue to play an important role | Theocratic government and eventually an Islamic global community |
| Future world order | Harmonious world order with end of conflicting classes | Democratic and peaceful world, particularly as democratic values spread | Faqih rules both temporally and spiritually until the coming of the twelfth Imman |

Note: These descriptions represent slightly oversimplified characterizations of beliefs that—although often violated in practice—provide a lens through which reality is viewed.

cluded as inheritors of Paradise along with Muslims, provided they believe in Allah and the last judgment and do good works."[20] References to notions of equality in yet other Islamic texts can be viewed as providing the underpinnings of democracy.

Clerics have often played important roles as mediators in various conflicts between groups. For example, it was Pope John Paul II who successfully mediated the Beagle Channel dispute that threatened war between Chile and Argentina in 1989; Catholic clerics played the critical mediating role in both El Salvador and Mozambique during the late 1980s and early 1990s; and Bishop Desmond Tutu was important to the process of ending apartheid in South Africa.[21]

## THE IMPACT OF COMPETING BELIEF SYSTEMS

Now that we have identified some of the functions of belief systems and examined the role they have played in several states, let us attempt to evaluate the general impact of these beliefs on foreign-policy behavior. Two conditions under which the ideological determinant will likely be instrumental in shaping foreign-policy choices have been identified.[22] The first consists of situations in which established values are widely challenged, as is the case especially during revolutionary times, for revolutions can be defined as events in which many people suddenly are willing to die for an idea. Second, ideological influences are likely to have a greater impact in those political systems that concentrate decision making in a very few individuals. As long as those individuals share a revolutionary ideology and wield the power of the state, they will be less constrained by other factors in the pursuit of their objectives.

A study of the experiences of China and Yugoslavia revealed that even after an ideology proves to be dysfunctional, leaders continue to adhere to it. The tendency in both instances, although against the national interest, was to "seize on preexisting ideas to guide them through times of high uncertainty and to allow them to legitimate and coordinate their actions."[23]

The role ideology plays in foreign policy depends on the specific beliefs inherent in the ideology itself. A democratic ideology will stress that a people should determine its own future. As we have seen, that does not preclude the possibility that democratic decision makers will use force in foreign policy—often with the justification that only force can eliminate a tyrannical ruler who stands in the way of a society's freedom—but, in general, constitutional democracies will select peaceful options and allow democratic choice for others.

---

[20]Donald E. Smith, "The Political Implications of Asian Religions," in *South Asian Politics and Religion,* ed. Donald E. Smith (Princeton: Princeton University Press, 1966), 16–17.

[21]The positive role played by various religions in peaceful settlement is noted in a number of case studies found in Douglas Johnston and Cynthia Sampson, eds., *Religion: The Missing Dimension of Statecraft* (New York: Oxford University Press, 1994).

[22]Werner Levi, "Ideology, Interests, and Foreign Policy," *International Studies Quarterly* 14 (March 1970), 1.

[23]Nina Halpern, "Creating Socialist Economies: Stalinist Political Economy and the Impact of Ideas," in *Ideas and Foreign Policy,* ed. Judith Goldstein and Robert O. Keohane (Ithaca, N.Y.: Cornell University Press, 1993), 110.

Advocates of crusading ideologies—such as communism, some fundamentalist Islamic movements, and French republicanism in the early stage of the French Revolution—are more likely to utilize aggressive means to extend their belief systems. Totalitarian ideologies are even used to rationalize killing and violence as the price a people must pay in order to enjoy a brighter future. The aggressiveness of Fascist leaders can be explained in part by the organismic view of the state perceived by **fascism,** which holds that a nation must expand or decline. Mussolini declared, in invading Ethiopia, that "Fascism sees in the imperialistic spirit a manifestation of its vitality."[24] Similarly, Hitler was concerned with gaining **Lebensraum,** or living space, for the people of his proclaimed "master race."

Ideologies that claim to have discovered all of the truths may make it difficult for those believing in such ideas to adapt to new circumstances. For example, the Islamic belief that the "gates of interpretation" were closed soon after the death of Muhammad may reduce flexibility in the foreign policy of Islamic states.[25]

The fact that different ideologies drive political movements in the world may lead to conflict out of the fear generated at the sight of people whose views and values seem so different. For the ideologue, the mere existence of another group with a different viewpoint can be psychologically threatening because it implies that one's own views may be wrong. The world of the cold war period was often described in terms of a struggle between communism and democracy. Although this was far too simplistic a depiction of all the issues that produced conflict among important state actors, the labels were quickly applied to rally support for one belief system over the other. Such support tended to feed upon itself, maintaining the gulf between East and West through fear and considerable mutual misperception.

What is perhaps most remarkable about belief systems is the extreme length to which national leaders often will go to pay lip service to a given belief even when the evidence contradicting it is overwhelming. U.S. leaders were guilty of this during the cold war when they classified the vilest of dictatorships as part of the "free world"; Hitler's apologists went so far as to define the Japanese as part of the favored Aryan race after the Berlin-Tokyo axis was forged; and Soviet leaders sometimes engaged in ideological gymnastics to keep their Marxist beliefs compatible with their real-world needs as they sought to redefine class enemies whenever temporary alliances were desirable.

Finally, whether the belief systems in question are religious or secular, their impact on foreign policy is directly related to the fervor with which they are held by members of the society in question. In this sense, the revolutionary zeal of Shiite Iran with the establishment of its Islamic Republic in 1979 is comparable to that of the citizens of the French Republic after the Revolution of 1789, or perhaps to that of the crusaders who sought to reestablish Christianity in the Holy Land during the eleventh to the thirteenth centuries. In contrast, Islamic teachings are scarcely a factor in Turkey's foreign policy today, any more than "republicanism" motivates con-

---

[24] Robert W. Tucker, *The Inequality of Nations* (New York: Basic Books, 1977), 24.

[25] Judith Goldstein and Robert O. Keohane, "Ideas and Foreign Policy: An Analytical Framework," in Goldstein and Keohane (eds.), *Ideas and Foreign Policy*, 53.

> ## Global Changes
>
> ### Changing Belief Systems
>
> [A] remarkable consensus concerning the legitimacy of liberal democracy as a system of government [has] emerged throughout the world over the past few years, as it conquered rival ideologies like hereditary monarchy, fascism, and most recently communism. More than that . . . liberal democracy may constitute the "end point of mankind's ideological evolution" and the "final form of human government," and as such constitute the end of history." (Francis Fukuyama, *The End of History and the Last Man* [New York: Macmillan, 1992], p. xi.)
>
> It is my hypothesis that the fundamental source of conflict in this new world will not be primarily ideological or primarily economic. The great divisions among humankind and the dominating source of conflict will be cultural. Nation states will remain the most powerful actors in world affairs, but the principal conflicts of global politics will occur between nations and groups of different civilizations. (Samuel Huntington, "The Clash of Civilizations?" *Foreign Affairs* 72 [summer 1993]: 22.)
>
> The notion of the creation of a universal civilization based upon Western values would appear to be quite unrealistic. A review of 100 comparative studies of values in different societies, conducted by the social-psychologist Harry C. Triandis, concluded that "the values that are most important in the West are least important worldwide." (Samuel Huntington, "The Clash of Civilizations?" *Foreign Affairs* 72 [summer 1993]: 41.)
>
> Whereas Fukuyama envisions a more benign world in the end of conflict over ideas, Huntington sees conflict continuing in the future as a result of clashes between eight civilizations: Western, Confucian, Japanese, Islamic, Hindu, Slavic-Orthodox, Latin American, and African.
>
> - Which of these civilizations are more likely to clash with one another? Since Huntington views such eruptions as most likely to occur between geographically contiguous groups or, as he puts it, along "cultural fault lines," where can we expect the flash points to occur?
> - How does one square the quote by Triandis—finding Western values to be the least significant ones in the world—with Fukuyama's notion of the triumph of liberal democracy throughout the world? Does liberal democracy mean more than just whether free elections are held? What characteristics of liberal democracy do you think might make it the wave of the future?

temporary French foreign policy or Christianity that of Great Britain, France, or Spain.

# THE FUTURE OF NATIONAL BELIEF SYSTEMS

Ideas, belief systems, and ideologies are the product of the complex, yet unique, social circumstances of particular times and places. Marx's views of the world would have been impossible without his encountering a newly powerful industrialism making "wage slaves" of millions of workers for the first time in Europe's history. Woodrow Wilson would not have succeeded in proposing an international league to keep the peace collectively had not much of the world just experienced a disastrous war as the result of the effort to find security through rival alliances. Similarly, late in our century, Gorbachev's glasnost responded to the growing impossibility in our time of keeping the Soviet public sealed off from the explosion of information technology that today provides the cutting edge of economic advance.

These are reminders that the belief systems that inspire people today are themselves evolving in response to new problems and new combinations of social and material forces. We should not expect the world of the twenty-first century to reflect the same ideologies as those dominating world politics at the moment—particularly not when we recall how quickly some of yesterday's important ideologies have faded from the scene today.

## NATIONALISM AND MULTICULTURALISM

Nationalism, the ideology that most clearly rationalizes the nation-state—and, thus, an international system of nation-states—looks increasingly potent at the end of the twentieth century. For example, since the end of the cold war, we have seen a resurgence of nationalist separatism across Eurasia, where Soviet imperialism had kept the lid on expression of such sentiments for decades.

However, challenging the future of nationalism are two long-term global trends that seem to work against efforts to divide the world into ever smaller units. First is the tendency of contemporary industrial and postindustrial economies to keep expanding, rather than contracting, making "nations" less viable as economic units. Second are the ways in which today's unprecedented communication methods give people access to information that may encourage their shared beliefs, rather than their sense of separation.

It is a fact of today's world that the most highly developed economies on earth are establishing new structures for integrating their economies with those of other nations. A North American Free Trade Agreement is intended to create one giant economy from those of Canada, the United States, and Mexico. In Europe, the European Union unites the economies of fifteen economic systems that once were separate.

Notice how this trend merges with that in communication. It is increasingly commonplace for an individual who has grown up in a traditional culture in Japan to manage, say, an automobile plant in South Carolina, or for an Italian to work in Germany assembling electronic equipment. The outlooks, attitudes, and values of both those individuals, as well as many others with whom they come in contact in their "foreign" locales, are being changed as a result. Moreover, both manager and worker are likely to turn to CNN or some other global network for identical information about events in Pakistan or Brazil. And they have counterparts in Brazil and Pakistan (a Korean manager, perhaps, and a Syrian worker) with access to information about Japan and Italy that also comes from one of the television giants and that is available to millions of others across the globe.

Despite these global changes, no one would suppose that the national identities of all these individuals is suddenly disappearing or that they now view themselves as global citizens with cosmopolitan and tolerant outlooks. But neither should we imagine that these persons are not undergoing changes in their perception and thinking that may well, in a generation or two of such new influences, distinguish many adults in the near future substantially from their parents and grandparents.

**Multiculturalism** is the usual label today for the greater mix of people, cultures, and, therefore, ideas that is increasingly characteristic of social life across much of the planet. The challenge facing today's young adults is to enrich themselves by drawing upon the most attractive influences and traditions of others at the same time they learn to share the values of their own cultures. Feminism, globalism, environmentalism—along with worldviews still to come—are increasingly influencing older outlooks, such as nationalism, in ways that we can still only dimly perceive.

## MORALITY AND FOREIGN POLICY

All belief systems are built upon the acceptance of ideas shared within a social group about the nature and purpose of life. When such beliefs clash, the result may be political conflict, which is likely to be more intense and uncompromising the greater each group's certainty that it alone is in the right. A sense of moral absolutism may lead to holy wars against those thought to be in the wrong, a phenomenon that has led to much of the violence that has characterized the history of international politics.

Yet humankind generally has thought that such moral absolutism should not be equated with true moral conduct. *Moral behavior* usually means to do what is right in a way that transcends what is merely expedient for one's group or political situation. But of course the trick is to distinguish the one from the other. The world's religions have created priests or other religious professionals charged with making impartial judgments about what is right and wrong on behalf of those who submit to the religious authority. In the secular realm, moral standards have been encoded into the laws and mores that govern social groups where, again, impartial judges help us make such determinations.

The question of determining what is moral conduct in foreign policy is a far more subjective matter. The very sovereign equality of states in the Westphalian system permits every government the right to judge such matters for itself, and then to act in accordance with that judgment, which is virtually certain to reflect its own strong self-interest. Thus, officials of the Johnson administration may have been convinced that their behavior was truly moral when they directed South Vietnamese civilians to be rounded up into strategic hamlets to protect them from the Viet Cong; so, too, may Leonid Brezhnev have supposed that when he ordered the Warsaw Pact to invade Czechoslovakia in 1968 he was acting in accordance with history's highest moral dictate: to prevent a socialist state from falling back into the camp of capitalism. The very sovereign equality of the Westphalian system encourages us to think that *my* country's behavior is moral; *your* country's stems from ideology that is false or immoral.

## IDEALISM, REALISM, AND MORAL FOREIGN POLICY

Moral considerations are also related to our earlier discussion of idealism and realism in the conduct of international politics. The relevance of the previous discussion to idealism should be clear enough, because ideologies and moral standards could not exist without ideas and ideals, and because the idealist tradition is charac-

**Normative Dilemmas**

## Questions Regarding the U.S. Decision to Intervene in Haiti, 1994

In September 1994, President Clinton announced that the United States would intervene with appropriate military force in the republic of Haiti to restore its duly elected president, Jean-Bertrand Aristide, to power. Aristide had fled his country three years earlier as the result of a military coup that had made General Raoul Cedras and his associates the effective government of the island nation. As the result of last-minute negotiations between Cedras and an American team consisting of former President Jimmy Carter, General Colin Powell, and Senator Sam Nunn, Haiti's military rulers agreed to relinquish power by October 15, when President Aristide would return. As a result, the American intervention in Haiti met virtually no resistance, and paved the way for Aristide's peaceable return.

Assume you are one of President Clinton's foreign-policy advisers who must help him decide, in late summer of 1994, whether or not to intervene in Haiti. Among the arguments opposing U.S. intervention are the following:

1. Military intervention would substantially risk the loss of American lives. Such a loss could not be justified to the American people because the question of how Haiti is ruled is not of significant national interest to the United States.
2. Military intervention risks initiating a long-term commitment to the occupation (or administration) of Haiti. That is too open-ended to allow for a clear policy outcome, and too expensive to be worth any possible gain.
3. The United States effectively occupied Haiti from 1915 to 1934 without demonstrably improving the quality of life for Haitians or producing democratic government there. The lesson may be that occupiers are unloved and that a new occupation will not advance the goal of democracy in the hemisphere.

Among the arguments supporting U.S. intervention are these:

1. The military leaders of Haiti have overthrown a democratically elected government and have refused to relinquish power as the result of nonmilitary sanctions (economic sanctions had been imposed more than a year earlier). The United States has an interest in promoting democratic rule and discouraging others from attempting illegal coups against legitimate governments.
2. The military leaders of Haiti have ruled brutally, murdering several thousand Haitians and imprisoning many others to maintain their position in power. A humanitarian intervention would be in accordance with international legal standards and would help bring an end to grave wrongs done to innocent Haitian people.
3. The United States has committed itself to bringing about an end to Haiti's military rule and the restoration of President Aristide. Its own credibility will not be taken seriously by other states if it does not follow up on its threats by pursuing its announced policy to its conclusion.

Decide which of these arguments (or others) you find most convincing, then try to analyze why they have persuaded you. What values and norms do you hold that you find reflected or advanced in the position you recommend? How do you account for the fact that those values are more important to you than those that would be advanced if you took a different position?

Next, determine how you assess the risk (that is, that something may go wrong or have unintended, negative consequences) in the policy choice you advocate. Has your calculation of probable risk influenced your policy recommendation? Do risk calculations help you choose between policies in which your values conflict with each other?

terized by the view that certain standards, principles, and norms transcend the interests of individual states and therefore ought to be supported by foreign-policy decision makers of any state.

The approach of realism, on the other hand, denies the ability of foreign-policy decision makers to apply a single, universal standard of right and wrong. Yet, that does not mean that realism lacks moral implications of its own. Rather, what is right for the realist is the effort to protect the interests of the state on whose behalf one acts, generally requiring some accommodation to the comparable interests of others. What the realist considers to be wrong is either blind submission to the will of other states or the attempt to impose one's own values on others through conquest. A "moral" realism is reluctant to advocate the use of a state's power to coerce others since it recognizes that other states have their own rights, duties, and interests. The existence of a multitude of states demands that every foreign policymaker respect the interests of others in an atmosphere of mutual tolerance.

We have seen that an extreme doctrine, fascism, is an ideology built upon a simplistic distortion of realism's concern with protecting the interests of the state. Fascism argues that the interests, and therefore the morality, of the organic, totalitarian, and fascist state are superior to those of all other groups, allowing the fascist to justify virtually any governmental exercise of power for "reasons of state." Clearly, given the horrors committed in the name of the supremacy of the state in our century, few today would advocate such extreme views. Yet, without embracing the fascist extreme, many self-styled realists still are frequently tempted to engage in coercive action for similar **reasons of state.** In foreign policy, that doctrine has been used to justify espionage, subversion, and the assassination of foreign officials. Those who hold such a view are only a step away from the notion that the use of the state's power to advance its interests abroad is always justified—that might actually makes right.

Yet our age is one in which we see increased evidence of social interests, and therefore policy goals, that are shared by people across nation-state lines. Exploration of what our growing interdependence means for the practice of international politics is one of the chief functions of this book, and we take it as a given that doctrines of state absolutism, which were always dubious, are even less acceptable in an age when the state is capable of wreaking unprecedented violence and destruction.

Nonetheless, international life still is organized so differently from life within the state that even if we subscribe to the view that a single, universal moral standard ought to govern us, most would argue that differences in the nature of the social systems within which one is operating may place different ethical demands upon the individual. As realism suggests, the foreign policymakers of a nation no doubt may still behave in accordance with different ethical imperatives in their official roles than when they interact privately with their friends, families, or fellow citizens.

For example, it is a requirement of good citizenship that a witness to a crime report it to the police and do whatever else is prudent to try to thwart the criminal. Yet in the world of state interactions, where no global police force exists, it may actually be more—not less—responsible for the government that is a witness to a "crime" committed by other nations to refrain altogether from intervening in their quarrel. As a result, a limited conflict can be prevented from spreading into a more general war. That is why neutrality and nonintervention have frequently been viewed as sound ethical doctrines for a nation-state even though they would less often be regarded as the correct moral standards for private citizens.

Nor is there apparently a single, uniform ethical demand placed upon all foreign-policy officials, regardless of the state they serve. The foreign ministers of great powers obviously have a much greater ability to try to coerce others into doing their bidding than do the foreign ministers of small, weak states. Great powers, in fact, seem always prone to view themselves as would-be police officers of the world, frequently identifying their state's own self-interest with that of the world community. When they attempt to enforce their view of what is right upon others, they act as judges in their own case, whereas the comparatively weak state has no such opportunity and must either submit to the stronger's interpretation of what is right or perhaps undertake a costly, or even futile, effort at resistance. The situation presents an ethical dilemma for the decision makers of both countries, although it is quite different for each: For what political and social values can the great power justify using its force, and will the use of force advance those values? What national interests can the small power afford to sacrifice through submission, and which ones are so vital that they demand even a self-destructive effort to resist the strong?

# Conclusion

The objectives of states have changed dramatically over the centuries, depending upon the evolving needs of societies over time. The Westphalian system itself was born out of the concern about security and survival. The state gradually assumed other objectives over the years as it became more concerned about economic well-being and sharing its presumed superior values and ideologies with others, by force if necessary. The objectives of states are, in reality, the objectives of individuals who act in the name of the state. They act in pursuit of goals that assume their own desire to better their state's situation in some way. That, in turn, implies that decision makers bring their sense of what ought to be to bear on policies they undertake.

Belief systems, consisting of the ideologies and historical traditions held by a given population, play a particularly important role in shaping the objectives of states. Without a coherent and strongly held set of beliefs about the nature and purpose of human society, no state would be motivated to act in international politics. Belief systems influence the perceptions decision makers hold about the world, they define the range of choices available to them, and they provide the means for rationalizing their choices. Without belief systems, there could be no continuity in a state's foreign policy and little sense of hope for its future.

We have examined several contemporary belief systems with values that transcend those of the nation-state. Marxism, liberal democracy, and various religious beliefs all have no doubt impelled some of their exponents to highly activist foreign policies. Yet what is at least as remarkable is that each of these belief systems has also shown itself to be highly adaptable to the demands of the Westphalian state system, often becoming the servant of nationalism.

Marxism is a more coherent ideology than are many belief systems that affect contemporary world politics, and it calls for an activist promotion of class interests across nation-state lines. Yet the evidence is overwhelming that it is the national beliefs and experiences of governments espousing Marxism that adapted the doctrine to serve their states, and that, for most of the twentieth century, Marxism probably was used more for rationalizing foreign-policy decisions than for determining them.

The impact of liberal-democratic beliefs on foreign policy is by no means simple to ascertain, both because there are many intellectual and historical roots to these beliefs and because they do not add up to a coherent guide to action in the foreign-policy arena. To take only the case of the United States, this tradition has produced conflicting and contradictory impulses toward withdrawal from the world in the form of isolationism (a tendency that might be at least partially ascribed to the kinds of restraints on government advocated by Locke), and interventionism, ostensibly to promote a democratic way of life abroad. Certainly, the U.S. interventionism that has been dominant since World War II can be attributed to a conviction that the world's most powerful democracy must resist authoritarian communism and, perhaps in the process, make the world safe for capitalism, too.

The case of Iran in the grip of a fundamentalist Islamic movement illustrates how religion as ideology can affect foreign policy. Fundamentalist Islamic thought seeks to go beyond the nation-state with its hope of uniting all people, much as Christianity did in other times and places. Yet, also as with Christianity, national differences within the religious community frequently have proved much stronger than the transnational appeal of a unified faith. As a result, national leaders of predominantly Muslim nations today often appear to exploit Islam for their own ends. The foreign policies of the Islamic world today appear about as diverse as were those of Christian Europe at the dawn of the Westphalian period.

The increasing multiculturalism that is the product of advanced communication, rapid transportation, and the demands of much modern economic life is affecting older ideologies in ways that are not yet clear. In the short run, it is responsible for a renewal of ethnic tensions as diverse groups of peoples are coming in ever greater contact with one another. For the longer run, multiculturalism may be changing experiences, attitudes, values, and beliefs in ways that could transcend nationalism as the focus of much human identification in the modern world.

# CHAPTER 5

# Foreign Policy Decision Making

In the previous chapter we sought to explain foreign policy from the perspective of the societal goals and belief systems of the state. We turn in this chapter to the decision-making process itself, focusing first of all on the impact of the individual decision maker on foreign policy, for ultimately it is human beings who give shape to the policies of the state. How in particular do the motivations and perceptions of individual decision makers affect foreign policy choice? Is it possible to make rational calculations in foreign policy?

Foreign policy behavior can also be explained from the perspective of the state level of analysis, as various kinds of political structures respond differently to their foreign policy challenges. The following are the sorts of questions we will be asking: Do democracies differ from dictatorships in the efficiency and effectiveness of their foreign policies? What roles do central decision makers and bureaucracies play in the foreign-policy process? What impact do public opinion and various subnational groups have upon the decision-making process?

## THE RATIONAL ACTOR MODEL

The most extensively used approach for explaining the behavior of foreign policy decision makers has been that of the **rational actor model** (sometimes called the *unitary model*). This is the approach often used by diplomatic historians to describe the foreign-policy interaction of different state leaders in response to one another. The decision-making units—whatever form they take—are viewed as unitary actors searching to maximize their goals. It is assumed that those goals, which are the product of normative choices, are clear to those who are pursuing them. With such an approach, little—if any—effort is made to understand the internal political and psychological forces affecting foreign policy choice. Instead, the analyst seeks to explain foreign policy as consisting of rational calculations in response to moves made by the other side (using a simple action-reaction model) or as a decision made only after a systematic and rational calculation of various alternatives. The objective is to maximize some form of utility.

There are certain advantages to using a rational or unitary actor model in seeking to understand foreign policy, and these are largely derived from the simplicity of the model itself. It provides an inexpensive approximation of reality in which analysts can attempt to consider what they would do if they were acting for the other state. This approach has been particularly useful in war game exercises, in which political and military experts play the role of the adversary. To the extent that such exercises can actually explain foreign-policy behavior, they have the very real virtue of parsimony—that is, they allow one to reduce complexity by focusing only on the fewest possible variables best able to explain a given phenomenon.

The most serious weakness of this approach is that it obviously assumes rational calculation on the part of decision makers, and that is an ideal situation that is seldom realized. Moreover, it may proceed from the premise that what is rational for one researcher or actor is rational for another in a similar circumstance. For example, during the Vietnam war, American decision makers and some analysts believed that an acceleration of bombing in North Vietnam would soon lead the North Vietnamese leadership to surrender, presumably because that would be the North's rational choice. Instead, such action tended only to increase the resolve of the North Vietnamese to defend their homeland. In 1994, the Clinton administration tried to pressure the North Korean government to reopen its nuclear facilities to international inspection. Although American officials held out the inducement of improved trade relationships with the West, they acknowledged that such a rational choice from their point of view might not be seen in that light by the isolated Communist government of Kim Il Sung and his son and successor, Kim Jong Il.

## THE INDIVIDUAL AND FOREIGN POLICY CHOICE

As the Westphalian state matured and became more democratized, some began to wonder whether particular individuals any longer made much difference in foreign-policy choice. No longer would it be appropriate to speak of the foreign policy of a Castlereagh or a Metternich, as diplomatic historians tended to do, for in the contemporary world there were too many international and domestic constraints on foreign policy, and too many individuals involved in the decisional process. Even so, when we consider the impact that Mikhail Gorbachev had upon both Soviet domestic and foreign policy, we realize that it would be premature to dismiss completely the role played by individuals in the making of policy. Even the personality predispositions of subordinates can make a difference in foreign-policy behavior. For example, the "gung-ho" attitude of Colonel Oliver North in the Iran-contra scandal led to an illegal reversal of official U.S. policy opposing the trade of arms for hostages.

> ## Conditions in Which Individuals Make a Difference
>
> - *When decisional latitude is high:* Since dictators are the least constrained by public and interest group opinion, one might expect that the personal predispositions and idiosyncracies of leaders would have their greatest impact upon the foreign policies of authoritarian states. The same would also be true of those states led by charismatic leaders.
> - *When a leader has a strong interest in foreign-policy matters:* Although President Ronald Reagan was largely a hands-off president, his preferences played an important role on those issues in which he became interested, as in the case of his almost obsessive support for the Strategic Defense Initiative. Few presidents, however, can compare with George Bush in terms of personal interest in foreign-policy matters—an interest that may have cost him the presidency when the electorate became primarily concerned with domestic economic issues.
> - *When involved in nonroutine situations:* Included here are decisions initiating or terminating major international undertakings such as war, alliance relationships, and aid programs—areas in which the habitual ways of doing things are not entrenched. Because crises are nonroutine situations, they also provide greater latitude and hence more opportunity for the idiosyncracies of the decision maker to be felt.
> - *In ambiguous, unanticipated, and remote situations:* Opposition forces will not have yet coalesced because of the ambiguous and unexpected nature of the issue, thus allowing leaders to make decisions primarily in accordance with their individual predispositions.
> - *When information is overloaded or too sparse:* Too much information can be as great an impediment to rational decision making as too little. Wading through a mass of information is a difficult task that is often short-circuited by paying attention only to preferred and expected signals while ignoring many contradictory inputs. On the other hand, a lack of information requires more reliance upon gut feelings in making a foreign-policy choice.

Those who approach the topic of international relations from the perspective of the individual level of analysis tend to focus upon what are called **idiosyncratic variables**—in other words, factors and characteristics of a particular individual. Of primary interest are the motivations and personality characteristics of the individuals involved in the decision-making process. In a sense all foreign policy is made at this level (although typically today involving a number of individuals), for states and international systems are merely abstractions; only human beings who reason and feel can make choices on foreign-policy matters.

While it is clear that individual variables can have varying effects upon foreign policy—depending on the circumstances—the critical issue to be examined is how specific psychological profiles and traits might impinge on foreign policy.

## PERSONALITY TRAITS AND FOREIGN POLICY

Researchers have sought to identify and classify various personality traits. The schemes they have developed are often based on complex personality testing in which respondents react to questionnaires, projective tests, and clinical evaluations.

Although such psychological testing is seldom done on foreign-policy decision makers directly, the available research allows one to ascertain how given personality types are likely to behave when they assume decision-making roles. One can perhaps get a better notion of just how foreign policy might be affected by comparing the authoritarian personality with that of the self-actualizer.

The **authoritarian personality** has been identified in the classic work of T.W. Adorno and colleagues, in which the F-scale (fascism scale) was developed.[1] Among the traits that have been noted in the authoritarian personality are a tendency to dominate subordinates, deference toward superiors, sensitivity to power relationships, a need to perceive the world in a highly structured fashion, excessive use of stereotypes, and adherence to whatever values are conventional in one's setting. Authoritarians also tend to be highly nationalistic and ethnocentric—characteristics that are closely related to support for war and aggression. Studies indicate that the authoritarian personality also prefers clear-cut choices. An examination of reactions to the Korean War of 1950–1953 showed the more authoritarian personality opting either for withdrawal of American forces or escalating the war by bombing China, while the less authoritarian personality was more likely to prefer peaceful settlement.[2]

Whereas the authoritarian personality is likely to have a negative impact upon the making of rational foreign policy, Abraham Maslow's **self-actualizer** has personality traits that are highly desirable for effective decision making.[3] For self-actualization to be achieved, certain basic needs must be met as the individual develops from infancy. These include physiological needs, safety or security, affection and a sense of belonging, and self-esteem. Most decision makers, except possibly those from the developing world, have been able to satisfy their basic physiological needs, which consist largely of sufficient food and shelter. Those, like former Prime Minister Golda Meir of Israel, who suffered physiological deprivation as children may be particularly sensitive to the needs of starving people—especially children. This may in part explain the extensive Israeli foreign aid program targeted toward Africa when Meir led Israel during the late 1960s. Such deprivation may also shed light on why so many leaders in the developing world live ostentatiously after achieving positions of power.

In addition to satisfying psychological needs, the self-actualizer has also developed self-esteem and a sense of security and belonging. Self-esteem is seen as particularly important if one hopes to achieve a more peaceful foreign policy. The distinguished political scientist, Harold Lasswell, concluded that wars and revolutions arose primarily as a means for a people and their leaders to discharge their collective insecurities.[4] People with low self-esteem tend to be most predisposed to nationalistic appeals, for by identifying with the state, the individual is able to raise his or her own self-esteem. Social-psychological research has also shown that individuals

---

[1] T. W. Adorno et al., *The Authoritarian Personality* (New York: Harper & Row, 1950).

[2] Robert E. Lane, "Political Personality and Electoral Choice," *American Political Science Review* 49 (March 1955): 173–190.

[3] Abraham Maslow, *Motivation and Personality* (New York: Harper and Row, 1954).

[4] Harold Lasswell, *World Politics and Personal Insecurities* (New York: McGraw-Hill, 1935), 255.

with a negative self-concept are highly responsive to information that calls attention to difficulties and complications in any policy problem.[5]

Individuals with high self-esteem, on the other hand, are more likely to trust others, and this may have some positive implications for the making of foreign policy. The person who feels pressure to control others will find it difficult to make the necessary conciliatory moves to achieve an agreement. President Woodrow Wilson, who demonstrated considerable lack of flexibility when personally challenged, could nevertheless be effective in situations in which he did not link success with his need for self-esteem and when he felt a minimum of personal involvement.[6]

Although those with low self-esteem often seem to be uncooperative or bellicose in their foreign-policy behavior, it has been hypothesized that people with high self-esteem may also sometimes be less than cooperative. The problem lies in the fact that such individuals may attempt to exploit those who are less secure, raising the probability that conflict will erupt. Those with high self-esteem also tend to take more risks than might be desirable.

Interviews with 126 State Department officials, in which respondents were asked to react to five scenarios involving international crises, confirmed that those measuring higher in self-esteem on psychological tests tended more often to oppose the use of force than those with lower self-esteem. But when the respondent had both high self-esteem (which alone would tend to make one oppose the use of force) and a strong ambition to feel active and personally influential, the net result was an unusually powerful tendency to advocate the use of force.[7]

A study of forty-five present and recent political leaders offered additional personality traits concerning aggressiveness.[8] Aggressive leaders were found to have low conceptual complexity (tended to think in black and white rather than in terms of gray), were distrustful of others, nationalistic, and tended to believe in their own abilities to control events.[9]

**ABERRANT PERSONALITIES** Although a number of personality traits may have a negative impact upon foreign-policy decision making, whenever **aberrant personalities** (whether due to illness or emotional instability) assume positions of leadership, serious problems affecting the stability of the international system may result. Cases of mental disturbance among leaders are more common than one would like to believe. For example, in the nineteenth century Lord Castlereagh of Great Britain committed suicide as a result of his deteriorating mental situation while serving as prime minister; and King

---

[5] Irving L. Janis, *Crucial Decisions* (New York: Free Press, 1989), 217.

[6] Michael Sullivan, *International Relations: Theories and Evidence* (Englewood Cliffs, N.J.: Prentice-Hall, 1976), 33.

[7] Lloyd S. Etheredge, "Personality and Foreign Policy," *Psychology Today*, March 1975, 37–42.

[8] Margaret G. Hermann, "Explaining Foreign Policy Behavior Using the Personal Characteristics of Political Leaders," *International Studies Quarterly* 24 (March 1980): 7–46; and idem, "Leaders and Foreign Policy Decision Making," in *Diplomacy, Force, and Leadership*, ed. Dan Caldwell and Timothy McKeown (Boulder, Colo.: Westview Press, 1993), 77–94.

[9] For further reading see Eric Singer and Valerie M. Hudson, eds., *Political Psychology and Foreign Policy* (Boulder, Colo.: Westview Press, 1992).

Ludwig of Bavaria demonstrated increasingly bizarre behavior during his reign. Serious illness clouded the war-time decision-making capabilities of such leaders as Winston Churchill, Joseph Stalin, and Franklin Roosevelt, while severe depression may have led both to the miscalculations made by Prime Minister Menachem Begin in Israel's 1982 invasion of Lebanon and to his increasingly paranoid behavior.[10]

Dealing with aberrant personalities in foreign-policy matters is extremely difficult. For example, it has been said that, given Soviet leader Joseph Stalin's extreme paranoia, it was virtually impossible to influence his behavior.[11] Every move, even a conciliatory one, was viewed by Stalin with extreme suspicion. As a result, when Stalin was alive, there may have been little alternative to the cold war—besides that of transforming it into a "hot" one.

**COMPATIBILITY OF PERSONALITIES** The personality traits of leaders, whether normal or aberrant, also significantly affect how leaders relate to one another. In an age of increased interaction among heads of state both at frequent summit meetings and through personal and rapid communications, the way in which their personalities mesh can influence interstate relations. Stalin's arrogance and aloofness undoubtedly affected the ability of other contemporary world leaders to relate to the Soviet Union, just as Nikita Khrushchev's insensitivity, erratic style, and bluntness bothered a number of leaders—particularly Mao Zedong and other Chinese leaders. In contrast, the style and personality of Mikhail Gorbachev not only led Margaret Thatcher to declare in 1985 that he was a Soviet leader with whom the West could deal, but it also made him popular with other leaders and the world's population in general.

The importance of personal feelings in foreign relationships was underscored by George Ball, undersecretary of state during the Kennedy administration, who related how he confronted President Ayub Kahn on a trip to Pakistan in 1963. Ball wished to know why Pakistan, a large, Asian, Muslim nation could not pursue better relations with Indonesia, another large, Asian, Muslim nation with whom it shared many interests. "Well, the answer's very simple," Ayub responded, "Sukarno [Indonesia's president] is such a shit."[12]

# MOTIVATION OF DECISION MAKERS

In order to obtain a better understanding of why certain foreign-policy decisions were made, it is necessary first to obtain some notion of what motivates foreign-policy decision makers and how they perceive the world. Determining such motivation is never an easy task. Indeed, decision makers themselves often are not certain why they behave as they do. Research on the motivations of decision makers is further complicated by the fact that psychological data about leaders are al-

---

[10]Jerrold M. Post and Robert S. Robins, *When Illness Strikes the King* (New Haven, Conn.: Yale University Press, 1993).

[11]Morton Schwartz, *The Foreign Policy of the U.S.S.R.: Domestic Factors* (Encino, Calif.: Dickenson, 1975), 193.

[12]George Caldwell, *Political Analysis for Intelligence* (Washington, D.C.: Central Intelligence Agency, Center for the Study of Intelligence, 1992), 3.

most impossible to obtain. Leaders are unlikely to submit to psychological testing and even less likely to be receptive to psychoanalysis. As a result, indirect means have to be utilized to obtain such information. These techniques include the content analyses of speeches, making inferences about traits and motivations from behavior, or drawing analogies and comparisons from experimental and interview studies of more accessible subjects. Despite the difficulties of direct observation, the considerable psychological research that has been conducted to date allows for some fairly accurate judgments regarding how leaders with certain needs and psychological predispositions are likely to behave when placed in positions of authority.

A key motivating force for many, if not most, political leaders is a *need for power*. Social-psychological studies suggest that those with very strong drives for power tend to desire leadership positions, want to dominate others, are argumentative, show little humanitarian concern, tend to be paranoid, and, at the same time, do not like to take risks.

For many leaders, the search for power may be an effort to compensate for deprivations experienced during childhood. As children they may have felt unloved because of a domineering mother or father. In some instances they did not relate well to their peers because of certain physical limitations. Hitler, for example, was "a small, unattractive, somewhat sickly child who was dwarfed by his dignified, uniformed father."[13] Woodrow Wilson also was sickly as a child and had an unhealthy relationship with his exacting father. This may not only have affected his power needs but could have "contributed to Wilson's breakdowns whenever his most intensely held commitments were challenged."[14]

The perceived need for power has been contrasted with two other basic psychological motivations—need for affiliation and need for achievement. The person who places *need for achievement* at the top of his or her motivational hierarchy tends to take moderate risks. Such a person is reluctant to sacrifice possible achievement of success by taking risks, but at the same time does not want to risk failure by not attempting to achieve a given goal. Prime Minister Margaret Thatcher of Great Britain, a self-described "conviction politician," probably represented this type.[15]

The person with a high *need for affiliation* constantly seeks the approval of significant others. In the political arena such a person has great difficulty firing advisers and may tend to be loyal to them long after such loyalty is useful. A person with this predisposition is likely to be less "imperialistic" because of a greater concern for other people. More negatively, indecisiveness in decision making may also be associated with the need to have the approval of others. A person with a high need for affiliation is hardly likely to fit the description of the self-directed person who is able to act on his or her own counsel. Some have suggested that the desire to be liked by all may have contributed to President Bill Clinton's seeming predisposition to flip-flop on issues.

---

[13] James MacGregor Burns, *Leadership* (New York: Harper and Row, 1978), 57.

[14] Betty Glad, "Contributions of Psychobiography," in *Handbook of Political Psychology*, ed. J. N. Knutson (San Francisco: Jossey-Bass, 1973), 299.

[15] Thatcher's ideological-driven personality may have impeded her ability to achieve results, since it made her less flexible in negotiation. See Margaret G. Hermann, "Leaders, Leadership, and Flexibility," *The Annals of the American Academy of Political and Social Science* 542 (November 1995): 152.

Prime Minister Margaret Thatcher of Great Britain and President Ronald Reagan reportedly enjoyed a very cordial relationship, June 1984.

A study of the effects of twentieth century American presidents' needs for power, affiliation, and achievement found that the power motive was associated with war and the failure to reach arms limitation agreements. Presidents with higher affiliation and achievement need scores, in contrast, were less likely to engage in war and more likely to support arms limitation agreements.[16]

## PERCEPTION AND FOREIGN POLICY CHOICE

Among other factors that can interfere with rational choice in the making of foreign policy is that of **perception**—or, perhaps more accurately, the problem of misperception. If one is analyzing a situation with a false sense of reality, no amount of clear thinking or logical and rational analysis will compensate for this deficiency. International events are perceived by decision makers on the basis of images they hold about the world. These images, of course, have been shaped over many years and are the product not only of the individual experiences of the decision maker but also of the broader myths and traditions that prevail in a society.

The impact that images and perceptions can have upon foreign policy can be seen in an analysis of public pronouncements about the Soviet Union made by John Foster Dulles, who served as secretary of state during the Eisenhower administration.[17] This study found Dulles to have had an unwavering view of the Soviet Union as an evil actor in the international system. This led him to perceive that the Soviet Union tended to decrease its hostility toward the United States only when Soviet capabilities decreased relative to those of the United States. Given that belief, Dulles therefore would always deem it appropriate for the United States to retain its own military superiority, even when the Soviet government was being conciliatory.

Of course, it is possible for decision makers to alter their views of reality radically when highly unexpected events occur. President Jimmy Carter noted how the Soviet incursion into Afghanistan in December 1979 shocked him into a more cautious stance toward the Soviet Union. He seemed to shift gears completely, asking for significant increases in military spending and shelving the SALT II Treaty. Indeed, it may be that those decision makers who feel most betrayed by the abrupt change of policies on the part of other leaders whom they had trusted will react the most negatively. Conversely, President Ronald Reagan showed remarkable change in his image of the Soviet Union, from viewing that country as the focus of evil in 1983 to a position of increasingly warm relations with Mikhail Gorbachev during five top-level meetings held during Reagan's last term in office.

**HISTORICAL EXPERIENCES** Considerable research has been conducted on the question of how decision makers develop their perceptions of foreign policy and, in par-

---

[16]David G. Winter and Abigail J. Stewart, "Content Analysis as a Technique for Assessing Political Leaders," in *A Psychological Examination of Political Leaders,* ed. Margaret G. Hermann (New York: Free Press, 1977), 60.

[17]Ole R. Holsti, "The Belief System and National Images," *Journal of Conflict Resolution* 6 (September 1962): 244–252.

ticular, what factors lead them sometimes to misperceive reality.[18] It has been discovered, for example, that one's perception of and reaction to current events is very much influenced by one's interpretation of the historical record. Prime Minister Harold Macmillan of Great Britain was said to have become obsessed with what he saw as the increased dangers of war inherent in the Berlin crisis in the late 1950s. While the crisis was ongoing, his journal entries included many analogies to World War I—a war that he believed had been quite preventable but had arisen as the result of a number of foolish miscalculations.

World War II and the events leading up to it provided the most important analogies for cold-war decision makers. Since the vast majority of postwar leaders were socialized during the 1930s, events such as the 1938 Munich Conference, which came to symbolize appeasement against Hitler, left an indelible impression on their minds. On a number of occasions Presidents Harry Truman and Lyndon Johnson called attention to the failure of Western appeasement policy as justification for their own tough stance against what they saw as Soviet and Chinese efforts to extend their control in Asia and Europe. In fact, once Hitler began his aggressive actions against Germany's neighbors, he continued them for several years with little resistance from the major international players. When the allies finally concluded that he must be stopped following the Nazi attack on Poland, World War II was the result. That experience no doubt gave rise to the *domino theory* in America's attitude toward Asia decades later, when those who had come of age in Hitler's time were now U.S. leaders. As it was, American decision makers assumed that communist successes, like Hitler's successes, would only cause communist leaders to continue their aggression; hence, they saw it as a necessity not to concede an inch. More recent official comparisons between Hitler and Saddam Hussein appear to have had similar effects on American responses to the Iraqi invasion of Kuwait when President George Bush adopted a policy of no compromise.

President Clinton and his generation of decision makers are more likely to see the defining event as the Vietnam War, the experiences of which are likely to make one very suspicious of the use of force, particularly in remote areas. This may help explain the reticence shown by that administration to military action in Bosnia and various parts of Africa.

It should be noted, though, that experience is not always a good teacher because of the problems involved in making the appropriate inferences. For example, the Aztecs who offered human sacrifices to obtain good harvests would not necessarily have ceased to do so if one year's harvest was bad. They might merely have increased the number of people they sacrificed, believing that the experience proved only that the previous sacrifice had been insufficient.[19]

---

[18]Among the best of these studies are Robert Jervis, *Perception and Misperception in International Politics* (Princeton, N.J.: Princeton University Press, 1976); Robert Jervis, Richard Ned Lebow, and Janice Gross Stein, *Psychology and Deterrence* (Baltimore: Johns Hopkins Press, 1989); John Stoessinger, *Why Nations Go to War,* 6th ed. (New York: St. Martin's Press, 1992); Yaacoc Y. I. Vertzberger, *The World in Their Minds* (Stanford, Calif.: Stanford University Press, 1990); Ralph K. White, *Nobody Wanted War: Misperception in Vietnam and Other Wars* (Garden City, N.Y.: Doubleday, 1970); and Ralph K. White, *Fearful Warriors* (New York: Free Press, 1984).

[19]J. D. Armstrong, *Revolutionary Diplomacy: Chinese Foreign Policy and the United Front Doctrine* (Berkeley: University of California Press, 1977), 18.

**PERSONAL EXPECTATIONS**  A person's *expectations* also tend to affect his or her perception of new events. Throughout the cold war, Western decision makers were inclined to suspect that communist subversion underlay every move against a pro-Western government, no matter what the domestic record of that government. Thus, President Johnson and his advisers concluded, on the flimsiest of evidence, that a 1965 rebellion designed to return Juan Bosch to power in the Dominican Republic was led by communists and thus justified military intervention by the United States. In sending in the Marines, Johnson declared that he was not going to sit around while Castro took over the hemisphere. Yet the evidence points to the fact that the revolt that led LBJ to intervene consisted primarily of students who were dissatisfied with the 1963 military overthrow of the democratically elected Bosch by General Wessin y Wessin.

Throughout the cold war, little evidence was required to lead Western decision makers to conclude that if a new regime were a communist one, it would behave in an aggressive fashion externally. Yet this was not necessarily the case, as suggested by the behavior of such self-described Marxist leaders as Salvador Allende of Chile and Robert Mugabe of Zimbabwe. The problem with such facile assumptions, of course, is that they sometimes become self-fulfilling prophecies. That was the case when Western decision makers reacted in a hostile and suspicious fashion to actors whose intentions they may very well have misperceived. Some think, for example, that Fidel Castro was "pushed" into the Soviet orbit by the treatment he received from the Eisenhower administration soon after Castro came to power. Conversely, the usual expectation of communist regimes was the mirror image of the Western one; namely, that a capitalist state is imperialistic and therefore not to be trusted. That could explain the refusal of the Soviet Union in 1948 to allow any of its fellow communist governments in Eastern Europe to participate in the Marshall Plan aid program offered by the United States.

In situations in which there are several plausible explanations for a given event, a person's expectations usually determine his or her perception. It has been suggested that one of the reasons why the German attack on Norway in 1940 surprised both that country and Britain—even though they had detected German ships moving toward Norway—was the expectation that the Germans were planning to break out into the Atlantic Ocean. The miscalculation in this case might also have been the result of a false analogy, as the affected decision makers thought that Hitler would declare war before attacking, as he had done in both Poland and Czechoslovakia. In fact, Hitler seems not to have been much concerned about possible international legal objections to a surprise attack, possibly because war had been declared against his government by the time of the Norwegian incursion.[20]

**COGNITIVE CONSISTENCY**  Regardless of the sources of one's images about reality, there is considerable psychological pressure to keep them consistent. As new information is received, an effort is made to interpret that information so that it will be compatible with existing images and beliefs. Social psychologists have identified a

---

[20]Donald R. Kinder and Janet A. Weiss, "In Lieu of Rationality," *Journal of Conflict Resolution* 22 (1978): 721.

number of ways in which an individual is able to reduce what is called **cognitive dissonance**, which occurs when seemingly conflicting information is received. One psychological mechanism utilized to reduce the amount of dissonant information is *selective perception*—when individuals focus on information that supports their basic predispositions while overlooking contradictory or conflicting information.

Persuasive evidence of this phenomenon is seen in a study of student responses to a questionnaire in which the respondents were asked to rate themselves and their international relations instructors on a seven-point liberal-conservative scale involving positions on such issues as the United Nations, nuclear disarmament, and foreign aid.[21] The results demonstrated that students tended to rate their instructors very much like themselves, so that a given instructor was seen as basically conservative by the more conservative students and liberal by the liberal students. The students apparently were selecting information from lectures and discussions that tended to bolster their previous opinions, and overlooked information that was contradictory. These findings suggest that the basic political predispositions of students on international issues are fairly rigid by the time they enter college. Indeed, studies of political socialization suggest that many political attitudes, whether national or international, are formed even before a child goes to school.

Whereas selective perception implies that a person simply fails to perceive a given piece of information, it is also possible to *deny* the information and thus retain cognitive consistency. Illustrative of this is the reaction of Soviet command headquarters upon the receipt of a message from the Soviet front line on June 22, 1941, indicating that Soviet troops were being fired upon. The response from Soviet headquarters in this instance was that the message sender must be insane, and the sender was reproached for not sending the message in code. An even more blatant example occurred when Hermann Göring, Hitler's second in command, was informed that an Allied fighter had been shot down over Aachen, proving that the Allies had produced a long-range fighter. Göring, responding on the basis of his own experiences as a fighter pilot, asserted that it was impossible. He went on to say, "I officially assert that American fighter planes did not reach Aachen. . . . I herewith give you an official order that they weren't there."[22]

A third way of coping when one receives contradictory information is to *compartmentalize* such information. An individual may focus on religious values on the day of worship while concentrating on a scientific approach at school. Atrocities committed by a leader against his or her own population may be compartmentalized by an allied leader who only wants to think positively about the allied partner. This phenomenon may help explain why American leaders did not react more indignantly in the 1960s and 1970s to the atrocities committed by the Shah of Iran, who was viewed as providing a major bulwark against the threat of communism in the Middle East. The ability of decision makers to accept and compartmentalize completely contradictory information was shown by the simultaneous acceptance throughout the cold war of the notion of "the free world" and the recognition that it was composed of a large number of states led by dictators.

---

[21] Jane S. Jensen, "Attitudinal Selectivity in International Relations Courses," *Teaching Political Science* 1 (April 1974): 240–252.

[22] Robert Jervis, *Perception and Misperception in International Politics*, 144.

Fourth, efforts may be made to *redefine* the dissonant information once it has been received. When the American secretary of the navy was told about the bombing of Pearl Harbor, his response was that it could not possibly be true and that the message must refer to the Philippines. In another example of redefinition, British generals in 1917 reacted to the news that few Germans were surrendering during a major British offensive by declaring: "We are killing the enemy, not capturing him."[23]

Finally, cognitive consistency can be enhanced by the tendency to *bolster* one's own choices or viewpoints. Decision makers spend considerable energy attempting to convince themselves of the correctness of whatever position they take, for the admission of error has a high psychological cost. Such leaders may have a particular fear of looking weak or foolish. If a response is required, they tend to exaggerate the favorable aspects of the position taken while minimizing its unfavorable consequences. Decision makers may also seek to minimize their own personal responsibility for the decision by asserting that they had no other choice.

**THE PROBLEM OF MISPERCEPTION** The attempt to maintain cognitive consistency often leads to misperception of international events, making rational choice exceedingly difficult. When information is discarded or reinterpreted simply for the sake of consistency, considerable error is likely to develop. But there are other psychological defense mechanisms that tend to distort a person's image of reality. One of these is the tendency to project one's own feelings onto an external object. People who feel a lot of anger, for example, tend to see others as angry and aggressive, while those who are more trusting view others in a similar fashion.

The predisposition to see one's own moves as motivated by magnanimous objectives, while viewing the same types of moves by the adversary as being maliciously motivated, may also lead to inaccurate interpretations. Throughout the cold war, when the United States intervened in the affairs of another state, the action was viewed by Americans as stabilizing; but if the Soviets did the same thing and with the same motivation, it was viewed as destabilizing.

Misperception about others may also arise from misreading communications as applying to oneself when they are really targeted for another audience. This is particularly true with respect to messages designed for a state's domestic public that are perceived and reacted to by an external actor. American decision makers, for example, appear to have resented the highly inflammatory comments about the United States made by the late Prime Minister Indira Gandhi, whose effort apparently was to divert attention from India's domestic problems by making the United States the scapegoat for some of those problems.

Perceptions about another state may also be distorted when lower-level officials act without official sanction, with the result that they may be countermanded by their superiors. The United States had persistent problems in obtaining an accurate reading about the intentions of Iranian revolutionary officials, beginning with the American embassy hostage situation in 1979 and continuing with the Iranian arms for hostage deal in 1986. Many different Iranian spokespersons were involved, not all of whom could speak for the nation and who often therefore had to be corrected by higher authorities, such as the Ayatollah Khomeini.

---

[23]Kinder and Weiss, "In Lieu of Rationality," 710.

As if such distortions of perception are not enough, one must also remember that foreign governments often intentionally seek to mislead the decision makers of another state. If a state's intentions and behaviors are not benign, as in the case of efforts to undermine a foreign government, considerable energy will be directed toward covering up such disinformation activities.

But whatever the explanation for the misperception of events and behaviors in the international system, it is clear that such misperception has often led to war and the escalation of conflicts. A detailed examination of some eleven wars led one analyst to conclude that misperception was the primary explanation of the conflict in each instance.[24] Clearly both Iraq's Saddam Hussein in attacking Kuwait and Argentina's General Leopoldo Galtieri in invading the Falklands (Malvinas) miscalculated respectively the probable responses of the United States and Britain.

# DOMESTIC STRUCTURES AND FOREIGN POLICY CHOICE

It is not only the psychological factors involving individual motivation and misperception that tend to undermine the explanatory value of the rational actor model. Its utility in accurately reflecting reality also is impeded by a number of domestic structural problems, resulting from the way decisions are made. Who has the authority to make decisions? To what information do they have access? With what competing domestic and international agendas must they contend? These are but a few of the sorts of considerations one needs to examine in order to explain foreign-policy behavior.

## DEMOCRATIC VERSUS AUTHORITARIAN REGIMES

Some writers have viewed democratic decision making as a major impediment to rational choice in foreign policy. It has been alleged, for example, that democracies are not as efficient as authoritarian and aristocratic regimes in making wise and timely choices in foreign policy. The French chronicler of American democracy, Alexis de Tocqueville, wrote almost one hundred and fifty years ago that the management of foreign affairs requires knowledge, secrecy, judgment, planning, and perseverance, qualities in which he saw autocratic systems as superior to democratic ones.[25] In this century, noted columnist Walter Lippmann criticized democratic foreign policy making on the ground that the mass public is generally uninformed about foreign policy and will always opt for taking the easy way out of situations that demand more assertive action.[26] The French analyst Raymond Aron also raised questions about democratic decision making because of what he saw as the danger of "conservative paralysis" and a corresponding inability to deal with pressing problems.[27]

---

[24]Stoessinger, *Why Nations Go to War*.

[25]Alexis de Tocqueville, *Democracy in America*, vol. 1 (New York: Knopf, 1945), 234–235.

[26]Walter Lippmann, *The Public Philosophy* (New York: Mentor, 1955), 23–24.

[27]Raymond Aron, *Peace and War: A Theory of International Relations,* trans. Richard Howard and Annette Baker Fox (Garden City, N.Y.: Doubleday, 1966), 67.

> ### Global Changes
>
> ## Changing Decision-Making Structures
>
> Throughout most of the Westphalian period, nondemocratic regimes have dominated the global scene. Only five states could be regarded as liberal democracies in 1848 and thirteen in 1900. According to Freedom House, by 1995 that number had grown to 117.
>
> - Is this trend toward democratization likely to continue? What sorts of things might cause a reversal in the growth of democracy?
>
> The relative power of executive and parliamentary bodies has fluctuated, but generally speaking the executive has primacy in making foreign-policy choices, particularly during periods of national emergency.
>
> - Which branch of government do you think should have the predominant role in the making of foreign policy and why?
>
> The impact of bureaucracies in the foreign-policy decision-making process has increased substantially as the number of agencies and personnel have proliferated severalfold in most states during the past few decades.
>
> - How is this growth likely to affect rational foreign policy?
>
> As the world becomes more complex and interdependent, the state has become only one of many actors involved in foreign policy. Interest groups, multinational corporations, and governments within the state as well as supranational groups increasingly have their own foreign policies as they deal directly with foreign governments and international organizations.
>
> - Are these trends to be applauded or condemned in terms of their impact on a more just and efficient global system?

If one is looking simply at the effectiveness and efficiency of foreign policy, there are several reasons to expect the more authoritarian decision-making structures to perform better than democratic ones. In the first place, the more authoritarian government ought to be able to *produce more rapid decisions,* since by definition it is not as responsive to a mass public and usually involves a smaller number of elites who need to be consulted, or at least considered, in the decision-making process. Moreover, less intraorganizational bargaining is required, since opposition from within the bureaucracy can be bypassed or crushed.

Second, the effectiveness and efficiency of the authoritarian regime is enhanced by the fact that it *can better ensure compliance* with its foreign-policy decisions, for a clear hierarchy of command exists, and the punishment for noncompliance may be harsh.

Third, the centralization of foreign-policy decision-making power enables the more authoritarian regime to *present a united front in its foreign policy,* as all spokespersons are expected to follow the party line. In contrast, owing to their pluralism, democracies often speak with several voices. This lack of unity might be particularly disadvantageous when a state is attempting to present a credible deterrent or even a promise of reward, only to find its position undercut by others in the foreign-policy establishment. In 1993–1994, deep divisions within the Clinton administration and the country at large over the issue of the use of force, for example, made it extremely difficult for the administration to develop a credible threat to the Serbians in their war against the Bosnian Muslims.

At the same time that an authoritarian regime can guarantee a more consistent external presentation of its foreign-policy views and thereby enhance the credibility of the message it desires to present, such a regime would seem to enjoy a fourth advantage in that it can *pursue a more adaptable foreign policy* to respond to changing conditions. Since, by definition, the authoritarian regime is less constrained by the mass public and the number of different groups it has to satisfy, it need not wait until the mood of either the elite or the public changes to make a shift in policy. President Franklin Roosevelt felt thwarted in his desire to involve the United States in the Allied cause prior to World War II as a result of isolationist sentiment. Various Democratic presidents who wanted to normalize relations with the People's Republic of China prior to President Richard Nixon's initiative in 1972 felt impeded by what they perceived as public opposition to such a move. In contrast, in the heyday of the Soviet Union, its leaders were able to make some radical departures in foreign policy with minimal internal repercussions, as when Khrushchev announced the Soviet bloc's peaceful coexistence with capitalism at the Twentieth Party Congress in 1956. The ease with which Mikhail Gorbachev was able in a short time to reorient Soviet foreign policy dramatically toward the West and to issue one unilateral initiative after another on arms control issues and other matters became the envy of many a democratic policymaker.

A fifth advantage for the authoritarian regime lies in its ability to *pursue contradictory policies at the same time,* if such a strategy is desirable for obtaining a given foreign-policy goal. During his tenure as secretary of defense in the Nixon administration, James Schlesinger complained that, in contrast to the United States, a "closed society like the Soviet Union has no difficulty in pursuing detente and simultaneously strengthening its defense efforts."[28] Publicizing **detente** in a more democratic regime, on the other hand, was likely to make it more difficult for the decision makers to convince the public that increased military spending was necessary.

Most of these advantages of authoritarian regimes derive from the alleged freedom of action that such a regime enjoys, given its minimal need to be responsive to the public and other interested groups. But one can perhaps exaggerate the amount of decisional latitude that dictators enjoy in the making of foreign policy. Although the foreign-policy elite is generally smaller in authoritarian than in democratic regimes, experts on the subject have increasingly noted that struggles similar to those involved in the political process of a democratic polity also occur at the top levels of single-party states, as well as among bureaucratic agencies and interest groups. In the spring of 1989, Western observers caught a glimpse of an elite struggle within the highest echelons of the Communist Party in China as the appropriate response to student demonstrations in Tiananmen Square was debated. After 1985, struggles within the leadership of the Soviet Union also became increasingly obvious with televised debates and a more substantial reporting of the increased opposition to Gorbachev, from both the right and the left.

---

[28]Cited in P. Williams and M. H. Smith, "The Conduct of Foreign Policy in Democratic and Authoritarian States," *Yearbook of World Affairs* (London: Stevens, 1976), 205.

Perhaps the most serious deficiency for the authoritarian regime lies in the fact that it may be severely *hampered when it comes to policy innovation.* Since its command and control structure is so centralized, and because of the tendency toward paranoia in such structures, authoritarian regimes often generate politicians who tend to accept whatever the dictator desires (or whatever the subordinates think the dictator desires). Initiative is lost in such a system, and there is no opportunity to explore a range of options. Reliance on heavily centralized structures—with their emphasis on secrecy and isolation from external criticism—also destroys the opportunity to tap fresh viewpoints and obtain new information. As a result, there is a tendency for the foreign policy of an authoritarian system to rely extensively on precedent, particularly with respect to minor issues that the higher-level bureaucracy is too busy to address and the lower levels of the bureaucracy have no authority upon which to decide. A rigid, rather than flexible, foreign policy tends to be the outcome of such structural arrangements.

## BUREAUCRACIES AND FOREIGN POLICY

Apart from the chief executive and his or her immediate advisers, the executive branch of government in most states is made up of thousands of bureaucrats, many of whom are involved in the making and implementation of foreign policy. Since chief executives and their immediate advisers are usually transitory, they are forced to rely heavily on the permanent bureaucracy for advice and cooperation in developing and implementing foreign policy. The bureaucracy has achieved the essential skills for dealing with foreign governments and must of necessity be deferred to in the conduct of such policy. Moreover, the bureaucracy collects the relevant information and makes decisions at each level as to what information and which issues will rise to the next level of decision making.

If expertise, experience, and control of information are not enough to give the bureaucracy a very important role in decision making, bureaucracies also become critical at yet another stage in the process: policy implementation. Policies do not implement themselves; they must be executed by subordinates. Through strategies of procrastination, not listening to instructions, or even intentional sabotage, many policies remain dormant. The chief decision maker often becomes preoccupied with other issues and as a result fails to monitor adequately the activities of the subordinates charged with executing a given decision. The frustration of President John Kennedy upon learning that his order to withdraw missiles from Turkey had not been carried out is merely one example of a fairly common phenomenon.[29] It is not enough simply to issue orders, as President Truman observed with regard to his successor, General Eisenhower. Truman was speculating on the problems Eisenhower might encounter as president, considering his military experience in which compliance with official orders is taken for granted. Even a president as strong as Franklin Roosevelt commented in the early days of World War II that the best that could be hoped for from his own State Department in the emerging conflict would be neutrality.[30]

[29]See Steve Smith and Michael Clarke, eds., *Foreign Policy Implementation* (London: George Allen and Unwin, 1985).

[30]John Kenneth Galbraith, *The Culture of Contentment* (Boston: Houghton Mifflin, 1992), 114.

**BUREAUCRATIC TRENDS** Before evaluating the role of the bureaucracy in the making of foreign policy, it might be useful to examine some current trends. The most obvious of these is the extensive increase in the size of the bureaucracy concerned with foreign-policy matters in most states. This can be seen in the overwhelming increase in the number of advisers and analysts within the **foreign office** itself. In the 1870s, the foreign office of a newly united and powerful Germany had only four permanent officials, roughly the scale at which all foreign offices operated at the turn of the century.[31] As late as 1892, the *New York Herald* was proposing the abolition of the State Department since it had so little business to conduct overseas. But times have changed dramatically; foreign-policy personnel for many governments now number in the thousands. Even small, developing states have established fairly large foreign-policy bureaucracies.

There has also been a substantial increase in the size and significance of *non–foreign-office bureaucracies* that deal with foreign-policy issues. In some systems, defense and economic ministries challenge the preeminent position of the foreign minister in the making of foreign policy. With increased global economic interaction, economic well-being is no longer determined by activities within the nation, but is very much affected by external economic activities. The power that the Organization of Petroleum Exporting Countries (OPEC) had during the 1970s to affect national economies throughout the world by setting the price of oil is but one illustration of the impact of external economic decisions. Others include governmental policies establishing tariffs, manipulating currency values, dumping products at cheaper prices abroad, and the like, all of which have serious repercussions for economic interests throughout the world. Consequently, in many countries experts on international economic issues are being added in large numbers to such departments as agriculture, treasury, commerce, and labor in an effort to look after their country's special interests. These experts also become involved in continuing negotiations abroad concerning a whole range of economic concerns and are being assigned to overseas embassies on a permanent basis.

The growth in the foreign-policy role of *defense ministries* has been even more impressive. Traditionally, defense ministries have assumed critical roles in foreign policy during wars and international crises, but continuing national security concerns in the cold war period increased both the power and size of defense establishments. Never before had military alliances such as NATO and the Warsaw Pact persisted during peacetime to the extent that they did, for more than four decades, after World War II. (Although the Warsaw Pact has been disbanded, NATO continues to play an important security role in Europe.) Nor have standing armies been as pervasive as they are today. Given the continuing concern about national security issues and the tremendous resources available to defense departments, they are able to make themselves heard by decision makers on many foreign-policy matters.

Bureaucratic decision making has also become increasingly specialized. Contemporary foreign policies have become extremely complex, involving political,

---

[31] Henry A. Kissinger, "Bureaucracy and Policymaking: The Effects of Insiders and Outsiders on the Policy Process," in *Readings in American Foreign Policy,* ed. Morton H. Halperin and Arnold Kanter (Boston: Little, Brown, 1973), 88.

economic, technological, environmental, and cultural factors; hence they require individuals with very specialized skills. While some generalists remain in foreign offices, the vast expansion of the foreign-policy bureaucracy has brought with it experts in issues involving much more than political and diplomatic affairs.

**LACK OF CREATIVITY AND DECISIVENESS** Distrust of foreign-policy bureaucracies has been rampant in governments throughout the world. As prime minister, Margaret Thatcher of Great Britain frequently vented her displeasure with the Foreign and Commonwealth Office, often seeking independent advice on foreign-policy matters. President Nixon shared the distrust of his chief foreign-policy adviser, Henry Kissinger, and accordingly adopted procedures to minimize the impact of the professional foreign-policy apparatus. But concern about the role of the bureaucracy is not merely a modern phenomenon; an eighteenth-century czar once suggested that it was not he but ten thousand clerks who ruled Russia.

Some of the criticism leveled at bureaucratic decision making is really directed at the issue of decision making by committee. Rational and strategic choice becomes particularly difficult when decisions are subjected to endless rounds of discussion by committees both within and between various agencies concerned with the making of foreign policy. Most governments rely heavily upon interdepartmental committees—which means that any foreign-policy decision is heavily compromised by the time it is made.

While he was a professor at Harvard University, Henry Kissinger wrote articles that were extremely critical of the role of the bureaucracy in the making of foreign policy.[32] He argued that because of the pervasive bureaucratization of American society, leaders have been socialized to act both timidly and in keeping with orthodox views, precluding creative responses to the demands of foreign policy. Moreover, he suggested that, owing to the tremendous energy required to prevail in bureaucratic politics, once a decision has been made, flexibility in international affairs is diminished because of the reluctance to hazard a hard-won domestic consensus.

**BUREAUCRATIC FRAGMENTATION OF POLICY** The growth of bureaucratic decision making has been criticized on the grounds that it tends to fragment foreign policy. With many different agencies involved, each with its standard ways of doing things as well as its jealousy and suspicion of outsiders, a coherent policy becomes difficult to achieve. Since foreign-policy issues affect many agencies, decision making is likely to proceed at a snail's pace if all relevant agencies are provided with the opportunity to "sign off" on a given report or decision. Failure to consult all the appropriate agencies, on the other hand, can sometimes lead to disastrous results, particularly if that agency has critical information or can make a difference in the implementation of policy.

The existence of several layers of decision making also means that information relevant to rational decision making can be scattered among many agencies, making a reconstruction of the broader picture almost impossible. Had the various pieces of

---

[32]Henry A. Kissinger, "Domestic Structure and Foreign Policy," in *American Foreign Policy*, ed. Henry A. Kissinger (New York: W. W. Norton, 1974), 44–52.

> ## Symptoms of Groupthink
>
> Based upon a careful study of a number of U.S. foreign-policy failures, Professor Irving L. Janis of Yale University concluded that the making of foreign policy at the highest levels of government involves a groupthink mentality that impedes "independent critical thinking, resulting in irrational and dehumanizing actions directed against outgroups." Among the symptoms of groupthink, Janis identified the following:
>
> - An illusion of group invulnerability, which creates excessive optimism and encourages the taking of extreme risks.
> - Collective rationalization to discount warnings that might lead the group to reconsider its assumptions.
> - An unquestioned belief in the group's inherent morality, causing its members to neglect the moral consequences of their acts.
> - Stereotyping of the enemy as evil, weak, or stupid.
> - Pressure against any member who challenges the stereotypes or assumptions of the group.
> - Self-censorship of deviations from apparent group consensus.
> - A shared illusion of unanimity among group members (silence is often viewed as approval).
> - The emergence of self-appointed "mindguards," who seek to make certain that others in the group do not deviate from the established norms and consensus.
>
> Janis saw the greatest threat to rational choice in small decision-making groups that are highly amicable and enjoy high *esprit de corps,* leading each member to reassure the others that a given course of action is desirable and that no further questions need be asked.
>
> **SOURCE:** Irving L. Janis, *Victims of Groupthink* (Boston: Houghton Mifflin, 1982).

information relevant to the surprise attack at Pearl Harbor in 1941 been available and coordinated in a single place, that intelligence failure might well have been averted.

Fragmented bureaucratic decision-making structures may also affect how other states relate to one another. Officials often seem to prefer to deal with governments that are more centralized, which adds to the predictability and the speed with which a given transaction can be executed. At the same time, fragmentation can enable external actors to gain bureaucratic allies to influence a given nation's policy or perhaps even to counter a decision that has already been made. Lobbying across national borders often involves bureaucracy-to-bureaucracy interaction in an effort to influence one's own foreign policy as well as that of the other state. For example, American bureaucrats within the Department of Agriculture regularly negotiate levels of grain exports with their counterparts in other countries rather than going through diplomatic channels.[33]

Another problem of bureaucratic decision-making is the fear that the narrow interests of an agency may be substituted for the broader national interest. Agencies are more concerned with their own special interests and tend to view problems only from that narrow perspective. Policies that will contribute to the budgetary prowess

---

[33] Raymond F. Hopkins, "Global Management Networks: The Internationalization of Domestic Bureaucracy," *International Social Science Journal* 30 (1978): 37.

and role of the bureau are likely to be favored. Indeed, there may be considerable mutual back-scratching, as in military budgetary requests, when each military service agrees either tacitly or formally not to challenge the pet weapons systems of the other in the expectation that its own projects will not be questioned. Obtaining unbiased recommendations becomes difficult for the central decision maker who depends upon bureaucratic advice in making foreign-policy choices.

Individual career interests may also interfere with providing the best advice. The desire to be promoted, or even to remain employed, places pressure on the individual not to rock the boat or object to the position of a superior. Bureaucrats consequently move slowly in uncharted territory until they are able to see which way the wind is blowing.

Failures in foreign policy, however, should not entirely be blamed upon the bureaucratic structures; indeed, such structures often facilitate rational foreign policy by providing specialization, hierarchical efficiency, institutional memory, and the consideration of a wide range of alternatives.[34] As the Tower committee, created in 1987 to investigate the Iran-contra fiasco, made abundantly clear, the failures in that case were not so much in the bureaucratic structures involved as in the people who operated those structures.[35] Among the bureaucratic irregularities noted were the bypassing of the State Department, the tendency of the National Security Council to exceed its terms of reference, the absence of congressional approval, and the devolution not only of operational responsibility, but also of command decisions to lower echelon officials, particularly Colonel Oliver North.

**HOW IMPORTANT ARE BUREAUCRACIES IN DECISION MAKING?** Although one can find arguments and illustrations both supporting and criticizing the bureaucracy in its role in making foreign policy, there is no consensus on whether bureaucracy is the most critical actor in explaining foreign-policy behavior. Several authorities have argued that bureaucracies are not the major architects of foreign policy in most nation-states, and the mere increase in the size and activities of bureaucracies is hardly proof of their impact on final choices. It might be noted that a number of those who regard the bureaucratic model as providing a major explanation for foreign-policy behavior also have been involved as practitioners in government. They may simply be following the natural inclination to exaggerate their own roles, much like Sir Francis Bacon's fly that sat on the axletree of the chariot wheel and declared, "What a dust do I raise."

Case studies based upon the role of bureaucratic politics tend to focus on issues that have exercised the bureaucracy, neglecting those that enjoy broader consensus or that are decided at higher levels without much bureaucratic input. By selecting such cases, one can document a great deal of bureaucratic bargaining and maneuvering. Whether such activities influenced or determined the final foreign-policy output, however, cannot be ascertained from an analysis of bureaucratic infighting.

[34]Charles W. Kegley Jr. and Eugene R. Wittkopf, *American Foreign Policy: Pattern and Process*, 4th ed. (New York: St. Martin's Press, 1991).

[35]John Tower, Edmund Muskie, and Brent Scowcroft, *The Tower Commission Report* (New York: Bantam Books, 1987).

The chief decision maker usually has the power to select his or her immediate advisers and to determine which advice will be accepted. Decision makers, like all people, have a tendency to listen to that advice that confirms their own predispositions. At the same time, bureaucrats have a vested interest in providing the kind of information and advice that they believe their employer desires. But when the chief decision maker delegates much of the details of policy making, as was the case with President Ronald Reagan, the bureaucracy may have greater influence. A number of observers have commented on Reagan's tendency to accept the latest advice provided. Bureaucracies would consequently compete to be the last to offer advice, sometimes resulting in embarrassing shifts in policy. In response to the 1982 Israeli invasion of Lebanon, for example, White House advisers recommended that the president instruct the American delegation to vote in favor of the Security Council decision condemning Israel, whereupon Secretary of State Alexander Haig, who had not participated in the decision, was successful in persuading Reagan to reverse his decision. When Haig notified State Department personnel to arrange the veto, however, he was countermanded by the deputy NSC head, Robert McFarlane. Further negotiations had to be held with the National Security Council, and it was only with minutes to spare that the American ambassador to the United Nations received instructions to ignore all previous instructions and to veto the resolution.[36]

[36] Alexander Haig, *Caveat* (New York: Macmillan, 1984), 339.

---

**Normative Dilemmas**

## What Foreign Policy Choices to Make?

In January 1995, after it was clear that Congress would not support a Mexican economic bailout, President Clinton found funds enabling him to guarantee billions of dollars to support the falling Mexican peso.

- What would you have done in a similar circumstance, having been advised by most economists, both in and out of government, that to do otherwise would risk considerable economic chaos that could reverberate throughout the global economy? Would you still support Mexico knowing that your presidential reelection would be jeopardized by political opponents exploiting the issue to an American public unalterably opposed to the action?

As president, a number of your major campaign contributors from businesses that have been unfairly hurt by China's disregard of U.S. patent and copyright laws have asked you to take strong economic retaliatory action against China. You are ambivalent on the issue, for you realize that only by encouraging continued economic interaction with China will there be any hope for moderating its repressive political regime.

- What would you do?

As a staff member of a major foreign-policy bureau, you have been asked to develop a set of recommendations to be forwarded to the president. As you prepare your report you realize that your view of what is in the national interest and that of your agency are at considerable variance.

- How would you draft your recommendations, knowing full well that your promotion and perhaps even your job is at stake, particularly since your superiors have been ordered to downsize the bureau?

**CLINTON, ARAFAT, RABIN**
Reprinted from *Foreign Policy* © 1995 Joseph Azar.

## THE ROLE OF PUBLIC OPINION

The indictment against democracies in the making of foreign policy rests in part upon a belief that matters of state should not be dependent upon the whims of public opinion. This concern is based upon the general lack of knowledge about international affairs. The *mass public*—those ill-informed about most current political issues—is generally viewed as representing from 75 to 90 percent of the adult population—even in an advanced nation like the United States. Readership of world news stories tends to be limited among the mass public, and with respect to news, domestic affairs tend to receive the most attention. Even at the height of World War II, the public's interest in domestic affairs was almost twice as intense as its interest in foreign affairs.[37]

**DECISIONAL LATITUDE**  Perhaps one need not be overly concerned, however, about an uninformed public pressing decision makers into unwise choices, for the evidence suggests that policymakers have considerable *decisional latitude* in making foreign policy. One multinational study concluded that in the short run, "changes in attitudes and shifts in policy are correlated at a relatively low level. And to the degree that a short-term relationship can be found, it is the public that responds to governmental action rather than vice versa."[38]

The latitude of decision makers is especially high during crises, since the public tends to rally around the flag at such times. Presidential popularity tends to peak with crises whether governmental policies are effective or not. Kennedy's popularity rose from 61 to 74 percent after the Cuban missile crisis, but it stood at an even higher 85 percent following the Bay of Pigs fiasco in 1961; Truman had an impressive 81 percent of the public supporting his commitment to South Korea in 1950, despite his low popularity at the time. But the most dramatic rise in presidential support occurred in response to the Iranian seizure of the United States Embassy in Teheran in November 1979. President Carter's popularity rating rose from 32 to 61 percent in one month—the most dramatic turnaround in a president's rating since Gallup polling began in the late 1930s.

A recent review of a number of case studies of public opinion and U.S. foreign policy suggests that its impact has been increasing in recent decades; indeed, it appears to be less volatile and more coherent than initially thought.[39] The frustration surrounding the Vietnam War made Congress and the American public reticent to support American involvement in the Angolan crisis in 1975, despite the urging of President Gerald Ford and Secretary of State Henry Kissinger. Public opinion also appears to have been a major factor in the hesitancy of the Clinton administration to use military force in the Bosnian crisis in 1993–1994, even though many in the government and in the media thought that strong action should be taken to stop Serbian aggression. Once a cease-fire was reached in November 1995, President Clinton had to overcome strong public opposition in committing 20,000 American troops to help police the agreement.

**MANIPULATION OF PUBLIC OPINION**  Political elites may attempt to manipulate public opinion for both domestic and international purposes. Just as political decision

---

[37] Ralph B. Levering, *The Public and American Foreign Policy, 1918–78* (New York: William Morrow, 1987), 32.

[38] Martin Abravanel and Barry B. Hughes, "The Relationship Between Public Opinion and Governmental Foreign Policy: A Cross-National Study," in *Sage International Yearbook of Foreign Policy Studies,* vol. 1, ed. Patrick J. McGowan (Beverly Hills, Calif.: Sage Publications, 1973), 126.

[39] Ole R. Holsti, "Public Opinion and Foreign Policy: Challenges to the Almond-Lippmann Consensus," *International Studies Quarterly* 36 (December 1992): 439–466.

makers have been known to use external conflict to divert attention from internal problems, they have sought to demonstrate their peacemaking capabilities in the international arena in order to bolster sagging public support. Both Presidents Nixon and Carter actively sought to play a peacemaking role while in office. Nixon attempted to counter his deteriorating position at the time of Watergate by holding a summit conference with the Soviet Union in 1974 in the hopes of concluding the SALT II agreement on strategic arms limits. Many have speculated that Carter gambled on success in the 1978 negotiations between Israeli Prime Minister Menachem Begin and Egyptian leader Anwar Sadat at Camp David as a way of improving his standing with the public, which was then lower than that of any president since Harry Truman.

Decision makers also attempt to influence public opinion in order to improve their bargaining position in the international arena. Demonstrating that the public is behind them on a given policy and suggesting that the state has no other options given the pressure of public opinion can be a useful strategy in international bargaining. Franklin Roosevelt employed this stratagem so extensively that Stalin began to use Soviet public opposition to justify Soviet foreign policies which were at variance with those of the United States. As the SALT II negotiations entered their final phase, the Carter administration exploited the Senate's hostility toward a SALT agreement as a way of inducing the Soviet Union to be more conciliatory in the negotiations.

In seeking to manipulate public opinion to support a given policy, decision makers need to be aware of latent predispositions that exist within a target population. Otherwise, efforts to mobilize that opinion might backfire. Illustrative of this problem are two instances in which the United States government sought to educate the American public on foreign-policy issues: its efforts to generate support for freer trade in the 1950s and to increase public support for foreign aid in 1961.[40] In both cases what might be called an "iceberg effect" resulted from the official educational efforts to sell a more liberal policy on these issues. The government efforts only addressed the informed and more liberal part of the public (the tip of the iceberg) without seeing the views of the uninformed mass public (the much larger portion of the submerged iceberg). As a result, efforts to lower the level of ignorance regarding foreign aid and trade served only to activate a large mass public whose latent attitudes were hostile to a more liberal policy. In 1978 the Carter administration risked a similar backlash in attempting to generate public support for the Panama Canal Treaties, which called for the eventual return of the canal to Panama—a policy overwhelmingly opposed by the latent attitudes of the mass public. Of course, if one's adversaries make a major point of an issue, as Ross Perot did in the case of the 1993 debate on American ratification of the North American Free Trade Agreement (NAFTA), an administration may have little choice but to engage in a counterattack in support of its own position.

---

[40]Raymond A. Bauer, Ithiel de Sola Pool, and Lewis A. Dexter, *American Business and Public Policy: The Politics of Foreign Trade* (Chicago: Aldine-Atherton, 1963); and James N. Rosenau, *National Leadership and Foreign Policy: A Case Study in the Mobilization of Support* (Princeton, N.J.: Princeton University Press, 1963).

# Decision Making in a World of Complex Interdependence

As we have noted, the rational or unitary approach to foreign-policy analysis provides an inadequate explanation of foreign-policy decision making, since it excludes the role played by individuals, their belief systems, and a variety of domestic political forces. The approach also neglects, in an increasingly interdependent world, the influences of a number of other global actors, such as other governments, international organizations, and a variety of nongovernmental linkage groups.

When responding specifically to another state, decision makers must take into account the probable reactions of yet other national governments. Will the proposed decision create an undesirable precedent that others might exploit? Will other states support or seek to punish the action taken? How will both friend and foe react? For example, in attempting to initiate military strikes against the Serbs in Bosnia and an end to the arms embargo against the Muslims, President Clinton had to be concerned not only with congressional and public opinion in the United States, but also with the attitudes of other governments in both NATO and the U.N. Security Council. Indeed, it was the opposition of other NATO and U.N. members that led the president to back away from his preferred position of increased air strikes and an end to the arms embargo—policies that were supported by both U.S. congressional and public opinion.

Nor can we confine our analysis to authoritative foreign-policy positions taken by other national governments, for this would exclude the important role played by both subnational and supranational nongovernmental actors in global politics. Increasingly, provincial governments and cities have developed their own foreign policies, conducting trade missions and negotiations with their foreign counterparts as well as with national governments.[41] Business corporations deal directly with foreign governments as well as with their foreign counterparts in areas that affect the relations between states. Interest groups and individuals with certain policy interests—such as improving the global environment, enhancing human rights treatment in foreign countries, or improving living conditions abroad—play an increasingly active role in affecting domestic and foreign policy.

Since most problems do not end at a nation's boundaries, all sorts of subnational groups deal directly and indirectly with foreign governments and particularly with their foreign counterparts. Ethnic groups in one state seek to influence not only the policies of the government in which they are located, but also those of governments with whom they have ethnic linkages.

At large international conferences, such as the 1992 global environmental conference in Rio de Janeiro, many governments welcomed foreign lawyers and technical experts interested in environmental issues to serve on their delegations. Such experts also lobbied and advised international civil servants assigned to the U.N. Environmental Program, which played the central role in directing the activities of

---

[41] Brian Hocking, *Localizing Foreign Policy: Non-Central Governments and Multilayered Diplomacy* (New York: St. Martin's Press, 1993).

the conference. Moreover, representatives of environmental groups were allowed to make presentations at the conference itself, belying the Westphalian notion that only states have standing in international fora.[42]

## CONCLUSION

Evidence has been presented in this chapter to suggest that the structure of the decision-making process and the people who make those decisions can affect foreign policy. A major debate surrounds the question of whether authoritarian or democratic structures have an advantage in the making of foreign policy. In terms of effectiveness and efficiency, there are reasons to believe in the superiority of authoritarian regimes, for such governments ought to be able to make decisions more rapidly, ensure domestic compliance with their decisions, and perhaps be more consistent in their foreign policies. Yet in practice, authoritarian regimes often are less effective in developing an innovative foreign policy because of subordinates' pervasive fear of raising questions of the top leadership.

In making foreign-policy choices, central decision makers rely heavily upon foreign-policy bureaucracies. But despite some very persuasive arguments and anecdotal evidence suggesting that the bureaucracy plays a key role at both the formulation and implementation stages, the central decision makers determine who their key advisers will be and which advice they will heed.

The impact of public opinion on foreign-policy choice has been found to be quite limited. Since mass public opinion is easily manipulated, its importance as a determinant of foreign policy is restricted even in democratic polities. At the same time, we know that democratically elected leaders often seem to have their eyes on the popularity polls when certain foreign-policy decisions are being made. The evidence does suggest, however, that there is little reason to fear that an uninformed and emotional public will interfere with rational foreign-policy calculation by forcing ill-conceived policies on reluctant decision makers.

In examining the behavior of states and what influences their foreign-policy choices, it would be shortsighted to focus only upon their foreign policies. Other actors share the world's stage in an increasingly interdependent global structure—and these can be found among groups and individuals both above and below the nation-state level. Today, interest groups, corporations, and international organizations are developing their own foreign policies and beginning to influence state actors directly.

---

[42]For further discussion see Karen A. Mingst, "Uncovering the Missing Links: Linkage Actors and Their Strategies in Foreign Policy Analysis," in *Foreign Policy Analysis,* ed. Laura Neack, Jeanne A. K. Hey, and Patrick J. Haney (Englewood Cliffs, N.J.: Prentice-Hall, 1995), 229–242; and James N. Rosenau, *Turbulence in World Politics* (Princeton, N.J.: Princeton University Press, 1990).

PART III

# Conflict and the Search for Security

The Westphalian system was born out of the perceived need to address the problem of violence that had confronted Europe during the sixteenth and early seventeenth centuries. The logic of the Peace of Westphalia was that by centralizing power in the hands of secular leaders, peace and security could be provided within defined territorial states. Under the new system, peace would result from independent sovereigns individually and collectively protecting their national power interests as they sought to enhance both tangible and intangible factors affecting that power.

A primary purpose for developing military power was to deter or protect against attack from other sovereign states. Military technology continued to evolve over the next three centuries, culminating in the development of nuclear weapons. The destructive capability of these and other modern weapons raised more horrifying prospects of the effects of warfare: not security, but its very negation. Therefore, throughout the cold war period, many grew concerned over the dangers of a potentially unstable nuclear deterrent situation. Although such fears have subsided somewhat with the end of the cold war, the spread of nuclear weapons to smaller and potentially less responsible antagonists (which could include nonsovereign actors) has intensified the search for alternative approaches to security, such as arms control and disarmament.

Nor has the end of the cold war brought an end to global violence, and for a number of reasons—related to the world system, the nature of the nation-state, and the frailty of human nature—we can expect violence to continue in the world.

Evidence continues to mount that military weapons and the use of force have been insufficient in providing security to societies; violence in one part of the global system increasingly affects other areas. Also evident is the erosion of the idea that sovereignty prohibits the political or military intervention of all in the internal affairs of other states. Increasingly, states justify such intervention in the name of some world order value. Such trends are turning up the volume on the question of whether a system of sovereign and independent states can still provide global security. For all these reasons, today it seems clear that a new appreciation has developed of the potential of negotiation for resolving and managing conflicts between peoples. ■

*Overleaf* • Rwandan refugees wait for relief food, 1994.

# CHAPTER 6

# Capabilities and Power

No concept is more central to understanding the search for security during the Westphalian period than that of power. Yet none has been more controversial because of the elusiveness of its meaning and, therefore, disagreement over how to measure it.

In its tangible capabilities, the United States is today no doubt the most powerful state that has ever existed. Yet it has been unable to prevail over small states like Cuba and North Vietnam; it was neutralized into seeming ineffectiveness in the 1979–1980 Iranian hostage situation; and it has often had small, fledgling states snub its attempts to influence votes in the United Nations. Similarly the former USSR, despite obvious power advantages, was forced to withdraw its troops from Afghanistan in 1989 as a result of a war it seemed unable to win. Months later, the Soviet Union's enormous military capability, in the form of troops and weapons stationed throughout its one-time "satellite" neighbors in Eastern Europe, failed to prevent those countries from leaving the Soviet orbit. Indeed, all of Moscow's military might proved impotent in stopping the collapse of communism throughout what had been its vast empire for many decades.

To explain why the weak can sometimes prevail against the strong, it is first necessary to examine what makes the concept of power so complex.

## DEFINING POWER

### THE MEANING OF POWER

A nation-state's capabilities consist of **tangible factors** that can be measured—such things as factories, guns, and natural resources—and such **intangible factors** as a dynamic leadership, a hard-working population, and a cohesive society. Both kinds of capabilities are essential to, though not identical with, what we usually mean by power in international politics. Power is generally thought of as the capacity to control or influence the behavior of others. Yet, much of the confusion with respect to this concept arises over the failure to distinguish between the capacity to act and the actual exercise of power. A state may enjoy such tangible assets of power as a strong industrial base, a large population, advanced technology, and important resources; it may also rank high in such intangible factors of power as high morale, effective leadership, and high educational levels. Yet this state may be unable or unwilling to translate these components into actual influence.

For example, one researcher who estimated the differential between **capability** and **influence** during the period from 1925 to 1930 suggested that if one looked only at capabilities, the following rank order would be appropriate: (1) the United States, (2) Germany, (3) Great Britain, (4) France, (5) Russia, (6) Italy, and (7) Japan. But when considering the nation's impact in world politics and the responses the state was able to evoke in seeking to change the behavior of others, the ranking would change to (1) France, (2) Great Britain, (3) Italy, (4) Germany, (5) Russia, (6) Japan, and (7) the United States.[1] In a similar vein, in mid-century, such leaders as Josip Broz Tito of Yugoslavia, Abdel Nasser of Egypt, Jawaharlal Nehru of India, and, a bit later, Fidel Castro of Cuba, probably exerted far more influence in the world than the limited capabilities of their respective states would appear to merit.

Some of these apparent anomalies can be explained by the fact that even in situations in which a state is clearly interested in exercising power, its capabilities are not always translated into influence. Several factors in turn help explain this failure. First, power involves a *perceptual relationship*. A state may have abundant capabilities that would allow it to prevail in most situations, but that capability must be perceived as both available and likely to be utilized before another actor will be influenced by it. Perceptions may often be in error as a result of a variety of blinders and communication problems, but it is one's perception of the capabilities and intent rather than what they "really" are that determines how states will act in response to the behavior of others.

Second, power is a *relative and reciprocal relationship*. A state may be able to exercise power over one state but not over another. It has been suggested, by way of partial explanation for this phenomenon, that a loss-of-power gradient might exist, making it more difficult to influence states at greater geographic distance than those close at hand. This inability to influence over great distances may be partly due to the lack of resolve, as a government is sure to be less concerned with marshaling all of its power in regions that it perceives as remote to its interests. This would certainly help explain the United States' failure in Vietnam, as well, perhaps, as China's failure to sustain great influence over much of the continent of Africa despite considerable activity there. Power is also relative and reciprocal in the sense that in any influence situation each party has some impact on the other. This is true even where one state is quite weak, for if the weak state can do nothing else, it may threaten to collapse and thereby provide a vacuum that a larger state's adversaries can penetrate.

---

[1]Kalevi J. Holsti, *International Politics*, 7th ed. (Englewood Cliffs, N.J.: Prentice-Hall, 1988), 146.

Third, power tends to be *issue oriented*. One state may have power over another on one issue but not with respect to others, for reward and punishment are not always the only factors that affect influence. A state may be influenced because it sees a given request as legitimate. A leader of one's alliance is more likely to be seen as making an appropriate demand for allied support than an actor without such authority. But even in this instance such a demand by the alliance leader is likely to be more effective if it involves issues affecting the common defense than if it is concerned with economic or ideological issues, which may be viewed as beyond the scope of the alliance commitment. If a government perceives an act as threatening its domestic sovereign rights, it is likely to be highly resistant to any effort to influence its behavior.

A state that is perceived as having expertise or as able to serve as an appropriate role model will also enjoy greater influence than the mere measure of its national power capabilities would suggest. For example, Singapore, with a population of less than 3 million, has exercised far more influence in international fora dealing with both political and economic matters than its small size would seem to merit. With among the highest economic growth rates in Asia, it serves as a leading example of economic development and the entrepreneurial spirit.

Fourth, power is *affected by one's expectations* in relation to another state. Before glasnost, the Soviet Union may actually have benefitted from expectations of a rather heavy-handed style. When its leaders made conciliatory moves, they may have received more credit from other actors than did the United States for making similar moves. Concessions tended to be expected of the latter and, therefore, were more likely to be taken for granted.

Fifth, a discrepancy between actual power capabilities and influence is likely to arise by virtue of the tendency of both decision makers and analysts to *ascribe higher power positions to groups and individuals that are perceived as aggressive.* Both the People's Republic of China and the Soviet Union, during their more assertive days in the 1950s and early 1960s, were generally seen in the West as more significant and powerful than their real capabilities at the time would have suggested. During the 1980s, the Reagan administration almost certainly exaggerated the capability of the Nicaraguan government as a result of the pervasive belief within the administration that the Sandinistas had aggressive intentions in Central America.

## THE ELUSIVENESS OF POWER

The elusiveness of power can be seen in the fairly rapid changes in the power fortunes of states during the relatively brief time that the nation-state system has been in existence. States have achieved and lost great power status over the years, as illustrated in Table 6.1. Considerable fluctuation in the power fortunes of the non-great powers has also been the rule—for example, 95 percent of the state units that existed in Europe in the year 1500 have now been obliterated, subdivided, or combined with other countries.[2]

---

[2]Richard N. Rosecrance, *The Rise of the Trading State* (New York: Basic Books, 1985), 130.

**TABLE 6.1** Great Powers in History

|  | 1700 | 1800 | 1875 | 1910 | 1935 | 1945 | 1995 |
|---|---|---|---|---|---|---|---|
| Turkey | X |  |  |  |  |  |  |
| Sweden | X |  |  |  |  |  |  |
| Netherlands | X |  |  |  |  |  |  |
| Spain | X |  |  |  |  |  |  |
| Austria (Austria-Hungary) | X | X | X | X |  |  |  |
| France | X | X | X | X | X |  |  |
| England (Great Britain) | X | X | X | X | X |  |  |
| Prussia (Germany) |  | X | X | X | X |  | X |
| Russia (Soviet Union) |  | X | X | X | X | X | X |
| Italy |  |  | X | X | X |  |  |
| Japan |  |  |  | X | X |  | X |
| United States |  |  |  | X | X | X | X |
| China |  |  |  |  |  |  | X |

SOURCES: Author's adaptation of "Great Powers, 1700–1979" by Kenneth N. Waltz, *Theory of International Politics* (Reading, Mass.: Addison-Wesley, 1979), 162. Reproduced with permission of The McGraw-Hill Companies.

From the perspective of the entire historical system, the half century in which the United States and the Soviet Union dominated the world stage was quite brief. In the last decade of this century, both the United States and the Soviet Union's chief successor state, Russia, retain overwhelming military power; but because of Russia's internal difficulties, only the United States is regarded as a superpower today.

However, the supremacy of the United States in the global economic system is increasingly threatened. On one side is the phenomenal growth of Japan and, on the other, the economic power of the European Union that now rivals, and may soon surpass, that of the United States. More surprising perhaps is the World Bank estimation that if China's prodigious growth continues, the combined economies of China, Hong Kong, and Taiwan will exceed the United States economy in less than a decade.[3] Since both China and two of the European Union members, France and the United Kingdom, have nuclear weapons and are in the process of extensive nuclear modernization programs, China and a united Europe could conceivably become military superpowers.

Power relationships can change, not just across centuries, but in a matter of a few years. During the 1970s, the OPEC states quickly acquired power through the effective exploitation of a single resource, while at the same time Japan and various Western European nations were squeezed by their own energy dependence. One result was a decline in support for Israel as these oil-dependent nations sought to placate Israel's oil-rich neighbors.

[3] Steven Greenhouse, "New Tally of World's Economies Catapults China into Third Place," *New York Times*, May 20, 1993, A1.

Domestic events also can unsettle the power position of a state, forcing leaders to withdraw from the world stage to concentrate upon domestic affairs. During the Chinese Great Cultural Revolution in the 1960s, domestic turmoil forced the Chinese leadership to recall most of its representatives from abroad. Similarly, the disintegration of the Soviet Union led to its demise as a superpower and removed its successor states from the active and influential global role that the USSR had played in the past.

## TANGIBLE SOURCES OF POWER

A nation's geopolitical features—such as the size of its population, military capabilities, economic resources, and industrial capacity—constitute tangible sources of power because they can be measured. In contrast, intangible factors, such as national morale, leadership capabilities, and the like, obviously are far more difficult to assess. Although the following sections examine the roles that the more tangible sources of power can play in global politics, it is important to remember that a state's power and influence always depend not just upon whether it uses its resources in a rational way in pursuit of its interests, but, most importantly, how its behavior is perceived by those it attempts to influence.

### GEOPOLITICAL SOURCES OF POWER

Certain states enjoy national power capabilities by virtue of the geographic attributes with which they are endowed or are able to obtain. States vary greatly in their territorial size, location, and topography—all of which can influence the power of the state as well as the role it is able to play in the international system.

Occupying a *large expanse of territory* not only is likely to increase the probability that the state will enjoy more resources, a larger population, and arable land; it will also provide an opportunity for temporary retreat and regroupment after an invasion, actions that eventually enabled the Russians to triumph against both the Napoleonic advance to the gates of Moscow in 1812 and the German invasion in 1941. Small states do not have such an advantage; and unless they have the benefit of a natural mountain fortress, like Switzerland, or are situated on a distant, resource-poor island, like Iceland, they can be conquered and occupied rapidly, as was shown by Hitler's invasion of the lowland countries of Europe. On the other hand, a large territory may present serious transportation problems for its government, making it difficult to integrate an efficient national economy. Japan's population of 125 million living in a small area probably contributed to its highly successful economic development. This provided a sufficient domestic market for industrial products as well as a "densely meshed communications network that worked to heighten the efficiency of economic activities."[4]

[4] Michio Royama, "Environmental Factors and Japan in the 1970s," in *Japan, America, and the Future World Order*, ed. Morton A. Kaplan and K. Mushakoji (New York: Free Press, 1976), 344.

The *location* of a state is also likely to affect both its power and its policies. Sea powers historically have been protected from invasion by the water surrounding them. Even the narrow British Channel provided Great Britain with protection from invasion by Hitler during World War II. More than a century earlier, Britain had the luxury of rising to the first rank of great powers without having to invest in a large standing army, thanks to its sea buffer. Although modern military technology, which allows destruction at great distances, has undermined the security of insular states, conquering and controlling such a state remains difficult to the extent that troops still must land on its beaches before it can be occupied.

The *topography* of a state likewise has power and security implications. Mountains traditionally have provided barriers to invasion. Since it is much simpler to move modernized armies through flat lands, certain invasion routes, such as those through Poland or across the lowlands of Western Europe, have been popular, frequently wreaking havoc upon the security of states located in those regions. The Soviet Union became aware of the drawbacks of inhospitable terrain in its 1979 incursion into Afghanistan, which helps explain its withdrawal from the country a decade later.

**GEOPOLITICAL THEORIES** During the early part of this century, a number of geopolitical writers developed theories regarding the implications of the geographic setting of a state for its global power position. Among these, American Admiral Alfred T. Mahan advocated expansion of sea power as the most useful approach to providing power and security.[5] Impressed by the example of Great Britain, he suggested that an emphasis upon sea power was particularly relevant for the United States, given its position on two oceans. U.S. Secretary of the Navy (later to become president) Theodore Roosevelt was particularly influenced by Mahan's arguments. Roosevelt worked to make the United States a major naval power, which he viewed as both requiring and justifying the nation's acquisition of the Philippines and other far-flung territories. Modern-day disciples of Mahan are found in the Navy and among those who would prefer to see the United States emphasize submarine-launched ballistic missiles over land-based ICBMs and bomber forces.

A different conception of geopolitical power was offered by the British writer Sir Halford Mackinder, who argued in 1904 that the critical element in world power was control of the heartland of the Eurasian continent, consisting largely of the European part of Russia.[6] Control of the heartland would enable a state to dominate the world island, which included Europe, Asia, and Africa, which in turn would lead to control of the entire world. By 1943, the new technology of air power caused Mackinder to revise his thesis; then he noted that his earlier dictum had assumed the continuation of military strategies built on armies and navies. Air power, on the other hand, had made the North Atlantic region both a potentially cohesive area and one with rapid access across the North Pole to the Soviet heartland, whose power it could easily counterbalance once Germany was defeated.[7]

[5] Alfred T. Mahan, *The Influence of Sea Power in History, 1660–1783* (Boston: Little, Brown, 1918).
[6] Sir Halford Mackinder, *Democratic Ideals and Realities* (New York: Norton, 1962).
[7] Colin S. Gray, *The Geopolitics of the Nuclear Era* (New York: Crane and Russak, 1977).

> ### Global Changes
>
> ## The Changing Face of Power
>
> "The explosive power amassed during the cold war was roughly 1,000 times the power used in all wars since the introduction of gunpowder six centuries ago" (Ruth Sivard, *World Military and Social Expenditures* [Washington, D.C.: World Priorities, 1989], p. 5).
>
> - How does this huge increase in destructive capability help explain the emphasis of the major nuclear powers on the threat, rather than the actual use, of such weapons?
>
> At the beginning of the Westphalian system, cannons, the longest-range weapon at the time, could fire a projectile three miles. Today, the range of missiles is over 8,000 miles, accurate within 300 feet.
>
> - Does this development make the territorial (or sovereign) state more or less secure from attack today?
>
> Economic disparity between people in different nations has increased considerably over the centuries. In 1850, early in the industrial period, the world's wealthiest countries were, on average, about twice as rich as the poorest. By 1950, the gap between richest and poorest was about ten to one. In 1960, it had grown to fifteen to one. It may reach nearly thirty to one by the end of the century.
>
> - To what extent is this widening gap reflected in the power positions of certain rich and poor countries in the world today?
>
> Power today is shared by a variety of nonstate actors that can do great harm to nation-states. For example, Moody's, a private firm that provides credit ratings, can cause millions if not billions of dollars to be shifted from one country to another simply by decreasing a country's investment rating.
>
> - In what ways do such developments attack traditional concepts of sovereignty?
>
> To communicate across the oceans in 1648 took several days. Now with fax, E-mail, telephones, satellite transmission, and so on, communication is instantaneous.
>
> - How does the communications revolution enhance a state's ability to utilize propaganda and persuasion to influence peoples at a remote distance?

Modifying Mackinder's 1904 thesis, Yale professor Nicholas J. Spykman suggested that control over Eurasia would more likely go to the rimland powers, such as Great Britain and Japan.[8] Spykman viewed the heartland as fundamentally limited in its ability to assume the dominant position because of the area's undesirable climate, its emphasis on agrarian productivity, and its general lack of resources.

As a final illustration of geopolitical theory, we note the work of Karl Haushofer, which was used to justify Nazi aggression.[9] Haushofer developed the notion of *Lebensraum,* defined in Chapter 4 as the need for living space. Borrowing from Social Darwinist ideas, he formulated the notion of an organic state that must either expand or wither. Although he influenced Nazi policy, as a geopolitician he opposed Hitler's invasion of the Soviet Union because of the Soviet advantage of defense in depth and its ability to relinquish space temporarily to gain time to regroup. Haushofer argued

---

[8] Nicholas J. Spykman, *The Geography of the Peace* (New York: Harcourt Brace Jovanovich, 1944).
[9] Andreas Dorpalen, *The World of General Haushofer* (New York: Farrar and Rinehart, 1942).

that world leadership should come from a combination of powers, consisting of the Soviet Union, Japan, China, and India, under German leadership.

One of the principal dangers in geopolitical theory is the tendency to mask the writer's own value preferences (e.g., Haushofer's view that the German nation had a "right" to expand) under the guise of objectively assessing the requirements of geography. While geography clearly influences human societies to behave in particular ways, foreign policies result from human choices and are by no means geographically predestined.

## POPULATION

A state's power is also a product of its population base. Human beings are important not only for providing military troops in case of war and using weapons in battle, but also for producing the machines that add to economic productivity more generally. States with relatively small populations can never aspire to great power status, regardless of how much they might try, for they can produce neither the economic nor the military base required. The historical trend shows clearly that, in each recent century, larger and larger political and economic units alone can be classed as leading powers; tiny Portugal had enough relative capability to act as a great power in the fifteenth century; the Netherlands could play that role in the seventeenth, as could Britain in the nineteenth. But by the late twentieth century, the gradually emerging, single great power of a united Europe included all of those nations and many more of their neighbors as well.

Numbers of people alone, however, do not provide the foundation for great power status. The power potential of a population also depends upon its **demographic profile** with regard to age and perhaps even sex. A population with a very large mix of children and young people will have serious disadvantages against a population with a more normal age distribution—a problem for developing countries. Some 45 percent of Africa's population, for example, was under the age of 15 in 1995—too young to be economically productive, and indeed, still largely dependent on an overburdened adult population.[10] Similar disadvantages might accrue to a state top-heavy with elderly citizens who must be supported by those still working. The principal reason the East German government built the Berlin Wall in 1961 was to try to stem the flood of working-age refugees who were leaving the country. The same concern in 1989 led the East German regime to liberalize its internal policies in an effort to discourage the further emigration of its most productive citizenry. States that have suffered much in war, as did the Soviet Union during World War II, may be left with a serious imbalance in the ratio of males to females, producing labor shortages as well as an inadequate base from which to conscript security forces. Although all states may benefit from greater use of all of their human resources, those suffering a serious sexual imbalance will find it particularly useful to make fuller employment of the talents and productive capacity of women in their efforts to increase capabilities.

---

[10]As projected by World Resources Institute, *World Resources, 1992–1993* (New York: Oxford University Press, 1992), 16.

Overpopulation can also be a problem, as many developing countries have found. Economic growth is hindered by having too many mouths to feed (particularly the nonproductive mouths of children). The mass starvation that may result from overpopulation is the most terrible outcome, but even short of that, malnutrition produces stunted lives and minds that inevitably mean a less capable and productive population. Under such circumstances, consumption demands are so high that capital savings, essential to economic growth, become an impossibility.

Still, assuming that a state's population is stable enough to keep it effectively productive and able to generate some surplus capital, larger populations are an asset over smaller ones. All of the states viewed as great powers over the past two centuries have had populations exceeding fifty million when including their colonies, placing them among the largest fifteen or so states. Recognizing that a large population is a factor in determining capabilities, the government of France from time to time in the past pursued tax and benefit policies designed to enhance its rather low birth rate, sometimes anxiously eyeing the greater population growth in neighboring Germany. For similar reasons, Mussolini's government offered bonuses to Italian women for having babies. More extreme were the population policies of the late Nicolae Ceausescu of Romania, policies that in part contributed to his downfall and execution at the end of 1989. The Romanian dictator had outlawed all abortions and contraceptives and required frequent medical examinations of every woman of childbearing age to make certain she had not had an abortion.

## ECONOMIC WEALTH

States vary considerably in the kind, amount, and variety of raw materials they possess for building a strong economic base. Some enjoy vast reserves of oil, as is the case with several Middle Eastern states; others, such as Australia, Canada, and South Africa, have substantial mineral resources that can be important for both industrial and military strength; still others, such as Argentina and New Zealand, are generally able to prosper because of their agricultural commodities and arable land.

Few, if any, states are completely self-sufficient in the kinds of resources needed to operate in a complex, industrial world. The most telling trend in modern economic life is toward greater resource interdependence of nations. Today, the United States and Russia, given their vast land mass and plentiful raw material supplies, come closest to being able to pursue policies of **autarky** (self-sufficiency), but even they are increasingly forced to import certain strategic materials.[11] The 1973 oil boycott effected by the **Organization of Petroleum Exporting Countries (OPEC)** underscored the vulnerability even of a superpower that earlier had been self-sufficient in oil. In the United States, long lines soon formed at service stations throughout the country as gasoline prices skyrocketed. Subsequent efforts at conservation did not end the U.S. dependence on foreign oil. At the time of the Iraqi invasion of Kuwait in 1990, fully half of the oil used in the United States came from abroad.

---

[11] Russia's problems in providing adequate consumer goods by the 1990s reflected a near breakdown in the system of distribution, resulting largely from the effort to replace the highly centralized system with a market economy, and not from any inherent inability to produce needed supplies.

**TABLE 6.2** Gross Domestic Product, 1994 and 2020 (Projection)

| 1994 GDP (IN BILLIONS OF U.S. DOLLARS) | PROJECTED RANK IN 2020 |
|---|---|
| 1. United States: 6,648 | 1. China |
| 2. Japan: 4,591 | 2. United States |
| 3. Germany: 2,046 | 3. Japan |
| 4. France: 1,330 | 4. India |
| 5. Italy: 1,025 | 5. Indonesia |
| 6. United Kingdom: 1,017 | 6. Germany |
| 7. Brazil: 640 | 7. Korea |
| 8. Canada: 543 | 8. France |
| 9. China: 508 | 9. Taiwan |
| 10. Spain: 483 | 10. Brazil |
| 11. Korea: 377 | 11. Italy |
| 12. Mexico: 375 | 12. Russia |
| 13. Australia: 332 | 13. United Kingdom |
| 14. Netherlands: 330 | 14. Mexico |
| 15. India: 291 | |
| 16. Argentina: 282 | |
| 17. Russia: 275 | |

SOURCES: World Bank, fax transmittal, October 25, 1995; projections for 2020 are cited in Paul Lewis, "An Eclipse of the Group of Seven," *New York Times,* May 1, 1995, D2.

**MEASURING WEALTH** The wealth of a state has often been used as one yardstick for ascertaining its relative power base. The most commonly used indicator for comparing the economic strength of states is the **gross domestic product (GDP),** which measures the total value of economic goods and services produced and marketed annually within a state. The economic strength of a state, however, is far from static; today's rank order of states in the world is likely to change somewhat in the next twenty-five years, as shown in Table 6.2. The changing fate of nations is well illustrated by the collapse of the Soviet Union that resulted in placing that former superpower, which is now represented by Russia, seventeenth in terms of gross domestic product in 1994.

A more accurate indicator of military potential considers a state's industrial base in addition to its GDP, for the production of military weapons requires a strong industrial capability. Today, postindustrial economies like that of the United States, whose service sector is growing while its manufacturing capability declines, may simultaneously be experiencing a decline in military potential. The large number of persons engaged in law, advertising, and other service industries are hardly equipped to turn quickly to the production of weapons in the event of war.

As an aspect of its wealth in a general sense, a state's financial capabilities constitute an important economic resource. Such capabilities derive from a government's ability to levy the taxes needed to pay for its foreign policy actions as well as from its creditworthiness. According to Yale professor Paul Kennedy, the European "financial revolution" of the late seventeenth and early eighteenth centuries, which provided a relatively sophisticated system of banking and credit, determined success or failure in war.[12] States most capable of borrowing to support their war effort were those most likely to prevail. In more democratic times, the ability of leaders to exercise power to engage in warfare is highly dependent upon the willingness of the population to support the effort. Without that, sufficient revenue will not be collected, nor will the state be viewed as creditworthy.

## MILITARY CAPABILITIES

Westphalian changes in weapons technology have been dramatic, most particularly during this century. As we have seen, the increased ability of Europeans to kill at a distance by using gunpowder contributed to the growth of the nation-state as the structure best able to provide security for very large communities. For several more centuries, firepower did not increase enormously. But in the twentieth century, the revolution in military technology that followed the invention of the airplane and, subsequently, nuclear missiles made nation-states permeable as never before. State boundaries became virtually impossible to defend against the new technologies.

Since the first atomic bomb was exploded over Hiroshima on August 6, 1945, advances in nuclear technology and delivery systems have only served to intensify the potential dangers. Whereas almost 100,000 people lost their lives on that fateful day in Hiroshima, nuclear weapons currently threaten the lives of hundreds of millions. We now measure their unprecedented explosive power in such novel terms as megatons (millions of tons of TNT equivalent). The Hiroshima bomb, estimated at about thirteen thousand tons (thirteen kilotons), was a mere toy compared to contemporary thermonuclear devices. The largest detonated to date measured more than fifty megatons. Such a device has about four thousand times the explosive power of the Hiroshima bomb.

**DELIVERY SYSTEMS** This immense increase in the destructive power of bombs and missile warheads is central to the nuclear revolution. But there also have been considerable improvements made in the ability to deliver these weapons to their targets with ever greater capacity, speed, and accuracy. Because nuclear warheads have been much reduced in size since 1945, more explosive power can be carried in each delivery system. The bombs dropped on Hiroshima and Nagasaki were so large and heavy that only one atomic weapon could be carried in a bomber. Today that same destructive capability could fit inside a small suitcase. The miniaturization of nuclear bombs became particularly significant in the missile age, for missiles have a more limited carrying capacity than bombers.

Missiles also have radically reduced the delivery time required to transport nuclear warheads to their destination. A maximum of thirty minutes' warning is available should a government with **intercontinental ballistic missiles (ICBMs)** choose

---

[12] Paul Kennedy, *The Rise and Fall of the Great Powers* (New York: Random House, 1987).

**FIGURE 6.1** | Nuclear Warheads

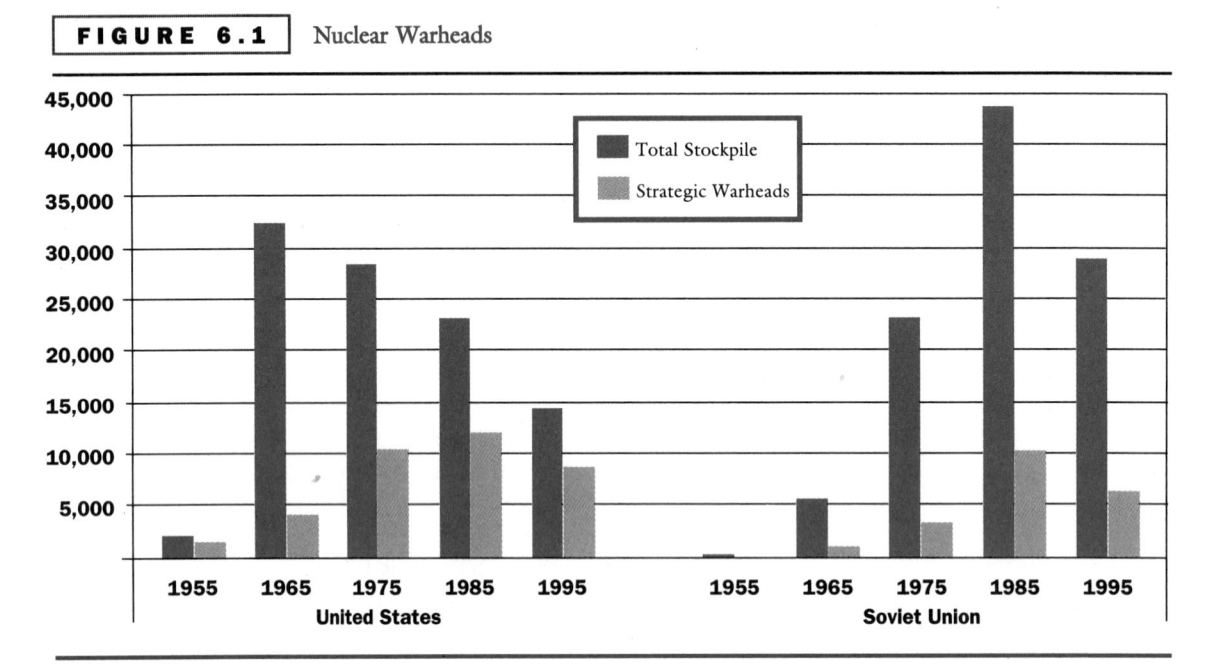

Includes warheads in active operational forces as well as those in reserve or in retirement awaiting dismantlement. USSR figures for 1995 represent Russia only and the total stockpile figures are based on 1994 data.
**SOURCES:** "Nuclear Notebook," *Bulletin of the Atomic Scientists* 50, no. 6 (November/December 1994): 59; "Fact File," *Arms Control Today* (November 1995), 30.

to strike another state, even when the latter is a continent away. Time can be cut to fifteen minutes or less if **submarine-launched ballistic missiles (SLBMs)** or **intermediate-range ballistic missiles (IRBMs)** positioned closer to the target are utilized.

The increased accuracy of missiles also contributes to the nuclear threat. Whereas early missiles were only accurate within a range of several miles, ICBMs are now able to hit within three hundred feet of their target. This pinpoint accuracy guarantees maximum destructiveness but at the same time makes ICBM launch sites highly vulnerable to attack, even when they are hardened with reinforced concrete and buried deep in the ground. The **cruise missile,** which is essentially a pilotless jet plane that can be launched from land, sea, or air, demonstrated its considerable accuracy delivering nonnuclear weapons in the Gulf War, as well as in the August 1995 NATO strikes in Bosnia. Given its relatively low cost and simple construction, it threatens to become the weapon of choice for many states in the future.

The development of the **multiple-independently targeted reentry vehicle (MIRV)** in the late 1960s also increased each vehicle's destructive capability by making it possible to deliver warheads to several different targets at the same time. For example, both the American MX and Soviet SS-24 ICBM arsenals during the 1980s carried up to ten warheads, each of which was capable of hitting a different city, military base, or industrial site.

Bombers still remain an important part of military arsenals, but to make them more effective and less vulnerable to antiaircraft fire, cruise missiles as well as **short-range attack missiles (SRAMs)** have been placed upon their wings, allowing the bomber to fire at a greater distance from the target.

**DIFFICULTIES IN COMPARING MILITARY STRENGTH** During the cold war, debate raged over the question of whether the Soviet Union or the United States was ahead in the arms race. The complex differences in their nuclear capabilities continued to bedevil the effort to determine military equivalence in arms reduction talks even at the end of the cold war. The fact that the two powers had emphasized different kinds of delivery systems made the problem especially difficult. The United States preferred bombers and submarine-launched missiles to intercontinental ballistic missiles. Whereas the Soviet Union had placed about 75 percent of its strategic force capability into ICBMs, the equivalent share for the United States was only 25 percent. The United States had maintained a substantial lead in cruise missiles and in the accuracy of various missile systems; in contrast, the Soviet Union's missiles were capable of delivering a substantially heavier "payload" of nuclear destruction.

The geopolitical balance between the two states was also unequal. The Soviet Union had to be concerned not only with adequate defense against the United States, but also with defense against American allies in Europe and Asia, as well as a sometimes hostile China. With the 1989 revolutions in Eastern Europe, the Soviet-sponsored Warsaw Pact suddenly ceased to be a credible threat to the West, thereby radically altering the power equation in Europe. These and many other considerations created ongoing complications for decision makers in Washington and Moscow about the size and type of military budget each should adopt. With the breakup of the Soviet Union, several of its former republics suddenly found themselves in control of nuclear weapons located on their soil, although their governments were quick to assert that they would not use or threaten to use them unilaterally; most of these weapons have been turned over to the Commonwealth of Independent States under Russian control or have been destroyed.

Measuring military capability is not much easier when it comes to nonnuclear weapons. Comparisons of overall defense budgets are misleading because of differing budgetary procedures. Soviet military budgets were always viewed in the West as undervalued since they excluded a number of items included in American defense budgets. When the Soviet Union finally revealed figures on its military spending for the first time in 1989, its estimates were about half what the United States had been projecting. Even comparing numbers of troops was difficult because divisions within the Soviet Union were considerably smaller than those in the United States. Equivalence also would have required that the United States count reserve and National Guard units in its force totals.[13]

An additional complication in measuring capability is that an attacker, who must normally operate with longer supply lines, typically needs larger numbers of forces

---

[13]Edward N. Luttwak, "The Missing Dimension of U.S. Defense Policy: Force, Perceptions, and Power," in *International Perceptions of the Superpower Military Balance,* ed. Donald C. Daniel (New York: Praeger, 1978), 23.

than a defender. Determining intent, a most subjective enterprise, is of critical importance to assessing relative capability. One should also try to account for qualitative differences in conventional weapons—an area in which the United States had decided advantages over the Soviet Union throughout the cold war period.

One also must ask whether it is crucial to have overwhelming numbers of weapons in existence at the onset of a war. After all, the United States demonstrated in both World Wars I and II that it could divert a civilian economy from a peacetime to a wartime footing in a relatively short period, and thus turn the tide of war. Rushing to production in advance of a conflict may only produce many obsolete weapons. Devoting more resources to technological development and to creating fewer but better weapons might prove to be the wisest strategy in the long run.

Focusing upon building a nation's military strength may also undermine a state's economic and political well-being, as Paul Kennedy argues. After discussing the rise and fall of such states as Holland, Spain, France, and Britain, Kennedy states that "by going to war, or by devoting a large share of the nation's 'manufacturing power' to expenditures upon 'unproductive' armaments, one runs the risk of eroding the national economic base, especially vis-à-vis states that are concentrating a greater share of their income upon productive investment for long-term growth."[14]

Obviously, no final answers are possible regarding the optimum military force structure. This will depend upon the sort of war a government anticipates. If a quick, decisive war is expected, then it will be more important to have weapons ready and deployed. But if the war is likely to continue over several years, having the economic potential to create a massive industrial and military mobilization will be more important.

Since power involves perception, some observers believe that the most important way to increase power is simply to increase aggregate military spending. This was the dominant view during the Reagan presidency; as a result, the United States in the 1980s seemed unconcerned about which specific military programs to emphasize. Rather, the Reagan administration doubled military spending simply in an effort to communicate resolve. Throughout the cold war, Soviet leaders also tended to regard it as critically important to their nation's power position to communicate a resolve stemming from their commitment to an ideology and a high morale.

# INTANGIBLE SOURCES OF POWER

Most neglected in the construction of indexes of power are the so-called intangible factors. It is not of much use to have the resources and the weapons required to exercise power in the world unless a government can also mobilize those resources. Such mobilization capabilities depend upon such things as the quality of leadership and government, the skills of a nation's diplomats, the educational and technological capabilities of the people and, ultimately, their morale.

To call such matters *intangible* means that they are at best difficult to measure. Yet some things that can be quantified at least bear upon these intangible factors. For example, one might use measures of governmental stability or public support as

---

[14] *The Rise and Fall of the Great Powers,* 539.

> **Normative Dilemmas**
>
> ## Normative Challenges to Realpolitik
>
> The following are quotations from Hans J. Morgenthau's classic statement of realist thinking in international relations, *Politics among Nations: The Struggle for Power and Peace* (6th ed., rev. Kenneth W. Thompson [New York: Knopf, 1985].)
>
> > International politics, like all politics, is a struggle for power. Whatever the ultimate aims of international politics, power is always the immediate aim. (p. 31.)
>
> > It would be useless and even self-destructive to free one or the other of the peoples of the earth from the desire for power while leaving it extant in others. (p. 38.)
>
> > [I]f we look at all nations, our own included, as political entities pursuing their respective interests defined in terms of power, we are able to do justice to all of them. (p. 13.)
>
> Should power always be the immediate aim of states in international relations? Does it matter what that power is to be used for—that is, what are the goals of the state? As long as any single political entity pursues power, is it necessary for all others to do the same? If so, what are likely to be the consequences? Can the system provide relative justice to all if all pursue power?

measured by public-opinion polls to get some notion of national morale. Leadership quality can be tapped in a similar fashion. What cannot be ascertained in advance is the degree to which a people will rally around a leader when attacked from outside. The world hardly expected the Soviet population to react with the unity it did to repel the Nazi invasion in World War II. Before the decisive turning point in the battle of Stalingrad in 1943, some even speculated that the Ukrainians and other non-Russians would defect to the invading Nazis. Instead, Hitler's ruthlessness only served to increase the determination of those groups to resist and to expel the invader in what was known in Soviet history as "the Great Patriotic War."

## LEADERSHIP AND DECISION-MAKING CAPABILITY

The effect that leadership qualities can have on foreign policy was amply demonstrated by President Charles de Gaulle of France, who pulled together an unstable French Republic in 1958 by centralizing the foreign policy decision-making process under his own authority.

Mikhail Gorbachev similarly showed what a single person could do to affect political developments, although he proved more adept on the global scene than within the Soviet Union. After he came to power in 1985, he argued increasingly that his country and the West should join as partners in fighting problems of environmental pollution, poverty, and the threat to all humanity contained in the superpowers' nuclear arsenals. He took much of the initiative in improving relations with the West, then watched while East European societies threw off Communist rule without Moscow's intervention. While his foreign policies did much to help end the cold war, his domestic policy failures, in contrast, led ultimately to the collapse of the Soviet Union—and Gorbachev's position within it.

During his first three years in office, President Bill Clinton was perceived by many to be less than effective as a leader in world affairs because of his inconsistency in pursuing foreign policy goals. Others were more charitable, viewing such behavior as a result of the end of the cold war. That uncertainty seemed to have pervaded American society at large, but it is typical for political leaders to be blamed for not asserting decisive leadership when foreign policy goals are in flux and domestic opinion is divided.

In a perverse variation on the power provided by strong and stable leadership, some leaders may actually gain strength in achieving foreign policy goals because their prospects for remaining in office are tenuous. For example, the European Union, concerned about a possible victory of a fundamentalist Islamic party, rewarded Turkey with a long-sought customs union agreement on the eve of the 1995 elections in the hope of strengthening the electoral prospects of Prime Minister Tansu Ciller. Political weaknesses in Japan, Russia, and China during the mid-1990s also led the United States to soft-peddle the pressure it exerted upon those three states in the hope of getting them to change some of their practices related to trade and human rights issues.[15]

## DIPLOMATIC AND STRATEGIC SKILLS

How the leader operates on the world stage also determines a state's ability to influence others in the international system. The success that President Gorbachev had in generating support among the peoples and governments of the world is a case in point. By offering a number of unilateral concessions in the military sector, and by establishing himself as trustworthy, he obtained agreements and goodwill from others; these included some of his bitterest ideological adversaries, such as Ronald Reagan and Margaret Thatcher. He, of course, was aided in the process by the stark comparison to his more blustery predecessors.

Still, boisterous threats may not be an entirely unproductive way to enhance a state's power. Mussolini's threats and bluster, for example, provided Italy with a degree of international deference greatly disproportionate to its capability.[16] The uncouth behavior of Nikita Khrushchev when he removed his shoe and pounded it on the table at the United Nations may likewise have given him some leverage; certainly, it showed him capable of an angry, irrevocable action that could lead to a nuclear Armageddon.

## INTELLIGENCE CAPABILITIES

Power also can be enhanced if one has superior intelligence services, enabling a state to collect information regarding the capabilities and intentions of other states. Allied diplomatic messages intercepted and decoded by American intelligence agencies in the closing months of World War II enabled the United States "to control

---

[15]Elaine Sciolino, "Why the U.S. Takes Guff from Weaklings," *New York Times,* July 2, 1995, E3.

[16]A. F. K. Organski and Jacek Kugler, *The War Ledger* (Chicago: University of Chicago Press, 1980), 31.

the debate, to pressure nations to agree to its positions and to write the U.N. Charter mostly according to its own blueprint."[17] In an earlier case, the United States succeeded in breaking the Japanese diplomatic code and learned of its fallback position on naval construction. This information helped the United States at the 1922 Washington Conference on Naval Limitation to force the Japanese to accept a greater limit on warship construction than they wanted.[18] Finally, it was through the use of electronic eavesdropping upon the participants that the British gained considerable mediating leverage in negotiations on Zimbabwe in 1979. Since the outcome was positive in this situation (moving Zimbabwe to legitimate independence), some might even approve such nefarious activity.

Advance knowledge of another state's plans enables a government to take countermeasures to reduce the action's predictable consequences. For example, if a proposed move is unpopular, domestically or internationally, public opinion and political opposition can be incited to prevent the policy from being carried out.

## COMMUNICATION CAPABILITIES

In the information age, communication capability has become an increasingly important intangible factor of power.[19] Given highly competitive belief systems, governments have attempted to generate international public support for their policies through the use of propaganda and cultural education, sometimes dubbed **public diplomacy**.[20] Targets of propaganda efforts include not only foreign government officials, but also important elites and mass publics abroad. One of the earliest and most effective of these efforts was directed by Louis XIII and Louis XIV in France during the seventeenth and early eighteenth centuries. By the end of the eighteenth century, French culture had been diffused throughout Europe and into Canada and the Levant; French had become the common language of nobles, ambassadors, and educated people. Foreign Minister Talleyrand would dismiss French ambassadors going to their posts with the words, "Make them love France."[21]

The capacity to engage in propaganda and educational efforts abroad has increased substantially over the course of the Westphalian period. One of the most important developments was the invention of the printing press, which allowed propagandists such as Martin Luther to spread his message by distributing stirring pamphlets throughout Europe. The development of radio, movies, and television in the twentieth century opened up new horizons for propagandists, enabling them to send messages over long distances without a physical presence. In the 1980s, the USSR was broadcasting nearly 2,100 hours per week in 80 languages, China 1,400

---

[17]William H. Honan, "War Decoding Helped U.S. to Form U.N.," *New York Times*, April 23, 1995, A4.

[18]Ibid.

[19]Howard H. Frederick, *Global Communication and International Relations* (Belmont, Calif.: Wadsworth, 1993), 234.

[20]Hans N. Tuch, *Communication with the World: U.S. Diplomacy Overseas* (New York: St. Martin's Press, 1990).

[21]J. M. Mitchell, *International Cultural Relations* (London: Allen and Unwin, 1986), 16, 22.

hours in 45 languages, the United States 2,000 hours in 45 languages, Germany 780 hours in 39 languages, and the BBC 720 hours in 37 languages. Even such small states as North Korea (593 hours weekly) and Nigeria (322 hours) joined the fray.[22]

The computer, satellite communication, E-mail, Internet, and fax machines have added to the capacity of governments and groups to be in constant communication, and thus to engage in efforts to propagandize and influence one another. The ability to beam radio and television programming worldwide via satellite is particularly important to the propaganda effort.

## NATIONAL MORALE

Having the support of the population, especially in a democratic state, is also important to the power that a state can project. Some have speculated that Saddam Hussein may have miscalculated American resolve in responding to his invasion of Kuwait in 1990 because of the fickleness of American public support for past military operations, especially Vietnam. The so-called Vietnam syndrome, after all, had prevented the U.S. administration from intervening in Angola in 1975, despite President Ford's wish to do so. Because of their limited size, location, and duration, the U.S. incursions into Grenada in 1983 and into Panama in 1989 provided insufficient evidence as to whether American public opinion would support massive troop movements to more distant lands. The initial response was one of popular support; three-quarters of the American people supported President Bush's sending troops to Saudi Arabia. But as the number of U.S. troops was doubled to more than 400,000, the public became momentarily more hesitant about the wisdom of the action.

## MISCELLANEOUS INTANGIBLES

A variety of other intangible factors can affect the power equation. For example, a dependent state is more likely to be influenced by that state upon which it relies for its economic well-being or for its military security. There may even be a generalized responsiveness between certain states based upon habit or preference, such as the mutual relations between the United States and Canada. This creates a predisposition on the part of each state to be responsive to the wishes of the other, which, of course, increases the likelihood of success for any attempts at influence.

Still other intangible factors may affect the power relationships of states. In some cases, the ability of a state to exercise power over another may be little more than a function of how much time and energy the influencer is able to spend on a problem. A state with a limited agenda and an intense interest in a single issue may have a considerable power advantage over a state with a broad agenda and far-flung interests. For example, Panama, with its relatively limited overseas interests, had considerable bargaining leverage over the United States in the renegotiation of the Panama Canal treaties in the 1960s and 1970s. Even cultural differences may affect the power balance, as one population may be more predisposed to working harder

---

[22]Garth S. Jowett and Victoria O'Donnel, *Propaganda and Persuasion*, 2d ed. (Thousand Oaks, Calif.: Sage Publications, 1992), 112.

than another, thus increasing productivity. States may also fail to exploit their full potential because of cultural or ideological factors. For example, Hitler failed to use women in the German industrial war effort, given his reactionary view of what constituted appropriate gender roles.

## MEASURING POWER

As a result of the many difficulties in conceptualizing power and the ever-changing power situation, it is small wonder that students of international relations have not been very successful in developing agreed-upon measures of power. A useful operational conceptualization has been provided by Yale professor Robert A. Dahl, who suggested that power is equal to the ability of A to get B to take action X, minus the probability that B would take action X anyway.[23] Despite the logic of such a formula, applying it to the world of international relations is a formidable task. The difficulties of measuring just what B would do in the absence of A's attempt to influence it are almost insurmountable.

Most efforts to measure power in the international system have focused on indicators of the capabilities of nation-states, including such tangible factors as population, industrial capability, and military budgets and forces. The difficulties of using such indicators, however, are shown by studies conducted in the 1960s that predicted that the People's Republic of China would become the most powerful actor in the world in the 1980s.[24] That hardly proved to be the case, as domestic turmoil and problems of overpopulation impeded China's capital savings and economic growth.

The most popular indicators used to measure industrialization and economic power have been gross domestic product, energy consumption, and iron and steel production. Events of recent decades, however, have demonstrated that high energy consumption can create power vulnerability, particularly if that consumption requires extensive importation of oil from abroad. In developing indices of economic power, some analysts also include measures of modernization such as urbanization, technology, and agricultural productivity.[25]

Former CIA analyst Ray S. Cline is one of the few authorities to use intangible variables in his power index, which includes his assessment of the national strategy and national will of states at the beginning of the 1980s.[26] The result of including these two intangibles in his formula was to reduce the power score for the United States by 30 percent, while increasing that of the Soviet Union by 20 percent. These scores for Soviet intangibles would obviously look quite different at the

---

[23] Robert A. Dahl, "The Concept of Power," *Behavioral Science* 2 (July 1957): 201–215.

[24] Karl W. Deutsch, *The Analysis of International Relations* (Englewood Cliffs, N.J.: Prentice-Hall, 1968), 23; and A. F. K. Organski, *World Politics,* 2d ed. (New York: Knopf, 1968).

[25] James Lee Ray and J. David Singer, "Measuring the Concentration of Power in the International System," in Measuring the Correlates of War, ed. J. David Singer and Paul F. Diehl (Ann Arbor: University of Michigan Press, 1990), 115–138.

[26] Ray S. Cline, The Power of Nations in the 1990s (New York: Free Press, 1989).

end of the decade. By then the Soviet Union's domestic difficulties, including the inability of its leadership to deliver its tangible potential, had contributed to a dramatic decline in how its power was perceived. Indeed, the fact that the Soviet Union collapsed so soon after Cline's estimates were made underscores the difficulties of any attempt to measure intangible factors of power.

# Conclusion

Power is an important yet elusive concept that plays a vital role in international behavior. Much of the confusion over the concept lies in the failure to distinguish between the capacity to act and the actual exercise of power, for the two do not always go together. Capability does not necessarily translate into influence, because perceptual factors may distort one's views of the capabilities and intentions of others. Power is also relative and reciprocal, and has varying effects on different issues, particularly since legitimacy, expertise, and serving as a role model also affect influence. Expectations about the intentions of others and the general tendency to equate aggressiveness with power have led to further disparity between power and influence. As a result, it is difficult to measure power directly; instead what is usually measured is national capability.

The sources of power for a state are many and varied. Among the tangible factors of power are geographical variables such as size and location, although the latter has become less important in the age of nuclear missiles. Population and economic resources are important in defining power, although overpopulation can impede economic growth. Ultimately, who prevails in a conflict situation may depend upon military arsenals, which are important both for deterrence and for fighting a war.

Measuring power becomes exceedingly difficult because so many things that determine the power of a state involve such intangibles as leadership capability, diplomatic reputation and skill, and the morale of a given population. Yet, even if it were possible to measure both the tangible and intangible sources of power, it would still be the case that only capabilities were being measured.

Capabilities, whether they represent economic resources, military weapons, or various political and psychological resources, become important as they are translated into instruments of power. In the next chapter we turn our attention to one such instrument—that of military force—which will be examined in terms of its deterrent value and its war-fighting potential. We shall also consider the effort to control the development of force in the international system.

# CHAPTER 7

# The Role of Military Force

From the realist perspective, the Westphalian system depends heavily upon military force to provide security for its separate units, those sovereign and independent states that operate in an anarchic environment. An overarching authority with the power to regulate violence is lacking in the international system. As a result, states acting independently and in cooperation with one another rely upon force—and the threat of force—to provide for their security.

The use of military power, however, is not limited to efforts to provide for the security of the state against internal and external enemies, for it serves a variety of other functions as well. Force has often been used as an instrument to support a state's greedy aspirations against another. Throughout history, some ruling groups have ordered their armies to encroach upon the territory of others, often using their own security needs as the excuse, but actually with the intent of claiming another's land and resources.

Force may also be used to deny gains to others in the international system, to weaken adversaries by engaging them in long, arduous wars, or perhaps simply to punish them for disliked behavior.

Additionally, military force has been viewed as a way of increasing a state's bargaining power and its ability to get attention from others in the international system. In the eighteenth century, Frederick the Great of Prussia summarized the way in which military power influenced diplomacy by asserting that "diplomacy without armaments is like music without instruments."[1]

Increasingly, force has also assumed an important role in balancing the power of others. In the age of potential mass destruction it has come to play a central role in deterrence through its use as a threat system. This function has become particularly vital today, since no effective defense exists against a nuclear missile attack.

---

[1] Cited in Geoffrey Blainey, *The Causes of War* (New York: Free Press, 1973), 108.

# Balance of Power

Concern with threats and increases in the power of other states has led governments to pursue **balance of power** politics. Balancing power assumes that independent states will react against any nation or group of nations that attempts to improve its own position and thereby upset the balance in a way that would threaten the security of others within the same system.

Writing in the eighteenth century, Scottish philosopher and historian David Hume suggested that the notion of the balance of power had existed throughout Western history, dating back to ancient Greece.[2] After noting the tendency of Greek city-states such as Athens and others to shift alignment in defense of weaker city-states, Hume sketched the process through the Roman Empire to his own time, viewing Britain as assuming the role of balancer in the eighteenth century. In the view of many, Britain continued that role during the following century as it sided with the weaker, threatened state in order to prevent any stronger state or group of states from dominating Europe.

But, as the world became more polarized during much of the twentieth century, a global balance of power system based upon a balancer and flexible alignment largely ceased to exist. The bipolarity of the cold war period was characterized by two alliance blocs whose membership remained highly fixed and inflexible. Still, balancing power remained relevant at the regional level, as, for example, in the Middle East, where France and Britain used balancing strategies to assume control over former Ottoman territories following World War I. Each made efforts to carve out administrative units that would preclude any single entity from obtaining a dominant position. The protectorates of Kuwait, Oman, and Yemen, among others, were established on the Arabian coast to lessen the prospect of dominance of the Arabian peninsula by larger entities, such as Saudi Arabia and Iraq.

Concern about maintaining a regional balance in the Middle East continued during the 1990–1991 Persian Gulf War. Great restraint was exercised by the U.S.-led coalition forces so as not to eliminate Iraq as a regional power, should Iraq be needed to balance other potentially dominating powers in the region, such as Iran and Syria.

## BALANCE OF POWER PREREQUISITES

Whether the balance of power system is global or regional, certain prerequisites are necessary for it to operate effectively. Among these is the obvious requirement that the independent actors within the system recognize that their fates are shared and that they have a vested interest in reacting to the efforts by any party to change the essential balance.

It is also essential that there be a limited number of major actors in the system with roughly equal power, who are willing to play a leadership role in maintaining the balance. Too many actors can complicate the decision-making process, which means that in any global balance of power system only the major actors take a lead-

---

[2] David Hume, "Of the Balance of Power," in *Classics of International Politics,* 2d ed., ed. John A. Vasquez (Englewood Cliffs, N.J.: Prentice-Hall, 1990), 273–276.

ing role. Too few dominant actors in the system, on the other hand, may lead to bipolarization and the replacement of the balance of power with a balance of terror.[3]

Balance of power operates most effectively when the participants share certain values and respect for the rules of the game. Thus it was far more successful in Europe during the eighteenth and nineteenth centuries than it has been at other times and in areas of greater diversity. The ideological divisions of the twentieth century have not been conducive to a stable balance of power system.

An effective balance of power system also requires that decision makers be able to pay attention to their vital power interests and not be led astray by such things as emotion, ideology, or domestic opinion in making their decisions. Nor should a state have permanent enemies or permanent friends, for this may interfere with the flexibility of alignment required for a stable balance of power.

Finally, for balance of power to function, decision makers must be willing to use force and the threat of force when necessary to restore balance. The balance of power theory does not claim that war will be entirely eliminated; rather it assumes a reduced likelihood of war whenever a stable balance of power is obtained. As such, balance of power is seen by theorists as the best hope for preserving sovereignty, since states are motivated to take action against those who would upset the balance by attacking others.

## MILITARY DETERRENCE

Following World War II, notions of a balance of terror replaced those of a balance of power. That is, the role that force played increasingly came to be thought of for its ability to deter. The change in thinking was in part due to a recognition that the development of horrendous weapons of mass destruction had made it vital not to repeat major wars, such as those that had occurred twice during the century. During the half century following the end of World War II, Europe enjoyed one of its longest periods of general peace since the days of the Roman Empire. Underpinning this peace were both the development of the very nuclear weapons that had generated so much fear, and the creation of well-armed military alliances, such as the North Atlantic Treaty Organization. Those developments combined to make the threat to use massive force, which is the central idea of deterrence, a nearly constant reality throughout the cold war period. Many saw the credibility of that threat as the explanation for what kept that military might from being used—in other words, as what ushered in the long peace.

Yet, as that peace endured, many became increasingly troubled over the questionable morality upon which it was built—that is, that governments based their security on the perpetual threat to engage in genocide on a nearly unimaginable scale. Whether or not it was deterrence that maintained the peace between East and West for more than forty years—and there is no way to prove that it did—huge stockpiles

---

[3]For a discussion of the relative advantages of bipolarity and multipolarity see Charles W. Kegley, Jr., and Gregory Raymond, *A Multipolar Peace: Great Power Politics in the Twenty-first Century* (New York: St. Martin's Press, 1994), 67–120.

of nuclear weapons were left at the end of the cold war that made this question particularly insistent: Would not the world be better off if a greater effort were made to control weapons than to continue to threaten enormous destruction with them?

An obvious goal of any national security policy is to provide a defense capable of repelling any attack that might occur. But a more hopeful approach to security would be to deter any potential attacker from striking in the first place. The idea underlying **deterrence** is simple enough: the purpose of vast destructive power, whether conventional or nuclear, is to prevent or deter one's enemy from some unacceptable foreign policy action by threatening retaliation if the opponent makes such a move. The term deterrence itself comes from the Latin word *terrer,* which means *to threaten.* For a threat to be effective, deterrent theorists tell us, both credibility and capability are essential. It is not enough simply to have the capability to unleash death and destruction upon the threatened state; it is at least as important to convince your adversary that you have the will to carry out the threat, despite the possible consequences such action may hold for you as well as your opponent.

## COMMUNICATING CREDIBLE THREATS

Perhaps most important for the credibility of one's threat is that it be *realistic.* That typically means it is crucial that the threat be clearly appropriate and proportionate to the action that one is seeking to deter. It is usually not credible to threaten consequences far larger than the injury done to the one issuing the threat. For example, how credible was Nikita Khrushchev's bombastic threat in 1956 to rain nuclear missiles on Paris and London if France and Britain failed to desist from their invasion of the Suez? The realism of the Soviet threat was questionable, first, on technical grounds, since the accuracy and carrying capability of missiles had yet to be proven, and, secondly, on the basis of sound policy considerations, since it seemed doubtful that Moscow would risk nuclear escalation over an issue that was not that significant for Soviet national interests.

Credibility also may be dependent upon a clear *consensus in support of carrying out a threat.* President Bill Clinton had great difficulty in issuing a credible threat in the Bosnian crisis during his first year in office. His repeated verbal threats to bomb artillery sites in Bosnia and to enforce a no-fly zone failed to deter Serbian aggression largely because of the opposition of the NATO allies and the lack of support for such an action on the part of the American public, within Congress, and even within his own administration.

Credibility is also a function of delivering a *consistent message* to the party one is seeking to deter. President Clinton again failed to do this during his early involvement in the Bosnian crisis, whipped as he was by divided forces on the issue within both the administration and the country. The failure may also have been in part related to his own ambivalence at a time when he was concerned to show that his primary objective was to solve some of the problems confronting the domestic economy, which had been the basis of his electoral success.

Governments also seek to enhance the credibility of deterrence by taking *demonstrative actions* to try to communicate the seriousness with which they regard an issue and how far they are willing to escalate the conflict. Such action might involve military war games, the introduction of gunboat diplomacy, or even the limited use of force at choice targets in an effort to communicate their seriousness. It was not difficult, for example, to ascertain why the United States held so many military exercises in Honduras in the 1980s at a time when the Reagan administration was seeking to frighten the Sandinista government in neighboring Nicaragua into greater compliance with Washington's own foreign policy goals.

Credibility may be enhanced by *burning one's bridges behind oneself* so that there are obvious consequences if one does not make good on a threat. Placing troops in harm's way becomes one possibility, as was the case when American troops were stationed on European soil after World War II and when Chiang Kai-shek positioned Nationalist Chinese troops on the offshore Chinese islands of Quemoy and Matsu to deter the Chinese Communist government from trying to seize them. Under such conditions there would be little choice other than to fight if attacked.

Deterrence is enhanced if one has a *history of carrying out past threats and responding quickly and decisively to acts of aggression*. U.S. intervention in the Persian Gulf War in 1990 certainly served to bolster American deterrent credibility in the immediate post–cold war world. On the other hand, the recurrent threats in Bosnia and Somalia in the early 1990s, coupled with the American failure to carry them out, served to undermine that credibility.

Because deterrence sometimes requires the pursuit of policies that are not completely rational, the effectiveness of the deterrent threat might actually be enhanced by demonstrating a propensity to *behave irrationally*. Nikita Khrushchev's pounding of his shoe on the table at the United Nations in 1960 was a possible case in point. Such leaders as Saddam Hussein, Mu`ammar al-Gadhafi, and Fidel Castro, who were seen in some quarters as fanatical and irrational, might therefore have been able to issue more credible threats than leaders who were viewed as more cautious and rational. Yet, they also risked being isolated as pariahs in the international system and were targeted for elimination because of the dangers they allegedly presented to the world.

Governments may increase credibility by *arguing the need to act in order to enhance future deterrence*. Decision makers might persuasively argue that it is necessary to carry out a threat regardless of how ill-conceived or potentially damaging to their own nation if failure to do so would mean that they might not be believed again, either by the party they were seeking to deter or by anyone else. In the final analysis, credibility depends upon whether or not one has made good on threats in the past. The worst possibility for future deterrent effectiveness is to be caught bluffing. For example, Bosnian Serbs were not deterred from overrunning U.N. "safe havens" in Bosnia in 1994 and 1995 after it was clear that U.N. forces there were neither large enough nor sufficiently armed to prevent those enclaves from being seized.

Finally, *having the necessary capability to carry out a threat* is also critical to believability. A state known to lack nuclear weapons, for example, can hardly make a valid nuclear threat. Nor is it likely that small nuclear powers with inadequate delivery capability will be able to mount credible threats against one of the major nuclear powers. They may, on the other hand, prove effective in deterring regional enemies, which helps explain why Israel, surrounded by many, far-larger Arab states, was so adamant to achieve a nuclear capability of its own.

"WATCH OUT OR WE'LL HURL ANOTHER THREAT AT YOU."
© 1993 Herblock/Creators Syndicate.

## DETERRENT STRATEGIES

Confronted with weapons of mass destruction, the superpowers expended great effort to develop military strategies in pursuit of the Westphalian logic of preserving the sovereignty of the state. In the late 1940s, some American military officials actually went so far as to suggest that the United States consider a **preventive strike** against the Soviet Union before the latter developed its own nuclear technology. The central problem with a preventive strike is that the government engaging in such a strike might find it extraordinarily difficult to convince the rest of the world that the use of force against a comparatively defenseless adversary is somehow justified. As Hitler's attacks on his much weaker neighbors in the late 1930s made clear, alleged preventive strikes may well be nothing more than rationalizations for aggression.

After 1949, when the Soviet Union obtained its own nuclear capability, talk in the United States of a preventive strike against the Soviets largely ended. But in 1954, when the United States was still far superior to the USSR in the nuclear field, the

Eisenhower administration introduced its **massive retaliation doctrine** that threatened the Soviet Union with the possibility of a massive nuclear strike even if the Soviets were only to engage in aggression through conventional means. Although the doctrine was introduced largely as a cost-saving measure, since nuclear weapons provide a "bigger bang for the buck," the policy was heavily criticized within the United States as a less than credible threat. Critics reasoned that the Soviets would be unlikely to believe that the United States would risk nuclear annihilation simply to defend against a minor Soviet border attack.

Campaigning for president in 1960, John F. Kennedy argued for placing renewed emphasis upon conventional force. Once he became president, he began to push for a policy of **flexible response** in which the adversary would be allowed to choose the instrument of attack, but the United States would retain a flexible arsenal of both conventional and nuclear weapons with which to respond.

Also during the 1960s, the policy of **mutual assured destruction (MAD)** was introduced. This strategy held that stability would be provided, making nuclear war highly unlikely if both sides had the assurance that they could retaliate and essentially destroy the other side even after a first strike. Retaliatory targets would primarily include so-called *soft assets* such as the enemy's population and industrial centers since they are easier to identify and destroy than military targets. Less emphasis would be given to retaliating against nuclear forces, for the goal under MAD was to gain stability by assuring that each side achieved an invulnerable retaliatory capability to deliver a devastating strike against the other's society.

---

### Global Changes

## The Changing Role of Military Force

At the beginning of the Westphalian period, military forces were based upon the use of mercenaries, often hired by dynastic leaders and consisting not only of nationals, but often foreigners as well. The rise of the mass citizen army and conscription began with Napoleon at the beginning of the nineteenth century to aid him in his imperial expansion.

- What do you think were the implications of such changes upon the severity and duration of war?

Historically, the use of force has involved a struggle between offense and defense. Defense became ascendent with the development of walled cities, until the invention of the cannon made them increasingly vulnerable to offensive attack. But this hardly compares with the superior position gained by the offense in the nuclear-missile age.

- Given such awesome offensive power, is the defense ever likely to achieve a preeminent position again? Would former President Reagan's proposed Strategic Defense Initiative (popularly dubbed "Star Wars") have provided defensive security for a state?

The invention of new and highly destructive weapons has often been viewed as increasing world safety on the assumption that rational human beings would be unwilling to use them and risk retaliation in kind. Thus the inventor of dynamite, Alfred Nobel, saw his creation as ending war. Prime Minister Stanley Baldwin asserted that the airplane would make war unthinkable. The same case has been made by many writers with respect to the threat of nuclear weapons.

- Why have such predictions proved to be wrong? Is it likely that the "ultimate weapon" will ever be developed?

The logic of MAD was that some finite level of nuclear capability serves as an effective deterrent, although the major nuclear powers disagreed on what that level should be. Clearly, a mere 5 to 10 percent of the strategic missiles currently in the stockpiles of Russia or the United States, if unleashed, could provide a holocaust such as the world has never experienced. A lower number of this sort seems a sensible estimate for so-called minimum or **finite deterrence.** President Jimmy Carter perhaps inadvertently put the matter in even more stark terms when he said in 1979 that "just one of our relatively invulnerable Poseidon submarines carries enough warheads to destroy every large and medium-size city in the Soviet Union."[4]

Many observers have never been comfortable with a strategy of mutual assured destruction, for the consequences of its failure would be so catastrophic. It has been proposed that more attention be given to **damage limiting strategies** so that if deterrence fails, the terrible consequences could at least be minimized. Damage can be limited through a strategy designed to destroy any remaining nuclear forces in the hands of an adversary or by building antiballistic missile systems capable of intercepting incoming missiles. Civil defense programs concerned with population relocation or the building of nuclear fallout shelters can also be thought of as damage limiting strategies.

Yet, the danger with such protective strategies points to the central dilemma of deterrence: They may actually provoke the adversary into that first strike those strategies are intended to deter. The reason is that weapons capabilities meant for damage limitation in the event of war also provide first-strike capabilities. The ability to destroy offensive capabilities may make the adversary very touchy on the trigger, just as efforts at missile or civil defense might make such an opponent feel that a first strike is being planned.

The previously noted nuclear doctrines primarily address the threat of conflict directly between the superpowers themselves and the fear that a devastating nuclear war might result. But since superpower conflicts during the cold war were usually fought out in proxy wars overseas, attention has also been given to the problem of deterring aggression against the territory of other states through a process known as **extended deterrence.** Making such a deterrent threat credible is difficult, for it is obviously easier to communicate a willingness to defend one's own territory than it is to make a convincing promise to defend a foreign territory from attack. That explains why, throughout the cold war, NATO allies particularly wanted to couple the U.S. deterrent to the defense of Europe by making American forces essentially hostages to Europe's fate.

## DOES MILITARY DETERRENCE PROVIDE SECURITY?

Since a major rationale for military weapons has been to deter war, several studies have sought to ascertain their effectiveness. Establishing a causal link between the existence of military weapons and the deterrence of war is virtually impossible. To prove that deterrence has worked, it is necessary to demonstrate that the opposing

---

[4] Theodore Draper, "How Not to Think about Nuclear War," *New York Review of Books* (July 15, 1982), 42.

power intended to attack but failed to do so only because of its adversary's weapons. Even if such a state does retreat in a confrontation, the explanation may be related to factors other than the deterrent threat. Such factors might include a recognition that the potential gain is not worth the possible negative repercussions of adverse world opinion, loss of support of allies, or the opposition of domestic groups whose moral sensitivities had been kindled.

Nonetheless, a study of eight postwar international crises revealed that as the American strategic and/or tactical preparedness was perceived by the Soviets to have increased, Soviet perception of American resolve also increased.[5] An examination of American attempts to influence the outcomes of fifteen major crises from 1946 to 1975—when both strategic nuclear and conventional force were threatened—revealed that in nearly every instance a favorable outcome, as defined in terms of U.S. objectives, was achieved within a span of six months.[6] Success dropped to three-quarters of the fifteen cases during the longer span of three years, suggesting that the deterrent effectiveness of a military threat declines over time.

When it comes to an extended deterrent, which is somewhat more difficult to make credible than an attack upon oneself, some evidence suggests that even here the threat of force may be successful. In a study involving sixty-seven cases of extended deterrent situations from 1902 to 1979 in which a defender was trying to protect a protege, success was recorded in almost two-thirds of the cases (forty-one out of sixty-seven).[7] The results, however, were challenged in another study that argued that many of the cases selected did not actually involve deterrent encounters. It is difficult to prove that the would-be attacker planned to use military force in the first place or that it resisted striking simply because of another state's deterrent threat.[8] Nevertheless, the fact that war did not occur in so many different crises suggests that part of the explanation may relate to the threat to use military force to defend against an attack. The end of the cold war has actually seen an increase in the number of militarized disputes throughout the world—military actions that during the cold war might have been deterred by the actions of the superpowers concerned lest a military outcome provide the adversary with an advantage.

## DESTABILIZING FACTORS IN DETERRENCE

Even though it might be argued that both nuclear and conventional deterrence can reduce the frequency of war, we should still be wary (not to say troubled on moral grounds) about relying upon huge military arsenals in our search for world peace.

---

[5] David C. Schwartz, "Decision Theories and Crisis Behavior: An Empirical Study of Nuclear Deterrence in International Political Crisis," *Orbis* (1967): 485.

[6] Barry M. Blechman and Stephen S. Kaplan, *Force Without War* (Washington, D.C.: Brookings Institution, 1978), 99–100.

[7] Paul K. Huth, *Extended Deterrence and the Prevention of War* (New Haven, Conn.: Yale University Press, 1988).

[8] Richard Ned Lebow and Janice Stein, "Deterrence: The Elusive Dependent Variable," *World Politics* XLII (April 1990): 336–369.

The stability of the deterrent system is threatened in a variety of ways, and particularly by factors that could lead to an **accidental war.** During the cold war, the United States experienced a large number of false warnings about an imminent Soviet missile attack. One study noted at least 147 false indications of attack during a single eighteen-month period.[9] In June 1980, American forces were on nuclear alert for six minutes because of a computer error created by the malfunction of a circuit chip worth 46 cents.

Accidental war might also occur because of psychological or physical problems among those who have charge of the weaponry. For example, in 1977 some 1,219 American military personnel with access to or responsibility for nuclear weapons were removed from duty because of mental disturbances, 265 for alcoholism, and 1365 for drug abuse.[10] Political leaders themselves have not been immune from psychological breakdown, and the intense pressures resulting from international crises and fear serve only to increase the dangers of nuclear hostilities.

In addition to the possibility of war erupting by accident, there is also the danger of **war by miscalculation.** If one side or another believes a military strike is imminent, the pressure could become irresistible to deliver the first blow. Such pressures would be particularly acute whenever one's own retaliatory capability is believed to be vulnerable. It is likely that Kaiser Wilhelm II of Germany struck at Russia in August 1914—thereby initiating World War I—in the expectation that Russia's mobilization meant that the tsar was preparing to order an attack on Germany.

The threat of miscalculation is increased in light of the rapid pace of technology. The fear that the opposition may be close to a scientific breakthrough that would give it some advantage is likely to cause particularly anxious moments. Most threatening today would be a system capable of intercepting the bulk of the incoming missiles and aircraft, thus signaling the end of deterrence. Since the incentive would be high to assure that the adversary never achieves such a capability, the state fearful of that prospect would be forced to consider seriously the possibility of a preemptive strike. Such considerations help explain Soviet anxiety during the Reagan administration over the U.S. pursuit of an antiballistic missile system with its Strategic Defense Initiative.

Miscalculation may result from concern about the viability of **command and control** structures in the nuclear age. John Steinbrunner of the Brookings Institution has estimated that fifty to one hundred nuclear weapons could disrupt the central nervous system of the American command structure, making it vulnerable even to smaller nuclear powers.[11] In October 1981, President Ronald Reagan announced a $20 billion effort to improve command and control structures—a task he viewed as having higher priority than building new weapons systems. The cost of the program was another indication of the ever-rising price of a deterrence system whose stability is constantly undermined by changes in numbers and kinds of military hardware.

---

[9] Ruth Sivard, *World Military and Social Expenditures* (Washington, D.C.: World Priorities, 1983).
[10] Tom Wicker, "War By Accident," *New York Times,* November 21, 1982.
[11] John Steinbrunner, "Nuclear Decapitation," *Foreign Policy* 18 (1981/1982): 18.

## STABILIZING DETERRENCE

Given the many factors affecting the stability of deterrence, the effort grew in the last years of the cold war to reduce the threat of a possible breakdown and, particularly, to avoid war by accident or miscalculation. To minimize the danger of unauthorized use of weapons, *command and control* functions were highly centralized in the governments of both superpowers. That remains the case for the United States today. The president is the only person empowered to initiate a nuclear strike, although contingencies are provided in case the president is incapacitated. At every step in the decision to use nuclear weapons, at least two individuals are made responsible for setting in motion a nuclear strike under the so-called *two key system*. This procedure has been designed to reduce the danger that someone in the chain of command could go berserk and initiate a nuclear attack.

The disintegration of the Soviet Union at the end of the cold war raised questions about the command and control of the Soviet arsenal, for nuclear weapons were left in the hands of four independent states: Russia, Ukraine, Belarus, and Kazakhstan. Assurances were given that only Russia would control the former Soviet nuclear arsenal; yet, the political instability in much of the region reminded the world of the dangers of accidental use, blackmail, or diversion of nuclear weapons when highly centralized control disappears.

If a delivery system is unleashed by accident or unauthorized action, or a nuclear bomb is accidentally dropped, it is critical for the decision makers to communicate promptly and effectively with each other to reduce the possibility that the incident could lead to war. To aid in such communication, the U.S. and Soviet governments agreed in 1963 to establish a **hot line** between Washington and Moscow. The hot line today does not actually provide leaders with direct telephone communication; rather, it uses modern computer technology to transmit nearly instantaneous exchanges of block messages and even maps. The idea of a hot line has caught on elsewhere: Sometime-rivals Greece and Turkey, Pakistan and India, and China and Russia, have all discussed or developed similar systems.

During the cold war, the mutual fears and suspicions of the nuclear superpowers led defense planners to develop what they called an **invulnerable retaliatory capability.** Its purpose was to enable the delay of a retaliatory response to a provocative action as long as possible, while remaining confident of the ability to retaliate at any time. Invulnerability could be enhanced in several ways, such as by utilizing mobile vehicles, as in the case of submarines; dispersing weapons intended for a retaliatory strike to make it more difficult to eliminate them; and hardening the weapons—in the case of missiles, placing them in underground silos with a thick cover made of reinforced concrete—to assure survivability from all but the most direct hit.

## CONTINUING THREATS TO NATIONAL SECURITY

With the end of the cold war, Russia and the United States no longer targeted nuclear weapons at each other. That has virtually eliminated the once overriding fear of a possible nuclear war between those two states. Instead, three other major threats to global security are receiving increased attention: nuclear proliferation, conventional arms spread, and international terrorism.

# NUCLEAR PROLIFERATION

Concern that weapons of mass destruction might fall into the hands of irresponsible leaders or even terrorist groups, has led to efforts to control **nuclear proliferation.** The breakup of the Soviet Union into its component republics raised serious concerns about how nuclear weapons were to be controlled. So too did the fact that Saddam Hussein had come so close to producing a nuclear bomb before his nuclear facilities and materials were confiscated—and hopefully destroyed—in the early 1990s.

A number of reasons invite concern about the spread of nuclear weapons to yet other nations. Increasing the number of nuclear powers simply adds to the statistical probability that one of them might unleash a nuclear war. In addition, there are probably more pressures on the leaders of new nuclear weapon states to use that capability than on the older nuclear powers. The latter know that they risk a holocaust if their use of even the least destructive nuclear weapons should lead to escalation. But leaders of states with only a limited nuclear capability run no such risk, at least not from exploding a few nuclear weapons on a regional enemy that lacks nuclear capability. Moreover, the offensive weapons of any fledgling nuclear state are likely to be less well protected, since the costs of making them invulnerable may be prohibitively expensive for a state that has had to spend so much of its limited resources simply to develop its nuclear capability. Finally, since many of the potential nuclear powers, particularly those among underdeveloped countries, are confronted with problems of internal political instability, it would require a very strong leader to resist domestic pressures to use the ultimate weapon in a serious conflict situation.

Having military weapons does not automatically increase a state's security in some clear and unambiguous way. The most that can be said, particularly for the most destructive weapons, is that they present a mixed blessing. While some argue that states seek to acquire nuclear weapons for their prestige value within their region, there is insufficient evidence that such acquisition actually does increase prestige. It is unlikely that the small number of nuclear weapons that a new nuclear state is able to stockpile can provide much leverage, particularly in relation to the huge nuclear arsenals of states such as the United States and Russia. Britain discovered that to be the case at the time of its invasion of Suez in 1956, when it was forced to back down as a result of pressure from Moscow and Washington.

A number of negative implications face a state that chooses to develop weapons of mass destruction. To the extent that other powers remain unalterably opposed to a state's acquisition of nuclear or chemical weapons, they might use force to destroy facilities designed to build such a capability. The Soviet Union seriously proposed a joint U.S.-Soviet strike against China's nuclear assets during the 1960s; Israel went a step further in 1981 when it bombed Iraq's Osirak nuclear research reactor, claiming that the reactor would provide the basis for an Iraqi nuclear bomb. Short of military action, outside states might also punish the potential nuclear weapon state through economic and political means. Fellow alliance members may have an incentive to withdraw commitments from a newly nuclearized state, for a continuing relationship only increases the probability that an ally would be drawn into a nuclear war not of its choosing. It is quite possible that an existing nuclear power might even go further by offering to protect, by nuclear means if necessary, any nonnuclear weapon state threatened by the newly acquired nuclear weapons.

Nuclear weapons may be of limited value as an instrument of influence for great powers as well. Soviet adventurism abroad was not clearly related to its rise as a nuclear superpower, but rather occurred earlier. The Soviet Union engaged in considerable risk-taking behavior during the Stalin period, despite the fact that the United States enjoyed an atomic monopoly until 1949 and considerable superiority for many years thereafter. Stalin sought to deny the efficacy of nuclear weapons by urging that such factors as morale, command, and the quality and quantity of conventional forces were more important. In many respects, Nikita Khrushchev appeared to behave more erratically and to be more willing to take risks than Leonid Brezhnev; yet the balance of forces was far more favorable to the Soviet Union during Brezhnev's tenure than when Khrushchev was in power.

## CONVENTIONAL ARMS SPREAD

With the end of the cold war, the threat of nuclear war between the superpowers has been replaced with concern about the spread of conventional armaments to other regions of the world, much of which occurred as a consequence of the cold war itself. As each bloc sought to aid its allies and to keep neutral states out of the camp of the opposition, global traffic in arms among governments reached as high as $63 billion in 1987.

The years following the end of the cold war have seen some decline in the global arms trade; arms deliveries from major suppliers were estimated to have fallen 55 percent between 1987 and 1991.[12] Another study showed major weapons exports for the United States to have declined by 42 percent over the period from 1983 to 1992 in real terms, with comparable figures of 86 percent for the USSR/Russia and 72 percent for France.[13] Increasingly, the predominant arms exporting states of Russia and the United States find new competitors in countries like Brazil, India, and Taiwan that offer not only such equipment as tanks, but also surface to surface missiles.

The spread of certain types of conventional capabilities presents real problems for security. For example, sophisticated delivery systems have been exported to all regions of the globe. Jet aircraft and missiles have found their way to rogue states such as North Korea, Iran, Libya, and earlier to Iraq. Recognizing the problem, several Western states in 1986 (joined later by the Soviet Union and China) formed the **Missile Technology Control Regime (MTCR)** to discourage the transfer of missile technology. The regime's efforts have not always been successful, as is evident

---

[12]Frederic S. Pearson, *The Global Spread of Arms: Political Economy of International Security* (Boulder, Colo.: Westview Press, 1994), 13.

[13]Kevin O'Prey, *The Arms Export Challenge* (Washington, D.C.: Brookings Institution, 1995), 5–6.

> ### Normative Dilemmas
>
> ## Dilemmas in the Use of Force
>
> The *security dilemma* refers to the situation in which one state increases its arms to add to its own security, but in the process only creates insecurity on the part of the adversary, who responds in kind by increasing its armaments. Both are made less secure.
>
> - What can be done to stop this vicious cycle?
>
> - Because states that possess nuclear weapons believe that they have a deterrent value, is there any moral justification for those states to demand that such weapons be denied to nonnuclear weapon states, who have just as much right to be concerned about their own security?
>
> - Is it morally appropriate to support a policy of mutual assured destruction in which one's own security is premised upon killing thousands, if not millions, of unarmed civilians?
>
> - Is it moral to provide protection for citizens from the dangers of nuclear blast and fallout, knowing that nuclear deterrence is more stable when one makes one's own citizens hostage to the bomb?
>
> - Would it have been moral during the period from 1992 to 1995 to supply arms to the Muslims in Bosnia-Herzegovina, who were widely viewed as victims of Serbian aggression, even though doing so might have exacerbated the violence and made the Muslims less compromising in negotiation? Or would the morally appropriate response have been to deny arms to all parties in the hope that the killing might end sooner, albeit with the victory of the stronger Serbian forces?

in the earlier transfer of Scud missiles to Iraq and subsequent transfers of missiles and missile technology to Pakistan and Iran.

How arms transfers can come back to haunt a state is illustrated by the U.S. transfer of Stinger missiles to rebel groups during the 1980s in an attempt to defeat the communist government of Afghanistan. Since these missiles, which can be carried by a single person and are capable of shooting down an airplane, were provided to Islamic fundamentalist groups in Afghanistan, there is concern that they might now be used in acts of terrorism. In an effort to forestall such an eventuality, the United States has attempted—with little success—to buy them back at three times the cost of production.[14]

Land mines are a particularly ominous threat resulting primarily from arms transfers. They are likely to remain a problem long after war (and the threat of war) subsides in a region. It was just such a land mine that caused the first American casualty in the NATO peacekeeping operation in Bosnia in December 1995. According to a recent State Department report, as many as 110 million antipersonnel land mines are scattered in 64 countries and continue to wound or kill 26,000 people annually. Although about 100,000 mines are cleared each year, officials estimate two to five million continue to be deployed.[15]

---

[14]William D. Hartung, *And Weapons for All* (New York: Harper Collins, 1994), 3.

[15]Sarah Walkling, "First CCW Review Conference Ends in Discord over Landmines," *Arms Control Today* (November 1995), 26.

## INTERNATIONAL TERRORISM

The terrorist bombing of the New York World Trade Center in February 1993 was a dramatic instance of a recurring threat to national and personal security. Terrorism has become a popular tool for certain ethnic, religious, and other politically revolutionary groups to force national authorities to pay more attention to them. Terrorists typically are frustrated by the lack of opportunity to negotiate their differences with a relevant authority peacefully; hence, part of the purpose of their terroristic acts is to call attention to their cause, hoping either to win the sympathy of foreign governments and peoples or, failing that, to frighten such governments into giving them what they want. Terrorism has also become an instrument used by weak states to get the attention of larger states, as illustrated by the state-sponsored terrorism of Cuba, Iran, Libya, North Korea, and Syria.

**Terrorism** has been defined in a variety of ways, but a Chinese philosopher more than 2,000 years ago captured its essence with this description: "Kill one, frighten 10,000."[16] Paul Wilkinson, a contemporary British authority on terrorism, defines it as the "systematic use of murder, injury, and destruction, or threat of same, to create a climate of terror, to publicize a cause, and to intimidate a wider target into conceding to the terrorists' aims."[17]

Terrorist acts have been known to immobilize nations, causing an inordinate preoccupation with a single issue, as in the case of the 444-day Iranian hostage situation during the Carter administration. They have caused economic harm to an entire business sector, as happened to the travel industry when many people, especially Americans, feared an outbreak of terrorism on the heels of the Persian Gulf crisis. And they have caused corporations to abandon operations in countries where the risk of terrorism has been perceived as unacceptably high.

**HISTORY OF TERRORISM** Terrorism is nothing new in international history. Since Greek and Roman times, such actions usually have been executed by groups unable to draw attention to their interests in any other way. An early example involved the Zealots, Jewish patriots opposed to Roman rule in Palestine, who used such means as assassinations and guerrilla attacks on Roman personnel and installations to try to advance their cause.

In addition to the use of terror by groups seeking power and attention, states and armies sometimes engage in terroristic acts. For example, General Sherman's march through Georgia entailed a deliberate strategy of terror, designed to demoralize the South during the American Civil War.[18] Similarly, the revolutionary government in France instituted a reign of terror from 1792 to 1794 during which 100,000 to 300,000 French citizens were detained, with many tortured and guillotined. That experience became a terrible model for the terrorism and resulting atrocities engaged in by authoritarian governments in the twentieth century.

---

[16]Richard Clutterbuck, *Kidnap, Hijack and Extortion* (New York: Macmillan, 1987), 6.

[17]Paul Wilkinson, *Terrorism and the Liberal State* (New York: New York University Press, 1986), 23–24.

[18]Russell F. Weigley, *The American Way of War* (Bloomington: Indiana University Press, 1973), 149.

**MISCONCEPTIONS ABOUT TERRORISM** Among the misconceptions many hold about terrorism is, first, that left-wing and revolutionary movements are primarily responsible for terroristic acts. The fact is that terrorists of the right, such as those who produced death squads in El Salvador during the 1980s, have been at least as commonplace throughout modern history. Second, some believe that terrorism only exists where people have genuine, legitimate grievances, and that if those grievances are removed, terror will cease. But the unfortunate fact is that no society is without those who are disaffected and alienated; yet, many of these individuals never become terrorists. Similarly, some believe that terrorists come primarily from among the downtrodden of society, who presumably have the least to lose, whereas studies indicate that they are recruited primarily from the upper middle classes.

However, as any discussion of this subject must note, because of differences in political goals and values, what one person calls a *terrorist* is frequently referred to by another as a **freedom fighter.** If one is sympathetic to their cause, one may be inclined to romanticize political terrorists, viewing them as modern-day Robin Hoods, more intelligent and less cruel than ordinary people. The most objective evidence suggests that one should generally be skeptical of such assumptions; a willingness to engage in terror reflects, at a minimum, such single-minded zeal for a cause that no room is left for toleration or any questioning of what its perpetrator sees as the overriding good for society.[19]

Spectacular events—such as the 1989 bombing of Pan Am Flight 103, which exploded over Lockerbie, Scotland, and killed all 270 persons on board—create the impression that terrorism is a major form of violence with which the world must cope today. Although terrorist incidents reached new highs in the 1980s, the number of fatalities as a result of international terrorist activities remains quite insignificant when compared with other forms of violent death, whether from such natural causes as earthquakes and typhoons or from criminal activities of the kind that plague American cities. In 7,343 international terrorist attacks recorded between 1975 and 1987, for example, only 416 U.S. citizens were killed, and most of these occurred in a few bloody instances, such as the 1983 bombing of the U.S. Marine headquarters in Beirut, Lebanon.[20]

Some believe that domestic terrorism is actually more of a threat than international terrorism, particularly for the United States. After all, the bombing of the federal center in Oklahoma City in April 1995, in which 168 lives were lost, involved action by domestic dissidents. The trend in terms of international terrorist events in recent years, on the other hand, has actually been downward, as illustrated in Figure 7.1.

As we approach the twenty-first century, certain trends in terrorist activities should still give us concern, particularly if these trends continue through the 1990s. Among these is the growth in state-sponsored terrorism. According to 1995 reports of the U.S. State Department, the main culprits include Iran, Iraq, Libya, Syria, Sudan, North Korea, and Cuba. The most flagrant of these seven states is Iran, which

---

[19] Walter Laquer, *The Age of Terrorism*, rev. ed. (Boston: Little, Brown, 1987), 5–8.

[20] Cecilia Albin, "The Politics of Terrorism: A Contemporary Survey," in *The Politics of Terrorism*, ed. Barry Rubin (Baltimore, Md.: Johns Hopkins University Press, 1988), 183–184.

**FIGURE 7.1** International Terrorist Incidents, 1975–1994

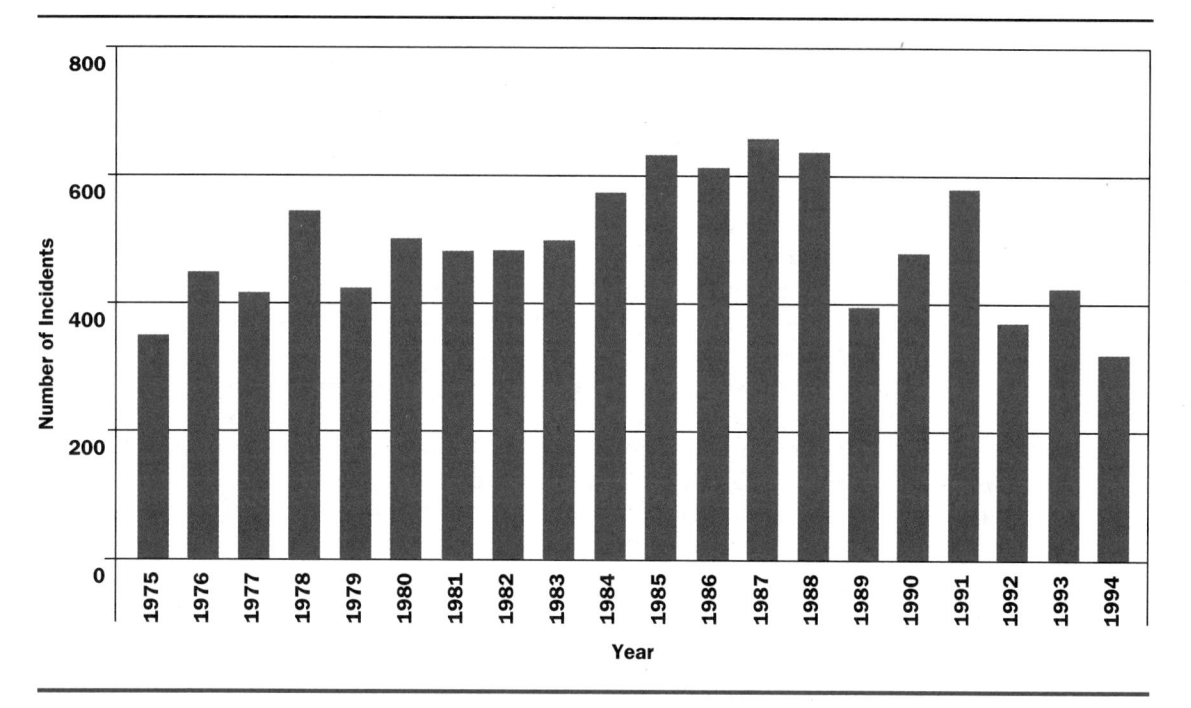

SOURCE: Department of State, *Patterns of Global Terrorism, 1994* (Washington, D.C.: Government Printing Office, 1995).

intelligence officials estimate provides up to $100 million a year to terrorists in a campaign to attack its enemies and derail Mideast peace talks.[21]

While only 1.4 percent of all international terrorist attacks between 1976 and 1983 were believed to have been state-sponsored, the figure increased to 15.3 percent between 1984 and 1987. Moreover, such attacks have been targeted increasingly against people rather than simply property. Only about 20 percent of all terrorists attacks were directed against persons in the early 1970s, but by the 1980s the figure had increased to 50 percent.[22]

It is also a matter of concern that the growing sophistication of weapons may aid the terrorist. Small amounts of explosives, which can be fitted into false bottoms of suitcases, can bring down huge aircraft. Terrorists now often have access to mortars and rockets. Even more threatening is the danger that terrorist groups might one day obtain chemical, biological, or nuclear weapons—a threat that became real with the lethal March 20, 1995 nerve-gas attack on the Tokyo subway system, which left eight dead and 600 hospitalized.

**DEALING WITH TERRORISTS** An oft-repeated notion says that one should not negotiate with terrorists, on grounds that rewarding terrorism will only increase its inci-

---

[21] Tim Weiner, "U.S. Lists Threats of Terrorism, Mainly from Iran," *New York Times,* May 12, 1995, A3.
[22] Ibid.

dence. Yet, in fact, governments often ignore this prescription when they are intent upon obtaining the release of hostages. President Carter's director of the Central Intelligence Agency, Stansfield Turner, has noted that every American president from Lyndon Johnson to George Bush has had a hostage problem in which he has been willing to deal with the terrorists.[23] Despite his strong pronouncements to the contrary, President Reagan demonstrated just such a willingness in the Iran-contra affair. Six separate arms sales were made to Iran beginning in late 1985 in the hope that they would facilitate the release of U.S. hostages in Lebanon. The American history of such attempts begins with President George Washington, who paid the Barbary pirates of North Africa one million dollars for the release of hostages, plus an additional annual tribute to discourage them from further terrorist actions.

Nor is it simply American presidents who have engaged in such behavior. Even Israel, while publicly denying that it "negotiates" with terrorists, has jailed large numbers of Palestinians with the express purpose of subsequently trading them for Israeli citizens or to achieve other objectives. Thus Israel was in a position to release some 700 Shia prisoners in return for the safety of a hijacked TWA aircraft that had been flown to Beirut in June 1985. With direct encouragement from the United States, hundreds more were released by Israel in 1991 with the hope not only of obtaining information and the possible return of downed Israeli pilots, but also to facilitate the return of Western hostages in Lebanon.

In an effort to reduce terrorism, governments also engage in a variety of defensive moves to protect citizens from possible terrorist attack. These include elaborate security checks at airports and other public places known to be potential targets, efforts to infiltrate terrorist organizations to obtain advance warning and to identify would-be terrorists, and surveillance of suspected terrorists by using highly sophisticated technology. Each such defensive move, however, produces inconveniences to many innocent people while restricting their freedom.

Governments may engage in direct threats and retaliation against terrorists or the states that support them to encourage the reduction of terrorist attacks. Indeed, retaliation may have some impact upon the incidence of terrorism. For example, after Israeli countermeasures against Jordan between 1967 and 1970, Jordan's King Hussein took action to force the PLO to flee from Jordanian soil. The result was that the number of terrorist attacks from Jordan's territory decreased from 3,400 during that four-year period to a mere 20 such incidents in the fifteen subsequent years. The U.S. aerial attack upon Libya in 1986, which killed Mu`ammar al-Gadhafi's daughter and came close to killing him, may also have been a major factor in reducing the level of international terrorism that could be attributed to Libyan assistance. Even so, the problem remains that, as one government's support of such activity is cut off, terrorists may find support or sanctuary from another state. Lebanon, for example, became the major launching point for Palestinian terrorism after access to Jordan was denied.

Since terrorism is an international problem, efforts have been taken by the international community to control it in much the same way that piracy was made an international crime in the early nineteenth century. International efforts to criminal-

---

[23]Stansfield Turner, *Terrorism and Democracy* (Boston: Houghton Mifflin, 1991).

ize terrorism have always suffered because of the role terrorism has played in nationalist and sometimes even in revolutionary democratic movements. Nevertheless, we can count a few legal successes along the way, including the adoption of the *Convention for the Prevention and Suppression of Terrorism* during the League of Nations era, and the unanimous 1985 adoption of a General Assembly resolution that condemned all acts of terrorism as "criminal." Conventions on the topic have also been adopted by the Organization of American States and the European Union, although each contains a number of loopholes.

More success has been obtained in getting states to agree to documents designed specifically to protect diplomats, and to decry the taking of hostages and the hijacking of airplanes. The Iranian seizure of the U.S. embassy in 1979, for example, provided the impetus for the adoption by the General Assembly of the *Convention against the Taking of Hostages*. Concern about hijacking led to a series of multilateral conventions to address the problem: Tokyo (1963); the Hague (1970); and Montreal (1971). But perhaps the most effective antihijacking agreement has been a bilateral "Memorandum of Understanding on Hijacking of Aircraft and Vessels and Other Offenses" signed by the United States and Cuba in 1973, which led to the return of a number of hijackers to the United States for prosecution.

## ARMS CONTROL AND DISARMAMENT

If personal and national security could be guaranteed through the threat or use of force, the trillions of dollars spent on global armament might well be worth the cost. But the fact is that with increased armament the world has become less secure, which suggests the need for alternatives. One such alternative is disarmament, whereby existing arms are reduced or eliminated; another is to adopt arms control measures designed to control further weapons development.

### THE RELATIONSHIP BETWEEN ARMS CONTROL AND DETERRENCE

Although there are sound reasons, both strategically and ethically, for preferring a world with a greater emphasis upon controlling armaments, arms control and deterrence are not entirely antithetical. Indeed, each approach may bolster the other. Deterrence has been vital to arms control because once arms control agreements are reached, the weapons that remain on each side and the ability to produce more such arms serves as a deterrent to those who might consider violating the agreement. Even if the agreement involves reducing arms to the level necessary only for internal defense, as envisioned in proposals for general and complete disarmament, deterrent forces would still be required, perhaps in the form of a United Nations police presence.

Arms control can contribute to the stability of deterrence by *regulating or eliminating destabilizing weapons systems*. As far as nuclear weapons are concerned, the most destabilizing are those capable of a first strike and those that threaten to provide a highly reliable defense against the other side's deterrent capability. In the early 1980s, concern arose in the United States about the dangers of a disarming first strike by the Soviet Union when some defense analysts asserted that the large Soviet land-based missiles presented a serious threat to the American ICBM. Similarly, the multiple-warhead MX missile was viewed by the Soviet Union, as it was by some critics in the West, as having first-strike potential because of its great accuracy and heavy payload. If the primary mission of a missile is simply one of retaliation, such capabilities are not required.

Arms control agreements may be used to *slow down the modernization of weapons systems,* thereby enhancing the stability of deterrence. Throughout the cold war, the fear that the other side might develop the ultimate defensive or offensive weapon fueled an arms race and created pressures for a preventive strike. Arms agreements in that period, although not terribly far-reaching, sought to reduce fears of a scientific breakthrough by placing restrictions on the testing of new weapons systems and by imposing limits on the size and power of any replacement delivery systems.

Arms control can be an important *adjunct to military planning* since quantitative arms agreements can lend predictability to arms competition. When no limits have been agreed upon as to the numbers of particular kinds of weapons adversaries may possess, military planners are more inclined to base their decisions on worst-case projections of their adversary's force. So, numerical limits can help restrain arms races and decrease military costs.[24]

Finally, it may be argued that arms control negotiations are useful for deterrent stability even when agreement is not reached, for the debates themselves *serve as learning centers,* enabling each side to communicate its concerns on defense issues. Since either party, even in the post–cold war world, can still destroy the other, regardless of what defensive measures are taken, it is important that such concerns be taken seriously.

## THE HISTORY OF ARMS CONTROL

Despite the thousands of meetings that have been held on the subject of arms control and disarmament, the world remains heavily armed. Although the cold war has ended, tens of thousands of nuclear weapons remain, primarily in the hands of the United States and Russia, as do conventional arms and troops numbering into the millions in the arsenals of the former cold warriors. The 1990 Iraqi invasion of Kuwait served to underscore the fact that huge arsenals also remain in the hands of third world states. These, too, pose a continuing threat to global stability—a threat made only that much more ominous with growing stockpiles of chemical and biological weapons and the potential of nuclear spread.

The cold war will probably be remembered as an era in which the nuclear antagonists were steadily engaged in arms control discussion but with limited results, as Table 7.1 suggests. Indeed, it took over four decades before agreement was reached on dismantling *any* weapons system; that occurred in December 1987 when the United States and the Soviet Union signed the **Intermediate Nuclear Force Treaty** calling for the elimination of existing land-based missiles with a range of between 500 and 5,000 kilometers. Although the **Biological Convention** of 1972 required

---

[24]Lisa D. Shaw, "Nuclear Arms Limitation Policy," in *Conflict and Arms Control,* ed. Paul R. Viotti (Boulder, Colo.: Westview Press, 1991).

**TABLE 7.1** | Major Arms-Control Agreements

| TREATY | PROVISIONS | DATE AS OF 1/1/95 | PARTIES |
|---|---|---|---|
| Geneva Protocol | Bans the use of gas or bacteriological weapons | 1925 | 132 |
| Antarctic Treaty | Demilitarizes the continent | 1959 | 42 |
| Limited Test Ban | Bans nuclear tests in the atmosphere, in outer space, or under water | 1963 | 124 |
| Outer Space Treaty | Demilitarizes space and the moon and other celestial bodies | 1967 | 94 |
| Latin American Nuclear Free Zone | Bans nuclear weapons from Latin America | 1967 | 29 |
| Nuclear Nonproliferation Treaty | Prohibits acquisition of nuclear weapons by nonnuclear states | 1968 | 170 |
| SALT I ABM Treaty | Limits deployment of antiballistic missile systems to two sites | 1972 | 2 |
| Interim Offensive Weapons Agreement | Five-year freeze on number of ICBMs and SLBMs | 1972 | 2 |
| Biological Weapons Convention | Prohibits development, production, and stockpiling of biological and toxin weapons | 1972 | 133 |
| Threshold Test Ban Treaty | Prohibits underground tests above 150 kilotons | 1974 | 2 |
| Environmental Modification Treaty | Prohibits military or other hostile uses of environmental modification techniques | 1977 | 62 |
| SALT II Offensive Weapons Agreement | Limits numbers of missiles and bombers on each side to between 2,250 and 2,400 | 1979 | Not in force |
| Intermediate-Range Nuclear Forces Treaty (INF) | Prohibits production and deployment of missiles with a range of 300 to 3,400 miles | 1987 | 2 |
| Chemical Arms Treaty | Requires U.S.-Soviet stockpiles of chemical arms be cut to 5,000 tons each by 2002 and that production be halted | 1990 | Not in force |
| Conventional Armed Forces in Europe | Limits number of conventional arms for NATO and Warsaw Pact | 1990 | 30 |
| START I Treaty | Reduces the number of warheads from 11,600 to 8,600 for the United States and from 10,222 to 6,500 for the USSR | 1991 | 5 |
| START II Treaty | Reduces number of nuclear warheads for each side to 3,000 to 3,500 by 2003 | 1992 | Not in force |

SOURCE: Multilateral agreements are reported in Stockholm International Peace Research Institute, *SIPRI Yearbook, 1995* (New York: Oxford University Press, 1995).

the destruction of existing biological weapons, these weapons were not an important part of Soviet and American arsenals, given the difficulty of controlling such weapons in a war situation.

Attempts to ban the use of obnoxious weapons systems date back to ancient times, when proscriptions against the poisoning of water sources and the burning of forests were established. Shocked by the million or so who died as a result of chemical weapons during World War I, a number of nations agreed to prohibit their future use by signing the **Geneva Protocol** of 1925. The Protocol appeared successful in deterring the use of chemical weapons during World War II; even Hitler did not resort to their use at a time in which his empire was crumbling around him. In the 1980s, however, the agreement did not stop one of its signatories, Iraq, from using chemical weapons in the Iran-Iraq war or against its own Kurdish citizens. Nor did the Geneva Protocol prevent Saddam Hussein from threatening their use during the Persian Gulf War during 1990–1991. The follow-up chemical arms treaty signed by some 154 nations in 1992 goes much further than the Geneva Protocol, which lacked any provisions for enforcement. The new treaty requires the destruction of each nation's chemical weapons arsenal and production facilities under strict international inspection.

Efforts to restrict the testing of weapons systems have been premised on the notion that such restrictions would discourage weapons development, since few states are likely to engage in massive production of systems that they have not adequately tested. Despite years of negotiation, a comprehensive nuclear test ban has yet to be achieved and the world has had to settle for the 1963 **Limited Test Ban Treaty** that prohibited nuclear testing in the atmosphere, outer space, and under water. But since testing was still allowed underground, the number of nuclear tests actually increased after the treaty was signed.

Despite pressure from the overwhelming majority of states and massive protests against resumption of tests (a global outcry arose when the French and Chinese both engaged in nuclear testing in the fall of 1995), a comprehensive nuclear test ban eluded the world for decades. It seemed nearer after President Clinton agreed that the United States would support a ban on all but extremely low-level testing whose purpose would be to verify the reliability of existing stockpiles. France also announced that, upon completion of its 1995 series, it would support a comprehensive ban on nuclear testing.

During the cold war, little progress was made in attempts to place treaty restrictions upon the production and possession of weapons. An important exception was the **Nuclear Nonproliferation Treaty (NPT),** which came into effect in 1970 and was renewed on an indefinite basis in 1995 with a membership of over 175 nations. Despite this progress, shortcomings remain. Some nonnuclear weapon states regard the NPT as an unequal treaty since it prohibits them from producing or receiving nuclear weapons while allowing nuclear weapon states to produce as many nuclear weapons as they desire. Moreover, Iraq and North Korea, for example, made considerable progress in nuclear weapons development despite having signed the NPT. Such states as India, Israel, and Pakistan all have refused to join the NPT and have themselves become nuclear weapon states.

Because of the difficulties in accounting for their past production, most efforts at restricting nuclear weapons have been directed toward delivery systems. Agreements reached during the **Strategic Arms Limitation Talks (SALT)** in the 1970s were largely cosmetic, freezing weapons at their existing levels. These included the 1972 **Interim Offensive Weapons Agreement,** which froze Soviet and U.S. ICBMs and SLBMs for five years, and the 1979 **SALT II treaty,** which added bombers to the strategic freeze. Although SALT II was never ratified, the United States and the Soviet Union continued to adhere to its restraints until the Reagan administration chose to exceed the 2,250 strategic force limit in 1987.

While that decision represented a setback for arms control, 1987 also saw the signing of the **Intermediate Nuclear Force Agreement (INF),** which called for greater reductions on the part of the Soviet Union than for the United States. The INF treaty was significant in that it was the first actually to abolish an entire class of weapons. Yet, ironically, the United States had no weapons of this type when the negotiations began in 1982, then proceeded to build them only to destroy them later.

Progress on conventional arms reductions in Europe proved far more difficult throughout the cold war. NATO entered such talks with the Warsaw Pact in 1973, largely as a response to threatened U.S. Senate efforts to reduce troops unilaterally in Europe. It took the end of the cold war and several unilateral Soviet conventional arms reductions, along with the collapse of the Warsaw Pact, before success was finally achieved with the signing of the **Conventional Forces in Europe (CFE) Agreement** in November 1990. With this agreement, the Soviet Union renounced its huge superiority in conventional arms and was required to destroy tens of thousands of tanks and other conventional armaments; for NATO, however, whose conventional forces were not substantially above the treaty's ceilings, only a few thousand conventional armaments needed to be destroyed.

## ARMS CONTROL AFTER THE COLD WAR

By the end of the 1980s, the cold war had clearly begun to wind down. Mikhail Gorbachev had taken a number of impressive unilateral initiatives to improve relations with the West. Then came the collapse of the Eastern bloc. In August 1991, an abortive coup in the Soviet Union hastened the demise of the Soviet empire, thereby ending the rule of the Communist Party after three-quarters of a century. The breakup of the Soviet Union confronted the world with new kinds of problems, calling into question the central security assumptions that had underpinned the cold war. Suddenly, governments that had been mortal enemies for decades appeared to be friends, while the superpower status of one of those erstwhile enemies vanished almost overnight. The decades-long ideological and military relationship between the two superpowers had been radically altered. Since these changes had occurred so quickly, the nuclear arsenals and deterrent strategies of the United States and the successor states to the Soviet Union remained in place. Decision makers and their publics had to rethink their security needs and determine how to transform their capabilities to make them appropriate to the new era.

The most fundamental questions revolved about the very logic of nuclear deterrence that had been built upon a sort of gruesome logic—that is, that an international balance of terror might be maintained as long as each antagonist believed the other's threat to commit mutual suicide. Deterrence—or at least nuclear deterrence—requires enmity if it is to be credible enough to keep the peace. With the cold war consigned to the history books, the security of previous adversaries would have to be built upon some other premise than that of their potentially suicidal mutual hostility. Yet a new doctrine to govern the nuclear weapons policies of the United States and Russia has been slow to emerge.

Nonetheless, the end of the cold war did fairly quickly produce the most significant cutbacks of nuclear arsenals in history. Just weeks before the August 1991 coup attempt against President Gorbachev, he and President Bush signed the **Strategic Arms Reduction Treaty (START I).** The treaty called for deep cuts of long-range bombers, submarines, and missiles by some 50 percent for the Soviet Union and 35 percent for the United States over a seven-year period.

Before action could be taken on this treaty, President Bush and Russian President Boris Yeltsin, who had succeeded Gorbachev with the demise of the Soviet Union, met in December 1992 to sign the **START II Treaty,** which required more than a two-thirds reduction in strategic nuclear warheads by 2003 (three years sooner if the United States helps Russia to pay for dismantling the weapons). This would leave the United States with 3,500 strategic nuclear warheads and Russia with between 3,000 and 3,500. Most significantly, the treaty would eliminate all Russian heavy MIRV missiles as well as the American MX MIRV missile, both of which systems threaten the stability of deterrence with their first-strike features and tremendous destructive capacity (see Table 7.2).

The end of the cold war has produced a profound shift in our nuclear consciousness, inasmuch as the two greatest nuclear powers no longer have their nuclear missiles directly targeted at each other. Nevertheless, the threat of nuclear war remains with us for the foreseeable future. Even if we assume that START commitments are met or somewhat exceeded, the destructive capacity of the remaining weapons will represent levels of destruction surpassing those that existed at the beginning of the Strategic Arms Limitation Talks (SALT) in 1969. Nor is detente in East-West relations likely to reduce the danger that the remaining weapons might be used by accident or miscalculation. The electoral power shown by hard-line Russian nationalists and Communists, the arrest in 1994 of an American CIA agent on charges of spying for the Yeltsin regime, and the growing animosity resulting from the Russian use of force in Chechnya all serve to remind us that East-West relations are not likely to be trouble-free. In fact, in April 1995 Defense Minister Pavel Grachev announced that Russia would not be able to fulfill its commitment under the Conventional Forces in Europe Agreement, arguing that his government would require additional arms to deal with situations like that in Chechnya.

Nor did an end to the cold war solve the problem of nuclear weapons in the hands of other states, some of which may be more tempted to use them than the superpowers were during the cold war. It may well be that as the United States and the successor states of the Soviet Union reduce their nuclear weapons, other states may be encouraged to increase theirs to enhance their competitive positions in relation to the United States and, especially, the former Soviet republics.

The breakup of the Soviet Union contributes to the proliferation problem in yet other ways. Smuggled fissionable materials from the former Soviet Union have been intercepted in a number of instances, particularly in Germany; former Soviet

| TABLE 7.2 | U.S. and Russian Nuclear Warheads |

| STRATEGIC WARHEADS | NUMBER AS OF JANUARY 1, 1996* | | START II PROJECTED NUMBER (2003) | |
| --- | --- | --- | --- | --- |
| | U.S. | Russia† | U.S. | Russia |
| ICBM RVs | 2,407 | 3,695 | 450–500 | 505–1,005 |
| SLBM RVs | 3,872 | 2,520 | 1,680 | 1,696 |
| Bomber weapons | 1,926 | 572 | 1,320 | 798 |
| Total warheads | 8,205 | 6,787 | 3,450–3,500 | 2,999–3,499 |

*Warhead figures for 1996 are based on START I counting rules. This results in bombers having fewer warheads attributed to them than they actually carry.
†Although most nuclear warheads from other former Soviet republics have been destroyed or turned over to Russia, some 1,838 additional nuclear warheads remain accountable to Ukraine, Kazakhstan, and Belarus until all such delivery systems are destroyed.
SOURCE: Adapted from Arms Control Association, *Fact Sheet* (Washington, D.C.: Arms Control Association, April 8, 1996). START II projected numbers are found in "Factfile," *Arms Control Today* 24, no. 10 (December 1994): 29.

nuclear scientists have been hired by such aspiring nuclear weapon states as Libya; and, squeezed by economic pressures, Russia agreed in 1995 to provide the state of Iran with several nuclear reactors, which will place Iran in a better position to meet its aspiration of becoming a nuclear weapon state.

The problem of preventing a determined government from acquiring nuclear weapons was underscored following the end of the Persian Gulf War. A condition of Iraq's cease-fire agreement with the United Nations was that Iraq cease all activities designed to give it a nuclear capability on its territory. Monitoring of such activities was provided by agents of the Security Council and the International Atomic Energy Agency, which in effect established an international police operation to ensure that Iraq complied with this commitment. Repeatedly, in the months that followed, the U.N.'s monitoring activities produced charges of Iraq's noncompliance, and of its continuing effort to proceed with its nuclear plans. These episodes suggest how much more difficult it may be to prevent nuclear proliferation where the international community has *not* imposed a directive, as it had on Iraq, requiring the elimination of the country's nuclear potential.

# CONCLUSION

Military force remains an important instrument of power in international politics, albeit perhaps not to the extent that realists would have us believe. Historically, military force has been used by groups for a variety of purposes, which include providing internal security; expanding territory; gaining independence; weakening, punishing, or balancing adversaries; denying gains to others; or as a way to gain bargaining leverage. In the age of mass destruction, increased attention has been given to deterrence by threatening would-be attackers with force.

Deterrence and arms control are both viewed as useful approaches to a more peaceful world. Some believe that the existence of a massive deterrent capability has provided post–World War II Europe with one of its longest periods of peace. Others, however, question this causal connection and believe that the stability of deterrence, whether conventional or nuclear, is delicate at best.

To the extent that the world continues to rely upon the threat of nuclear and conventional deterrence, attention must be given to ways to make the deterrent system safer. These include measures to improve command and control functions as well as communication systems. Deterrent stability also depends upon the creation of less vulnerable weapons systems to discourage first strikes and the control of those weapons that are especially destabilizing. Strategic defense systems, rather than contributing to deterrent stability, might have exactly the opposite effect.

Although the end of the cold war provides some hope for arms control and disarmament as approaches to peace, the record to date is hardly impressive. The United States and Russia still have nuclear overkill capabilities and additional states threaten to join the nuclear club. Even though agreement on unprecedented cuts in nuclear stockpiles has been achieved, there is every reason to expect that reduced—but still highly destructive—arsenals will be maintained for their presumed deterrent effect. Huge stockpiles of conventional weapons exist in all parts of the world, little affected by efforts to control them. As a result, it appears that the world will continue to rely for some time on military threat systems for its presumed security.

# CHAPTER 8

# Explaining Violence

As has been the case throughout history, the threat of violence committed by one person or group against another continues to challenge our security. That threat has become increasingly serious, given the modern weapons technologies (discussed in Chapter 7) designed to kill thousands, if not millions, of human beings in a single attack. In this chapter our concern will be to examine trends in the use of violence in both interstate and intrastate warfare, followed by explanations of why violence continues to play such a prominent role in human affairs.

## HISTORY OF GLOBAL VIOLENCE

### HISTORICAL TRENDS

Our capability for instant global communication makes people aware as never before of intergroup violence throughout the world today. But it is not entirely clear whether the incidence of violence in the international system is increasing or decreasing. The answer varies depending upon what periods and which states are being compared. In general, the statistical evidence suggests that individual nations are much less likely to engage in war at present than they were in earlier centuries. According to one study, the odds of a country's becoming involved in an international war in any given year declined steadily from 1 in 43 in the mid-1800s, to 1 in 62 around the turn of the century, and to only 1 in 166 since World War II.[1]

---

[1] Melvin Small and J. David Singer, "Conflict in the International System, 1816–1977," in *Challenges to America: United States Foreign Policy in the 1980s*, ed. Charles W. Kegley Jr. and Patrick J. McGowan (Beverly Hills, Calif.: Sage Publications, 1979), 89–116.

Yet those figures may be misleading. Because of the greater number of sovereign states in the world in recent times compared to one hundred or more years ago, warfare may still be prevalent even when most states in the world are not involved. One study, for example, found an average of between thirty-five and forty major ongoing wars annually during the period from 1986 to 1991, despite the winding down of the cold war.[2] The rise in the number of sovereign states and the shrinking of the time required to send weapons to distant lands have also meant that the number of states involved in the average war has increased considerably over the centuries, "from two or three in the fifteenth and sixteenth centuries to six or ten in the eighteenth century, to 30 in the First World War and over 40 in the Second. . . ."[3]

What is clear is that the amount of death and destruction from interstate war has increased dramatically as war has become more globalized and as the killing capabilities of our weapons have increased astronomically. Data collected for 118 international wars during the 165-year period from 1815 to 1980 revealed combat fatalities totaling some 31 million. Almost half, or 15 million, occurred during the six years of World War II. World War I accounted for 9 million casualties, Korea for 2 million, and Vietnam 1.2 million.[4] Together, these twentieth century wars accounted for more than eight times as many casualties as occurred in the century from 1815 to 1914.

Another recent pattern in violence more reminiscent of much of the nineteenth century has been the predominance of deaths caused by intrastate war rather than interstate war (see Table 8.1). Since 1975, for example, intrastate battle deaths have surpassed those of interstate deaths by more than 400 percent.

More disturbing for the security and dignity of humankind, the number of civilians killed in war also appears to be on the increase. During World War I, 8.4 million soldiers and 1.4 million civilians were killed. But in World War II the proportions were the reverse, with 16.9 million troop and 34.3 million civilian fatalities.[5] Even the almost surgically "clean" Gulf War of 1991—clean in the sense that the U.S. coalition's "smart" bombs produced very few direct civilian casualties—fought over Iraq's cities, nonetheless had a devastating longer term impact on the lives of innocent citizens of Iraq. It is estimated that about 70,000 Iraqi civilians died in the months after the war's conclusion because of the fact that U.S. bombs had destroyed the nation's power grid; that action crippled public health by knocking out water purification and distribution stations, sewage treatment facilities, health-care systems, and refrigeration. Other wars after the close of the cold war era—in the Caucasus, the former Yugoslavia, Rwanda, Somalia, the Sudan, and elsewhere—all claimed a disproportionate amount of suffering and death among displaced civilian populations, as well.

---

[2] Peter Walensteen, "Conflict Resolution after the Cold War: Five Implications," in *Resolving Intra-National Conflicts: A Strengthened Role for Intergovernmental Organizations* (Atlanta, Ga.: The Carter Center of Emory University, 1993), 34. A major war is defined as one involving at least 1,000 battle deaths.

[3] Evan Luard, *The Blunted Sword* (New York: New Amsterdam, 1988), 57.

[4] Melvin Small and J. David Singer, *Resort to Arms* (Beverly Hills, Calif.: Sage Publications, 1982).

[5] Frances A. Beer, *Peace Against War* (San Francisco, Calif.: W. H. Freeman, 1981), 37–38.

**TABLE 8.1** Battle Deaths in War, 1815–1994

| YEARS | INTERSTATE DEATHS | EXTRA-SYSTEMIC DEATHS* | INTRASTATE DEATHS |
|---|---|---|---|
| 1815–1824 | 1,000 | 38,500 | 141,000 |
| 1825–1849 | 159,100 | 557,000 | 562,150 |
| 1850–1874 | 862,100 | 338,575 | 2,922,025 |
| 1875–1899 | 346,332 | 554,424 | 182,400 |
| 1900–1924 | 9,069,303 | 334,576 | 920,925 |
| 1925–1949 | 16,515,552 | 645,849 | 2,683,135 |
| 1950–1974 | 3,185,755 | 280,000 | 3,143,213 |
| 1975–1994 | 1,395,369 | 10,000 | 5,926,732 |
| Totals | 31,534,511 | 2,758,924 | 16,481,580 |

*Extra-systemic deaths refer to casualties resulting from war with and between political entities that are not in the Westphalian system, as in the case of colonial wars.
**SOURCE:** Courtesy of Professors J. David Singer and Meredith Sarkees, Correlates of War Project, University of Michigan, April 1996 data base.

Despite such depressing statistics, the world is not necessarily becoming a more violent place. Small and Singer conclude that "whether we look at the number of wars, their severity, or their magnitude, there is no significant trend upward or downwards over the past 165 years."[6] Compared with earlier centuries, most modern wars end before massive destruction occurs. The Seven Years' War (1756–1763) engaged in by Frederick the Great, for example, resulted in the loss of one-ninth of the Prussian population.[7] Most estimates conclude that the Thirty Years' War (1618–1648), which provided the impetus for the Westphalian system, ended with the loss of one-third of Germany's population. A much higher figure has been cited by the French historian, Jean Perre, who suggested that Germany's population declined as a result of that war from thirteen million to four million, or by a total of 69 percent.[8]

In contrast, modern democratic societies have had no history of interstate warfare with one another. That is what is most hopeful about the apparent advances of democratic government at the end of the cold war. Much of the world seems to be moving farther from the threat of large-scale interstate violence. Global wars like the two that occurred in the first half of the twentieth century appear much less likely today.

Nonetheless, small-scale and regional and civil wars continue to threaten the world (see Table 8.2). That was made dramatically clear in the Iraqi invasion of

---

[6]Small and Singer, *Resort to Arms*, 141.

[7]Evan Luard, *War in International Society* (New Haven, Conn.: Yale University Press, 1987), 51.

[8]Kalevi J. Holsti, *Peace and War: Armed Conflicts and International Order, 1648–1989* (New York: Cambridge University Press, 1991), 28. One should use caution in interpreting statistics of war. Some relate only to battle deaths, others to related deaths from war—including disease and famine, which often accompany such violence. Most studies making comparisons between states and times try to be consistent in such measurements.

### TABLE 8.2  Armed Conflicts by Type and Region, 1945–1995

| TYPE | WESTERN EUROPE | SOVIET BLOC | AFRICA | MIDDLE EAST | ASIA | LATIN AMERICA | TOTAL |
|---|---|---|---|---|---|---|---|
| State versus state/intervention | 0 | 3 | 21 | 12 | 15 | 7 | 58 |
| Internal resistance/secession | 1 | 8 | 27 | 9 | 17 | 1 | 63 |
| Internal ideological | 1 | 4 | 14 | 9 | 18 | 20 | 66 |
| Total armed conflict | 2 | 15 | 62 | 30 | 50 | 28 | 187 |
| Total internal | 2 | 12 | 41 | 18 | 35 | 21 | 129 |
| Percentage internal | 100% | 80% | 66% | 60% | 70% | 75% | 69% |

SOURCE: Adapted from K. J. Holsti, "War, Peace, and the State of the State," *International Political Science Review* 16 (October 1995): 321.

Kuwait in the summer of 1990 and in the less publicized border war between Peru and Ecuador in the spring of 1995. Wars between Arab states and Israel, between India and Pakistan, and perhaps among neighboring states elsewhere, remain a real possibility. If the pattern of the past half century holds, these threats are greatest in the southern hemisphere.

Nor can cold wars and periods of tension between powerful states be viewed simply as historical curiosities. Growing hostilities between the United States and Russia over such issues as nuclear sales to Iran and the Russian invasion of Chechnya remind us that it is quite possible to return to periods of intense military rivalry.

## CIVIL VIOLENCE

Although it may be that interstate war will be somewhat diminished in the future, the same cannot be said with respect to violence within states, which accounted for over two-thirds of the wars between 1945 and 1995. If anything, the trend toward internal violence is only likely to accelerate with the end of the cold war. Ironically, civil violence tends to be exacerbated with the development of more political openness in multinational countries, as the cases of the former Soviet Union and Yugoslavia illustrate. Moreover, the heterogeneous makeup of many less developed countries, resulting from the artificially drawn imperial boundaries, is likely to remain a major source for conflict as well (see Table 8.3 later in this chapter for a listing of minorities in various parts of the world).

Violence between such groups has a tendency to spill over national boundaries, particularly if ethnic and religious minorities identifying with one of the disputing parties resides in a bordering country. At minimum, those who want to escape the violence are likely to seek refuge across borders. Ethnic and religious groups are also likely to pressure their own governments to help alleviate the problems encountered by members of their group in the state experiencing conflict. In 1993, the German parliament passed legislation restricting Germany's liberal immigration policies. That action was prompted by the social problems—including ultranationalist violence against foreigners—that had erupted in Germany as it sought to absorb huge numbers of persons displaced by fighting in the Balkans, and by the economic upheaval of other East European countries after the cold war.

Finally, violence in human society is not merely a function of war, defined as organized violence between sovereign (or would-be sovereign) groups. In fact, during the twentieth century, according to a 1989 estimate, at least three times as many lives have been lost as a result of slaughter and genocide undertaken by governments against their own citizens as by military activity.[9] The mass murder of civilians has been accomplished "almost never by heavy weapons but relying on hunger, exposure, barbed wire and forced labor to kill the bulk, executing the rest by shooting them with small arms, rolling over them in trucks (China circa 1950), gassing, or smashing skulls with wooden clubs as in Cambodia."[10] Perhaps the most tragic trend in the history of the nation-state has been the growth in the sovereign government's capacity and willingness to kill the very populations sovereign authority was designed to protect.

## ANALYSIS OF VIOLENT CONFLICT

To fathom why such violence occurs in the global system, we need only examine Iraq's invasion of Kuwait in 1990 and the Persian Gulf War that followed. It might be useful to group those explanations on the basis of the level of analysis: systemic, national and societal, and individual explanations.

From the *systemic level* perspective, Iraq had reason to be concerned about the balance of power, particularly given its many enemies within the region and the fact that the Soviet Union was no longer in a position to support Iraq militarily. As a result of such security concerns, Iraq has long been involved in arms races with its immediate neighbors, as well as with Israel. These races were not only costly, but they also tended to increase tension and fear within the region.

Those viewing the issue from the *national or societal level* blamed Saddam's aggression upon Iraq's deteriorating economy, a political structure made more vulnerable by an exhausting eight-year war with Iran, and declining oil prices. Kuwait, with its rich oil reserves, was a particularly attractive target for a country experiencing domestic difficulties. Others called attention to internal structural conditions in

---

[9]According to Paul Seabury and Angelo Codevilla, some 35 million (including 25 million civilians) were killed in interstate military operations during the century, but at least 100 million died at the hands of their own police forces [*War: Ends and Means,* (New York: Basic Books, 1989), 7]. Figures produced by another researcher, which included strategic bombing of innocent urban populations, increased the estimated number of nonbattle-related deaths in this century to 169 million [Rudolph J. Rummel, *Death by Government* (New Brunswick, N.J.: Transaction Books, 1994), 4].

[10]Paul Seabury and Angelo Codevilla, *War: Ends and Means,* 7.

Iraq, including the prominent role played by the military in influencing the decision to invade.

Some, like President George Bush, sought to personalize the conflict by blaming it directly upon Iraq's Saddam Hussein, whom Bush described as a madman, equating him with Hitler. Others who focused on the *individual level* to explain the conflict suggested that the initial invasion of Kuwait was the result of misperception and miscalculation by Saddam. According to some analysts, Saddam was influenced in part by signals the American ambassador to Iraq sent on the eve of the aggression; she suggested that the United States had no opinion on how Arab states should settle their conflicts among themselves.

These are sketches of explanations for a particular war from three levels of analysis. Now, recognizing the probable multicausality of all wars in the international system, we shall investigate some of the alleged causes in more detail.

# SYSTEMIC EXPLANATIONS

To illustrate how global and regional systemic conditions can influence the choice of war or peace, we shall explore two such causal explanations. The first concerns how power is organized within the system, particularly in terms of whether it is bipolarized or multipolarized. The second explanation addresses arms racing, which many see as leading directly or indirectly to war.

## POLARITY AND WAR

From the systemic perspective, the issues that have received the most attention have been the effect of the distribution of power upon how states behave and how this behavior in turn affects the stability of the system. Some experts have been highly impressed with the fact that the cold war period provided what some call the **long postwar peace,** during which the world saw the most extensive period without war in Europe since the time of the Roman empire.[11] This long peace was viewed as resulting from the **bipolarity** of the two superpowers and their respective alliances—both sides, given their highly destructive capabilities, feared that a direct attack could mean mutual suicide.

**BIPOLARITY AND WAR**  In arguing the utility of bipolarity over multipolar international systems, one scholar observed that where only two world powers exist, all other actors tend to divide as allies of one or the other.[12] Therefore, both sides will tend to involve themselves in happenings throughout the globe in order to deny any advantages to the adversary—which, in turn, will provide stability. Under such condi-

---

[11] Charles W. Kegley Jr., ed., *The Long Postwar Peace: Contending Explanations and Projections* (New York: Harper Collins, 1991).

[12] Kenneth N. Waltz, "The Stability of a Bipolar World," *Daedalus* (summer 1964): 881–909.

tions, each world power becomes concerned about even the smallest changes in the balance. In the process, leading states in a bipolar system tend to develop schemes for coping with recurrent crises and will have a particular interest in restraining crisis-provoking behavior on the part of their allies.

With preponderant power in a global bipolar system, minor shifts in the power balance will not be as decisive and consequently not as destabilizing as in a multipolar system. The American defeat in Vietnam and the Soviet failure in Afghanistan did little to change the essential balance, nor did those losses have the domino effect that many had predicted. Indeed, the power of each of these superpowers might even have been enhanced by their respective losses, since they meant an end to conflicts that had proved to be a serious drain upon the resources of the intervening states.

Finally, others have suggested that "the uncomplicated edifice of bipolarity makes it easier to establish the game rules that facilitate reaching international accords; hence war is less likely since these rules and accords help regularize behavior and institutionalize cooperation."[13] Examples abound in the U.S.–Soviet nuclear rivalry, during which the two adversaries sought to stabilize the relationship by adopting the hotline to facilitate communication, open skies to reassure each other, and non-interference in each other's national technical means of inspection.

Conversely, it might be argued that bipolarity may actually increase, rather than reduce, the motivation of the two most powerful rivals to expand their influence, owing to the concern of each to establish ever-wider buffer areas out of fear of the other polar power. During the height of the cold war, both the United States and the Soviet Union sought to establish as many client states as possible, and each often seemed unwilling to accept the other's sphere of influence. It is questionable whether real peace can be achieved when nations are involved in recurring crises, any one of which may slip over the threshold into violence.

It is probably not accurate to suggest that polarized powers will intervene if only minor changes in the balance are threatened. Superpowers may realize that they themselves enjoy such overwhelming force that it is unnecessary for them to be concerned about the effect of minor power shifts upon their own security. Instead of intervention, they may prefer retrenchment, particularly since—as was the case in the most recent period of bipolarity—intervention risked leading the superpower into a nuclear war simply for the defense of an ally.

**MULTIPOLARITY AND WAR** An international system organized around several different poles of power has been viewed by advocates of **multipolarity** as providing more foreign policy options for a state than a bipolar system. With several actors involved, it becomes possible for a state to rely upon shifting alliance patterns to realize its objectives, making the use of force less important in obtaining goals. Furthermore, multipolar participants are less likely to be suspicious of other actors in the system. No longer must relations be viewed solely as a zero sum game in which everything gained by one participant is viewed as a loss to the other. Similarly, in a multipolar system

---

[13]Charles W. Kegley Jr. and Gregory A. Raymond, *A Multipolar Peace?* (New York: St. Martin's Press, 1994), 49.

not every increase in armaments by an adversary need be seen as directed against oneself. Such an increase may, after all, be a response to the threat posed by other potential enemies of the arming state.[14]

Conflict behavior might be dampened somewhat within a multipolar system as a result of the fact that where many actors interact, the amount of attention each pays to any specific state or conflict situation will, of necessity, be diminished. Moreover, with many independent participants—instead of two entrenched adversaries in a bipolar world—it becomes more likely that a state or a group of states will be willing and able to serve in the role of a third party. A third party can help mediate differences between the major adversaries or perhaps even come to the military assistance of a state that becomes a victim of aggression.

Finally, the existence of several great powers in the system may encourage **risk-averse behavior** by states who are uncertain about the strength and composition of their opponents. Several different alignments would be possible in a multipolar system.

A **preponderance of power** in the hands of a status quo state or group of states may actually lessen the likelihood that other states will resort to the use of force. In a situation in which preponderant power exists, weaker states are less likely to fight, knowing that they cannot win, and the more powerful can often achieve their needs and interests without warfare. For example, the dominance of Rome after it had extended its empire provided a period of peace and stability in Europe almost unequaled in human history. Britain's preponderant power in the eighteenth and nineteenth centuries, which allowed it to dominate the seas, was in no small measure related to the relative peace that prevailed during those two centuries.[15] And the dominant position of the Soviet Union over the Soviet bloc yielded far less violence than we see today in the republics of the former USSR and in Eastern Europe. From this perspective it may not be an equal balance that one should seek for stability but rather a preponderance of a status quo power. In the post–cold war period, some no doubt see the United States as playing this kind of stabilizing role in the world.

Rather than viewing the power balance as static, A. F. K. Organski argues that what he calls a **power transition model** shows major wars to be most likely when power transitions are underway. When the dominant power feels that it may be surpassed, it has an incentive to strike a challenger preemptively, hoping to forestall the inevitable. It is also when the two are moving toward equality that the challenger, perhaps due to the overconfidence stemming from its rapid growth or the desire to speed up the process, attacks in an effort to achieve victory. In fact, the faster the rate of the power transition, the greater the probability of war, according to Organski.[16]

The empirical results from studies of the effects of polarity on war have been inconclusive, both because scholars have examined different periods and have used

---

[14]Karl W. Deutsch and J. David Singer, "Multipolar Power Systems and International Stability," *World Politics* 16 (April 1964): 390–406; and Kegley and Raymond, *A Multipolar Peace?*

[15]Such preponderance, however, did not prevent, but more likely facilitated, British imperial ambitions in Africa and Asia.

[16]A. F. K. Organski, *World Politics* (New York: Knopf, 1968); and A. F. K. Organski and Jacek Kugler, *The War Ledger* (Chicago: University of Chicago Press, 1980).

different methods of analysis.[17] Other factors, such as whether or not systemic power is concentrated among the satisfied or the dissatisfied states, may affect the findings. Thus, when power was concentrated among dissatisfied states, war was found to be more likely.[18] Kegley and Raymond also found that, in relating patterns of alignment to polarization, "Alliances can weaken the prospects for a multipolar peace if they combine capabilities in a highly polarized manner, or if they become so diffuse that it is difficult to ascertain who is in league with whom from one issue to the next. On the other hand, alliances can contribute to the chances of achieving peace in a multipolar world if they retain a modest amount of flexibility."[19]

## ARMS RACES

Some students of war believe that the reciprocal competition for arms, rather than the power balance itself, is a major explanation of why wars occur. That the military spending of one state is related to that of its adversaries is suggested by the work of the British statistician, Lewis F. Richardson, who collected data on defense spending during the nineteenth and twentieth centuries.[20] His findings indicate that military spending tended to be reciprocated, particularly when controlling for the levels of grievance (which led to increased military spending), and fatigue (which resulted in less spending).

Yet, efforts to apply Richardson-type arms race models to United States-Soviet military programs have generally been unproductive. Increases in the defense spending of one side did not seem to lead to increased spending on the part of the other. In fact, some research has shown that a state's own past armament behavior has been a better predictor than that of its adversary.[21] More recent research by Ido Oren found that in the cases of the U.S.–Soviet and Indo–Pakistani arms races, the adversaries often responded to arms increases on the part of the adversary by decreasing their own arms levels, and vice versa.[22] He explained this seeming anomaly, in part, by the fact that states worry more about what they perceive to be the intentions of the adversary rather than its capabilities. Pakistan, for example, failed to follow the Indian lead of lower defense spending, following the 1965 war, out of fear

---

[17] For summaries of these studies, see Greg Cashman, *What Causes War?* (New York: Lexington Books, 1993), 245–253; and Kegley and Raymond, *A Multipolar Peace?* 67–120.

[18] Richard J. Stoll and Michael Champion, "Capability Concentration, Alliance Bonding, and Conflict among the Major Powers," in *Polarity and War*, ed. Alan N. Sabrosky (Boulder, Colo.: Westview Press, 1985), 67–94.

[19] Kegley and Raymond, *A Multipolar Peace?* 119.

[20] Lewis F. Richardson, *Statistics of Deadly Quarrels* (New York: New York Times Books, 1960).

[21] W. Ladd Hollist, "An Analysis of Arms Processes in the United States and the Soviet Union," *International Studies Quarterly* 21 (September 1977): 503–528; and Charles W. Ostrom, "Evaluating Alternative Foreign Policy Decision-Making Models," *Journal of Conflict Resolution* 21 (June 1977): 235–266.

[22] Ido Oren, "The True Arms Race Puzzle Is Not Zero Reaction Coefficients but Negative Ones" (paper presented at the American Political Science Association Convention, Chicago, September 1992); and Ido Oren, "The Indo–Pakistan Arms Competition." *Journal of Conflict Resolution* 38 (June 1994): 185–214.

that India would become more bellicose if it were to develop nuclear weapons in the future.

To the extent that arms races do occur, the critical question is the extent to which they contribute to decisions to engage in armed conflict. A study of wars that have occurred since 1945 argues that arms races and hot wars have been largely independent of each other.[23] The Korean War, for example, did not begin with an arms race; if anything, it started during a period of disarmament as the United States sought to bring its troops home as rapidly as possible after World War II. That arms races do not always end in war is also suggested by the long naval race between England and France (1840–1904), during which the expected violence did not occur. Despite one of the most heated arms races in history in the decades following World War II, that period brought a dramatic decline in the frequency with which military confrontations involving major powers escalated into war.[24]

Empirical studies have been divided on whether arms races necessarily lead to escalation and increased violence. One researcher found, for example, that of twenty-six great power military confrontations coded during the nineteenth and twentieth centuries that escalated to war, twenty-three were preceded by arms races. On the other hand, of seventy-two great power confrontations that did not result in war, only five were preceded by arms races.[25] Most other analysts, however, have found that less than a majority of arms race disputes have led to war, but in those instances where they have, the following conditions stand out: (a) a change in the balance of power between two rivals occurs as a result of the arms race; (b) a "revolutionary" power is winning the arms race, rather than a status quo power; and (c) the arms race is unable to attain an equilibrium.[26] In yet another study, it was discovered that historically those arms races involving large shifts in the military balance were likely to slip into war, particularly when the initiator was willing to take risks and the target sought to avoid danger at all costs.[27]

In a classic analysis, Samuel Huntington argued that the likelihood of war emanating from an arms race varies inversely with the length of time of its existence.[28] In effect, war is more likely in the early stage of an arms race, before peaceful patterns of action have been established. Huntington goes on to suggest that **quantitative arms races** are more likely to lead to war than **qualitative arms races** in which the emphasis is placed upon improving the capability rather than the numbers of weapons. Quantitative races are seen by Huntington as producing the greater inequalities, since a state slips further and further behind with little chance of catching up as the other side keeps increasing its number of armaments. Qualitative races

---

[23]John C. Lambelet, "Do Arms Races Lead to War?" *Journal of Peace Research* 12 (1975): 123–128.

[24]Melvin Small and J. David Singer, "Conflict in the International System, 1816–1977," in *Challenges to America: United States Foreign Policy in the 1980s*, ed. Charles W. Kegley, Jr. and Patrick J. McGowan (Beverly Hills, Calif.: Sage Publications, 1979), 89–116.

[25]Michael D. Wallace, "Early Warning Indicators from the Correlates of War Project," in *To Augur Well*, ed. J. David Singer (Beverly Hills, Calif.: Sage Publications, 1979), 17–36.

[26]Greg Cashman, *What Causes War?* 182.

[27]James D. Morrow, "A Twist of Truth: A Reexamination of the Effects of Arms Races on the Occurrence of War," *Journal of Conflict Resolution* 33 (September 1989): 500–529.

[28]Samuel Huntington, "Arms Races: Prerequisites and Results," *Public Policy* (1958): 41–86.

> ### Global Changes
>
> ## What Nations Fight About: Issues and International Conflict
>
> Based upon a careful analysis of 177 wars since the beginning of the Westphalian system in 1648, Professor Kalevi Holsti of the University of British Columbia reached the following conclusions about the underlying issues leading to the various wars:
>
>> Quarrels involving control, access to, and/or ownership of physical space figured in about one-half of all wars between Westphalia and the outbreak of the First World War, but since Napoleon's defeat there has been a gradual decline in the prominence of this issue, both as a percentage of all conflict-producing issues and as a source of war. (p. 307)
>>
>> The search for statehood has commanded the international agenda since the late eighteenth century, and in two of the periods (1815–1914 and since 1945) it has been more often associated with war and armed intervention than any other issue. (p. 311)
>>
>> While power and dynastic factors were involved [in the Thirty Years' War], contests over religious and derivative civil principles drove the war from the beginning. . . . The succeeding century and one-half was marked by conflicts over more concrete issues. . . . Only in the late eighteenth century did the consensus and uniformity begin to break down. . . . The importance of [ideological] issues has been particularly notable in the post-1945 period where, combined, they have been involved in 42 percent of the wars. (p. 313)
>>
>> There has been a general decline in the significance of economic issues as sources of international conflict. Between Westphalia and the Congress of Vienna they figured significantly in the etiology of wars. The nineteenth century, in contrast, saw the rapid diminution of armed contests over commercial and navigational issues. They revived during the 1930s thanks to Japanese and Italian imperialism, only to decline again after the Second World War. Contests over resources continue to appear and may become significant issues of the future. (pp. 316–317)
>>
>> The high frequency of wars in the nineteenth century in which [sympathy for ethnic and religious kin] played a role is partly accounted for by the persecution of Orthodox Christians and fellow Slavs in the Ottoman Empire and the frequent championing of the cause by Russia. There has been a decline of sympathy issues since the First World War, but they have been one source of conflict in more than one-fifth of the post-1945 wars. (pp. 317–318)
>
> As the quotations above indicate, the issues over which states fight have varied considerably over the centuries. As we move into the twenty-first century, which of those issues are more likely to be a source of conflict, and why?
>
> SOURCE: Kalevi J. Holsti, *Peace and War: Armed Conflicts and International Order, 1648–1989* (New York: Cambridge University Press, 1991). © Cambridge University Press, 1991. Reprinted with the permission of Cambridge University Press.

are seen as tending toward equality, for scientific advantages usually do not last very long in a highly technological world. Since quantitative arms races generally involve a greater financial burden and produce greater tension, powerful pressures are created to end the arms race through either war or disarmament. Unfortunately, war has been a more frequent outcome than disarmament.

## NATIONAL AND SOCIETAL EXPLANATIONS

Rather than viewing war as simply the result of the anarchic international system and competing power structures, others view the nature of the national units themselves as an important cause of war. In the following sections we will look at some of the national-level explanations of war, beginning with the impact of nationalism and ethnicity. Research suggesting that war is related to particular types of government and their stability will then be explored, followed by an analysis of the role that the so-called military-industrial complex plays in fostering conflict.

# NATIONALISM

Recent interethnic violence in the former Yugoslavia and the declaration of independent national republics in what was the Soviet Union remind us that nationalism continues as yet another societal explanation of conflict and war. The origin of nationalism is fairly recent, with some scholars dating its rise only to the time of the French Revolution. According to one scholar, "There is scant evidence to support the notion that nationalism in European history represented the aspirations of distinct ethnic or cultural groups to form their own states."[29] In fact, there were many instances of states crowning foreign aristocrats rather than one of their own. For example, the Belgian nationalist movement, after separating from the Dutch, offered the crown to the royal house of France and then to a German prince. Norway, after separating from Sweden in 1905, voted to have a Danish prince as its king. Nationalist movements in the nineteenth and early twentieth centuries often called not for nation-states but rather for the re-creation of European historical empires like those of Byzantium, Rome, and Charlemagne.

**POSITIVE AND NEGATIVE ASPECTS OF NATIONALISM** As a force in world politics, nationalism has had both positive and negative effects upon foreign relations. President Woodrow Wilson saw **national self-determination** as a way to peace. He reasoned that if people were given the opportunity to choose their own destiny, they would opt for democracy, which he viewed as more peace-loving than authoritarian regimes (a position for which there is some evidence, as we shall see). Nevertheless, it is clear that nationalism, particularly when aroused by an authoritarian government, often leads to war, for it results in chauvinistic exaggeration, **ethnocentrism,** and **xenophobia.** It was French nationalism that aided Napoleon in his effort to create an empire. Similarly, it was German nationalism that provided support to Hitler nearly one hundred and fifty years later. Competing nationalisms within a state have often led to civil war, and, in an effort to instill a sense of nationalism and national identity, leaders have sometimes engaged in external conflict behavior, as noted in our subsequent discussion of the relationship between domestic instability and war.

Nationalism may provide some stability to a region by unifying a given population under a national banner. But in order to be stabilizing, the national identity must be coterminous with the boundaries of the nation-state, and in many instances this has not been the case. According to one study that examined only politically active ethnic and religious groups, more than one-sixth of the global population (17.3 percent, or 915 million people) in 1990 belonged to some 233 minority groups sit-

---

[29]Sandra Halperin, "State Autonomy versus Nationalism: Historical Reconsideration of the Evolution of State Power," in *Transcending the State-Global Divide,* ed. Ronen Palan and Barry Gills (Boulder, Colo.: Lynne Rienner, 1993), 64.

**TABLE 8.3** Minorities at Risk of Conflict, 1990

| WORLD REGION | COUNTRIES WITH MINORITIES AT RISK | NUMBER OF MINORITY GROUPS AT RISK | TOTAL (THOUSANDS) | PERCENTAGE OF REGIONAL POPULATION |
| --- | --- | --- | --- | --- |
| Western democracies and Japan (21) | 15 | 24 | 84,023 | 10.8% |
| Eastern Europe and the USSR (9) | 5 | 32 | 153,658 | 35.0% |
| Asia (21) | 15 | 43 | 273,064 | 10.2% |
| North Africa and the Middle East (19) | 12 | 31 | 118,205 | 28.8% |
| Africa (south of the Sahara) | 29 | 74 | 237,023 | 42.3% |
| Latin America and the Caribbean (21) | 17 | 29 | 49,371 | 11.0% |
| Total (127) | 93 | 233 | 915,344 | 17.3% |

**SOURCE:** Ted Robert Gurr, *Minorities at Risk* (Washington: United States Institute of Peace Press, 1993), 11.

uated within larger communities.[30] As Table 8.3 shows, their numbers varied considerably from region to region; Africa leads with 42.3 percent of its population consisting of minorities.

Despite the recent achievement of independence for several national groups in the former Soviet Union and in Yugoslavia, a number of large minorities in the world still lack a state of their own; consequently, they remain a major source of conflict. Many of these minority groups can be found in India in addition to those in Africa and other parts of Asia. Several countries in western Europe and Canada similarly include national groups that from time to time have asserted their rights to independent status.

The problem of communal violence is unlikely to be resolved simply by providing each large minority with its own state, for remaining in those new states are often yet other minority groups. Some minorities, such as the Russians and Serbs who are found in large numbers in many of the new republics of the former Soviet Union and Yugoslavia, respectively, might well expect their more powerful ethnic brethren to intervene on their behalf whenever circumstances seem to warrant such intervention. We have seen that danger grow since the end of the cold war; historical rivalries between ethnically based states have reasserted themselves in ever more violent forms, no longer controlled by a superior force, as in the cases of the former Soviet and Yugoslavian states.

[30] Ted Robert Gurr, *Minorities at Risk* (Washington, D.C.: United States Institute of Peace Press, 1993), 315.

## DEMOCRACY AND WAR

Of greater normative interest is the question of whether one type of regime is more warlike than another. For a variety of reasons, authoritarian or single-party states have been viewed by many, particularly in the West, as more willing and able to initiate war than liberal, democratic ones. Authoritarian states appear better able to mobilize great military power; they are predisposed to unlimited action and to engage in war for economic reasons; and they are better able to exploit situations, for they can decide to go to war without the approval of the people.[31]

But other examples seem to suggest that liberal and democratic governments have frequently been less than peace loving and more bellicose in their behavior than others. Democratic Athens was said to have engaged in more foreign conquests than its authoritarian rival, Sparta, while the Japanese, despite the prevalence of militarism within their society, lived in isolation and consequently peace for many centuries. In more recent times, dictatorships such as Tito's Yugoslavia and Franco's Spain were highly peaceful in their external relations, whereas the liberal regimes of Britain and France pursued vigorous imperialist policies, particularly during the nineteenth century. But as the British scholar Evan Luard reminds us, the expansionism of Britain and France did not necessarily result from their growing democratic governments.[32] Rather, rising prosperity, commerce, and a sense of adventure may have induced both the development of democracy and the desire for expansion.

Among the many empirically based studies comparing the belligerent behavior of democratic and authoritarian states, Quincy Wright, in his monumental study of war covering the past six centuries, concluded that democracies are generally slower to move into war; yet once they have done so, they fight as vigorously or more vigorously than authoritarian regimes.[33] Perhaps a major factor in the hesitancy with which a democratic regime goes to war is related to concern that an opposition party will rally the public against the war, particularly if the war threatens to be a long, inconclusive one. In an extensive review of findings from a number of empirical studies covering both the nineteenth and twentieth centuries, Rudolph Rummel concluded that on balance libertarian (democratic) regimes have been less conflictful than authoritarian regimes.[34] Some have questioned his conclusions, arguing that the proposition does not hold for earlier periods or with respect to colonial and imperialistic wars.[35] Nonetheless, one trend is apparent, and that is that war *between* liberal democracies has been virtually nonexistent, particularly in the twentieth century.[36]

---

[31] Alastair Buchan, *War in Modern Society* (New York: Harper and Row, 1968), 21–24.

[32] Evan Luard, *Types of International Society* (New York: Free Press, 1976), 122.

[33] Quincy Wright, *A Study of War*, 2d ed. (Chicago: University of Chicago Press, 1965), 842.

[34] Rudolph J. Rummel, "Libertarianism and International Violence," *Journal of Conflict Resolution* 27 (March 1983): 27–71.

[35] Steve Chan, "Mirror, Mirror on the Wall . . . Are the Freer Countries More Pacific?" *Journal of Conflict Resolution* 28 (December 1984): 617–648.

[36] Among the many studies making this point are Bruce M. Russett, *Grasping the Democratic Peace* (Princeton, N.J.: Princeton University Press, 1993); and Rudolph J. Rummel, *Death by Government* (New Brunswick, N.J.: Transaction Publishers, 1994).

To the extent that democracies are associated with peace, there is considerable hope for the future. In 1800 only the United States and Switzerland could be classified as liberal democracies; a century later, some thirteen states met the criteria. But with the spread of liberal democracy from Eastern Europe to Central and Latin America and more recently to Africa, Freedom House in its annual January 1996 survey found a total of 117 states—or 61 percent of the world's 191 countries—to be formal democracies, up from 42 percent a decade earlier.[37]

Should we begin to regard war as "obsolete and subrationally unthinkable?" Is it likely to go the way of slavery and dueling in civilized societies as states democratize? Perhaps not yet. We must remember that a number of authoritarian regimes still remain in the world, and even those that have liberalized can return to a more authoritarian and aggressive posture. Aggressive behavior on the part of the former Yugoslavian states, most of which have held free elections in the 1990s, serves to remind us of the problem.

**WHY ARE DEMOCRACIES MORE PEACEFUL?** A democratic polity may assume a more peaceful orientation in foreign policy because democratic values tend to instill a belief in the importance of compromise as a way of solving problems. Research has shown, for example, that the more open or democratic political systems tend to utilize international courts and third party mediation more frequently than the more closed systems.[38]

Another explanation of why democracies may be less likely to go to war relates to the role played by a public that is unwilling to risk lives for a cause that its leadership perceives as appropriate. In a democracy, such views must be taken seriously or the leadership might suffer the electoral consequences.

Also increasingly removed from the repertoire of the democratic regimes are actions taken by their governments that are perceived as inhumane or immoral by the public. For example, there was a public outcry in the United States in response to the exposure of the Central Intelligence Agency's participation in assassination plots against foreign leaders—an outcry that became particularly vociferous after the 1973 death of President Salvador Allende of Chile. Or, when the U.N. Security Council agreed to an economic boycott of Iraq for its invasion of Kuwait in 1990, its members soon responded to public opinion by agreeing to make an exception of essential food imports to feed the civilian population.

## DOMESTIC INSTABILITY

Another extensively researched proposition concerning the causes of war holds that political elites engage in external conflict to divert attention from internal societal problems. The historian Geoffrey Blainey found that from 1815 to 1939, at least

---

[37]Adrian Karatnycky, "Democracy and Despotism: Bipolarism Renewed?" *Freedom Review* 27 (January/February 1996): 1.

[38]William D. Coplin and J. Martin Rochester, "The Permanent Court of International Justice, the International Court of Justice, the League of Nations, and the United Nations," *American Political Science Review* 66 (June 1972): 529–550; and William J. Dixon, "Democracy and the Management of International Conflict," *Journal of Conflict Resolution* 37 (March 1993): 42–68.

thirty-one wars—or just over half the conflicts occurring during that time—had been immediately preceded by serious disturbances within one of the fighting nations.[39] One writer even went so far as to assert that he doubted whether a totalitarian dictatorship could exist without taunting or attacking a foreign scapegoat.[40]

History is replete with illustrations in which leaders have been alleged to have taken aggressive action to divert attention from domestic problems. Louis Napoleon may have continued the Crimean War partly to deflect French attention from discontent at home. Russia perhaps used the Russo-Japanese war in 1904 for similar reasons. On the eve of the clash at Fort Sumter that began the American Civil War, incoming Secretary of State William Seward was reputed to have suggested to President Abraham Lincoln that the president provoke a war with France or Spain in order to hold the country together. In 1971, Prime Minister Ali Bhutto of Pakistan was accused of using the conflict with India to distract the Bengalis from their demands for independence; the Argentinean invasion of the Falklands in 1982 was said to have been similarly motivated.

During the 1960s, statistical research was conducted that examined the linkage between the domestic political instability and external conflict measures on a global basis to determine whether or not there was a positive correlation between the two, as the hypothesis would suggest.[41] Using such indicators as demonstrations, riots, and coups within states and such foreign conflict measures as protests, threats, and numbers killed, these studies found little relationship between domestic and international conflict behavior.

However, cross-national studies like those of Rudolph J. Rummel and Raymond Tanter in the 1960s—which uncovered minimal, if any, correlation between measures of internal instability and external conflict behavior—may have overlooked some important relationships that are lost because of the aggregation of data concerning the behavior of large numbers of states. Subsequent studies have shown that the predicted relationship does hold for certain kinds of states in specific circumstances, even though it may not translate to the world at large. A reanalysis of the very data collected by Rummel and Tanter found differences among states categorized as personalist (military-type authoritarian regimes), polyarchic (democratic), and centrist (communist type governments). Personalist states, located largely in Latin America, showed a statistically significant relationship between internal conflict indicators and diplomatic conflict behavior such as protests and threats, but the same was not true for more violent forms of external conflict. The findings also indicated that centrist (authoritarian) regimes were more likely than polyarchic (democratic) regimes to engage in external conflict during periods of domestic turmoil and revolution.[42]

---

[39]Geoffrey Blainey, *The Causes of War*, 2d ed. (New York: Free Press, 1986), 71.

[40]Wright, *A Study of War*, 272.

[41]Rudolph J. Rummel, "Dimensions of Conflict Behavior Within and Between Nations," *General Systems Yearbook* 8 (1963): 1–50; and Raymond Tanter, "Dimensions of Conflict Behavior Within and Between Nations, 1958–1960," *Journal of Conflict Resolution* 10 (March 1966): 41–64.

[42]Jonathan Wilkenfeld, "Domestic and Foreign Conflict," in *Conflict Behavior and Linkage Politics*, ed. Jonathan Wilkenfeld (New York: McKay, 1973), 107–123.

Kahil/Al-Sharq al-Awsat/Jidda, Saudi Arabia.

Studies also suggest that less developed countries are more likely than more highly developed ones to engage in external conflict in order to divert attention from internal problems. An examination of the behavior of eighty-four states from 1948 to 1962 discovered that less developed states were more likely than developed ones to show their frustrations in external aggression.[43] Given the general lack of economic resources, skills, and technology for them to draw upon, leaders of some developing countries may be particularly likely to be tempted by the short-term expedient of finding a scapegoat for their economic and political problems.

## THE MILITARY-INDUSTRIAL COMPLEX

In Chapter 11 we shall consider the arguments for and against the notion that capitalism—or capitalists—are responsible for imperialism and war. Here we look only at theories suggesting that responsibility falls on a particular segment of the business

---

[43]Ivo K. Feierabend and Rosalind L. Feierabend, "Level of Development and International Behavior," in *Foreign Policy and the Developing Nation*, ed. Richard Butwell (Lexington: University of Kentucky Press, 1969), 135–188.

world—that of munitions manufacturers who allegedly conspire with the military for a larger part of the total governmental budget.

A number of writers have asserted that although business as a whole may not favor external violence—and has much to gain from peace—weapons manufacturers have an economic interest in war or the threat of war. This view was especially popular in the period following World War I, when George Bernard Shaw wrote biting satirical plays, such as *Major Barbara*, condemning the arms manufacturers; the Nye Commission was established by the United States Senate to inquire into the profits of the arms industry; and Charles A. Beard expounded on his "devil theory" of war, accusing economic interests of driving the United States dangerously close to war.[44]

Concern about a **military-industrial complex** was also raised by the sociologist C. Wright Mills early in the cold war period in his treatise on the so-called power elite.[45] Mills's power elite included three groups of leaders (military, business, and political) whom he regarded as involved in a conspiratorial effort to keep military spending high. Collusion among these three groups was facilitated by the interchangeability of roles as military officers became government officials (Eisenhower and de Gaulle) and also assumed roles in the corporate hierarchy. During the 1980s— that is, thirty years after Mills's treatise—over 3,000 retired military officers of high rank went on to work for major U.S. defense contracting corporations, and close to 70 percent of every federal dollar allotted to research and development went to military research and development.[46] An earlier estimate suggested that 50 to 60 percent of military retirees who were employed in the United States had positions that were directly dependent on the defense and aerospace industries.[47]

That economic considerations tend to influence decisions related to military procurement is suggested by research on contracting procedures used in the American aerospace industry during the 1960s. Aerospace contracts tended to be awarded either to contractors who were in deep financial trouble, as in the case of Lockheed, or to contractors who were just completing a previous contract. The primary motivation appeared to be keeping the lines of production open rather than any rational calculation of need, or even of cost-effectiveness.[48]

Further evidence of the influence of a military-industrial complex in the United States might be inferred from the extensive cost overruns of most weapons systems. Some claim that military and industrial interests conspire to negotiate military procurement contracts with low bids, only to have the costs rise sharply once Congress, the bureaucracy, and the American public are committed to the program. Weapons systems have also been developed to a level of sophistication far beyond what is necessary to fulfill their mission, and are used to replace weapons that are far from obsolete. Given the destructive capability of nuclear weapons, for example, does it

---

[44]Charles A. Beard, *The Devil Theory of War* (New York: Vanguard Press, 1936).

[45]C. Wright Mills, *The Power Elite* (New York: Oxford University Press, 1956).

[46]Frederic S. Pearson and J. Martin Rochester, *International Relations,* 3d ed. (New York: McGraw-Hill, 1992), 288.

[47]Albert Biderman, "Retired Soldiers Within and Without the Military-Industrial Complex," in *The Military-Industrial Complex,* ed. Sam C. Sarkesian (Thousand Oaks, Calif.: Sage Publications, 1972), 113.

[48]James R. Kurth, "Why We Buy the Weapons We Do," *Foreign Policy* 11 (1973): 33–56.

> **Normative Dilemmas**
>
> ## Moral Considerations in the Use of Violence
>
> - Was it morally proper for the United States to have dropped atomic bombs on Japan in the hope of ending the war more quickly and thus avoid the costly casualties that armed invasion would inevitably have produced for American troops?
>
> - Would it have been more moral to have assassinated Saddam Hussein, assuming that his replacement would have been far more moderate, than to have wrought the tremendous death and destruction that the Persian Gulf War produced?
>
> - If, as the evidence seems to suggest, states become more cohesive by scapegoating other governments for their internal problems, what is wrong with engaging in such activity if it can be done at minimal cost, and internal peace and unity is thus provided for one's own state?
>
> - War may be said to play a positive role in creating new states when previously suppressed peoples demand and properly receive a state of their own. Similarly, war may sometimes decrease the number of conflicting sovereigns, replacing them with a more powerful sovereign able to provide security and stability to an entire region. Do either of these results justify the resort to violence?

really make much difference whether the accuracy of a nuclear missile is improved by one hundred feet? Billions of dollars have been spent for just such purposes.

The power of military-industrial interests, at least in the United States, is also illustrated by the fact that military expenditures have been reduced only marginally with the ending of the cold war. According to a former Reagan administration Department of Defense official, ". . . the United States will pay $15 billion more for defense this year [1995], in inflation-adjusted dollars, than it did in 1980 at the height of the Cold War."[49]

As the East-West confrontation has subsided, military-industrial interests in the United States and elsewhere have become increasingly concerned about selling military weapons abroad. Although such sales threaten to exacerbate global conflict, the matter is not that simple, for two reasons. First, in what has been referred to as the **dove's dilemma,** efforts to restrict the arms trade may only increase the incentive for potential recipients to develop their own military production facilities, as has been the case for states like Argentina, Brazil, Chile, India, and South Africa. Such states are then subject to continuing pressure from their own newly created industrial-military complex to expand arms production and to become arms exporters themselves. Under such conditions the supply of global armaments would increase substantially beyond what would have been the case had arms shipments been allowed. Second, failure to bolster the military capability of the weaker state in a regional conflict through arms trade may only encourage aggression on the part of its enemy. Those supporting the lifting of the arms embargo against the Bosnian Muslims before the peace accord late in 1995 did so on grounds that it might serve to deter continued Serbian aggression and restore a balance of power in the region.

---

[49]Lawrence J. Korb, "An Overstuffed Armed Forces," *Foreign Affairs* 74 (November/December 1995): 23.

# Individual Explanations

Ultimately, of course, it is human beings who must make the decision to go to war; but why do they do it, particularly given the risks and the inhumanity of it all? Some would say aggression is innate for human beings, just as it is seen as an inherent characteristic for all animals. Others object to that notion, suggesting that aggression is learned behavior. Aggression is also seen as rooted in psychological explanations, involving such factors as misperception and fear.

## HUMAN AGGRESSION

Some authorities believe that human behavior can best be explained by biological traits that predispose human beings to behave in certain ways. Of most relevance to international relations is the assertion that humans are inherently aggressive. Many who hold such a view about human aggressiveness have extrapolated from their studies of the animal world, where they have seen aggression as an innate drive, about as basic as those for hunger and sex.[50]

**VIOLENCE AS AN INNATE RESPONSE**  The view that violence is caused by **innate aggression** inherent in the biological makeup of human beings has been challenged by various experts on the grounds that unjustified analogies are being made from discrete bits of animal and human behavior to describe all human behavior.[51] It is argued that even if such biological theories of aggression are accurate, they fail to account for the considerable variation in aggression to be found among individuals and groups.

By probing people's unconscious motivations, Sigmund Freud came to believe that human aggressiveness was an inherent biological trait.[52] In addition to life forces, Freud saw a set of drives whose primary function is the destruction of the individual—for individuals, in his view, are driven to try to return to an inanimate state. He saw these drives, which he called death instincts, as causing overt aggressiveness against self and others. Freud even went so far as to assert that the atrocities of war are more "natural" than the civilized behavior of peace.

**THE FRUSTRATION-AGGRESSION HYPOTHESIS**  The **frustration-aggression hypothesis,** first presented in 1939 by Yale researcher John Dollard and associates, illustrates yet another instinctual theory, premised as it is on the notion that aggression is always a consequence of frustration.[53] Although Dollard and others later modified

---

[50]Konrad Lorenz, *On Aggression,* trans. Marjorie Kerr Wilson (New York: Bantam Books, 1967); Robert Ardrey, *The Territorial Imperative* (New York: Atheneum, 1966); and Desmond Morris, *The Naked Ape* (New York: McGraw-Hill, 1967).

[51]Gerda Siann, *Accounting for Aggression* (Boston: Allen and Unwin, 1985), 88–89.

[52]Sigmund Freud, *Freud: On War, Sex, and Neurosis* (New York: Arts and Science Press, 1947).

[53]John Dollard et al., *Frustration and Aggression* (New Haven, Conn.: Yale University Press, 1939).

their assertion that aggression was inevitable, they maintained that aggressive behavior was still the most likely response to frustration.

Subsequent research has suggested that people do not react aggressively to frustration if they believe that interference with their goal attainment (the cause of their frustration) is justified by the rules.[54] Nor are they likely to react violently when they see failure to obtain their goals as related to fate or to their own behavior. The frustration-aggression hypothesis has also been questioned because of the many other possible responses to frustration besides aggression. For example, frustration may lead to such responses as apathy, withdrawal, repression, resignation, self-transcendence, or merely righteous indignation.

Throughout history, groups have shown great ingenuity in coping with frustration without resorting to violence. An example is the native American tribes located in the southeastern part of the United States, who competed with one another in lacrosse but forestalled the development of perpetual schisms by requiring that competitors change opponents after a certain number of losses. Some Eskimo groups worked out their frustrations by singing satirical songs to each other with judges in attendance, and indigenous peoples in Colombia were able to settle their disputes by beating a rock or a large tree with a stick while cursing their antagonist. The victor was the one whose stick did not break in the process. Despite such examples, and research that seems to question the validity of the hypothesis that frustration produces aggression, it may still be that the greater the frustration, the more aggressively an individual or group is likely to behave, assuming a predisposition to aggressive responses as a way of coping with frustration.

**VIOLENCE AS LEARNED BEHAVIOR** Rather than seeing human behavior as biologically determined, some authorities believe that aggressive behavior is primarily learned. Indeed, the frustration-aggression hypothesis itself is often viewed from the perspective of learned behavior, since different individuals and societies develop varying ways of responding to frustration. Research conducted by Albert Bandura supports the notion that if people are aggressive when their goal attainment is blocked, it is only because such behavior has been rewarded in the past.[55]

Aggression, like other behavior, apparently can be learned by observation. After a careful review of the vast literature on aggression, one authority concluded that "there is compelling evidence that hostile and aggressive behaviors can be acquired, maintained, strengthened, weakened, and eliminated through modelling."[56] This is why many authorities are concerned about the extensive violence shown on television and in movies.

Socializing experiences in child rearing appear to be particularly relevant in explaining aggressive behavior. A study of ninety preindustrial societies found that the

---

[54]Leonard Berkowitz, "What Happened to the Frustration-Aggression Hypotheses?" *American Behavioral Scientist* 21 (May/June 1978): 696–697.

[55]Albert Bandura, *Aggression: A Social Learning Analysis* (Englewood Cliffs, N.J.: Prentice-Hall, 1973).

[56]Dolf Zillman, *Hostility and Aggression* (Hillsdale, N.J.: Lawrence Erlbaum, 1979), 239. (Emphasis in original).

least violent societies were those in which socialization was affectionate and child abuse minimal.[57] Those societies characterized by close contact between males and children were found to exhibit lower levels of warfare.[58] The learning aspect of violence is also suggested by research that shows that individuals who blame and threaten others in everyday conflicts also favor blaming and threatening other nations in international conflicts.[59]

# MISPERCEPTION AND MISCALCULATION

Regardless of whether aggression is an innate or a learned response, it is clear that conflict has often arisen as a result of misperception or miscalculation. As noted in Chapter 5, a number of factors impede the accurate perception of reality, raising serious questions about the prospects of making rational foreign policy decisions. Such misperceptions are of interest to those concerned about the causes of war since they have often been shown to lead to an escalation of conflict. The late Karl W. Deutsch of Harvard University once estimated that 50 to 60 percent of decisions to go to war have been related to misperceptions and misjudgments regarding the intentions and capabilities of other nations.[60] An examination of some eleven wars led another analyst to conclude that misperception was the primary explanation of the conflict in each instance.[61]

Studies of the origins of World War I—remembered as "the war nobody wanted"—have paid considerable attention to the problem of misperception as a cause of war. Among the many instances of misperception cited are the failure to appreciate Russia's concern about the Austrian government's possible humiliation and subjugation of Serbia, the failure of Germany to perceive accurately Britain's likely response to a German attack upon France, and the many miscalculations made regarding the meaning and implications of various military mobilizations. It has been said that President John F. Kennedy was so deeply impressed by Barbara Tuchman's chronicling of the multiple misperceptions of the leaders on the eve of World War I in her best-selling *The Guns of August* that he took special pains to assure that such miscommunication problems would not lead to a similar fate during the 1962 Cuban missile crisis.

**REASONS FOR MISPERCEPTION** A major explanation as to why such misperception arises can be found in the generally poor processing of information that occurs during crises. Studies of sixteen international crises in the last century revealed that only

---

[57] Marc Howard Ross, "Social Structure, Psychocultural Dispositions, and Violent Conflict: Extensions from a Cross-Cultural Study," in *Aggression and Peacefulness in Humans and Other Primates*, ed. James Silverberg and J. Patrick Gray (New York: Oxford University Press, 1992), 271–290.

[58] James Silverberg and J. Patrick Gray, "Violence and Peacefulness as Behavioral Potentialities of Primates," in *Aggression and Peacefulness in Humans and Other Primates*, ed. Silverberg and Gray (New York: Oxford University Press, 1992), 29.

[59] Lloyd S. Etheredge, *A World of Men* (Cambridge, Mass.: M.I.T. Press, 1978), 11.

[60] Lloyd S. Etheredge, "Personality and Foreign Policy," *Psychology Today* (March 1975): 38.

[61] John Stoessinger, *Why Nations Go to War*, 7th ed. (New York: St. Martin's Press, 1993).

## McNamara's Vietnam Mea Culpa

Thirty years after Vietnam, former Secretary of Defense Robert S. McNamara published his reflections on the war in which he admitted that he and his colleagues miscalculated the situation in Vietnam and provided 11 major causes for that failure:

1. We misjudged . . . the geopolitical intentions of our adversaries . . . and we exaggerated the dangers to the United States of their actions.
2. We viewed the people and the leaders of South Vietnam in terms of our own experience. We saw in them a thirst for—and a determination to fight for—freedom and democracy. We totally misjudged the political forces within the country.
3. We underestimated the power of nationalism to motivate a people (in this case, the North Vietnamese and Vietcong) to fight and die for their beliefs and values. . . .
4. Our misjudgments of friend and foe alike reflected our profound ignorance of the history, culture, and politics of the people in the area. . . .
5. We failed then—as we have since—to recognize the limitations of modern, high-technology military equipment, forces, and doctrine in confronting unconventional, highly motivated people's movements.
6. We failed to draw Congress and the American people into a full and frank discussion and debate of the pros and cons of a large-scale U.S. military involvement in Southeast Asia before we initiated the action.
7. After the action got under way and unanticipated events forced us off our planned course, we failed to retain popular support in part because we did not explain fully what was happening. . . .
8. We did not recognize that neither our people nor our leaders are omniscient. Where our own security is not directly at stake, our judgment of what is in another people's or country's best interest should be put to the test of open discussion in international forums. We do not have the God-given right to shape every nation in our own image or as we choose.
9. We did not hold to the principle that U.S. military action—other than in response to direct threats to our own security—should be carried out only in conjunction with multinational forces supported fully (and not merely cosmetically) by the international community.
10. We failed to recognize that in international affairs, as in other aspects of life, there may be problems for which there are no immediate solutions.
11. Underlying many of these errors lay our failure to organize the top echelons of the executive branch to deal effectively with the extraordinarily complex range of political and military issues, involving the great risks and costs—including, above all else, loss of life—associated with the application of military force under substantial constraints over a long period of time.

**SOURCE:** Robert S. McNamara, *In Retrospect: The Tragedy and Lessons of Vietnam* (New York: Times Books, 1995), 321–323. Copyright © 1995 by Robert S. McNamara. Reprinted by permission of Times Books, a division of Random House, Inc.

---

40 percent of the messages between governments were correctly interpreted; the remainder were either incorrectly interpreted (51 percent) or distorted in transmission (9 percent).[62]

Failures in deterrence have often been blamed upon misperception and communication problems. In initiating his invasion of the Falkland Islands in 1982,

---

[62]Glenn H. Snyder and Paul Diesing, *Conflict among Nations* (Princeton, N.J.: Princeton University Press, 1977), 316.

General Leopoldo Galtieri of Argentina completely misread the likelihood of a British military response. In a subsequent interview, General Galtieri suggested that he regarded such a reaction by Britain as "scarcely possible and totally improbable."[63] A study of deterrent threats issued by the People's Republic of China during the Korean War underscored the American failure to perceive that U.S. troop movements would produce a military response from the Chinese, who saw the moves as threatening their borders.[64]

Just as a misreading of the likely response of an adversary might induce a military strike because of the failure to deter, misperception leading to excessive fear about the probability of an attack can also be a cause of war as a state seeks to gain the advantages of a first strike. The psychologist Ralph K. White, in a careful study of international conflict, provides considerable evidence for the proposition that exaggerated fear has been the major cause of war in the twentieth century.[65]

The fears generated by the assassination of Archduke Ferdinand in Serbia leading to World War I, the emotional public response in the United States to the Japanese attack on Pearl Harbor, and the heightening of reciprocal fears in the Middle East on the eve of the several Arab-Israeli wars are indicative of how easy it is for anxious warriors to slip into overt conflict.

# CONCLUSION

Despite advances in the destructiveness of weapons and the purposeful targeting of civilians in modern warfare, the world appears no more violent than it has in previous centuries. In many respects, the record is better. The number of wars and casualties from war seems not to have kept pace with the increased number of states or the rise in population. In fact, interstate war has shown some decline; the wars of the future appear likely to be civil wars—wars within the state between ethnic and religious groups. Instead of interstate war, most violent deaths have actually involved governments killing their own citizens or groups competing for power within a state.

Theories of war from the systemic, societal, and individual perspectives abound. At the systemic level, conflict can be seen as the result of states competing in an anarchical world. From this viewpoint, war is the result of changing power balances as one state or another sees the stability of the system and its position within it threatened. Such threat perceptions are only enhanced by the tendency to engage in arms races, which end either in war or disarmament.

From the societal or national perspective, war is seen as arising primarily from domestic instability, nationalistic fervor, or economic forces. Decision makers have often found it useful to engage in foreign adventures as a way to pull their nation

---

[63]Richard Ned Lebow, "Miscalculation in the South Atlantic: The Origins of the Falklands War," in *Psychology and Deterrence,* ed. Robert Jervis (Baltimore, Md.: Johns Hopkins University Press, 1985), 110.

[64]Allen S. Whiting, *The Chinese Calculation of Deterrence* (Ann Arbor: University of Michigan Press, 1975), 220.

[65]Ralph K. White, *Fearful Warriors* (New York: Free Press, 1984).

together by seeking a scapegoat for domestic problems. In appealing to a real or imaginary enemy, leaders have exploited the sense of national feeling among the population, resulting in the development of a highly expressive nationalism and ethnocentrism. Divisions of national identity, however, have also induced conflict within a state when such identity is not coterminous with the territorial boundaries of a state.

Marxists and many commentators from third world countries see conflict largely in terms of imperialism and exploitation by capitalist states motivated by greed. Within the capitalist state, the so-called military-industrial complex has come under bitter attack, accused of fostering conflict as a way of increasing its wealth through an arms buildup and military sales abroad.

Since it is ultimately individuals who make the decisions to clash, much effort has been made to try to understand whether such conflict is innately driven or whether it tends to be a learned response. Given the great variability in human aggression, most writers conclude that violence is not an innate characteristic. Contributing also to conflict are the many misunderstandings that tend to arise among human beings.

To think of all war as caused by any single factor is erroneous, as any historical study of the complexities of the origin of a particular war will reveal. Of one thing we can be certain: conflict between human beings is endemic. But there is another revelation in human history that suggests that warfare is not the only way to deal with conflict. There is reason to hope that global violence can be controlled, and perhaps even to believe that some of the root causes of conflict can be removed.

# CHAPTER 9

# Diplomacy and Negotiation

Although realists consider power and force to be central to their notion of how to provide national and international security, idealists and others believe diplomacy and negotiation to be the preferable approach to security. Diplomacy and negotiation play an important role in communication between peoples and can be particularly critical in reducing misunderstanding and misperception. Diplomacy can help resolve political disputes before they erupt into violence—but if violence does occur, diplomacy remains vital to the successful negotiation of peace treaties and cease-fire agreements. The importance of diplomacy, however, is not lost upon the realist who considers it to be critical in the process of building coalitions in support of the balance of power.

One of the oldest and most basic instruments of foreign policy is that of diplomacy, which has been defined by Elmer Plischke as "the political process by which political entities, generally states, conduct official relations with one another within the international environment."[1] Throughout recorded history, wherever separate political entities have interacted, they have thereby engaged in diplomacy, for in the broadest sense, diplomacy occurs where independent units find the need to communicate with one another on matters of common concern.

# THE HISTORY OF DIPLOMACY

## PRE-WESTPHALIAN DIPLOMACY

In prehistoric times, tribal social groups used special envoys to convey messages to one another. The early record shows that these emissaries between small

---

[1] Elmer Plischke, *Modern Diplomacy: The Art and the Artisans* (Washington, D.C.: American Enterprise Institute, 1979), 33.

clans were restricted, for the most part, to the role of delivering messages and then returning with a response.[2]

But as social units grew larger and more complex, emissaries gradually acquired other functions. In the already highly civilized Indian subcontinent of the fourth century B.C., the political philosopher Kautilya described the duties of diplomats as including transmittal of the points of view of their governments, preserving treaties, defending the objectives of their states by the use of threats, spreading dissension, collecting intelligence, and persuading those in the host country of the correctness of their own state's views.

In classical Greek and Roman times, envoys from foreign states were even heard in audiences before the Senate, creating a demand for prominent orators for the job. The status of ambassadors was particularly great at the height of Rome's power, when representatives from foreign states were provided with luxurious living accommodations and given magnificent gifts. Such treatment was designed to help increase the prestige of Rome in the eyes of those foreign emissaries so favored. Dynastic China also observed pomp and ceremony in receiving envoys from abroad.

The use of permanent, resident emissaries did not develop until the fifteenth century in Italy, where a host of tiny units, especially the trading city-states such as Venice and Genoa, engaged in economic diplomacy. The increased activity among these states and the ambiguity of their relationship with the papacy—which was both a territorial rival and the common spiritual ruler—made the use of resident ambassadors a necessity. Rather than serving simply as messengers, these representatives were increasingly given the power to negotiate and to speak for the prince.

## EARLY WESTPHALIAN AND MODERN DIPLOMACY

The nature of diplomacy has changed dramatically since the beginning of the Westphalian system in about the middle of the seventeenth century. Diplomacy during its early European period of the eighteenth and nineteenth centuries involved states that shared common values and standards of conduct. Diplomats came from the same upper-class society, held similar values, and spoke Latin in their interaction (replaced with French in the eighteenth century). What they shared often extended even to their personal relationships, given their frequent intermarriage. Contrast this with the modern age in which the diversity of culture, values, language, and ideology across a global system makes diplomatic interaction far more difficult. Today, the United Nations conducts its business in no fewer than six official languages; and the much smaller European Union, in spite of its growing political and economic unity, operates with eleven official languages.

Recent decades have seen a trend toward the recruitment of diplomats professionally trained for their positions. Family connections are not so important as they once were to diplomatic careers, nor is social class or university affiliation as critical as previously—at least not in the more advanced democracies. Even so, in the case of the United States, the effort to create a more professional diplomatic corps has been countered by the tendency of some American presidents to appoint wealthy

---

[2]Jose C. de Magalhaes, *The Pure Concept of Diplomacy* (Westport, Conn.: Greenwood Press, 1988).

friends and campaign contributors to important embassies even when these individuals appear to have minimal qualifications. The specter of an ambassador on the way to an overseas assignment who does not even know the name of the Prime Minister of the state, as happened with respect to an Eisenhower appointee to Ceylon (today's Sri Lanka), continues to haunt.

Compared to earlier centuries, the ethical standards of diplomats have improved somewhat. Diplomats of the sixteenth and seventeenth centuries were known to have bribed courtiers, stimulated and financed rebellions, encouraged opposition parties, spied, lied, and stolen.[3] To the extent such activities continue today, they are largely carried out by intelligence agencies rather than professional diplomats, no doubt because governments with a stake in the status quo have come to recognize that they are better served when their diplomats behave reliably. In recent years, it is governments with the most radical demands to make on the international establishment that often blur the distinction between acceptable, above-board conduct and that which is covert and illicit.

Ambassadors previously enjoyed much more latitude in making decisions in the field than is the case today, when they have become largely instructed representatives of their government. In earlier times, high-level diplomats often had the power to commit their state in dealings with other states; now, however, they must consult their governments on the smallest issues, including even the speeches they deliver and the minor votes they cast. The official title of **ambassador plenipotentiary,** which refers to the full representative powers of ambassadors, clearly has lost much of its meaning.

During the early Westphalian period, the great European powers held the prime responsibility for diplomatic activity. But with decolonization, these states have had to share the world stage with a large number of new states from Africa, Asia, and Latin America. The vast expansion of international organizations, from the various U.N.-related bodies to the many regionally based ones, has provided these smaller states with an arena in which to make known their often divergent views.

**STRAINS IN MODERN DIPLOMACY** With the enormous increase in the number of participants in the international system, there also has been a decline in the use of closed, behind-the-scenes negotiation. Much diplomacy is now carried out in such public forums as the United Nations in contrast to the secret diplomacy, usually of only a few great powers, of the classical period. But today, even in bilateral negotiations, proposals are often presented to the news media, either intentionally or through unauthorized leaks, before they are shown to the opposition. Occasionally, major proposals are even unveiled in public addresses by government officials before the other nation receives the proposal. Relations between the United States and the Soviet Union were strained by the Carter administration's public exposure of its March 1977 strategic arms proposal prior to presenting it to the SALT II negotiation process. The action was a violation of what had become the norm and resulted in setting back the negotiations by several months.

---

[3]Harold Nicholson, *Diplomacy,* 3d ed. (New York: Oxford University Press, 1969), 20.

Foreign offices throughout the world have expanded in size over the course of the Westphalian system. In the 1780s the British foreign office had only a dozen clerks and the secretary; France's included only ninety members in 1870; and the Hapsburg Empire's foreign ministry had only fifty-one members in 1890.[4] Now such offices in large states number in the thousands.

Recent decades, in particular, have witnessed exponential growth in the sheer amount of diplomatic interaction, thanks mainly to the vast increase in the number of matters requiring global or regional solutions. These include not only security issues, but economic, social, and environmental ones as well, because trade, travel, and many other transnational connections have grown enormously in the twentieth century. Some U.S. embassies are now larger than the entire State Department was prior to World War II. Part of this growth is obviously related to the enhanced stature of the United States in world politics, but it is also a function of the increased complexity of global politics. Coping with such a broad agenda also has required the creation of large missions assigned to international organizations in such cities as Geneva, Vienna, and New York.

The complexity of modern diplomacy has also been affected by the tremendous increase in the number of newly independent states that became participants in the international system after World War II. In 1800, the United States had diplomatic relations with only five countries; in 1900 the number stood at forty-one, but at present it has representation in over 180 countries.

Given the extent of this diplomatic activity, it is little wonder that the number of agreements major states have negotiated has reached monumental levels. The United States and the United Kingdom are each party to about 10,000 agreements, and all the world's treaties registered with the United Nations fill more than one thousand volumes.[5] Many of those agreements are exceedingly complex, requiring hundreds of meetings and often several years to hammer out the details—as demonstrated, for example, in negotiations on trade liberalization and arms control.

**DIPLOMATIC REPRESENTATION** Many of the norms related to diplomatic representation and behavior have their roots in premodern history; but they have evolved in recent centuries to serve the logic of the Westphalian nation-state system with its emphasis on the rights and duties of equal sovereigns. Governments that want to establish relations with other states usually do so through a process in which mutual **diplomatic recognition** is established. Such recognition may either be *de jure* (as a matter of law) or *de facto* (in fact). The former is regarded as the more formal or legally significant form of recognition, and it implies greater goodwill between the parties. De facto recognition acknowledges the factual existence of a state and government but does not ordinarily include a full-scale diplomatic exchange. It is the form most often used in situations in which the recognizing government disapproves of a governmental succession, either because of the means used to gain power or its objection to the ideology of the new regime. It took the United States twenty-three

---

[4]M. S. Anderson, *The Rise of Modern Diplomacy* (New York: Longman, 1993), 77, 110.
[5]Adam Watson, *Diplomacy* (Philadelphia: Institute for the Study of Human Issues, 1986), 42.

years before extending de facto recognition to the People's Republic of China, with President Richard Nixon's visit there in 1972, and another seven years before full de jure recognition was given. Earlier in the century, sixteen years expired after the success of the Bolshevik Revolution before the United States extended full diplomatic recognition to the USSR in 1933.

Even when one government disapproves of another, it may nonetheless find it useful to extend recognition, particularly if the government in question effectively controls a particular territory and population. In a highly interdependent world, most governments cannot escape the necessity of having to deal from time to time even with those they oppose. The nation-state system itself virtually demands such interaction. That was apparent in an unusual period before 1972 when the U.S. government and representatives from the People's Republic of China met repeatedly through a subterfuge to deal with common problems; contacts were made in Warsaw, through the U.S. and Chinese ambassadors to Poland.

A long rupture in relations between governments generally makes it more difficult to normalize relations once a decision to recognize a government is made. There will have been too much emotional investment in antipathy—not just between the governments, but their populations as well—over the years. Issues such as financial claims and counter claims will not have been resolved, creating a tremendous burden once recognition is accepted as the way for the governments to deal with one another. For example, many divisive issues remain between the United States and Vietnam, despite the U.S. decision to recognize the Vietnamese government after two decades. U.S. recognition of Cuba is unlikely to occur until Castro is removed from power.

Formal rules have developed regarding the rights and responsibilities of those engaged in diplomatic activity. At Vienna in 1815, four categories of diplomats were established, ranging from ambassadors to extraordinary envoys to resident ministers and, lastly, chargés d'affaires. In order to minimize pressure that might be exerted upon diplomats from a foreign state, such representatives are provided **diplomatic immunity** exempting them from civil and criminal prosecution by the state to which they are sent. Clearly, freedom from such pressures is crucial to their ability to do their job—that is, to communicate without fear of reprisal with the governments to which they are accredited.

Diplomats nevertheless are expected to behave in accordance with the laws and regulations of the host state; the flouting of its laws might lead to a diplomat's being declared **persona non grata**—in other words, not wanted by the host and thus forced to leave. Predictably, such declarations are rare between friendly governments and most frequent between ideological enemies. In a number of instances during the cold war, alleged spies were expelled either from a Soviet or Western embassy, after which the other side retaliated with similar expulsions. Following one such series of expulsions in 1987, the Soviet Union, much to the consternation of the United States, refused to allow the American embassy in Moscow to employ Soviet citizens in even menial jobs at the embassy.

**FUNCTIONS OF DIPLOMATS** Diplomats play a variety of roles for their states in dealing with other nations in the international system. First and foremost, they are *rep-*

*resentatives, empowered to convey messages and to speak on behalf of their government.* Individual citizens operating on their own traditionally have been expected to refrain from such activities. The penalties for doing so used to be particularly severe. As late as the fifteenth century, Venice threatened with banishment or even death those Venetians who dared to interact with any member of a foreign legation.[6]

Second, diplomats also play an important role in *collecting information about the host state and its policies.* Much of this information is simply gleaned from the newspapers, television, and other media sources in the country, as well as from talking to various local political groups and citizenry. Despite the popular messages of spy novels, such espionage activities appear to play a relatively minor role in intelligence-gathering. During the cold war intelligence officers from the CIA and the KGB operated in U.S. and Soviet embassies throughout the world; but, with the breakup of the Soviet Union, the KGB was abolished in its traditional form and the role of the CIA is being reassessed in the United States. Some now suggest that the CIA should play a role in investigating such things as international industrial espionage.

Most young professional diplomats begin their careers in a third role, by *providing consular services.* In this role they issue visas for foreigners to travel to their country as well as provide a variety of services to their own nationals visiting the country to which they are assigned. Special attention is given to those citizens who get in trouble with the law while traveling abroad. Increasingly, embassy personnel are playing a role in facilitating exports and other business interests of their citizens.

Fourth, diplomatic representatives assigned abroad also *engage in negotiation,* although the latitude provided in this role has, for the most part, been usurped by higher-level governmental officials. Those who envision the role of today's ambassador, who is the highest-level official abroad, as one of astute diplomatic bargaining with foreign counterparts is likely to be disappointed. More likely an ambassador's function will involve simply carrying messages between governments. Often when an important issue requires resolution, a special negotiator, sometimes even as high as the foreign secretary, will be utilized instead of the ambassador in the field.

Finally, much of the energy of the high-level diplomat will, of necessity, be directed toward *managing an embassy* composed of representatives from the numerous bureaucracies assigned to it. These personnel include military attaches, intelligence officers, and even staff from a number of domestic departments. Although the foreign ambassador theoretically is given authority over all who serve in an embassy, the reality is that individual departments continue to pursue their separate agendas abroad just as they do at home, making the control of them an uncertain matter.

# THE NEGOTIATING STRUCTURE

Negotiations between the designated representatives of states are structured in many different ways. They involve officials at several levels of authority, with varying degrees of secrecy, a wide array in the number of participants, and may be con-

---

[6]Nicholson, *Diplomacy,* 6.

> ### Global Changes
>
> ## Diplomacy and Negotiation
>
> Britain as late as 1914 maintained only nine embassies, and two of these were outside Europe and had only recently been upgraded from legations (Washington in 1889 and Tokyo in 1905).* Today embassies established by great powers usually number more than 150, as most diplomatic representations have been upgraded.
>
> - What problems do you foresee in the conduct of foreign policy with the vast expansion of diplomatic activity beyond the European sphere?
>
> Attention to protocol and diplomatic niceties has declined over the centuries. In 1661, relations were ruptured and almost fifty men were killed or wounded when the Spanish ambassador, in what would seem to be a minor infraction of protocol, tried to precede the coach of the French ambassador at a procession in London.† Today, spokespersons and even presidents label their counterparts as liars and cheaters, as President Reagan did in his 1983 speech in which he called the Soviet Union an evil empire. In an earlier age such a pronouncement might have been cause for war.
>
> - Is it necessary or even desirable that diplomacy return to the surface civility of earlier centuries?
>
> In the early Westphalian system, ambassadors represented their sovereigns and assumed much of the negotiating responsibilities. Today heads of government do a considerable amount of negotiation in face-to-face summit meetings and by telephone.
>
> - What are the possible advantages and disadvantages of the new forms of diplomacy made possible by technological advances?
>
> The 1992 Earth Summit in Rio de Janeiro involved some 4,000 official delegates and 30,000 unofficial delegates. In contrast, the Westphalia peace negotiations of 1648, which largely founded the nation-state system, involved only dozens of delegates.
>
> - What are some of the advantages and disadvantages of increased numbers of negotiators and of the inclusion of nonstate actors in the diplomatic process?
>
> ---
>
> *M. S. Anderson, *The Rise of Modern Diplomacy* (New York: Longman, 1993), 105.
> †Ibid., 63.

ducted bilaterally or multilaterally. Each of these structural characteristics will affect the way negotiations are carried out; as might be expected, some states will prefer one arrangement over another.

## SUMMIT NEGOTIATIONS

International negotiations held at the highest levels of government are referred to as **summit diplomacy,** involving the direct participation of the heads of state and/or government. Some writers view any meetings between foreign ministers as representing summitry, but this viewpoint tends to be the exception rather than the rule. As already noted, summitry has come to be used more frequently since political leaders today travel rapidly by air. President Woodrow Wilson was the first American president to be formally received in a foreign capital and the first to negotiate personally at an international conference. He sailed to France in 1919 to participate in the Versailles negotiations that ended World War I and created the League of Nations. Some seventy years later, President George Bush became the most traveled of presidents, visiting twenty-eight countries in his first two years in office. Not until 1906 did an American secretary of state, Elihu Root, attend a conference outside the United States, whereas by the second half of the century, secretaries were constantly on the road. In a period of twenty-three months, James Baker, secretary of state in the Bush administration, met no fewer than twenty-five separate times with his Soviet counterpart, Eduard Shevardnadze, in various sites throughout the world. And this, of course, represented interaction with only one state's representative.

Much attention has been given to the increased use of summitry by the United States and the Soviet Union, but many other states now also use summit diplomacy to discuss all sorts of issues. Among the largest such gatherings of governmental leaders are the recurring meetings of the Nonaligned Bloc, whose tenth meeting was held in Indonesia in 1993 with over one hundred states participating.[7] The **Group of Seven (G-7),** comprising heads of government from seven industrial states—the United States, France, Germany, the United Kingdom, Italy, Japan, and Canada—also meets once or twice a year.[8] The G-7 agenda now includes not only economic issues, as in its first meeting in 1975, but political and security issues as well. Governmental heads also meet from time to time under the aegis of various international organizations such as NATO and the European Union. In the largest of these, more than 120 heads of government attended the Earth Summit in Rio de Janeiro in 1992 and similar numbers attended its follow-up conference in Berlin three years later. Added to this are the many face-to-face meetings that occur during official state visits to various capitals throughout the world. State funerals have become a particularly useful time for leaders to engage in a form of ad hoc summitry.

Summitry has begun to play a more important role in resolving regional conflicts as well. Examples include the *Contadora Group,* comprised of the leaders of four Latin American countries, and the *Esquipulus Group* of five Central American presidents. Both groups, named after the places in which they held their first meetings, have sought to mediate conflict in Central America, just as several summit meetings held by a number of African presidents have dealt with regional conflicts in Angola, Namibia, Zimbabwe, and South Africa. These regional summits and the participating presidents unquestionably played very crucial roles with respect to the signing of the accords on Angola and Namibia in December 1988, and in setting the stage for the democratically elected Violeta Barrios de Chamorro to replace the Marxist Daniel Ortega in Nicaragua in April 1990.

U.S. presidents have also successfully hosted summit meetings to encourage parties engaged in lengthy wars to resolve their differences. Thus President Jimmy Carter summoned Anwar Sadat of Egypt and Menachem Begin of Israel to Maryland, where the Camp David Accords were signed in 1978; and President Bill Clinton hosted the presidents of Croatia, Serbia, and Bosnia, thus creating the environment

---

[7]With the end of the cold war alignments, nonalignment would seem to have lost its meaning. As a result, conference participants began to view the Nonaligned Conference as a United Nations of the developing world that would concentrate upon dispute resolution among its membership.

[8]In recent years Russia has participated in the G-7 meetings, but not as a full-fledged member.

for negotiating the cease-fire agreement on Bosnia that was signed in Dayton, Ohio, in November 1995.

**DANGERS OF SUMMITRY** Despite its prominence, summit diplomacy has provoked controversy for centuries. In the fifteenth century, Philippe de Commynes criticized direct meetings between kings, which he thought likely to worsen rather than improve relations. According to de Commynes, if kings wished to remain friends, they should refrain from direct contact entirely. Centuries later, Dean Rusk, who served as secretary of state during the Kennedy and Johnson administrations, suggested that negotiation at the summit "diverts time and energy from exactly the point at which we can spare it least."[9]

Nonetheless, most would agree that in the modern era, summit negotiations at the chief executive or foreign minister level have certain advantages over lower-level negotiations. They communicate serious intentions on the part of the participating states, for top-level decision makers are seen as being unwilling to risk their personal prestige unless agreement seems within reach. Higher-level officials also can commit their state more resolutely than lower-level negotiators, adding to the credibility of the positions taken and increasing the likelihood that the agreement will not be rejected. Such negotiations are also likely to progress more rapidly because of the ability of leaders to commit the state without relying upon instructions from the home government for every move.

But even today, the use of summit negotiations is not without its hazards. The knowledge of the high-level officials meeting at the summit may be fairly limited when it comes to complex technical and political issues. Criticism has been leveled against both Henry Kissinger and Richard Nixon for some of the positions they took at the Moscow summit conference in 1972 in their frantic attempt to complete the negotiation of the SALT I arms control agreements. Nixon's acceptance of one Soviet proposal would have precluded development of the major U.S. strategic missile—the Minuteman III—although, as it happened, the problem was recognized in time and appropriate adjustments were made.

More serious were events that occurred at the Reykjavik summit in October 1986, when President Ronald Reagan, in private meetings with Mikhail Gorbachev and without input from the joint chiefs of staff, came close to committing the United States to a position in which all nuclear weapons would be abolished. Such an abolition was unalterably opposed at the time by most U.S. military analysts, who had long argued that nuclear weapons were necessary to deter the Soviet Union from aggression and were particularly needed to offset Soviet superiority in conventional arms. Only at the last moment did the deal fail, when Gorbachev insisted that complete nuclear disarmament required the United States to scrap its Strategic Defense Initiative. Reagan vehemently refused to comply because of his perhaps naive view that an effective shield could be built against incoming nuclear missiles.

Rather than smoothing relations, personal contact between leaders may actually worsen them. Hostility may be particularly severe when summits fail, as happened when the Four-Power summit set for Paris in 1960 was canceled after a U-2 spy plane of the United States was shot down over the Soviet Union. Personal ani-

---

[9]Dean Rusk, "The President," *Foreign Affairs* 38 (April 1960): 365.

mosity may even develop in meetings between allied leaders, as occurred in the relations between President Carter and Chancellor Helmut Schmidt of the Federal Republic of Germany, and between Prime Minister Margaret Thatcher of Britain and President Francois Mitterrand of France.

Alternatively, problems may result from the development of too warm a relationship between leaders, blinding them to the broader national interests of their own states. Since foreign leaders are often received with great honor and adulation, it may be difficult for them to be critical of their hosts when perhaps they should be. The fondness that Henry Kissinger developed for Zhou Enlai through years of personal interaction, and the warmth that President Bush held toward the Chinese given his earlier position as American ambassador to China in the 1970s, may help to explain why the two were not more critical of the Chinese suppression of the democratic movement in China in the spring of 1989.

Although a political leader is able to make more credible concessions and adjustments to a nation's bargaining position than can a diplomatic official, this is not without its costs. One of the advantages of operating through diplomatic representatives is that a state may always contradict offers made by a representative, whereas the national and personal prestige of a leader is at stake if the offer is made by officials at the summit.

Since summit meetings have often been scheduled for domestic political reasons rather than for foreign policy objectives, there has been a tendency to oversell them to the home audiences. Disillusionment results when achievement falls short of inflated expectations. It may be that part of the hostile reaction to the 1975–1976 detente in the United States was a product of the overly optimistic accounts of the earlier Nixon-Brezhnev summits, which had depicted detente as ushering in a long period of more peaceful relations that did not come to pass.

Some suggest that in order to minimize the problems of rushed agreements to meet a summit deadline and to increase their perceived normalcy, summit meetings should simply be routinely scheduled at least on an annual basis. Such a pattern was established between the United States and the Soviet Union beginning in the Nixon years. Although President Reagan refused to engage in summitry during his first term in office, he made up for lost time by holding five summits with Gorbachev in his second term. President Bush approached the issue of U.S.-Soviet summitry very cautiously at first, but eventually acceded to Soviet, European, and domestic pressures by agreeing to meet Gorbachev on a ship in the Mediterranean in December 1989. Subsequent meetings between these two presidents were held routinely about twice a year as long as Gorbachev remained in office. Despite a desire to focus upon domestic issues, President Clinton met Russian President Boris Yeltsin in Vancouver, Canada, less than three months after assuming office, and the pattern of frequent U.S. and Russian meetings is now well established.

## OPEN AND CLOSED DIPLOMACY

Idealists such as President Wilson, who were critical of secret deals made by states before and during World War I, favored more openness in relations between states. Advocates of openness have come to look upon the use of **parliamentary diplomacy,** such as that practiced at the United Nations, as the preferred way of do-

**"TURN UP THE CLOUD MACHINE."**
Henry Kissinger conducts the summit meeting between Leonid Brezhnev and Richard Nixon, 1972.
Don Wright/Reprinted by permission: Tribune Media Service.

ing the world's business. Such openness has the advantage of educating the world's citizenry and serves to reduce suspicions that decisions are being made contrary to the people's interests. But perhaps more importantly, parliamentary diplomacy can be used to put pressure upon disputing parties to resolve their differences. From the perspective of the state itself, parliamentary meetings provide an opportunity to test a given foreign policy before putting it into effect. By presenting an issue in a hypothetical way before a forum of nations, a state, in effect, is able to float a trial balloon before committing itself to an irrevocable act.

**PROBLEMS OF OPEN DIPLOMACY** Open diplomacy nonetheless can have negative consequences, not the least of which is the tendency of such negotiations to become merely forums for issuing propaganda. A content analysis of U.S. and Soviet speeches made between 1954 and 1957, for example, revealed that the hostility and propaganda scores of the two states were substantially higher in the large, open meetings of the First Committee of the U.N. General Assembly and the Disarmament Commission than in the smaller, more closed Disarmament Subcommittee holding sessions at about the same time.[10]

Compromise is likely to be impeded in public negotiations because of the de-

---

[10]Lloyd Jensen, *Bargaining for National Security: The Postwar Disarmament Negotiations* (Columbia: University of South Carolina Press, 1988), 43–44.

sire to save face and to not appear as ready to give in to negotiators from other states. Other political leaders outside the negotiations who observe excessively compromising behavior are likely to demand as much when they themselves enter into negotiations with the facile compromiser. Concern about domestic public reaction may provide an additional constraint upon compromise; negotiators do not wish to let their domestic political enemies accuse them of being soft in negotiations and not standing up for the national interest. For example, during the limited test ban negotiations in the early 1960s, the Kennedy administration reduced its demand from twenty to six on-site inspections annually. That provided ammunition for opponents of the proposed treaty to argue that the Soviet Union had failed to compromise on the issue, for the Soviets were willing to allow only zero to three annual inspections. What the critics overlooked was the fact that the technology of inspection had improved to such a degree that the six inspections were probably as effective as twenty would have been using techniques available earlier.

On the other hand, it has been argued that concessions presented publicly will be more acceptable to domestic publics than those that have the flavor of closed-door conspiracies. As Daniel Druckman writes, "the same compromise arrived at secretly may not look nearly so bad if arrived at openly."[11] Suspicions are always likely to arise among publics who are given an incomplete picture of the negotiations and how far the other side moved on an issue.

What is needed in negotiation is a proper balance between secrecy and openness. For maximum flexibility, negotiations should be conducted in secret, but the results should be publicized. Secret negotiations should not be used to exclude important governmental actors with vital information and experience. A case in point involved the absence of representatives of the State and Defense Departments at critical points in the decision-making process in the 1986 Iran arms-for-hostages deal. The end result was the ill-fated Iran-contra scandal in which profits from the sale of arms were diverted to aid rebels in Nicaragua in contravention of U.S. law.

## BILATERAL AND MULTILATERAL NEGOTIATION

A wide variety of fora are used in the negotiating process. When two governments meet in face-to-face negotiations, we speak of bilateral negotiations. A third party might be added as a **mediator** in what remain essentially bilateral negotiations. That was the role played by President Carter in the Camp David meetings: His function was to assist the parties at the negotiations in reaching a settlement. Multilateral negotiations, on the other hand, usually involve large numbers of states. The optimum size of a negotiating body, and whether or not to negotiate bilaterally or multilaterally, ultimately comes down to the specific issue one is attempting to resolve. If a given party can veto a particular solution agreed to by others, it is desirable to have that party included rather than excluded. Negotiations to establish multilateral assistance for the developing world obviously must involve those states with the resources for providing aid. Negotiations on environmental issues ought to include as

---

[11]Daniel Druckman, "Human Factors in International Negotiations: Social Psychological Aspects of International Conflict," *Sage Professional Paper in International Studies,* 02-020 (Beverly Hills, Calif.: Sage Publishers, 1973), 50.

many states as possible from among those responsible, or potentially so, for the environmental damage.

If an issue involves a political dispute between only two parties, direct negotiations between them might prove to be preferable to multilateral negotiation on the subject. Article 33 of the United Nations Charter recognizes the primacy that ought to be given to the disputants themselves; it states that, if at all possible, issues should be resolved regionally before submitting them to the United Nations. If the conflicting parties are not speaking to each other, as has often been the case between Israel and its Arab neighbors, multilateral forums might make more sense. But ultimately the two parties will have to face each other if a negotiated settlement is to be reached.

**NEGATIVE ASPECTS OF LARGE-GROUP NEGOTIATIONS** The larger size of multilateral negotiating bodies changes the process as well as, perhaps, even the prospects of reaching a consensus. More interests and values will have to be taken into account and compromised when large numbers of states participate. The result will often be vague or watered-down agreements in order to accommodate the diversity of interests involved. The larger the number of governments participating, the more likely it is that it will take longer to reach an agreement, as each interest seeks to be heard.

With a large number of states taking part in multilateral negotiations, considerable complexity is introduced, often impeding concession making. For example, in commodity trade negotiations the **Group of 77** (actually now comprising more than 120 less developed countries) found it difficult to compromise once they had agreed upon a unified position. The group had become a prisoner of the bargaining structure, as it recognized that any attempt to modify the group position would destroy the hard-won consensus.

**POSITIVE ASPECTS OF LARGE-GROUP NEGOTIATIONS** On the plus side, multilateral negotiations open the opportunity for certain states to play a mediating role, which can be helpful in the conflict resolution process. States without so much at stake are in a position to be more objective and can frequently make suggestions for breaking a deadlock. A participant in the debates on the nuclear test ban has noted how a proposed compromise submitted by the eight neutral nations in the Eighteen Nation Disarmament Conference was a major factor prompting the Soviet Union to accept on-site inspection.[12] It has also been suggested that the addition of neutral participants in these negotiations made the Soviet negotiators use more reasonable arguments than they had in earlier conferences.[13] Since the Soviet government at the time was courting support from neutral states, it must have recognized that such support would not be forthcoming if its negotiators assumed unreasonable positions.

Suggestions and proposed compromises are easier to accept if they come from a third party in a multilateral framework rather than from an individual state opposed to one's actions. The United Kingdom, France, and Israel found it much easier to

---

[12] Arthur Lall, *Negotiating Disarmament* (Ithaca, N.Y.: Cornell University Center for International Studies, 1964), 22–23.

[13] Fred Iklé, *How Nations Negotiate* (New York: Harper and Row, 1964), 20.

accept recommendations emanating from the collective membership of the United Nations to cease their invasion in the Suez in 1956 than they did to buckle to the pressure being exerted by Washington—this in spite of the fact that the three governments shared friendly relations with the United States. The United Nations action in this case enabled the three intervening countries to retain their bargaining reputation and to adopt the face-saving position that they had merely acted until the United Nations could take over.

## CHALLENGES TO THE WESTPHALIAN NEGOTIATING STRUCTURE

The Westphalian model, which was based upon negotiation primarily between sovereigns and their representatives, is changing dramatically as nonstate actors increasingly become involved in negotiation. At times individuals, either by their own initiative or by request of foreign governments, have become participants in external negotiations. For example, the Reverend Jesse Jackson negotiated the release of a downed U.S. flier in Syria in 1984 as well as the release of a number of American hostages held by Saddam Hussein in 1990, and President Carter involved himself increasingly in private diplomacy in various regions of the world after leaving office. Carter's private diplomatic efforts led to cease-fire agreements in Bosnia and Sudan, among others; they also provided the basis for restoring the Aristide government to power in Haiti in 1994 and for an agreement with the North Korean government on restricting nuclear weapon developments.

It may be that these departures from traditional practice mark the unique ability of individuals with celebrity status in an age of instant mass communication to undertake what diplomats may sometimes be constrained from doing precisely because of their official positions. Whether such departures indicate the beginning of a worldwide trend, however, remains to be seen.

Another trend has been the increased activities of nongovernmental organizations, particularly in multilateral conferences, as illustrated most dramatically in the 1992 Rio Earth Summit. Representatives from environmental groups interacted extensively with official delegates at the conference, often serving as unpaid advisers and staff members for foreign delegations that lacked the expertise. Opportunities were provided for the NGO representatives to speak before groups of delegates and to provide extensive background information for the conference.

The state-centric model of the Westphalian system also misses the bargaining that goes on between nongovernmental interests that coalesce across national boundaries as they seek to devise acceptable schemes of cooperation. These can then be sold to the respective governments, thus forming the basis for an intergovernmental agreement.

Finally, increased complexity has been introduced in modern negotiation as a result of the growth of democracy. Democratic values make it all the more important to take account of domestic interests in negotiating international agreements. This is particularly true when parliamentary bodies have the power to ratify treaties. Many negotiators have noted that their most difficult negotiations have been with domestic interests. For example, Robert Strauss said of the Tokyo Round of trade

> **Normative Dilemmas**
>
> ## Dilemmas in Negotiation
>
> **Toughness**
>
> Being tough in negotiation is likely to improve your outcome, but showing toughness will lead the other side to respond in kind, making agreement less likely. Which risk should one take?
>
> **Fairness**
>
> Should divisions of goods be made on the basis of equity, in which a party receives a proportional share based upon its capabilities, or should impartial justice prevail in which each party shares and shares alike?
>
> **The Best Becoming Enemy of the Good**
>
> Was President Clinton correct in opposing the Vance-Owen plan for Bosnia, which would have created a loose confederation of ten provinces, on the grounds that the plan would reward Serbian aggression by giving them control over much of the territory they had obtained by force? (In the process he made the Bosnian Muslims less compromising by encouraging the hope for a better deal with U.S. help, which was never forthcoming. Ultimately, the Muslim-dominated Bosnian government had to settle for the 1995 Dayton cease-fire agreement that gave 49 percent of Bosnia to the Serbs while retaining the fictional viability of Bosnia's recognized external boundaries.)

negotiations: "During my tenure as Special Trade Representative, I spent as much time negotiating with domestic constituents (both industry and labor) and members of the U.S. Congress as I did negotiating with our foreign trading partners."[14]

# THE NEGOTIATING PROCESS

Procedural questions—such as what issues are to be discussed and who is to participate—usually must be decided before substantive negotiations begin; yet, such questions can sometimes create real difficulty. In 1648, it took six months simply to determine the order of entering and seating among the numerous delegations participating in the Conference at Westphalia that ended the Thirty Years' War. Status issues remain important even in the age of modern negotiation, as is illustrated by the much-discussed decision in the 1945 Yalta Conference to have Stalin, Churchill, and Roosevelt enter the negotiating room at the same time from three separate entrances. In the 1970s, months were also consumed in trying to determine the shape of the table in the Vietnamese negotiations, because of concern about what certain seating arrangements would mean for the implied status of such nonstate actors as the Viet Cong. These "merely" symbolic issues become important because they represent matters of serious political substance.

---

[14] Cited in Robert D. Putnam, "Diplomacy and Domestic Politics: The Logic of Two-Level Games," in *Double-Edged Diplomacy*, ed. Peter B. Evans et al. (Berkeley: University of California Press, 1993), 436.

# NONAGREEMENT OBJECTIVES OF NEGOTIATION

In any negotiation, significant time is usually given to a discussion of agenda items, and their order on the agenda, since the choices made can have an important effect upon the outcome. Part of the difficulty in reaching agreement on the agenda or any other issue is related to the fact that states engage in negotiations for a variety of reasons—frequently the least of which is to reach an agreement. For example, negotiations may be used for *propaganda purposes* or to *embarrass the opposition* by openly challenging it to show its goodwill at the bargaining table. Attempts may also be made to *divide one's opposition*—a tactic used by Mikhail Gorbachev when he proposed a number of arms control measures that he knew would divide the Western alliance system, particularly with respect to issues involving nuclear forces in Europe. Similarly, by the simple act of tabling carefully constructed arms proposals, the Soviets exploited deep divisions over the Strategic Defense Initiative, not only within NATO but also within the United States.

Governments may also engage in negotiations to *obtain information about the capabilities and intentions of their adversaries*. Important inferences can be made simply by carefully studying the sorts of proposals made by the other side. The inclusion of certain weapons systems in a disarmament scheme, for example, might indicate that a state is having trouble developing the system or perhaps has a better system on the drawing board. Bush administration proposals reviving a 1955 plan for open skies over American and Soviet territories, as well as its earlier insistence that verification procedures be developed *before* agreement on arms reductions, may have provided ammunition for the Soviet argument that the United States was more interested in obtaining intelligence on Soviet military capabilities than it was in achieving arms reductions.

Negotiations may be held *to enhance domestic political support for a leadership*, and have little to do with foreign policy objectives, much less with the desire to reach an agreement. Political leaders have often found that one of the surest ways to greater popularity is to engage in a summit meeting or to take other actions in support of negotiation in order to appear to be peacemakers. Summits remain popular with the public even when relations are strained between states, for they offer the hope of more peaceful relations. Prior to the first Reagan-Gorbachev summit in November 1985, American polls showed that 82 percent of the public believed summits were a good idea. Yet, at the same time, 66 percent of the respondents believed that Soviet decision makers could not be trusted.[15]

# SETTING THE AGENDA

Despite the ulterior motives that some parties may have for engaging in negotiation, the continued contact may facilitate the solving of new problems as they arise. Negotiations might also prove useful simply from the perspective of the old cliche that "as long as they are talking they are not fighting."

---

[15]Coral Bell, "Negotiation from Strength the Second Time Around," *The National Interest* 11 (1988): 55.

Assuming that reaching agreement is viewed as a primary objective of negotiation, scholars disagree on how broadly the negotiating agenda should be drawn. A number of years ago, Harvard professor Roger Fisher suggested that issues should be broken down so that negotiators can gain experience and success in solving small issues.[16] In a process that he called **fractionation,** little issues would not be linked to larger ones but would instead be settled on their own merits. By solving a number of smaller differences, the success and goodwill obtained would then make it easier to deal with the tougher issues.

To a certain degree, Secretary of State Henry Kissinger used a strategy of fractionation in his Middle East **shuttle diplomacy** in the mid-1970s. Kissinger favored an incremental approach, based upon transitional arrangements that would permit eventual resolution. Issues where movement seemed problematic were quickly withdrawn. The process resulted in agreement on cease-fire arrangements with Egypt in January 1974 and with Syria in May 1974, followed by the Egyptian-Israeli agreement in September 1975. For the first time since the June 1967 war, Israel had agreed to withdraw from certain occupied territories.[17]

Since the incremental approach failed to produce a lasting peace in the Middle East, President Carter opted for a comprehensive solution. The realities of the situation, however, soon forced Carter to accept the more limited Israeli-Egyptian peace treaty of 1979 in which Israel relinquished its occupation of the Sinai. The more significant part of the 1978 **Camp David Accords,** which called for a solution to the Palestinian problem and the occupied West Bank, remained little more than a vision for many years, despite several rounds of Mideast negotiations begun under Secretary of State James Baker during the Bush administration. A breakthrough was finally achieved in the historic Washington meeting between Israeli Prime Minister Yitzhak Rabin and Palestinian Liberation Organization (PLO) leader Yasir Arafat in 1993. This success underscored the utility of incrementalism, since it was many years in the making. But it also demonstrated that a small state, in this case Norway, and even private citizens, such as professors in Norway and the Middle East, can play an important role in the negotiation process.[18] Without the secret talks in Norway, which began at the private level and expanded to official involvement, the trust and basic framework necessary for the Washington agreement probably would never have happened.

# THE ROLE OF CONCESSIONS

Research suggests that if negotiations are to be successful, concessions must be made on each side. For an agreement to be regarded as fair, relatively equal satisfaction (and dissatisfaction) should be felt by both parties. Napoleon Bonaparte rec-

---

[16]Roger Fisher, "Fractionating Conflict," in *International Conflict and Behavioral Science,* ed. Roger Fisher (New York: Basic Books, 1964), 91–109.

[17]Janice G. Stein, "Structure and Tactics of Mediation: Kissinger and Carter in the Middle East," *Negotiation Journal* 1 (1985): 337–338.

[18]Jane Corbin, *The Norway Channel: The Secret Talks that Led to the Middle East Accord* (New York: Atlantic Monthly, 1994).

> ## The Giraffe and the Hippopotamus
> ## Who Talked Past Each Other
>
> A giraffe and a hippopotamus who had long been adversaries were being urged by the tick-birds who lived on their backs to work out ways to get along in a more harmonious fashion. The two great beasts began their discussions by the side of a watering hole.
>
> "I propose," said the giraffe, stretching himself up to his full height, "that all animals with large jaws and sharp teeth refrain from opening their mouths in the presence of others. This will be an excellent measure for building confidence."
>
> The giraffe's words floated into the air above the hippopotamus's head. The giraffe was pleased with his proposal and chuckled quietly.
>
> The hippopotamus then said, "I propose that we agree to a universal halt to all hostile kicking. Indeed, in the interests of peace I declare that I will not lash out at other animals with my legs."
>
> The hippopotamus's words passed under the stomach of the giraffe. The hippopotamus was also pleased with what he had said and grunted in satisfaction.
>
> After a short time the two animals began to get fidgety and impatient. The Hippopotamus opened his huge jaws and took a bite out of the giraffe's thigh. The giraffe swung his front leg with all his strength and kicked the hippopotamus in the stomach.
>
> A tick-bird who had watched the incident from close by said to the other tick-birds, "These big guys have made a mess of it again. Lopsided proposals may make you feel good—but not for very long."
>
> **SOURCE:** Alan Neidle, *Fables for the Nuclear Age* (New York: Paragon House, 1989), 134–135.

ognized this when he suggested that "to negotiate is not to do as one likes."[19] The satisfaction gained from driving a hard bargain may be of limited value, for the agreement reached may require renegotiation, as in the case of the Panama Canal treaty of 1903, which was renegotiated some seventy years later in response to long-standing Panamanian dissatisfaction with the original treaty that infringed upon Panama's sovereignty. Or it may produce continuing resentment, as in the instance of the treaty that gives control of the Cuban base, Guantanamo, to the United States.

The importance of reciprocal concessions in facilitating progress toward agreement is illustrated by several studies of the post–World War II disarmament negotiations. An examination of United States and Soviet concession scores on disarmament issues from 1946 to 1960, as well as during the SALT decade (1969–1979), revealed considerable reciprocation. In the earlier period, a positive correlation of 0.57 was found between the weighted concession scores of the two states, and an equally significant correlation of 0.49 was discovered when the data for the SALT period were analyzed. Similar findings were found in coding concession scores for the Nuclear Test Ban talks that began in 1958 and eventually culminated in the Limited Test Ban Treaty in 1963.[20]

---

[19]Thomas A. Bailey, *The Art of Diplomacy* (New York: Appleton-Century-Crofts, 1968), 145.

[20]All correlations were significant at the 0.01 level, having a probability of less than one in one hundred of occurring by chance. Such high statistical correlations in concession-making behavior suggest that to encourage flexibility from the other side, one must make concessions. See Jensen, *Bargaining for National Security*, 64, 67.

In these negotiations, as the two superpowers approached agreement on a comprehensive nuclear test ban in April 1960, the Soviet Union began to express avoidance behavior, as illustrated by a number of retractions from its earlier concessions on verification and other issues. Such behavior suggests that states may be able to make numerous concessions while results seem remote, perhaps in the expectation that the other side will never accept them. As the prospects for agreement increase, however, they will begin to show avoidance tendencies and may even make a number of retractions as they begin to worry about committing themselves. As a consequence of this backtracking, the world had to settle for a limited ban that allowed nuclear testing to be conducted underground but not in the atmosphere, outer space, or under water.

Despite the seeming necessity of making concessions if agreement is to be obtained, concessions are not without their hazards. A propensity to be conciliatory may be interpreted by the adversary as an indication that a given issue is not particularly important for the other state. There is also the danger that an adversary may misconstrue the move as an indication of weakness, in which case the adversary is likely to demand even more.[21] Experimental studies in which students engaged in bargaining exercises designed by Sidney Siegel and Louis Fouraker suggested that it is better to begin with hard positions rather than conciliatory moves, since the latter tend only to raise the expectations and consequently the demands of the other side.[22]

Some students of negotiation have rejected the reciprocal concession model altogether as a major factor in determining the outcome of negotiations. One study based on interviews conducted with some fifty ambassadorial-level diplomats at the United Nations, for example, concluded that the negotiation process involves finding an appropriate formula and attempting to implement it.[23] Minimal attention tends to be paid to the number of concessions made by the other side. Instead, concessions are largely symbolic moves made most often toward the end of the negotiations, when agreement is in sight and it is believed that the other side is likely to accept the package being offered. Interviews with participants in the Kennedy Round of trade negotiations during the 1960s demonstrated that there was minimal awareness of the concessions made by other participants. To the extent that concessions were made by the negotiators, they seemed to be responding more to their general free trade philosophy than they were to the behavior of other negotiators. The respondents, in fact, seemed unable to agree on when breakthroughs occurred in the trade negotiations or whether concessions were even made.[24]

Increasing attention has been given to the means of achieving a win-win outcome in international negotiations rather than emphasizing how much one side can

---

[21]Charles Lockhart, *Bargaining in International Conflicts* (New York: Columbia University Press, 1979), 125.

[22]Sidney Siegel and L. E. Fouraker, *Bargaining and Group Decision Making* (New York: McGraw-Hill, 1960).

[23]I. William Zartman and Maureen R. Berman, *The Practical Negotiator* (New Haven Conn.: Yale University Press, 1982).

[24]Gilbert R. Winham, "Complexity in International Negotiations," in *Negotiations: Social-Psychological Perspectives,* ed. Daniel Druckman (Beverly Hills, Calif.: Sage Publications, 1977), 347–366.

## Don't Bargain over Positions

PROBLEM:

POSITIONAL BARGAINING—
WHICH GAME SHOULD YOU PLAY?

SOLUTION:
CHANGE THE GAME—
NEGOTIATE ON THE MERITS

| Soft | Hard | Principled |
|---|---|---|
| Participants are friends | Participants are adversaries | Participants are problem-solvers |
| The goal is agreement | The goal is victory | The goal is a wise outcome reached efficiently and amicably |
| Make concessions to cultivate the relationship | Demand concessions as a condition of the relationship | Separate the people from the problem |
| Be soft on the people and the problem | Be hard on the problem and the people | Be soft on the people, hard on the problem |
| Trust others | Distrust others | Proceed independent of trust |
| Change your position easily | Dig in to your position | Focus on interests, not positions |
| Make offers | Make threats | Explore interests |
| Disclose your bottom line | Mislead as to your bottom line | Avoid having a bottom line |
| Accept one-sided losses to reach agreement | Demand one-sided gains as the price of agreement | Invent options for mutual gain |
| Search for the single answer: the one *they* will accept | Search for the single answer: the one *you* will accept | Develop multiple options to choose from; decide later |
| Insist on agreement | Insist on your position | Insist on using objective criteria |
| Try to avoid a contest of will | Try to win a contest of will | Try to reach a result based on standards independent of will |
| Yield to pressure | Apply pressure | Reason and be open to reasons; yield to principle, not pressure |

**SOURCE:** Positional bargaining chart from Roger Fisher, William Ury, and Bruce Patton *Getting To Yes* (New York: Penguin Books, 1983), 13. Copyright © 1981, 1991 by Roger Fisher and William Ury. Reprinted by permission of Houghton Mifflin Company. All rights reserved.

gain over the other. In their book, *Getting to Yes,* Roger Fisher and William Ury outline what they call principled negotiation, which is designed to produce solutions in which all parties can gain.[25] From their perspective, less emphasis should be given to presenting emotionally involved positions and more attention should be directed toward attempting to solve a problem jointly.

Others hold that establishing better relationships between peoples in different states is more important than negotiations and that in the absence of improved relations, state-to-state negotiations would be counterproductive. For example, Harold H. Saunders, a former U.S. diplomat with considerable experience in Mideast ne-

---

[25] Roger Fisher and William Ury, *Getting to Yes* (Boston: Houghton Mifflin, 1981).

gotiations, argued that Israeli-Palestinian negotiations were futile in the mid-1980s.[26] He preferred instead that attention be directed to improving relationships between the two communities. Central to this notion is the view that peoples, both at the official and the unofficial levels, should engage in discussions and problem solving rather than making threats and trying to negotiate one-sided agreements. It was just such informal meetings between Palestinians and Israelis in Norway and elsewhere that set the stage for beginning official talks on Palestinian autonomy in 1993 that would ultimately prove to be successful.

## NATIONAL NEGOTIATING STYLES

Although differing structures and processes have an impact upon negotiating behavior, perhaps the most important factor to explain negotiating outcomes is negotiating style: the distinctive characteristics and attitudes of the different national groups engaged in negotiation. Considerable variation appears to exist among states in such important negotiating traits as open-mindedness and flexibility, both of which can facilitate compromise. A study of Greek interaction with Americans found that whereas Americans, given their democratic values, were willing to settle for an outcome "in between two positions," the Greeks tended to view compromise as a clear defeat. Initial demands were considered justified and immutable by the Greeks.[27] According to one assessment, experts are in general agreement that few societies find give-and-take more distasteful than the Japanese.[28] Other studies have found Arabs to be less compromising than Americans, kibbutz children of Israel to be more compromising than city children, and Mexican village children to be more cooperative than both their counterparts in the United States and those from Mexican urban centers.[29]

Patience is a useful trait in negotiation, and is a quality that seems to vary considerably as the result of culture. Israelis have been found to be far more patient than Americans. Russians are also far more patient—a fact that a former Strategic Arms Reduction Talks (START) negotiator suggests is reflected in the games that are popular in each country: "The Russians play chess; we play video games. They like the well-thought through results of step-by-step reasoning; we like the instant results of electronic machines."[30] The impatience of American negotiators was illustrated in a detailed examination of seven early rounds of postwar disarmament negotiations that showed the Soviet Union making some 75 percent of its concession score during the *last third* compared with the United States' score of 82 percent during the *first*

---

[26]Harold Saunders, "We Need a Larger Theory of Negotiation," *Negotiation Journal* 3 (1985): 349–362.

[27]E. D. Davis and H. C. Triandis, *An Exploratory Study of Intercultural Negotiations,* technical report 26 (Urbana: University of Illinois Group Effectiveness Research Laboratory, 1965).

[28]Raymond Cohen, *Negotiating across Cultures* (Washington, D.C.: United States Institute of Peace Press, 1991), 91.

[29]Daniel Druckman, "The Person, Role, and Situation in International Negotiations," in *A Psychological Examination of Political Leaders,* ed. G. Margaret Hermann (New York: Free Press, 1977), 409–456.

[30]Edward L. Rowny, "Ten 'Commandments' for Negotiating with the Soviet Union," *New York Times,* January 12, 1986.

*third* of the same talks.[31] Although the figures were not as divergent with respect to the SALT negotiations during the decade of the 1970s, the United States continued to make its concessions early. Gorbachev may have changed the pattern with his flurry of unilateral concessions on arms control and other matters beginning in 1986; still, these concessions were initially dismissed by the United States on the grounds that they only reflected severe economic problems in the Soviet Union rather than a genuine desire to improve relations.

Other cultural differences between Americans and Russians have also been viewed as having given the Russians certain advantages in negotiations, at least during the cold war period. Henry Kissinger, for example, believed that American pragmatism and empirical-orientation doomed the United States "to an essentially reactive policy that improvises a counter to every Soviet move, while the Soviet emphasis on theory gives them the certainty to act, to maneuver, to run risks.[32] Critics of American negotiating behavior sometimes suggested that, as a result of the democratic approach to reaching consensus, the United States, in making its proposals, constrained itself by making compromises before presenting its proposals in international forums.[33]

Both the Japanese and the Russians are seen as extremely conscious of rank in their dealings with each other and also with other nations. They are more concerned about formality and rank equivalence in diplomatic negotiation than are American diplomats, who come from a more egalitarian and democratic culture. Preoccupation with one's status, as opposed to focusing on resolving the immediate issues that divide the parties, can only result in failed negotiations.

The Far Eastern concern for saving face may also impede compromise, as seems apparent in the Japanese negotiating style. According to an examination of Japanese negotiating behavior in eighteen case studies covering the period from 1895 to 1941, the Japanese have a "cultural proclivity to oppose rather than propose, to be passive rather than active, defensive rather than offensive, and evasive rather than forthright.[34] Making commitments appears to be particularly difficult, and if compromise is forced upon Japanese negotiators, they are likely to argue that the situation impelled them to act as they did.

Cultural differences go a long way toward explaining some of the difficulties experienced by Western societies when negotiating with the Arabs. The anthropologist Edward T. Hall has identified some of the cultural problems that arise when what he calls *high context* and *low context* countries negotiate with each other.[35] For example, in the high context Arab culture, emphasis is given to the total environment in deciding whether or not to trust the other party. Information about a negotiating partner tends to be gleaned from body language, particularly from the eyes (which could help explain the tendency for such Arab leaders as Yasir Arafat to

---

[31] Jensen, *Bargaining for National Security*, 69–70.

[32] Henry A. Kissinger, *Nuclear Weapons and Foreign Policy* (New York: Harper and Row, 1957), 425.

[33] Iklé, *How Nations Negotiate*, 149.

[34] Michael Blaker, *Japanese International Negotiating Style* (New York: Columbia University Press, 1977), 60.

[35] Edward T. Hall, *Beyond Culture* (Garden City, N.Y.: Doubleday, 1976).

> ## The "Asian Way" of Diplomacy
>
> - Initially, the Asian negotiating style seeks to maximize informal discussions. Decisions emerge only after a long process of equalitarian socializing.
> - When Westerners fail to reach an agreement, the immediate conclusion tends to be that one party was too greedy. For practitioners of the Asian way, discussion develops decisions by consensus building, not by horse-trading.
> - Decision making in the Asian way is incremental. While Westerners keep their eyes on the prize, Asians tend to be fascinated by the process.
> - Westerners tend to cite problems of technical feasibility in order to say "No" to idealistic aspirations. Asian leaders generally seek to commit to abstract principles first, leaving questions of implementation to subordinates.
> - The Asian way promotes unique solutions over universal nostrums. The goal is "Asian solutions to Asian problems."
> - The Asian way stresses the need for solidarity through multilateral cooperation with other Asian countries.
>
> SOURCE: Michael Haas, *The Asian Way to Peace: A Story of Regional Cooperation* (New York: Praeger, 1989), ch. 1.

often wear dark glasses). Less attention is paid to the verbal context and to written agreements in such societies. In low context societies such as those of the United States and especially the north European states, emphasis is given to writing everything out—hence the frequent use of highly formal contracts and the employment of lawyers in all sorts of transactions. As a result of such differences in the two cultures, high context societies believe it desirable to develop trust over a long period of time, but this conflicts with the lack of patience in the low context state that in turn can lead to considerable misunderstanding.

Yet we must conclude on a strong cautionary note about the impact of national styles. What may seem to be national differences may in fact be related more to the policy goals of elites than they are to a nation's culture. The changes in the leadership of the Soviet Union that brought about the end of the cold war clearly illustrate this point. Mikhail Gorbachev, Boris Yeltsin, and Eduard Shevardnadze more nearly resembled their contemporary Western counterparts as negotiators and statesmen than they did their own Russian and Georgian predecessors—men like Brezhnev, Khrushchev, and Stalin. It may be that deeply embedded differences in cultural traits matter far less in a world where elites are inevitably acquainted with one another, if only through the medium of television, in ways that could not be imagined in previous centuries. Most importantly, where the policy goals of negotiators are truly congruent, differences in their cultural styles evidently are of little importance.

## THE NEGOTIATING STYLE OF WOMEN

With the increasing importance of women on the political and diplomatic scene, it is appropriate to ask whether or not the sex of the negotiator is likely to make a difference in the conduct and outcome of negotiations. Evidence suggests that women tend to be more concerned with long-term relationships and with seeking win-win solutions. They are less predisposed toward positional bargaining and more likely to share ideas rather than engage in challenges and debates. They also tend to be more sensitive to signals sent by others, preferring to view the negotiating process as sharing thoughts and building upon what others say.[36] In some respects the female approach to negotiation is more akin to that proposed by supporters of a problem-solving approach, which many have come to view as the preferable procedure for dispute settlement.

While the image of women is one of being "soft" negotiators, the behavior of such former prime ministers as Golda Meir and Margaret Thatcher raises questions about the accuracy of this image—they were both seen as tough and intransigent. Interviews with women who have been successful in negotiation reveal that in order to be accepted and to move toward a more collegial position, women must first present themselves as tough.[37] Also, because of differences in expectations of "proper" sex roles, the same behavior may be viewed in women as aggressive, while seen in men as a sign of leadership.

As was the case with differences in national negotiating styles, there is considerable variability among women. Nonetheless, when it comes to questions of war and peace, women perhaps show greater sensitivity to the issue of killing. Both Thatcher and Meir brooded heavily over the killing of "their boys," as they put it, during the Falklands and the Yom Kippur Wars. Despite the heavy burden of the affairs of state during these crises, both wrote long letters to the families of those killed.

Prime Minister Indira Gandhi undoubtedly showed greater restraint than most male leaders would have during the Bangladesh War in 1971 that led to the breakup of the state of Pakistan. The war sent eight million East Bengali refugees streaming over the border into India, further straining its fragile economy. Despite domestic political pressures, Gandhi responded with military force only *after* several months of continuing civil violence in East Bengal and Pakistan had sought to expand the war in a frontal attack upon Indian territory.

Even if the responses of men and women to conflict and violence were not appreciably different, women politicians have often placed peacemaking activities higher on their agendas than their male counterparts. In Sri Lanka, President Chandrika Kumaranatunga, the daughter of the world's first woman prime minister, Sirimavo Bandaranaike, made the continuing violence of the Tamil rebels the primary issue of her 1994 campaign. Similarly, Philippine President Corazon Aquino made negotiation with the rebels a central objective during her administration. And it was the government of Prime Minister Gro Brundtland of Norway that served as facilitator for the secret negotiations between the Palestinians and Israelis, leading eventually to the historic agreement signed in Washington in 1993 between Yasir Arafat and Yitzhak Rabin.

---

[36] Deborah M. Kolb, *Is It Her Voice or Her Place That Makes a Difference?* (Kingston, Ontario: Queens University, Industrial Relations Centre, 1992).

[37] Ibid., 9, 11.

## NEGOTIATING AN END TO VIOLENCE

Whether it has been due to an improvement in the structure or the process of negotiation or perhaps simply the changed international environment with the end of the cold war, a number of lengthy and extremely violent civil and international wars have ended recently with negotiated settlements.[38] These include, among others, peace settlements of civil wars in Angola (1991 and 1994), Cambodia (1991), El Salvador (1992), Mozambique (1992), Liberia (1995), and Bosnia (1995).

In addition, after many years of refusal even to talk, a number of serious negotiations have recently been undertaken. These include those involving Syria and Israel, which began in Maryland under U.S. auspices in December 1995, and talks on Northern Ireland in which Sinn Fein, the political arm of the Irish Republican Army, has joined Britain and the Irish Republic at the negotiating table.

What is remarkable about many of these negotiations is that they virtually negate the Westphalian principle that makes sovereignty or control over one's own internal affairs all important. Rather, the international community has asserted its right to intervene in cases of civil violence in other countries. In such interventions, outside powers have used diplomacy, economic sanctions, foreign aid, and even military force in an effort to compel the combatants to resolve their differences at the negotiating table. Outside powers and international organizations have also furthered the peace process by supervising elections, monitoring compliance of disarmament schedules, and guaranteeing observance of the provisions of the agreement by placing military forces on the ground. An example is the 1995 accord that called for the emplacement of 60,000 NATO troops in Bosnia, including 20,000 from the United States. Peace has increasingly come to be viewed as indivisible; security is no longer strictly national, but is now seen as global in scope.

## CONCLUSION

The logic of the Westphalian system, with its emphasis upon sovereign and independent actors, has always required international interaction, for few problems end at a nation's borders. Such interaction in turn has resulted in the creation of certain norms involving recognition policy and appropriate forms of diplomatic behavior.

The nature of diplomacy and negotiation has been far from static, leading some to begrudge what they believe to be a decline of diplomacy in the twentieth century, whereas others view those changes more benignly as the natural result of a far larger, more open, and heterogeneous system. Diplomacy is no longer European-centered nor restricted to the upper classes. Ambassadors have lost some of their

---

[38] For some of the many studies involving the negotiation of violent disputes, see Jack Child, *The Central American Peace Process, 1983–1991* (Boulder, Colo.: Lynne Rienner Publishers, 1992); Chester A. Crocker, *High Noon in Southern Africa: Making Peace in a Rough Neighborhood* (New York: W. W. Norton, 1992); Roy Licklider, ed., *Stopping the Killing: How Civil Wars End* (New York: New York University Press, 1993); Paul Pillar, *Negotiating Peace: War Termination as a Bargaining Process* (Princeton, N.J.: Princeton University Press, 1983); and I. William Zartman, ed., *Elusive Peace: Negotiating an End to Civil Wars* (Washington, D.C.: Brookings Institution, 1995).

power, sometimes becoming little more than messengers. Protocol has declined in importance and negotiations are often conducted in open and noisy forums.

The reasons for such changes are not difficult to fathom. Among these are improved technology and communication, the vast increase in the number and cultural diversity of sovereign states, an increased emphasis upon ideology during the twentieth century, and a greatly expanded international agenda created by a more interdependent world that must cope with a variety of economic, environmental, and technological problems.

To contend with these various problems, a number of different negotiating structures have been developed. These include summitry and both open and closed meetings, as well as bilateral and multilateral negotiations. It is impossible to generalize as to which form is likely to be the more successful, for this depends upon the specific circumstances. Perhaps more important than the precise structure of the negotiations is the attitude and the responsiveness of the negotiating parties. Negotiations are held for many reasons, not always with the objective of reaching an agreement. The seriousness with which a party approaches negotiation is perhaps best shown by its willingness to make reciprocal concessions.

There is some evidence (although here anecdotal) for what might be called a national negotiating style. Cultures vary considerably in terms of the human traits they produce, such as open-mindedness and flexibility. Some cultures are more patient than others, which might provide certain advantages in negotiation. Dealing with people who think differently from oneself can be frustrating, but for negotiations to be successful, negotiators must make an effort to understand their counterparts' particular national styles.

The state-centric model of Westphalia cannot account for the increasing roles of individuals other than state officials who today play important roles in diplomacy. Prominent private citizens, nongovernmental experts, and others who represent some "public" interest are beginning to assume unprecedented importance in the effort to resolve conflicts in the world. This trend, which is almost certain to grow more prominent in the years ahead, may reflect the growing inadequacy of the state system to treat all of the issues whose dimensions are global today.

PART IV

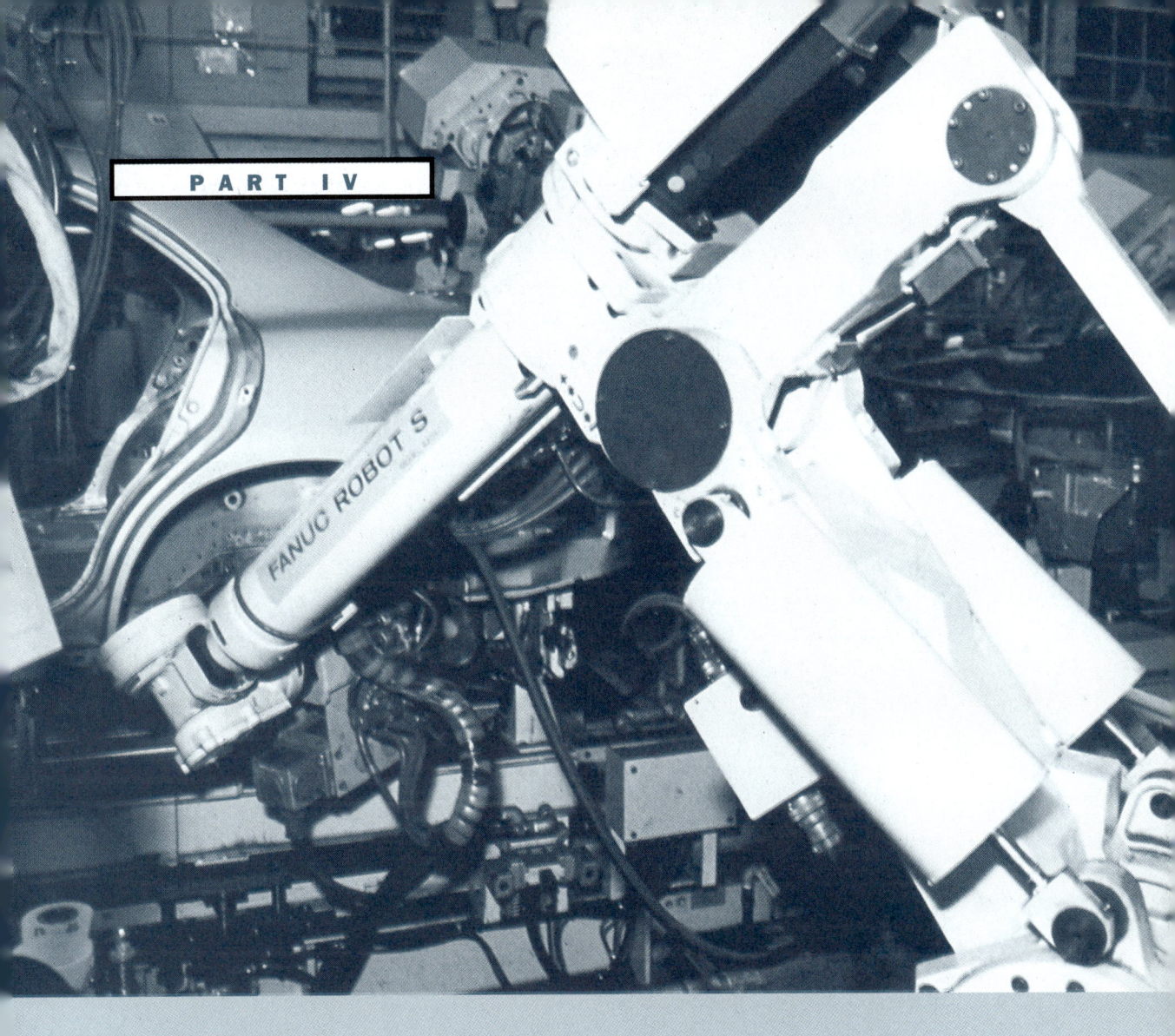

# The World Political Economy

The protection that governments provide against threats both from within and without their state's borders has always been the most basic function of the state. It remains so today. Increasingly, however, nation-states are paying more attention to other issues that affect their well-being; economic welfare is one such important issue. Today, citizens look to governments to protect their interests in an increasingly interdependent economic world. As a result, more emphasis than ever before is being placed on the study of the international political economy, which stands at the nexus of politics and economics—each of which influences the other in ever more complex ways.

For the first time in human history, the world today may be creating a single, global economy. In many respects, our economic interconnectedness across the planet is proceeding with little regard for the principle of separation upon which the nation-state system is built.

Chapter 10 explores the development of a world economy. It examines the factors working both for and against globalism in the marketplace and the dynamics of a world monetary system, which makes possible the expanding global market. Chapter 11 turns back to the state's perspective on economic life. Here we examine the economic tools of statecraft, as well as the impact on states of growing economic interdependence. Chapter 12 considers the problems of poverty worldwide and the effort to alleviate it through economic development. But it is increasingly clear that the quest for wealth as pursued in the modern world is also endangering the health of the planet. Chapter 13, therefore, concludes this section by investigating the impact of economic life across the globe upon the resources of our common home. ∎

*Overleaf* • **A robot assembles a car at a Nissan factory in Japan.**

CHAPTER 10

# The Development of a World Economy

France's province of Alsace has known hard economic times in this century. Situated across the Rhine from Germany, it was devastated in two world wars. As recently as the 1980s, this coal-producing region experienced a severe recession from the changing and declining coal and steel industry. Then something happened that turned this depressed province into one of the most dynamic in Europe in the 1990s. High-tech Japanese firms—Ricoh, Sharp, Sony, and Yamaha—built plants and created thousands of new jobs in Alsace. Why? It began when Alsatian officials courted Japanese companies in an imaginative way. They decided to subsidize a Japanese television show about a young Japanese businessman who married an Alsatian girl and settled in that province. The show was a hit in Japan. When Japanese executives began really to explore Alsace, they discovered that its location made it an ideal site for the European market, and their major investments followed.[1]

## PERSPECTIVES ON INTERNATIONAL POLITICAL ECONOMY

Just as studies of state security have been driven by various theoretical perspectives, so are the ideas and actions that constitute the international political economy informed by theory. Three such perspectives and their variations have been particularly prominent during the period of the Westphalian system: mercantilism,

---

[1] William Drozdiak, "Alsace: From Devastation to Dynamism," *The Washington Post*, September 24, 1994, A18.

liberalism, and Marxism.[2] Each presents a distinctive view as to the proper relationship between the efforts to acquire wealth and the role of the state.

# MERCANTILISM

**Mercantilism,** an economic doctrine that emerged in the seventeenth century, emphasizes that the creation of wealth should be the primary goal of a state, which should strive to enhance its own power and security at the expense of others. Like realism *(realpolitik),* mercantilism makes the state the central unit of analysis. It shares with realism the goal of guaranteeing state sovereignty, but emphasizes the acquisition and use of economic power rather than political and military power. The acquisition of wealth, for a mercantilist, is a largely zero-sum proposition—that is, greater wealth for one state means relatively less for another.

Under mercantilism, state instruments were to be used to achieve economic objectives—primarily by placing restrictions upon international trade. Mercantilists believed that it was better to export than to import, in order to create monetary reserves that could be used for war-making purposes supporting the security interests of the state. Those countries involved in international commerce were consequently hobbled by the tariffs and other protectionist measures imposed by outside states, as well as by the attempts of others to engage in imperial expansionism in pursuit of wealth.

Mercantilists like Alexander Hamilton, one of the American founding fathers, and Friedrich List, the nineteenth-century German political economist, believed the role of government should be to enhance domestic industrial production by protecting and subsidizing it, thus making it more competitive with the products of other countries. To be successful, the state would find it necessary to engage in mercantilist or restrictive trade practices.

Descendants of those mercantilists are more likely to be called **neomercantilists** today. In a world where there is vastly more trade than two hundred years ago, they see some utility in state restrictions on the trading system. Often, neomercantilists argue that such restrictions are necessary to counter what they see as the unfair trade practices of other states, to create what is often called a *level playing field*. Many neomercantilists also advocate a more active government role in supporting various national industries that have the potential of competing effectively in global markets.[3]

---

[2]For a more complete discussion of various economic perspectives, see David N. Balaan and Michael Veseth, *Introduction to International Political Economy* (Upper Saddle River, N.J.: Prentice-Hall, 1996); Robert A. Isaak, *Managing World Economic Change,* 2d ed. (Upper Saddle River, N.J.: Prentice-Hall, 1995); Jeffrey A. Frieden and David A. Lake, eds., *International Political Economy: Perspectives on Global Power and Wealth* (New York: St. Martin's Press, 1995); Robert Gilpin, *The Political Economy of International Relations* (Princeton, N.J.: Princeton University Press, 1987); and Thomas D. Lairson and David Skidmore, *International Political Economy: The Struggle for Power and Wealth* (Fort Worth, Texas: Harcourt Brace, 1993).

[3]For further reading from the perspective of neomercantilism and strategic trade policy, see Robert Kuttner, *The End of Laissez-Faire* (New York: Knopf, 1991); Clyde V. Prestowitz, *Trading Places* (New York: Basic Books, 1989); and John Zysman and Laura Tyson, *American Industry in International Competition: Government Policies and Corporate Strategies* (Ithaca, N.Y.: Cornell University Press, 1983).

## ECONOMIC LIBERALISM

Whereas mercantilism emphasized the role of the state in supporting the economic system on behalf of its own interests, classical **liberalism** downplays the state's economic involvement. It rejects the heavy hand of government and interference in the economy, believing that unfettered private market forces (or a laissez-faire system) will operate more effectively and efficiently than a state following mercantilist doctrines. Although some individuals and groups will lose out because they are unable to compete, liberals expect the end result to be greater efficiency and growth in markets as a whole. The traditional economic liberal today is often viewed as a political conservative, who wants a minimum role for government and maximum freedom for the individual.

Some liberals of both the nineteenth and twentieth centuries recognize that government is needed to regulate the excesses of a capitalist system based entirely upon the search for profit. Profits might be made at the expense of the environment, the worker, or the shareholder, all of which may require governmental regulation to assure fairness and the greater good. Even such early liberals as John Stuart Mill (1806–1873) believed that it would be necessary to take limited actions to correct market failures or to provide social progress in such areas as education, health, and social welfare. Such governmental activities were seen as aiding productivity in the long run.

Liberalism views economic enterprise as inherently a non zero-sum activity. That is, wealth for some is expected to help create greater wealth for others, whether these are the employees one hires or one's trading partners. When it comes to the international political economy, therefore, liberalism tends to see the entire system, and not the individual state, as the crucial unit.

**HEGEMONIC STABILITY THEORY** An important variation on the liberal theme has been that of **hegemonic stability theory.** According to this theory the effective functioning of a liberal global economy requires the existence of a hegemon, some nation or group of nations with considerable power and interest in perpetuating a liberal and stable trading regime.[4] The hegemon must accept that it is in its own interest to provide for certain *public goods*—even though, in doing so, that hegemonic actor must contribute more resources and energy than others who will also benefit from the enterprise.

Thus, the post–World War II United States was seen as an economic hegemon willing to pay a disproportionate share to assure such goals as peace, free trade, and a sound monetary system. It was the United States that played a leadership role in the reconstruction of Europe following World War II, absorbed the prime responsibility and by far the greatest expenditure in defending against the Soviet threat, and played a critical role in supporting institutions that helped nations in balance of payments difficulties and that were plagued by currency problems. Instead of acting

---

[4]Robert O. Keohane, "The Theory of Hegemonic Stability and Change in International Economic Regimes, 1967–1977," in *Change in the International System,* ed. Ole R. Holsti, Randolph M. Siverson, and Alexander L. George (Boulder, Colo.: Westview Press, 1980).

upon mere altruism, the United States expected to benefit disproportionately from the advantages inherent in global economic growth.

## MARXISM

**Marxism** originated as a critique of liberalism. It sought to correct some of the injustices and inequalities that resulted from an unfettered liberalism that had not improved the material well-being of the masses, even though a minority had become rich under capitalism. Writing in the mid-nineteenth century, Karl Marx (1818–1883) insisted that the advanced industrial societies of his time were characterized by a conflict between the dominant, capitalist class (bourgeoisie) and the workers (proletariat) whom the capitalists exploited for profit. The doctrine of laissez-faire, in the view of Marx and his followers, had allowed the capitalist minority to assume ownership of the means of production for their own private gain, rather than for the good of society. The state or political structure was the vehicle that had been created and used by the dominant class to perpetuate its own selfish interests.

Marx viewed capitalism as playing an important historical role in destroying slavery and feudalism and in setting the stage for economic and industrial growth, both of which he favored. Nonetheless, he believed that capitalism itself should and would be destroyed because of the social injustices it generated.

Like the early liberals, Marx focused most of his attention upon economic life within industrial states. But his analysis was decidedly non-statist. He argued that once the industrial proletariat of the world recognized their true mutual interest in ending capitalism, workers would unite across national lines to overthrow liberalism and, ultimately, the bourgeois state itself. In this sense, Marxism is a systems-level theory that anticipated the end of the nation-state system.

## THE RISE OF THE TRADING STATE

The Westphalian system has evolved from the mercantilist emphasis upon the use of economic policy to advance the political and military well-being of the state to one in which liberal doctrines for achieving power through trade are dominant. This evolution has been outlined in a study by UCLA professor Richard Rosecrance.[5] According to his argument, the early Westphalian state was territorially based and its leaders were mainly concerned with assuring its power and security. The sovereign tended often to concentrate on acquiring additional territory, thereby gaining more revenue, more resources, and, accordingly, more wealth. Expansion was also motivated by the desire to assure access to markets of highly valued goods, such as gold, silver, spices, sugar, indigo, and tobacco.

The mercantilist era began to fade in the late eighteenth century with the rise of capitalism, wherein commerce was seen to be mutually enriching rather than a zero-sum activity. Gradually, a few states began to recognize that their economic well-

---

[5] Richard Rosecrance, *The Rise of the Trading State* (New York: Basic Books, 1985).

being might be enhanced by engaging in trade with other states instead of simply attempting to dominate markets and resources by imperialist measures. As early as 1713, Holland gave up its great power ambitions, as did Sweden and Denmark shortly thereafter, in order to concentrate upon trading. The beginning of the Industrial Revolution early in the nineteenth century facilitated the process. Without conquering new territories, states could enhance their economies simply by trading with other states for the raw materials required for manufacture. Great Britain became the world's leading power in the nineteenth century primarily as a trading, rather than a military, state.

The tendency to rely less upon the political-military dimension of the state and more upon trade was also furthered, according to Rosecrance, by the shattering twentieth century experiences of World Wars I and II. These wars helped demonstrate that warfare no longer served a very rational purpose and that means other than territorial conquest should be used to achieve security objectives. As nationalism grew, it became more difficult for imperialist states to control their colonial populations. The burden of defense expenditures also was becoming apparent to many. Moreover, defeat in war could lead to a fervent desire for revenge, which was how Germany's defeat in World War I apparently set the stage for a second world war twenty years later. Conversely, defeated adversaries who were helped to recover (instead of being humiliated) could quickly become economic competitors with their conquerors, as post–World War II experience proved. Japan and Germany are clear examples of the power of the trading state today; their wealth increased dramatically once they abandoned their empires, dismantled the enormous military sectors it took to keep those empires under control, and sought their fortunes in the world through trade.

In general, what Rosecrance describes as the trading state is one guided by the doctrine of liberalism. Whatever one calls it, its central economic premise is that unfettered trade creates wealth, which further increases both wealth and trade. We can see clear evidence of the rise of the trading state by examining the phenomenal upward curve in foreign trade over the course of the Westphalian system. That increase has been particularly dramatic since the Industrial Revolution in the late nineteenth century. As Figure 10.1 illustrates, total global exports, adjusted in constant U.S. dollars, grew little and remained relatively low during the first two centuries after Westphalia, but began to rise somewhat after 1870. Yet this acceleration in the growth of trade seems modest compared with events of the post–World War II era: Total exports have demonstrated a more than tenfold increase. Indeed, in just seven years (1985–1992), exports almost doubled.

# THE DEBATE OVER FREE TRADE

The phenomenal growth in international trade, along with efforts to liberalize trade through organizations like the North American Free Trade Area and various GATT agreements, have intensified the mercantilist-liberal debate over free trade. Positions on this issue ultimately revolve around beliefs concerning how freer trade will affect both individual economic well-being and that of the nation.

**FIGURE 10.1** World Export Volume, 1720–1992 (in billions of 1973 U.S. Dollars)

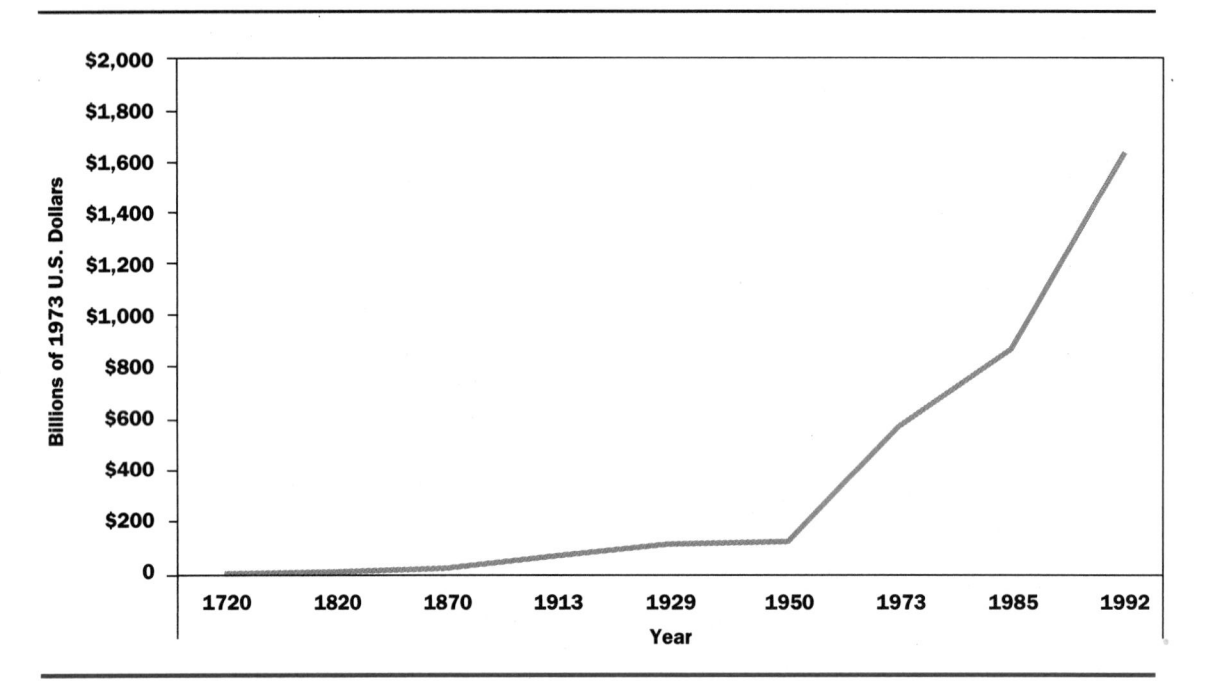

SOURCE: Angus Maddison, "World Economic Performance since 1870," in *Interactions in the World Economy,* ed. Carl-Ludwig Holtfrerich (New York: New York University Press, 1989), 230. The figure for 1992 was adjusted to 1973 prices on the basis of data from the United Nations, *World Economic and Social Survey, 1994* (New York: United Nations, 1994), 270.

## ARGUMENTS FOR FREE TRADE

Free trade has had many influential proponents for nearly two centuries. Associated with doctrines of **economic liberalism**—that is, laissez-faire capitalism as applied to world economic relationships—much of the discussion of trade policies in the world today still reflects the classical arguments for free trade as presented by such nineteenth century economists as David Ricardo. One of Ricardo's important contributions was to emphasize the *advantages of economic specialization.* The argument is that if each state concentrates on those products where its society is most competitive, all will benefit. The theory holds that a state should focus upon those products or activities in which it possesses a comparative advantage—even though it might produce many other goods quite efficiently.

Applying Ricardo's argument today could mean, for example, that since the United States has proven itself to be one of the most efficient producers of agricultural products in the world through its huge agribusinesses, it would make sense for the United States to concentrate upon agriculture, and leave to other states the production of such goods as VCRs, or perhaps even automobiles. The United States could then utilize revenue received from the overseas sales of its agricultural products to purchase even more cars or VCRs at a cheaper price than it could produce such products domestically.

Proponents of free trade also point out that *protectionism is extremely costly for the consumer*. Not only is the consumer forced as a result of protectionist measures to pay higher prices because of the lack of competition, he or she is also restricted in the variety of products that can be purchased. It has been estimated that the cost of agricultural protectionism in the countries belonging to the **Organization for Economic Cooperation and Development (OECD)** amounted to $350 billion in 1992—costs that would have to be born by consumers in the form of higher prices or governmental outlays.[6] A recent study estimates that tariffs, quotas, and other barriers to imports cost American consumers $73.3 billion a year, or about $1,000 for the average family.[7]

Although trade restrictions may save a few domestic jobs, they do so at considerable cost. The previous study, for example, found that for twenty-one protected industries, some 1.8 million jobs were saved as a result of U.S. trade barriers—but at a cost of $170,000 per job.[8] This amounted to six times the average annual pay of manufacturing industry jobs, which suggests that it would be far cheaper simply to pay the workers the going wage and let them do nothing.

Whether in the form of subsidies or trade barriers, *protectionism interferes with competition in the world marketplace, thus allowing inefficient companies to remain in operation*. If producers are not subject to market forces, they have little incentive to modernize their plants or reduce their expenses in order to improve their efficiency. It is often the inefficient companies that resort to political pressure on government to protect them through restraints upon trade or financial assistance. Special interests with political clout are thus able to prevail over the common good.

Also, *protectionism can create costly and time-consuming paperwork* mandated by government to determine whether an importer or an exporter is evading protectionist legislation through transshipment by way of other countries. Partly due to concerns about such evasion, "the average transaction in international trade was said in 1985 to require 35 documents and 360 copies, a state of affairs likely to please few besides the paper industry."[9]

Free traders argue that a state seeking to restrict imports from abroad is likely to find that *other states will retaliate against protectionist measures;* thus no one is benefitted. The world depression of the thirties was exacerbated when other states began to retaliate against the United States for raising high tariff barriers with passage of the Smoot-Hawley bill in 1930. In this way, states engaged in a **beggar-thy-neighbor** policy, in which one state sought to enhance its own domestic welfare by promoting trade surpluses at the expense of other countries. In the 1930s, world trade dropped by a full 50 percent, to the detriment of all.

---

[6] World Bank, *Global Economic Prospects and the Developing Countries* (Washington, D.C.: World Bank, 1994), 42.

[7] Gary Clyde Hufbauer and Kimberley A. Elliott, *Measuring the Costs of Protection in the United States* (Washington, D.C.: Institute for International Economics, 1994).

[8] Ibid. In the case of ten of those industries, the cost for saving a single job came to more than $400,000.

[9] Norman S. Fieleke, *The International Economy* (Boston: Ballinger, 1989), 46.

"TAKE IT OFF! TAKE IT ALL OFF!"
OLIPHANT © 1971 UNIVERSAL PRESS SYNDICATE. Reprinted with permission. All Rights Reserved.

Finally, the most sweeping argument made by liberal free traders is that *free trade promotes world cooperation and peace* by making nation-states more interdependent. If states need each other for raw materials and markets, it follows in this view, they will be less likely to risk damage to that relationship by engaging in war. Furthermore, by being involved in a trading relationship, they will develop habits of cooperation that might be transferred to their political relationship.

## THE CASE FOR PROTECTIONISM

In spite of the recommendations for free trade, self-interested reasons have always been found for supporting mercantilist restrictions. In the contemporary period, perhaps the most persuasive of these is the argument by many **less developed countries (LDCs)** that they need to *protect their infant industries.* Since the ability to compete in the production of many industrial goods is dependent upon developing an economy of scale—that is, sufficient production to reduce the price per unit to a competitive level—a new, fledgling company will be unable to do so unless it is protected from the more cheaply produced goods from abroad. Initially, preferential trading agreements—in which the developed countries agree to give preference by not applying restrictions to the importation of goods from a less developed country—may be required. During the 1970s, both the United States and the European Community extended such preferences to a wide range of less developed countries. Their goal was to assist the LDC economic development efforts by providing easier access to northern markets.

Protectionism is also viewed as important in order to *provide product diversification.* States that are largely dependent upon the sale of one or a few commodities are subject to the whims of the market or nature, and can consequently find their incomes dropping precipitously from one year to the next. Dozens of countries—such as the oil-rich states of the Middle East, such coffee producers as Colombia and Uganda, as well as those with extensive copper reserves such as Zaire and Zambia—find themselves in situations in which a single product represents the bulk of their export sector.[10] The ability of other states to withhold weapons parts also creates an incentive for a state to develop its own arms manufacturing capability. Part of Saddam Hussein's poor showing in the Persian Gulf crisis was related to the fact that the U.S.-led coalition had stopped the flow of military parts to Iraq as part of its larger embargo against that state.

Within developed countries, proponents of protectionism also call attention to the need to *protect jobs and wages at home.* The textile industry in North America and Western Europe has been especially critical of the loss of jobs to countries in the developing world where wages barely meet subsistence levels and where the lack of governmental regulations and environmental restraints makes it possible to produce clothing at far cheaper prices than in the North.

Those advocating protectionist measures also call attention to the need to protect certain industries and goods in order to *guarantee national security interests.* For example, the shipping industry argues that it deserves to be subsidized and its competitors restricted since, in case of a war, the merchant marine will be needed to transport war materials and perhaps even troops. Similar national security arguments have been made in connection with state control of communication systems and banking.

Some suggest that tariffs and other protectionist measures may actually *facilitate global economic development by increasing foreign investment.* The argument is that multinational corporations, restrained by protectionist measures from exporting their goods to a certain state, may simply evade such restrictions by building factories in that country in order to sell their products locally. To the extent that such activities occur in LDCs, the end result is presumably of considerable benefit in improving the economic well-being of those societies.

Finally, restrictions may be placed upon the free flow of goods, as in the case of efforts to *keep certain products out of enemy hands* through export controls or to *punish a nation for undesirable actions.* For example, economic sanctions were placed against South Africa because of its apartheid policy and against Iraq as a result of its invasion of Kuwait. Protectionism might even serve an environmental purpose, as illustrated by legislation proposed by the Clinton administration that authorizes the American president to impose duties on imports from countries found to have environmental standards less rigorous than those on American industry.

## RESTRAINTS UPON TRADE

Mercantilist measures involving restraints upon international trade and the use of unfair trade practices assume a number of guises. The most obvious of these is a **tariff,** which is a tax based upon the value of the imported product. By imposing

---

[10]Overall, there has been some improvement in the situation, making a number of LDCs less dependent upon commodities for their foreign exchange. In 1975, commodity exports of less developed countries, as a whole, stood at more than 80 percent of their total exports. But by 1992 that figure was down to 40 percent. For Africa, however, the figure remained above 80 percent, down from 94 percent in 1975. United Nations, *UNCTAD Commodity Yearbook, 1994* (New York: United Nations, 1994), 82–83.

such a tax, a state makes those goods produced within its own boundaries more competitive with imported goods. If the tax is high enough, few consumers will choose to buy the imported product, however much they might like to have it. Another restraint is the **import quota,** which limits or even forbids the importation of a particular good.

Though tariffs and quotas usually pertain to the importation of products, they can also be applied to exports. For the most part, such restrictions are imposed only upon trading with a national enemy or for particular products that a government may not want to see exported, such as armaments or materials that would aid in the production of weapons of mass destruction.

Aside from these direct and obvious restrictions upon the free flow of goods among states, many governments have shown great skill in developing what are called **nontariff barriers** that also serve to restrain trade or to provide certain advantages to one state over another. Governments may provide **subsidies** for the production of certain goods to make them competitive in world markets. Farmers enjoy considerable political power in many states and have consequently been particularly successful in obtaining subsidies. In Japan, for example, total transfers to agriculture in 1992 amounted to $74 billion, or $600 for every man, woman, and child in the country. In the United States, subsidies on sugar alone added $3 billion to consumer bills.[11]

Other less obvious nontariff measures may also have an impact upon the flow of trade. For example, some European states levied burdensome use taxes upon heavy cars, aware that such taxes would most likely discourage buyers of American automobiles that traditionally have been heavier than their European counterparts. Even some kinds of restrictions upon advertising can imperil the free flow of trade: For instance, the French government has placed restrictions upon the advertising of hard liquor, which tends to be produced abroad, but not upon wine, which is a major domestic industry in France.

Many nontariff barriers are established on the grounds that they are necessary for the protection of health and safety of a state's citizens. By changing certain health and safety specifications, a state may catch foreign producers off guard, impeding their ability to compete when they cannot meet the new specifications in a timely manner. Sometimes serious differences of opinion may surface as to whether a given restriction really is relevant as a health or safety measure. During the Bush administration a major crisis arose over the importation of beef into the European Union, whose officials regarded it as unhealthy because U.S. cattle were fed hormones to facilitate growth. Although American farmers and the Food and Drug Administration rejected arguments that this practice posed a health hazard, the EU restriction held.

The flow of goods may also be slowed, if not prohibited, by licensing and inspection requirements. Import licenses can be denied to some, or the impediments of inspection can be made so onerous and time-consuming that foreign exporters simply give up. At one time, France had only one individual assigned to inspect every single VCR coming into the country. Needless to say, few such products were admitted as a result of the constant backlog.

---

[11]World Bank, *Global Economic Prospects and the Developing Countries* (Washington, D.C.: World Bank, 1994), 19.

Government procurement regulations may likewise undermine the free flow of goods by stipulating that all such procurement be based upon local production. Such requirements are particularly likely to be applied in the defense sector; governments often seek to lessen their dependence upon foreign defense suppliers, who might be able to affect national security interests by denying them parts or additional weapons. Cities and other subdivisions of the state from time to time impose similar domestic procurement requirements in their efforts to protect local jobs.

Governments and private firms may also engage in unfair trade practices by **dumping products** in foreign markets at a lower cost than that required to make a profit, or at least at a cost significantly lower than the product's price in the producer's market. By selling the product at a lower price than domestically produced goods, the market share for the imported product will be increased. While this can be very beneficial for the importing state's consumers, domestic producers may eventually be driven out of business—with a resulting loss of jobs and the prospect of increased prices due to the greater market control obtained through dumping practices.

## CREATING A LIBERAL TRADING ENVIRONMENT

The protectionist wars of the 1930s, in which a variety of such restrictive practices were used, contributed to the global depression of that period. This return to neomercantilist practices, whereby states tried to obtain economic advantages over others, resulted in losses for all. As that fact came to be appreciated, the United States passed the **Reciprocal Trade Agreements Act** in 1934, authorizing the president to negotiate the reduction of tariff barriers on a reciprocal basis with other states by up to 50 percent.

After World War II, memories of the protectionist experiences of the Great Depression prompted a vigorous effort to create a liberal trading environment worldwide. The most far-reaching initial attempt was to create a treaty-based International Trade Organization (ITO). That effort failed, however, because of the opposition of the United States government; President Truman refused to submit the treaty to the Senate for ratification, arguing that ITO would unduly restrict the United States while providing exceptions for other countries. The United States pressed instead to establish a much smaller and more limited organization, the **General Agreement on Tariffs and Trade (GATT).** Inaugurated in 1948, GATT became the vehicle for freeing trade among a minority of countries. Initially, GATT had 23 members, almost all of whom were among the richest and most highly developed states of the time.

### GATT AND THE WORLD TRADE ORGANIZATION

Although GATT began as a "rich countries' club," it more than quadrupled its membership over the next forty-five years, proving itself to be an effective vehicle in the effort to reduce barriers to global trade. That was in spite of the fact that GATT lacked the extensive institutionalization and powers envisioned in the aborted International Trade Organization.

Initially, GATT consisted largely of annual meetings of the membership, during which the participants would negotiate reciprocal reductions of tariff barriers primarily on a bilateral basis. Once agreeing to reduced tariffs, those tariff levels would automatically be extended to a number of other states according to the GATT agreement provisions for **most favored nation (MFN)** treatment. This simply means that a state must extend its lowest tariffs to all other states in the system that are covered by MFN status.

Since the most favored nation status can be highly beneficial to a state interested in increasing its international trade, MFN treatment can sometimes provoke political controversy. This was the case in 1991 when the Bush administration debated extending such treatment to the Soviet Union for the first time since World War II in an effort to encourage continued liberalization of the Soviet economy and political system. Even more controversial was the extension of MFN status to China, particularly after the Chinese government ruthlessly suppressed the pro-democracy movement at Tiananmen Square in 1989.

The GATT process became more multilateralized with the **Kennedy Round of Trade Negotiations** (1963–1967), which resulted in widespread tariff reductions of 35 percent, covering some 80 percent of the dutiable trade of industrial states, the signing of an antidumping code, and an international grain agreement. This was followed by the **Tokyo Round of Trade Negotiations** (1974–1979), which reduced tariff barriers by perhaps an additional 40 percent. Some success was also obtained during this round on dealing with the problem of nontariff impediments to trade.

As global recession set in during the 1980s, a number of states resorted to protectionist measures just as they had in the 1930s: Voluntary export restraints and other nontariff barriers to trade were instituted. Since GATT had been highly successful in reducing tariffs, states looked for other measures to discourage imports. A 1985 study, for example, found that some 27 percent of the total imports of industrial countries were covered by nontariff barriers.[12] Yet despite the successes of the Kennedy and Tokyo Rounds, some 40 to 50 percent of world trade remained outside the GATT framework. Prominent exclusions from GATT included intellectual property and such services as transportation, banking, and insurance.

In an effort to reverse the protectionist trend, the **Uruguay Round of Trade Negotiations** began in 1987, ending some seven years later with about 120 nations signing the new trade agreement. Among the important new features coming out of the Uruguay Round was the creation of the **World Trade Organization (WTO)**, which became operative on January 1, 1995.

The new organization, which will probably replace the GATT structure, has greater power to settle trade disputes related to the new trading rules. Under WTO, quotas on textiles and clothing will be phased out over a ten-year period; the protection of agriculture through subsidies and other means will be restricted; and tougher international trade laws will be designed to prevent patent abuse, copyright theft, and counterfeiting. The negotiators met the overall goal of cutting import tariffs by more than one-third on thousands of products. Failure to have reached such

---

[12] Robert E. Baldwin, *Trade Policy in a Changing World Economy* (Chicago: University of Chicago Press, 1989), 233.

an agreement, according to a World Bank estimate, would have meant a lost opportunity to expand world trade by about $200–300 billion annually (in 1992 dollars). The gains anticipated for the developing countries from agricultural liberalization alone were calculated to be $20–60 billion a year.[13]

Although the several rounds of GATT negotiations have accomplished much in liberalizing international trade, particularly with respect to tariffs that were reduced from an average of 40 percent of the price of each product in 1947 to around 5 percent in 1992,[14] there is still more to be done. The agenda for the future includes completing negotiations on such key functions as financial services, telecommunications, and transportation. Additional work is needed for further liberalization of trade in agriculture, steel, and civil aviation.

## ENHANCING REGIONAL TRADE

When looking at global trading patterns, one is struck first of all by the tendency of many regions to trade largely among themselves. Thus the United States is the largest single trading partner for both Canada and Mexico; trade among European nations accounts for more than one-third of their total merchandise trade; and intra-Asian trade is also somewhat higher than with any specific outside region.[15]

There are a number of reasons why one might expect regional trade to be favored over global trade, including the lower transportation costs that would be involved and, in many cases, a tendency to trade with those with whom one shares greater cultural affinity. But the explanation also lies in the creation of a number of regional trading institutions with a variety of powers and functions, starting with a **free trade area.** Such arrangements typically require the full elimination of tariff and nontariff barriers among the parties, but each member may retain its own tariff schedule. In practice, these arrangements need not apply to all products; some may be excluded. A **customs union,** on the other hand, is a free trade area in which barriers to the free flow of goods are abolished, but a common external tariff is put into effect. If the liberalization extends to the free flow of such other factors as labor and capital, it is known as a **common market.** A full **economic union** requires agreement on provisions for harmonizing certain economic policies, particularly those dealing with macroeconomic and regulatory issues. Such a union may entail considerable erosion of sovereignty.

**THE EUROPEAN COMMON MARKET** The experience of Europe in the postwar period illustrates some of the differences among trading institutions. In 1952, France, Germany, Italy, and the so-called Benelux states (Belgium, Netherlands, and Luxembourg) took the first step in forming a customs union when they established the European Coal and Steel Community (ECSC). That was followed in 1957 with the much more ambitious European Economic Community (EEC), which created a

---

[13]World Bank, *Global Economic Prospects and the Developing Countries* (Washington, D.C.: World Bank, 1994), 19.

[14]Hal Kane, "Trade Continues Steep Rise," in *Vital Signs,* ed. Lester Brown et al. (New York: W. W. Norton, 1993), 74.

[15]"WTO: The New Age of Trade," *New York Times,* April 15, 1994, A23.

> ### Global Changes
>
> ## Changing Economic Patterns
>
> **E**conomic hegemons: Some within the liberal perspective argue that the success of the international economic system is dependent upon the existence of a major hegemon willing to use its power and resources to support a liberal trading environment and to assure economic growth. Three such hegemons have been identified during the Westphalian era: the United Provinces (Holland) in the eighteenth century; Great Britain in the nineteenth; and the United States in the post–World War II era.
>
> - Why do some argue that the United States is a hegemon in decline? If a new hegemon is to arise, who will it likely be?
>
> World trade: Based upon countries for which data are available, world exports as a percentage of global GNP increased from 4.6 percent in 1830 to 11.4 percent in 1913, followed by a sharp decline in the 1930s. By 1950 the figure stood at 8 percent.* The upward trend continues. Exports as a percentage of world GNP reached almost 20 percent in 1992.
>
> - Some believe that the nineteenth century increase in world trade was primarily related to the industrial revolution. Why might this have been so? The decline between the two world wars was in part attributable to the global depression. Why? Finally, what factors might help explain the increase in global trade in the post–World War II era?
>
> Tariffs: Tariffs were a major source of U.S. government revenue until the early twentieth century, at which time they became primarily instruments for trade protection. Following World War II, negotiations have led to substantial reductions in tariff levels throughout the world. These have fallen from 40 percent in 1947 to an average of 5 percent in 1992, and many products now have no tariff whatsoever.†
>
> - With the demise of meaningful tariffs on most products, what impediments remain to the liberal ideal of an unfettered global economy?
>
> Monetary system: Despite the 40 percent decline in the value of the dollar against the Japanese yen over the period 1989–1994, and similar trends with respect to other hard currencies, the dollar remains the world's most important currency. Nearly half of the world's trade is priced in dollars; more than 40 percent of all currency exchange transactions involve the dollar; 44 percent of the world's horde of stocks, bonds, bank deposits, and other instruments are denominated in dollars; and in places throughout the world when local currency values are threatened by inflation or political uncertainty, dollars serve as a store of value and are readily accepted.‡
>
> - What factors will determine whether or not the dollar remains in its preeminent position as a world currency? Will the decision of the European Union to create a single currency—to be called the *Euro*—by 1999 undercut the dominant position of the dollar?
>
> ---
>
> *Stephen D. Krasner, "Economic Interdependence and Statehood," in *States in a Changing World*, Robert H. Jackson and Alan James, eds. (Oxford: Clarendon Press, 1993), 311.
>
> †International Monetary Fund, *World Economic Outlook* (Washington: International Monetary Fund, 1993), 70.
>
> ‡Jeffrey A. Frankel, "Still the Lingua Franca: The Exaggerated Death of the Dollar," *Foreign Affairs* 74 (July/August 1995), 10–11.

customs union covering all products. By the end of 1992, EEC's successor organization, the European Community (EC)—with a membership of twelve after the addition of Great Britain, Denmark, Greece, Ireland, Portugal, and Spain—reached its goal of a single integrated common market when all remaining impediments to the free flow of goods, labor, and capital were abolished. The European Community had also made considerable progress on agreeing to a common foreign and defense policy. Armed with new powers under the Maastricht Treaty, which provided additional authority with respect to coordinating economic and social policy in the community, the EC in 1994 changed its name to the European Union to reflect its new status. Austria, Finland, and Sweden also joined the EU, for a total of fifteen members.

**THE EUROPEAN FREE TRADE ASSOCIATION** Because of concerns in the late 1950s about the EEC's potential effect upon their sovereignty, several other European states established the competing European Free Trade Association (EFTA), which allowed them to retain their individual tariff schedules. EFTA's purpose was to heighten trade among its initial seven members: Austria, Denmark, Norway, Portugal, Sweden, Switzerland, and the United Kingdom. After establishing a free trading area among its membership, EFTA negotiated free trade agreements with the EEC as well as with Finland and Iceland. But given the greater success of the European Community in establishing a broad common market, former EFTA members Austria, Denmark, Portugal, Sweden, and the United Kingdom have joined the European Union.

**THE NORTH AMERICAN FREE TRADE AGREEMENT** In January 1989, the United States and Canada, which exchange more goods and services than any other two countries in the world, entered into a bilateral free trade agreement. This agreement was expanded into the **North American Free Trade Agreement (NAFTA)** in 1992 with the addition of Mexico. NAFTA provides improved market access in many sectors, such as agriculture, automotive products, textiles, and apparel. Intellectual property and investment, which are usually not covered in such agreements, are also included in the new trade rules. After some political pressure, supplementary agreements were negotiated on labor and environmental standards due to concern that the lower standards required by Mexico would place it in a competitive advantage.

A primary issue raised during the U.S. debate on the ratification of NAFTA was the extent to which jobs would go to Mexico, where wages are lower than in the United States. Yet, in reality, such jobs could be relocated regardless of a trade agreement. Without NAFTA, there would have been an incentive for American automobile manufacturers to build plants in Mexico in order to evade the 20 percent tariff that Mexico collected for imported cars.

According to Stefan H. Robock, professor emeritus at Columbia University, the notion that jobs are exported overseas by more liberal trading agreements is largely a myth.

> American companies' overseas manufacturing operations are mostly in high wage, advanced countries, accounting in 1992 for 84 percent of the operations' sales and 64 percent of their work forces. And the host country and other foreign countries are their principal markets. In 1992, only 10 percent of total sales from these operations were to the United States, and only 25 percent of sales from operations in less-developed countries.[16]

---

[16] Stefan H. Robock, "Jobs Exported Overseas? It's a Myth," *New York Times*, November 27, 1994, F11.

## THE FUTURE OF REGIONAL TRADE AGREEMENTS

After the successful conclusion of the GATT negotiations and NAFTA, the Clinton administration took an aggressive approach to expanding regional economic arrangements that would go further than the Uruguay Round agreement in eliminating barriers to trade. In November 1994, the eighteen members of the **Asia-Pacific Economic Cooperation (APEC)** group committed themselves to "free and open trade and investment" by 2020. This would create the world's largest free trading area, involving countries situated on four continents that now account for one-half of the world economy and more than 40 percent of its trade. A month later, a summit of Latin American leaders meeting in Miami agreed to begin negotiations for a Western Hemispheric free trade agreement. Negotiations have begun for the accession of Chile to NAFTA and discussions with other Latin American states are expected to follow. And in June 1995, the Clinton administration agreed to a European proposal to begin discussions on a free trade area that would encompass all of the North Atlantic region.

It should be noted, however, that movement toward regional free trade areas is not always perceived as positive by those who would like to see freer global trade. GATT has accepted the legitimacy of such trading blocs if they are not more restrictive toward the outside than the individual members had previously been, and if the areas covered by the free trade arrangement include most sectors of the economy. Still, regional blocs have been able, through protectionist measures, to redirect trade at the expense of outside nations as the experience of the European Union and its predecessors have shown. Whereas in 1958 only 29.6 percent of imports and 30.1 percent of exports by Common Market countries came from other member nations, by 1970 those figures stood at 48.4 percent and 48.9 percent, respectively.[17]

Moreover, since a common external trade policy is required in a customs union, restrictions on trade will be forced upon those member states that would reject the protectionist measure if allowed to make their own choices. Given weighted voting and veto rights in some cases, the French, for example, were able to delay the Uruguay Round negotiations and to obtain community acquiescence for some of their preferred protectionist policies in the final agreement. It was only after considerable pressure that the large majority of the EC was able to persuade the French to accept a compromise on the issue of agricultural subsidies and thus to remove one of the last sticking points.

On the other hand, it can be argued that regional agreements, as long as they do not create warring trading blocs, can actually speed the creation of freer global trade. C. Fred Bergsten, the president of the Institute for International Economics, notes that if Europe sees the Asians and the United States negotiate regional trade agreements, it will seek to match or even better the terms. It was fear of just such an arrangement, he contends, that persuaded the Europeans to complete the long-stalled negotiations in the Uruguay Round.[18]

---

[17]Robert A. Isaak, *Managing World Economic Change,* 2d ed. (Englewood Cliffs, N.J.: Prentice-Hall, 1995), 85.

[18]David E. Sanger, "Trade Agreement Ends Long Debate, but Not Conflicts," *New York Times,* December 4, 1994, A1.

## CONFLICT OVER BILATERAL TRADE

Despite the considerable progress made in developing cooperative global and regional trading arrangements, various bilateral trading relationships have been quite acrimonious. A prime example of this has been the U.S.-Japanese trading system, spurred on as the result of huge American trade deficits with Japan, which reached $66 billion in 1995.

The United States has blamed this imbalance upon the protectionist policies of the Japanese—policies that the United States initially did not challenge because of its greater concern to help Japan develop economically after the devastation resulting from World War II; the United States wanted its former enemy to become strong enough to help deter the threat of communism.

During the early postwar years, Japanese exports were not competitive with American products and were viewed as shoddy in comparison. The situation changed in the 1960s, particularly in the auto industry, when Japan developed a reputation for building high-quality products. These automobiles became even more popular as the result of the oil crisis in the 1970s, when smaller cars were favored over the gas-guzzling American models. American consumers found the same quality control and cheaper prices in Japan's manufacture of electronic equipment, so much so that most televisions and VCRs began to come from Japan and, subsequently, from other parts of Asia.

In attempting to deal with the trade imbalance, the United States has negotiated a number of bilateral agreements with Japan that sought to open up Japanese markets or to limit Japanese exports to the United States. The negotiating pattern was established early; threats and acrimony would peak and then, at the last moment, the Japanese would acquiesce to American demands. In practice, however, Japan then would often fail to meet the specific requirements outlined, and more recently has begun to challenge the pressure exerted by the United States.

Some of the agreements reached with Japan, particularly those in the auto and textile industries, have involved what is called **managed trade** (rather than free trade). Popular among these are the so-called **voluntary export restraints (VERs),** according to which a state agrees to limit the quantity of products it exports. To refer to VERs as voluntary, however, is somewhat of an oxymoron, for it is usually made clear to the exporting state that if it fails to establish such restraints against its own exporters, it can expect far more restrictive trade legislation in the future. Not only are limitations established for the amount of exports, but some agreements have provided a target for importing an increased number of units of a product from its trading partner in what is called **voluntary import expansion.** But again, it is hardly voluntary. Unlike VERs, however, voluntary import expansion may actually help competition within domestic markets by bringing in more products to be sold.

Tariffs, for the most part, no longer inhibit American access to the Japanese market. Instead, a variety of nontariff measures have made it difficult to penetrate that market. Among these are agricultural subsidies to Japanese farmers, which in 1993 amounted to over $70 billion.[19] Subsidies on rice, for example, raised the cost of that food item to the Japanese consumer to six times that of its world price.[20]

---

[19] Hobart Rowen, "Exposing Trade Curb's Real Costs," *Washington Post,* August 22, 1993, H1.
[20] World Bank, *Global Economic Prospects and the Developing Countries,* p. 42.

Nontariff barriers are a major reason why the Jeep Cherokee, which in 1995 cost $19,100 in an American showroom, increased 64 percent, to $31,372, when sold in Japan. These barriers include the costs of checking for compliance with 238 Japanese regulations, cuts taken by distributors, costs related to complex distributor-dealer relationships, as well as additional inspections and reinspections for work already done.[21]

American officials have made other arguments to suggest that Japan plays unfairly when it comes to trade and commercial activities. These include charges regarding restrictions and complications in bidding on construction jobs in Japan and governmental procurement in general. (Many state governments in the United States and the national government itself place restrictions on such construction when public expenditures are involved.)

In an effort to provide what American officials like to call a level playing field, the United States has increasingly resorted to retaliatory threats against Japan. For example, in May 1995, President Clinton, in accordance with the authority given him under U.S. trade legislation (known as **Super 301**), threatened to impose a 100 percent tariff on thirteen Japanese luxury automobiles. The threat was designed to persuade Japan to import more American-made automobiles and to reform its insular business practice based upon the **keiretsu system.** The latter involves a system of interlocking ties between manufacturers and their suppliers and distributors, resulting in discrimination against foreign suppliers of automobile components. The extension of keiretsu to Japanese automobile plants in the United States has been particularly troubling to the U.S. automobile industry.

If American automobile manufacturers are successful in establishing targets for exports to Japan, other producers will certainly demand comparable treatment. Moreover, other nations, concerned about an increased American share of the Japanese auto market at their expense, will not remain idle. For instance, immediately upon the issuance of the Super 301 threat, the European Union took the issue to the World Trade Organization and also negotiated an agreement with Japan that would make it easier to export European automobiles to Japan.

Some are concerned that such unilateral actions can set the stage for a trade war between states, particularly since politicians have discovered that making threats and counterthreats involving trade restrictions increases their political popularity. So far, retaliation against the use of 301 legislation has been fairly muted. Despite ninety-one investigations based upon the legislation between 1975 and 1993, only nine resulted in U.S. retaliatory action against alleged unfair trade practices, and in only two of those nine cases did tit-for-tat retaliation by the foreign state result.[22] Indeed, the threat of economic retaliation is very much like a nuclear threat, in that it is hoped that the mere threat will change the behavior or deter the other side. To actually use the weapon would likely result in severe punishment to both sides.

Unilateral retaliation against alleged unfair trade practices violates the letter and spirit of the GATT and NAFTA agreements, which provide dispute settlement procedures involving panels of trade experts to rule on complaints. Obtaining compli-

---

[21] Sheryl WuDunn, "An Uphill Journey to Japan," *New York Times*, May 16, 1995, D1.

[22] Kimberley Elliott and Gary Hufbauer, "Sanctions Might Work," *Washington Post*, January 20, 1994.

> **Normative Dilemmas**
>
> ## Winners and Losers in Trade Agreements
>
> The environment as loser: In one of its first decisions, the new World Trade Organization, in January 1996, upheld a challenge to U.S. environmental standards for imported gasoline. WTO officials agreed with the Venezuelan government that American requirements for cleaner burning fuel discriminated against imports.
>
> - Should the United States modify its environmental laws simply because outsiders think such laws go further than they believe is necessary for good health?
>
> Jobs versus lower prices: The jobs of some 1.1 million U.S. textile workers are protected by various quotas and tariffs, which will gradually be removed under current free trade agreements. This protectionism adds 50 percent to the wholesale price of clothing, according to Louis E. Boone and David L. Kurtz.*
>
> - Whose interests should be respected in this case: workers or consumers? Should some protectionism remain to protect jobs of textile workers?
>
> National values versus the free flow of information and entertainment: A major factor in French reticence to complete the Uruguay Round agreement revolved around its concern to protect the French film industry and to keep out foreign films, which it saw as undermining French culture. Similar reservations have been voiced by Canadians upset about U.S. cultural dominance in that country. Germany has threatened to ban internet providers who allow pornography to get onto the World Wide Web.
>
> - How far should a nation be able to go to restrict the flow of information and entertainment that it believes to be against its national values? Will such restrictions simply become an excuse for trade protectionism?
>
> ---
>
> *Louis E. Boone and David L. Kurtz, *Contemporary Business*, 7th ed. (Fort Worth, Texas: Harcourt Brace, 1993), 82.

ance with GATT panel recommendations has been relatively successful as these sorts of dispute resolution procedures go. One study of 117 cases discovered failure of compliance in only eight to ten cases.[23] The strengthened dispute resolution procedures provided by the World Trade Organization forbid states from being able to veto their own punishment, as had been the case in GATT. Therefore, one might expect the compliance record to improve.

# THE INTERNATIONAL MONETARY SYSTEM

With the world's increasing involvement in international trade, procedures have been developed to pay for goods exported abroad. The primitive system of bartering one good for another has long since given way to economies based upon money and credit.

---

[23] John H. Jackson, *The World Trading System* (Cambridge, Mass.: MIT Press, 1989), 101.

## CURRENCY VALUATION

National currencies come with a variety of names, such as dollars, yen, marks, and the like. Their values fluctuate considerably, either because of governmental action or market forces. **Fixed exchange rates** are established by governments so that the state might trade one currency for another or purchase goods from abroad based upon its relative currency valuation. By manipulating such exchange rates, a government can affect the flow of goods between states by making it possible to purchase more or fewer products from abroad for the same amount of currency.

Alternatively, a state might simply operate within a **floating exchange rate** system, in which global market forces determine the value of the currency as it fluctuates from day to day. Relying upon market forces to determine currency values, however, produces considerable financial instability, creating an element of economic risk. Such instability and risk, in turn, tend to discourage international trade and introduce considerable speculation into the global financial markets.

Efforts have been made to stabilize the fluctuation in the value of various currencies. In the 1870s and again after each of the world wars, many states pegged the value of their currencies to that of gold. In 1934, for example, the United States placed the value of gold at $35 an ounce—a level that it continued to support until 1971 when foreign traders threatened U.S. gold reserves by demanding payment in gold rather than dollars.

Despite the lip service paid to the gold standard, governments often failed to adhere to it. States in balance of payments or other economic difficulties would remove themselves from the standard unilaterally, or would peg their currency at unrealistic rates. Many felt that the competitive devaluation of currencies in the interwar period (1919–1939) was a major factor inducing protectionist measures; that is, some states sought to protect their domestic industries from imported goods made cheaper in those countries that had resorted to devaluation. In an effort to correct some of these problems, a conference was held at Bretton Woods, New Hampshire, in 1944, during which the **International Monetary Fund (IMF)** and the **International Bank for Reconstruction and Development (IBRD)** were created.

Under the **Bretton Woods system,** governments were expected to peg their currencies to gold or to the American dollar, which the United States would support at $35 per ounce. The IMF had to approve the initial exchange rates as well as most subsequent changes. The agreement also required that **convertible currencies** be provided, so that governments or individuals could pay their bills in a variety of currencies or sell them to a central governmental bank, which in turn could convert the monies to its own currency, or even to gold.

Currencies that were easily convertible because of governmental policy or a general faith in the ability of the government to make good on its notes came to be known as **hard currencies.** Today, these include U.S. dollars, German marks, and Japanese yen, among a good many others. **Soft currencies** with restricted convertibility and limited value due to the adverse economic conditions of the state producing the currency, on the other hand, are found largely in less developed countries and in such state-controlled economies as that of the former Soviet Union. The successor Soviet republics have moved in the direction of currency convertibility;

U.S. Acting Secretary of State Dean Acheson addresses representatives to the Bretton Woods conference, 1944.

however, the costs of those moves were very high in the early 1990s, particularly in terms of the inflationary effect. Their monthly rates of inflation have moved, on occasion, into double digits.

## THE BALANCE OF PAYMENTS PROBLEM

**Balance of payments** refers to the total flow of money into and out of a country. Its largest component includes revenue from trade, but the balance is also affected by income from services and the flow of capital. Table 10.1 provides a summary of the major credits and debits involved in calculating the balance of payments. When debits are greater than credits, a state has an *unfavorable* balance of payments and may be considered to be in balance of payments difficulty.

To correct an unfavorable balance, a country must seek credits or take action to increase its exports or decrease its imports. In an effort to discourage protection-

### TABLE 10.1   The Balance of Payments Ledger

| CREDITS | DEBITS |
| --- | --- |
| Exports of goods | Imports of goods |
| Foreign travel of others to your country (military, tourists) | Citizen's travel to other countries |
| Foreign purchases of your services | Your purchase of foreign services |
| Gifts and other transfers from abroad (foreign aid, charity) | Gifts to those abroad |
| Capital imports (loans to and investments in your country) | Capital exports |
| Profits, interest from foreign investments | Profits, interest paid to foreign investors |
| Government receipts from foreign expenditures | Government overseas activity |

SOURCE: Leland B. Yeager, *International Monetary Relations* (New York: Harper and Row, 1976), as cited in John T. Rourke, *International Politics on the World Stage,* 2d ed. (Guilford, Conn.: Dushkin Publishing Group, 1989), 366–367.

ist approaches and the threat of unilateral currency devaluations as a way of correcting balance of payments problems, the IMF may provide short-term credits called **Special Drawing Rights (SDRs).** SDRs—which consist of reserves based upon the average value of several currencies—may be used by states to pay the central banks of other states with which they have an unfavorable balance of payments. Such reserves have been viewed by many potential recipients as both inadequate and requiring excessively stringent changes in economic policy. As a result, some states have sought credit support elsewhere.

It was primarily balance of payments difficulties that led to the U.S. rejection of parts of the Bretton Woods system in 1971, when the United States was confronted for the first time in the twentieth century with a balance of trade deficit. Pressured by France and other states that were beginning to demand payment in gold, President Nixon suspended convertibility of the dollar and imposed a 10 percent import surcharge. Since then, the United States has refused to return to a fixed rate system as envisioned in the Bretton Woods agreements. Nonetheless, the institutions created at Bretton Woods remain viable and active; today they play a critical role in economic development activities (as will be shown in Chapter 12).

## CURRENCY VALUES AND THE FLOW OF TRADE AND INVESTMENT

States with chronic balance of payments problems will generally find that the value of their currency in the world markets will decrease in relation to those of states that are doing better economically. Consequently, a floating currency mechanism theoretically will have a tendency to improve the balance of payments position by moving a state toward a favorable balance of trade. This is because currency devaluation makes one's products cheaper and more competitive in global markets. Devaluation will also encourage increased tourism and investment in that state as outsiders, searching for a good deal, find that the exchange rate is favorable—that is, that it lets them buy more for their money.

Japan's experience suggests that the theory does not always work in practice. The yen rose some 40 percent against the U.S. dollar from 1989 to 1994, making Japanese exports that much more expensive. Despite predictions that the yen's rise would soon price Japanese exports out of world markets, Japan enjoyed an upsurge in exports, which increased by 41 percent in that same five-year period.[24]

It would seem logical that because the value of the dollar has declined substantially in relation to other currencies—making it cheaper to buy American products—the U.S. balance of trade should improve substantially. But this has not been the case for several reasons: (1) it takes two or three years for economic forces to respond to changing prices; (2) Americans may still travel or buy foreign products even at the much higher costs, simply because they want the experience or the products remain of high value to them; (3) such countries as China and Taiwan, which peg their currency to the value of the dollar, will not realize any advantage to a falling dollar since their purchasing power remains the same relative to that of the dollar; (4) consumers from other states may not be able to take advantage of lower American prices with the fall of the dollar due to restrictions placed upon imports by their governments; and (5) foreign companies, in order to retain their share of the market, may simply absorb the additional costs to meet the lower U.S. prices rather than increase the prices of their products.[25]

## CONSOLIDATING CAPITALISM

Some see the contemporary world as moving increasingly toward a global capitalist economy. For evidence, they cite the sudden collapse of Marxism and the sorts of reforms undertaken in both the international trading and financial systems discussed in this chapter. Harvard economist Jeffrey R. Sachs, who became a major adviser to economic reform movements in the former Soviet bloc, asserts that while such a unified global economy has not yet arrived, "countries with a combined population of roughly 3.5 billion people have undertaken reforms to adopt the institutions of the capitalist system."[26] He lists in particular the following global achievements of recent years: the creation of open international trade; the adoption of currency convertibility; private ownership operating as the main engine of economic growth; corporate ownership as the dominant organizational form of large enterprises; openness to foreign investment; and the growth of membership in such in-

---

[24] Eamonn Fingleton, "Japan's Invisible Leviathan," *Foreign Affairs* 74 (March/April, 1995): 69.

[25] Dean H. Hanink, *The International Economy: A Geographical Perspective* (New York: John Wiley, 1994), 123.

[26] Jeffrey R. Sachs, "Consolidating Capitalism," *Foreign Policy* 98 (spring 1995): 51.

ternational economic institutions as the World Bank and the World Trade Organization.

These achievements may or may not lead to a permanent consensus on the policies that should control economic life. Nonetheless, the movement toward a global economic system in recent years is one of the most remarkable developments of the century's last decades. We have arrived at a moment, as another prominent scholar has put it, when "the logic of 'anarchy'—the fragmentation of the world into sovereign states—is checked by the logic of, and a broad consensus on, an open global economy."[27]

# CONCLUSION

Three major perspectives have dominated the discussion of international political economy: mercantilism, liberalism, and Marxism. The debate over the international trading and monetary system has been largely conducted between mercantilists and liberals. Mercantilists believe that economic instruments should serve the state and that a state's wealth and power is enhanced by an emphasis upon exports; economic liberals see great virtue in unfettered trade. From the Marxist view, the capitalist state is seen as an instrument allowing one class to exploit another; once capitalism is overthrown, capitalists as an exploitive class will no longer dominate, and greater economic justice should prevail. Today, Marxism has been eclipsed as an important governmental ideology. Few now adhere to the extreme views of any of these perspectives—hence, the tendency to add the term *neo* to describe contemporary variants on each of these doctrines.

With the end of the cold war and the prospect that weapons of mass destruction will play a diminished role in the foreign policies of great powers, military and security issues have assumed less importance while economic and developmental issues have become increasingly significant. Domestic economic issues have important effects on the international political economy. As economic conditions deteriorate within a state when jobs are lost and the trade balance suffers, governments often engage in a variety of protectionist activities. These include the use of tariffs and quotas, voluntary export restraints, and a variety of nontariff barriers. Regulations related to licensing, inspection, and procurement have also been used to restrict the free flow of goods.

In spite of these pressures favoring protectionist activities, the classical arguments for free trade presented in the nineteenth century remain powerful today, notwithstanding the fact that arguments for exceptions to the general rule still abound. Proponents of trade freedom insist that all participants are likely to gain by specializing in the production of those goods in which they enjoy a comparative advantage. Consumers, economic competition, and even world peace might likewise benefit from a liberal trading environment.

---

[27] Stanley Hoffmann, "What Should We Do in the World?" *The Atlantic Monthly*, October 1989, 86.

Protectionists, who tend to become more vocal during periods of recession, argue for the need to protect and develop infant industries, to preserve jobs and higher wages, and to provide diversification and protection for national security interests. In recent years, they have essentially changed the debate from one of free trade to one of fair trade.

It is clear that in an interdependent global economy, some institutionalization is required to regulate the system, stimulate trade, and facilitate economic interactions. A variety of institutional devices play this function; these include everything from regional customs unions and free trade areas to such global organizations as the International Monetary Fund and the new World Trade Organization.

# CHAPTER 11

# Economic Statecraft and Private Economic Activity

The growth of a global economy has increased economic interdependence and the need to assure continued access to markets, capital, and raw materials. As a result, states have unprecedented opportunities to utilize economic leverage to reward or punish—and thereby influence—other states to do their economic and political bidding. In this chapter, we first explore how economic statecraft has influenced the behavior of other actors in the international system, often serving political rather than strictly economic ends. Then we shall examine the tension between states and private market forces in the global system that can sometimes undermine the objectives of economic and political statecraft. There follows an analysis of the growing interdependence among the various global economic actors and its impact upon global peace and stability.

## ECONOMIC STATECRAFT

The state can utilize a variety of economic instruments to enhance its prospects of achieving economic and political objectives. Three such instruments of economic power have been particularly prominent during the course of the Westphalian system: economic imperialism, foreign aid, and economic sanctions. Determining the effectiveness of these various instruments, however, is not merely an empirical issue, for normative considerations are at stake as well. As with any policy instrument, some groups will benefit, while others will lose. Moreover, some values such as national security may be served, whereas others, such as economic growth or economic equality, might be sacrificed.

# ECONOMIC IMPERIALISM

Many observers have seen economic factors as a primary explanation for war and imperialism. Although a state might be hesitant to undertake imperialistic actions if its resources are limited, those same deficiencies may lead it to covet the economic assets of others, providing sufficient incentive for an attack, regardless of the risks. Plato was an early proponent of such a view, and went so far as to suggest that a republic should remain poor so that it would not be worth attacking. He argued that the rich society—what he called the "luxurious state"—would incite only envy in others and require constant expenditures for its own defense. The historical record is indeed one in which groups suffering from an inadequate hunting or harvest season have simply added to their food supplies by conquering neighboring peoples.

**CAPITALISM AND IMPERIALISM** Although Marxist and radical writers have often linked capitalism and imperialism, the latter has ancient roots, and much predates the rise of capitalism. Nonetheless, capitalism became the dominant mode of economic organization for the European states that engaged in imperialism in recent centuries. As a result, a number of writers living at the height of the European imperialist period sought to explain imperialism through an analysis of capitalist economics.

Writing in the mid-nineteenth century, Karl Marx set the stage for such analysis with his insistence that the advanced industrial societies of his time were characterized by a conflict between the dominant, capitalist class and the workers whom the capitalists exploited for profit. While Marx devoted most of his attention to analyzing these relationships within capitalist states rather than those between such states and poorer nations, others applied some of his notions to explain imperialist actions taken by capitalistic interests in other regions of the world.

Although not a Marxist, the British theorist John A. Hobson, writing in 1902, blamed British imperialism on special-interest groups composed of capitalists.[1] Borrowing from Marx, Hobson asserted that imperialism resulted when capitalist groups, confronted with an oversupply of capital and goods, sought to obtain markets abroad in order to divest themselves of that surplus. The economically underdeveloped areas of the world, given their chronic lack of capital and untapped markets, provided particularly useful investment opportunities. For Hobson, the imperialism of the British government became "a vast system of outdoor relief for the upper classes."

Unlike Marx, however, Hobson did not recommend eliminating capitalism and the capitalist; instead, he believed the means of purchase should be placed in the hands of the workers in order to reduce the incentive for economic imperialism. This could be accomplished by strengthening unions and by using the taxing powers of government to place more money at the disposal of the domestic consumer.

Vladimir Lenin, writing shortly before the Bolshevik revolution that brought him to power in Russia in 1917, adapted the notions of Marx and Hobson. He

---

[1] John A. Hobson, *Imperialism: A Study* (Ann Arbor: University of Michigan Press, 1963).

viewed imperialism as not only the highest, but also the last, stage of capitalism.[2] Like his predecessors, Lenin believed that capitalist states engaged in imperialist activities in order to acquire markets for the oversupply of capital and goods that capitalist economies inevitably produced. But this imperialism brought them into conflict with other capitalist states, and the end result was war among imperialist nations, as evidenced—according to Lenin—in the outbreak of World War I.

Apart from the fact that imperialism long predated the rise of capitalism, there are other reasons for rejecting the notion that capitalism necessarily leads to imperialism or conflict among imperialist states. Most of the wars that have occurred in the age of mature capitalism were not fought primarily for economic reasons. Examples include the Austro-Prussian War, the Franco-German War, the Crimean War, the Spanish-American War, all of which were fought between 1866 and 1898, and, in the twentieth century, the Russo-Japanese War, and the two World Wars.[3] According to another authority, "Wars have occurred during periods of capitalistic dominance, but they have been least frequent in the areas most completely organized under that system."[4]

The 1990 Iraqi invasion of Kuwait by Saddam Hussein, who lusted after its rich oil fields, is but the latest chapter illustrating that noncapitalist states (Iraq's economy is centrally planned) are just as likely to engage in economic imperialism. Yet even if one were to find that only capitalist states behaved in an imperialistic fashion, this would not necessarily mean that capitalism induces imperialism. Such imperialism may be due simply to the tendency of great powers to increase their power, and in modern times most of the powerful states have been capitalist.

# FOREIGN AID AS AN INSTRUMENT OF POLICY

**Foreign aid** is a broad term that covers every kind of grant or loan given from one government to another, including both military and economic assistance. As might be expected, liberal and Marxist writers have quite different conceptions of the role and even the morality of such foreign aid. For the liberal, foreign aid used for economic developmental purposes can be viewed as a way of enhancing the liberal trading order. By helping another state increase its economic well-being, an economic aid donor or investor can help create potential customers for its exports. In fact, legislation is often adopted requiring purchases to be made in the donor state as a condition of using such aid.

So, for liberals, foreign aid that is provided primarily with the expectation of increasing one's own export opportunities is hardly altruistic. Marxists see foreign aid from capitalist donor states in more sinister terms, as designed primarily to keep less developed states perpetually nonindustrialized and dependent. Marxists often argue that much foreign aid and investment from capitalist states goes into extractive

---

[2] Vladimir Lenin, *Imperialism: The Highest Stage of Capitalism* (New York: International Publishers, 1939 [orig. pub. 1916]).

[3] Hans Morgenthau and Kenneth W. Thompson, *Politics among Nations*, 6th ed. (New York: Knopf, 1985), 64.

[4] Quincy Wright, *A Study of War*, 2d ed. (Chicago: University of Chicago Press, 1965), 1163–1164.

industries such as minerals and other raw materials, while little, if any, is provided for building new industries.

In any case, foreign aid more often has had a political purpose than an economic one. The use of monetary rewards as an instrument of policy has a long history. Rulers of ancient China and other empires demanded tribute from their vassals—a practice meant in part to demonstrate the essential inequality of the actors. Since the beginning of the Westphalian period, economic payoffs have been used to influence the behavior of sovereign equals. According to one authority, "It was quite common in the eighteenth century to offer statesmen large gratuities in return for an alliance or the favorable conclusion of negotiations."[5]

The reasons for giving or withholding aid to a foreign government are many and varied. In its most crass sense, such aid has been used as a form of bribery. An American-brokered agreement in 1994 created an international consortium to provide North Korea with $4 billion in energy aid and light-water reactors in exchange for that government's dismantling of its nuclear weapons program. The United States provided millions of dollars in aid to Syria, Egypt, Jordan, and Israel following the Yom Kippur War in 1973 in an effort to induce agreement on cease-fire arrangements. Since the signing of the Camp David Accords in 1978 by Egypt and Israel, the United States, in a continuing effort to bring peace to the Middle East, has spent well over $75 billion in aid to just those two states. Egypt and Israel still account for almost 40 percent of the total U.S. aid budget.[6]

The threat or actual withdrawal of foreign aid has similarly been used to force states to follow their donors' biddings. In 1948, the United States threatened to cut off Marshall Plan aid to the Netherlands if the Dutch failed to make a settlement with Indonesian nationalists and relinquish control over what was then its colony. In 1956, the United States withdrew economic support from Britain, France, and Israel in an effort to discourage their aggression in the Suez. In other cases, South Korea, in order to receive U.S. aid in the 1950s, had to agree to restrict textile exports to the United States; aid to Peru and Bolivia was made contingent upon cooperation with the U.S. drug enforcement program;[7] and the Reagan administration announced that it would consider a state's voting record in the United Nations in assessing that state's qualifications for American foreign assistance.[8] The U.S. Congress has also mandated that economic and military assistance be withheld from any nonnuclear weapons state that appears to be pursuing policies designed to achieve a nuclear capability. Pakistan from time to time has been subjected to such restrictions as a result of its interest in developing nuclear weapons.

Aid has also been used in an effort to affect internal political and economic behavior within foreign states. In recent years, the United States has made aid contingent upon a state's human rights record (although during the cold war congres-

---

[5] Robert G. Wesson, *State Systems* (New York: Free Press, 1978), 121.

[6] Elaine Sciolino, "Call It Aid or a Bribe, It's the Price of Peace," *New York Times,* March 26, 1995, sec. 4, 3.

[7] Raymond Vernon and Debora L. Spar, *Beyond Globalism* (New York: Free Press, 1989).

[8] Charles W. Kegley, Jr., and Steven W. Hook, "U.S. Foreign Aid and U.N. Voting," *International Studies Quarterly* 35 (September 1991): 295–312.

sionally mandated restrictions were often secondary to national security interests). In the post–cold war era, foreign assistance to the former Soviet Union and Eastern bloc economies has required that these states change their basic economic structures from state ownership to privatization and from state-controlled economies to free market economies. Economic assistance to Russia has also been used as an instrument of foreign policy, as American officials threatened to stop aid to Russia in 1994 because of continued Russian involvement in spying operations directed against the United States. A principal example was the Aldrich Ames case, in which an employee of the CIA passed on information to Russia after the cold war had ended. In the 1990s, Americans concerned about Russian behavior in Bosnia and its use of force against the secessionist region of Chechnya likewise favored the use of economic assistance as leverage to punish Russia for engaging in policies and actions they opposed.

The fact that most foreign assistance is provided bilaterally rather than multilaterally is another indication that states view it primarily as an instrument of foreign policy rather than as an act of beneficence. Between 1981 and 1991, for example, 72.8 percent of grants and loans were given bilaterally, compared with 27.8 percent multilaterally.[9] By serving as the direct donor, the state is able to use aid—including the threat of its withdrawal—as an instrument of influence.

Whatever the motivation for states engaging in foreign aid, the trend in the United States in recent years has been one of declining military and economic assistance. Many Americans began to question the granting of assistance to former Warsaw Pact countries at a time when U.S. unemployment was high, even though aid was meant to help post-Communist governments weather their difficult economic conditions and thereby progress toward democracy and a free market economy. In fact, the United States now ranks next to last among the top twenty donor states in terms of the percentage of GDP directed toward foreign economic assistance. Even in absolute terms, Japan's foreign economic aid has now surpassed that of the United States.

# ECONOMIC SANCTIONS

If foreign aid is a "carrot" extended to would-be friends, economic sanctions are a "stick" intended to coerce recalcitrant governments into behavior that is acceptable to the state or states that institute such sanctions. With growing economic interdependence, economic sanctions have become an increasingly important weapon of state policy.

**TYPES OF ECONOMIC SANCTIONS** A government may impose an **embargo** on trade, which limits or restricts the flow of goods into a given country, or a **boycott,** which entails the refusal to purchase goods from a country. In a time of extensive overseas investment, the **freezing of foreign assets** located within a state is yet another economic weapon. Under this increasingly preferred sanction, neither the government

---

[9]United Nations, *World Economic Survey, 1993* (New York: United Nations, 1993), 244.

Mario/*Economic Times*, New Delhi.

nor its citizens may cash in those investments or receive interest or profits on the assets located in the country instituting the freeze.

The power of selective embargoes was illustrated by the Arab oil embargo in 1973–1974. Following the Yom Kippur War between Egypt and Israel in 1973, Arab leaders began to grasp the potential of oil for pressuring other states to accept their position with regard to Arab-Israeli issues. Their actual and threatened embargoes proved instrumental in obtaining the necessary votes to force Israel out of the International Labor Organization and influenced the passage of various U.N. General Assembly resolutions equating Zionism with racism. The leverage was so strong in the summer of 1980 that a highly inflammatory anti-Zionist resolution was passed in the General Assembly with only seven negative votes.[10]

Despite the extensive aid provided by Israel to African states during the 1960s, the Arab states were able to induce a number of African leaders to take an increasingly negative position toward Israel. They did so by offering these leaders both oil and financial aid funded by increased oil prices. Arab leaders also used their economic power vis-à-vis Western corporations by **blacklisting** Israel: Any corporation involved in trade with Israel would be forbidden from business activities with the far more populous Arab states.

The use of trade as a political instrument to pressure states to behave in a desired way is not limited to oil-producing states. France, for example, threatened to use the trade weapon against New Zealand in the conflict over the 1985 French sinking of the *Rainbow Warrior* in New Zealand waters. The ship had been sent to the South Pacific by Greenpeace in an effort to thwart French nuclear tests in the area.[11]

Economic sanctions have been used to influence or punish states that have been viewed as violating generally accepted international norms. Such sanctions were imposed by the League of Nations on Italy during 1935–1936 to protest the latter's invasion of Ethiopia. Decades later, the United Nations sought to impose similar sanctions on Rhodesia and the Republic of South Africa because of those governments' mistreatment of their black majorities. Other efforts of this sort have included the United States' attempts to impose economic restrictions upon Cuba and the more recent use of economic sanctions against the Soviet Union as a result of its 1979 invasion of Afghanistan, against Iran in connection with the seizure of the American Embassy in Teheran, and against Iraq and Serbia following their respective aggressions against Kuwait and Bosnia in the early 1990s.

**EFFECTIVENESS OF SANCTIONS** Despite their frequent use, the results of economic sanctions have hardly been satisfactory. The League's efforts against Mussolini's 1935–1936 aggression were undermined by the decision of the United States, which was not a member of the League, to make up for some of the shortfall by increasing U.S. trade with Italy. The French and the British were also ambivalent about pressing sanctions on Italy too vigorously for fear of pushing Mussolini too close to Hitler.

Since World War II, multinational corporations and other governments have sometimes softened the impact of economic sanctions against such states as Rhodesia (now Zimbabwe) and South Africa.[12] When the United States sought to apply economic sanctions against Cuba, Castro merely turned to the Soviet Union, which responded with massive economic assistance. The United States also found it difficult to obtain support among other nations for its 1979–1980 economic boycotts of the Soviet Union and Iran. Yet it is significant that, as the cold war ebbed, the United

---

[10] That vote was rescinded in 1991.

[11] Edmunds Dell, *The Politics of Economic Interdependence* (New York: St. Martin's Press, 1987), 98.

[12] Yet, in the latter case, as more states and companies joined the sanctions, the economic isolation led F. W. De Klerk to change the South African political system, leading to the election of Nelson Mandela.

States had considerably more success in obtaining global support for applying economic sanctions to Iraq as a result of the 1990 Iraqi invasion of Kuwait. The sanctions in this case were estimated to have been more than 95 percent effective in restraining both imports and exports to Iraq. They led the Iraqi GDP to fall some 50 percent in less than six months. Similar economic damage was done to Serbia in response to its aggressive policies in Bosnia. Even so, the war raged on for several years, though many concluded that Serbia's president, Slobodan Milosevic, eventually turned peacemaker because of the pain for Serbia of the economic sanctions.

Overall, however, economic sanctions have been relatively ineffective—as was shown in a study of more than one hundred sanction episodes in which the United States has been involved since 1922.[13] According to this study, success was achieved in only 34 percent of the cases; yet even this rate is higher than has often been found in other studies.[14] Related research has shown that, after two years following the application of sanctions, trade has returned to its previous levels whether or not the situation that had induced the sanction had been rectified.[15]

After surveying the limitations of trade embargoes and boycotts as instruments of economic pressure, one authority suggested that subtle, concerted use of economic weapons—such as the reduction of investments, delays in delivering spare parts, snags in licensing, decreasing loans and grants, and a refusal to refinance existing debts—tend to be more effective than trade boycotts and embargoes in influencing the policies of other states.[16] A problem with the more extreme economic sanctions is that they tend to unify the targeted population, increasing its sense of nationalistic fervor and resistance rather than coercing its compliance with the sanctioning states' wishes.

Special difficulties arise for pluralistic states that attempt to use trade as a weapon, because various governmental bureaucracies and interest groups are often divided on the issue. For example, food was not exploited as a weapon in September 1977 in several sensitive negotiations with the Soviet Union; American officials involved in the negotiations discovered only by reading the newspapers that the Department of Agriculture had unilaterally agreed to sell seven million more tons of grain than were called for in United States–Soviet agreements.[17] It was pressure from agricultural interests, not a resolution of the conflict, that led President Reagan to lift food sanctions that his predecessor, Jimmy Carter, had applied following the 1979 Soviet incursion into Afghanistan. Problems in applying sanctions also arise from the fact that many private corporations are beyond the reach of their national govern-

---

[13] Gary Clyde Hufbauer, Jeffrey J. Schott, and Kimberley A. Elliott, *Economic Sanctions Reconsidered* (Washington: Institute for International Economics, 1990).

[14] See Margaret Doxey, *International Sanctions in Contemporary Perspective* (New York: St. Martin's Press, 1987); and James M. Lindsay, "Trade Sanctions as Policy Instruments," *International Studies Quarterly* 30 (June 1986): 153–173.

[15] Peter Wallensteen, "Characteristics of Economic Sanctions," *Journal of Peace Research* 5, No. 3 (1968): 248–267.

[16] R. S. Olson, "Economic Coercion in World Politics," *World Politics* 31 (July 1979): 471–494.

[17] Samuel Huntington, "Trade Technology and Leverage: Economic Diplomacy," *Foreign Policy* (fall 1978): 75.

> **Normative Dilemmas**
>
> ## Questions Concerning the Use of Economic Leverage
>
> - If economic sanctions hurt the people you are trying to help, as was the case in South Africa in the 1980s where sanctions threatened to increase black unemployment, should one still insist upon their use? Are sanctions applied to *all* parties in a conflict justified when only one party is generally seen as the aggressor, as was the case with Serb groups in their conflict with Bosnian Muslims in the 1990s?
>
> - Legislation has been passed or proposed that would deny foreign aid in each of the following circumstances. Which, if any, would you view as justifiable reasons for refusing to aid a country finding itself in dire economic circumstances?
>
> (1) The government is not democratic.
>
> (2) The government has a history of considerable corruption..
>
> (3) The government, given its serious problem of overpopulation, allows abortions and family planning that are morally repugnant policies to you.
>
> (4) The government spends what you regard to be excessive amounts on its military.
>
> (5) The government appears to be trying to develop an atomic device.

ments. Subsidiaries of such companies may sell products that the parent company would be prevented from selling. Even during World War II, it has been alleged, some armaments were sold to Germany by American firms.

Generally speaking, the **less developed country (LDC)** will be more vulnerable to economic sanctions than the developed state. The LDC's economy is less diversified, so it stands to lose more when its trade is disrupted. If the state produces basically only one commodity, as is the case for many LDCs, curtailing the export of that product could threaten economic ruin. The less developed country lacks the opportunity available to the more developed state of decreasing its vulnerability by stockpiling or producing synthetic substitutes. The unavailability of marketing skills and the lack of diplomatic personnel capable of stimulating exports also make the developing state more vulnerable to market disruptions.

Still, the less developed state may have certain advantages in resisting economic coercion from outside. Its very poverty may reflect a subsistence-level self-sufficiency that can make it more resistant to economic deprivation. It has been suggested that a modern state such as Belgium would have been less able to resist American pressure than did Vietnam during its war with the United States; Belgium's comparably sophisticated economy could be readily disrupted by destroying or impeding one of its sectors, because the parts are so interdependent.[18] In advanced economies like Belgium's, where separate pieces of complex machines and other products are manufactured in many locations, dislocation of the transportation system or interference with the production of a single necessary item can shut down an entire industry.

---

[18]Franklin B. Weinstein, *Indonesian Foreign Policy and the Dilemma of Dependence* (Ithaca, N.Y.: Cornell University Press, 1976), 27.

## THE ROLE OF THE PRIVATE SECTOR

In the previous chapter we saw how such various international institutions as the International Monetary Fund, the World Trade Organization, and the European Union intrude upon the sovereignty of the state from above. At the same time, the state's ability to affect the international economic system is undermined by a number of private interests and economic forces, including international banks, investment firms, and multinational corporations. These developments led the distinguished economist Charles Kindleberger to assert in 1969 that "the nation-state is just about through as an economic unit."[19]

As might be expected, views of the proper role of the state versus **private market forces** vary depending upon one's economic perspective. It might be argued with only slight oversimplification that, from a mercantilist perspective, private economic interests are subordinate to state interests, whereas from the Marxist perspective, the state is merely the tool of the dominant economic class. Liberalism, with its emphasis upon laissez-faire, on the other hand, seeks to minimize state interference and involvement in the global economy, allowing domestic and global market forces to determine the distribution of economic wealth.

Whereas the liberal perspective considers private investment as benefitting both capitalist and developing economies alike, the Marxist believes such investment exploits the less developed state. For the mercantilist, investment serves the important function of helping the less developed country purchase industrial goods from the capitalist state, thus enhancing the latter's economy by adding to its wealth.

## MARKET FORCES VERSUS THE STATE

Several economic trends indicate that private economic actors are coming to play an increasingly important role in the global economy at the expense of state governmental institutions. One such trend, noted in the last chapter, has been the growth of a capitalist world economy with increased privatization of economic functions.

Perhaps no sector of the economy has become more global in its orientation and operations than that of finance. According to Stephen Kobrin of the Wharton Business School, "Government control over flows of funds and thus the values of currencies or monetary policy is very limited."[20] Electronic financial networks allow the transfer of funds at a moment's notice throughout the world. The international flow of funds has also been enhanced by branch offices positioned worldwide, with cities like London playing host to over 500 foreign banks.[21] Stock exchanges throughout the world have become interdependent with a large part of their capitalization coming from foreign sources. As a result of this increased financial activity, cross-

---

[19] Cited in Andrew Walter, *World Power and World Money* (Hertfordshire, England: Harvester Wheatsheaf, 1991), 12.

[20] Cited in Ethan B. Kapstein, *Governing the Global Economy: International Finance and the State* (Cambridge, Mass.: Harvard University Press, 1994), 4.

[21] Ibid.

**FIGURE 11.1** Largest Suppliers and Users of Net Capital Flows, 1989–1993 (Current exchange rates; as shares of net capital flows)

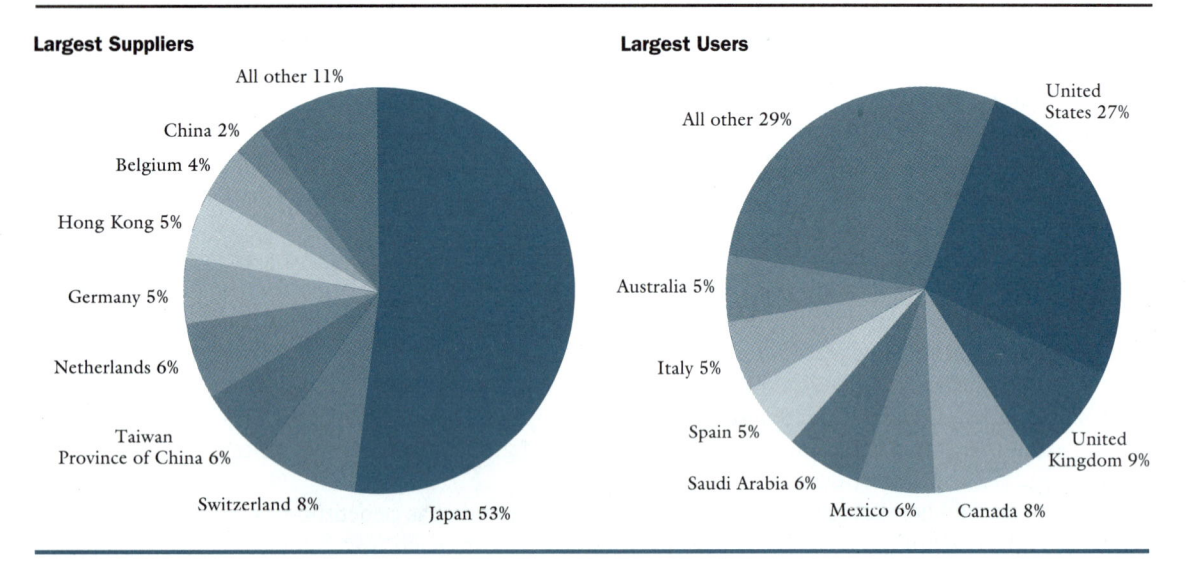

**Largest Suppliers**
- All other 11%
- China 2%
- Belgium 4%
- Hong Kong 5%
- Germany 5%
- Netherlands 6%
- Taiwan Province of China 6%
- Switzerland 8%
- Japan 53%

**Largest Users**
- All other 29%
- Australia 5%
- Italy 5%
- Spain 5%
- Saudi Arabia 6%
- Mexico 6%
- Canada 8%
- United Kingdom 9%
- United States 27%

SOURCE: International Monetary Fund, "Saving in a Growing World Economy," *World Economic Outlook* (Washington, DC: International Monetary Fund, May, 1995), 83.

border financial transactions of most industrial countries stood at less than 10 percent of their GDP in 1980, but had increased to well over 100 percent by 1992.[22]

Figure 11.1 provides a breakdown of those states that supplied the greatest net capital to the world (much of it private) and those that were the largest users of such resources during the period 1989 to 1993. As is clearly shown, Japan, accounting for 53 percent of the world's net capital supply, has become the world's greatest creditor nation, while the United States is the largest debtor or user of outside capital. Other things being equal, it is the creditor state that is likely to have more leverage in international affairs.

International businesses increasingly deal directly with one another in a number of activities without state involvement. Thus they negotiate joint economic ventures, utilize private rather than state dispute settlement procedures, and engage increasingly in a barter economy rather than using currency.

## THE MULTINATIONAL CORPORATION

Among private international economic actors, the multinational corporation (MNC) has received the most attention. Some have viewed the multinational corporation as a major challenger to the power and sovereignty of the nation-state, in

---

[22] International Monetary Fund, "Saving in a Growing World Economy," *World Economic Outlook* (Washington, DC: International Monetary Fund, May 1995), 80.

> ### Global Changes
>
> ## Change in Global Investment and Finance
>
> The globalization of finance: Lending by international banks grew from $40 billion in 1975 to well over $300 billion by 1990, while international bond lending rose almost tenfold, from $19 billion to over $170 billion during the same period.*
>
> - What are the implications of such trends for the creditor and for the debtor?
>
> Foreign direct investment: Although one can chart a significant rise in global foreign direct investment, the level of investment today can be exaggerated compared to yesteryear. Japanese investments abroad pale compared to those of Britain, which in 1914 had invested a quarter of its savings abroad.
>
> By contrast, Japanese companies and individuals created fears that they were "buying America," when they invested 10 percent of their savings outside Japan during the 1980s.
>
> - Why has greater concern been expressed in the United States about Japanese foreign investment than about European investment, which is somewhat more extensive? Should foreign investment be welcomed or discouraged?
>
> *Andrew Walter, *World Power and World Money* (Hertfordshire, England: Harvester Wheatsheaf, 1991), 198.

part due to the resources that it controls. As noted in Chapter 3, a number of MNCs' assets and sales are greater than the gross domestic product of a majority of the world's nations-states.

Given the resources at their command, multinational corporations have assumed a role as dominant actors in the global system, often operating at the expense of both home and host governments. By controlling enormous assets cross-nationally, decisions made by MNCs—such as where to buy and sell products and where to make foreign investments—have a major impact on the balance of payments and the value of currency in various nation-states. It has even been suggested that the currency speculations engaged in by MNCs, many of which are American, undermined the dollar and led to its devaluation in February 1973.

MNCs typically are headquartered in a highly developed country, whose citizens provide the staff, managerial skills, and technology to run them. Even so, MNCs and their subsidiaries have been known on many occasions to disregard the foreign policy directives of their home governments. For example, after the briefest hesitation, a subsidiary of Gulf Oil turned several hundred million dollars over to the winning side in the 1976 Angolan civil war even though the U. S. government had opposed the new Marxist regime that had come into power. More serious digressions from governmental policy occurred among American subsidiaries that sold arms to Hitler in the 1930s. During the 1973–1974 oil boycott, some American multinational oil companies failed to provide needed petroleum supplies to the United States, preferring instead to sell their products in more profitable markets.

With the extensive resources they control, multinational corporations have also been exceedingly active in extracting special considerations from their home governments in support of their overseas activities. Neo-Marxist writers have tended to portray the relationship between big business and government as one in which politicians, owing to bribery and political pressures, simply do the bidding of business interests. Foreign policy is seen by such writers largely as an effort to protect and further the economic interests of the MNCs and other businesses at home and abroad.

Some governments have been more active than others in meshing governmental and corporate interests. In socialist states, with their state-owned enterprises, the two policy levels are virtually inseparable. Among democratic states, the Japanese government has been exceedingly active in protecting and furthering overseas interests of business, as have Britain and France. The U.S. government, in contrast, has been more passive on this score. Nevertheless, there are a number of instances in which the United States intervened on behalf of specific economic interests. During the early decades of this century, the United Fruit Company clearly gained from the intervention of the U.S. Marines in several of the republics of Central America; indeed, United Fruit wielded such power over Central American economies that it influenced national politics in that region on several occasions. The U.S. government was also instrumental during the 1920s in obtaining access for American oil companies in Middle Eastern areas reserved for France and Britain. More recently, the government has placed considerable pressure on Japan in an effort to force that country to restrict exports to the United States unilaterally, as noted earlier.

From the Marxist and neo-Marxist perspectives, capitalist corporations, whether helped by government or not, tend to exploit smaller, less developed countries. But these smaller states have some leverage of their own. As a result of the global spread of MNCs, a small country can shop around for a foreign business that is more compatible with its needs and is willing to share more of the profits. Over time, as expertise and skill in managing its own resources improves, the small state has been able to negotiate more balanced concessionary agreements with huge business conglomerates, allowing it to share more of the profits.

## The Impact of Economic Interdependence

As might be expected, those with different economic perspectives disagree as to whether global economic interdependence—tying together both state and nonstate economic actors—is beneficial for humankind. Liberals tend to view the dependence of one state upon another in a positive way as long as it is not exploitative, arguing that such interdependence may have a dampening effect upon international conflict. The British political philosopher John Stuart Mill, an early proponent of free trade, was enthusiastic about the prospect that increasing trade would create the conditions for peace. Mill wrote, "It is commerce which is rapidly rendering war obsolete, by strengthening and multiplying the personal interests which are in opposition to it."[23]

Experience in recent decades is far from conclusive proof of Mill's thesis. Instead of providing peace, increased interdependence may actually produce more issues upon

---

[23] Cited in Dell, *The Politics of Economic Interdependence*, 196.

# Does the Multinational Corporation Exploit Less Developed Countries?

## The Case against the MNC

- Multinational corporations decapitalize less developed countries by taking out more money in profits than they invest in those nations. Rather than bringing in more investment, many MNCs simply borrow from local creditors, thus depleting capital resources that might have been used by indigenous business.
- The technological dependence created by outside business interests means that a country is less likely to develop its own innovative capacity. While borrowing technology can be useful in the short term, it does not help long-term national economic growth.
- MNCs, through advertising, encourage consumerism and the importation of luxury goods, thus undermining domestic investment that is vital to economic growth.
- MNCs can counter mercantilist restrictions on trade by establishing subsidiaries abroad. In effect, this allows them to jump the trade barriers, continuing production and collecting profits at the expense of the LDC in its own territory.
- MNCs intentionally discourage industrialization, for this would mean increased competition for themselves. They prefer instead that LDCs concentrate on supplying raw materials and cheap labor.
- Since LDCs are often forced to concentrate on primary materials for export, they are denied the added value derived from processing their own resources. The finished product must be imported, which adds the costs of processing, transportation, and profits to the final bill.
- The profit orientation of MNCs works against progressive social and political values. MNCs may be persuaded to do business with oppressive or racist regimes.
- MNCs are creating an elite class of wealthy individuals in developing countries much like their rich counterparts in the North, but who are increasingly detached from the impoverished masses within their own societies.

## The Case for the MNC

- Multinational corporations provide considerable investment of their own and many investors do not regard indigenous business interests as good risks anyway.
- The MNC can provide increased employment opportunities to the LDC citizenry. Employment of nationals is often made a requirement for a firm to do business in a foreign state.
- MNCs have an interest in improving the well-being and education of their indigenous work force and in contributing to the LDC infrastructure (such as roads and transportation).
- MNCs share the latest technology through their subsidiaries that, in many cases, do not have to pay research and development costs.
- As a result of sharing technology and a common work culture across national lines, MNCs tend to equalize working conditions, standards, attitudes, and values transnationally.
- Foreign subsidiaries of MNCs facilitate import substitution. This can be very important for states with severe balance of payments problems.
- MNCs generate royalties and taxable income, which provide needed revenues for the LDC.
- MNCs encourage the creation of a more peaceful international environment, because they cannot thrive if they have to cope with prolonged warfare, trade barriers, and the like.

---

which to differ. Economic interdependence may only heighten antagonism, particularly on the part of a state that feels itself forced into an unequal economic relationship. Neo-mercantilist arguments then may receive a sympathetic hearing. For example, the growth in economic interdependence between Canada and the United

States has tended to increase Canadian nationalism; as American investments have intruded into the Canadian economy, Canadians have feared the loss of economic and, ultimately, political and cultural control.

The economic interdependence of global economies can lead to friction among political leaders as they conflict with one another with respect to what would seem to be strictly internal economic matters. Despite Margaret Thatcher's warm personal relationship with President Reagan, the British prime minister bitterly criticized his administration for running huge budget deficits, for this tended to increase the interest rates the British had to pay in a global economy. Successive American presidents condemned Germany for not reducing German interest rates (a policy adopted due to obsessive German fears of inflation resulting from its experiences during the 1930s). And President Clinton encouraged Japan to stimulate its own economy by reducing taxes and offering other incentives in the hope that Japan would buy more American products.

That economic interdependence might increase the likelihood of war is shown by a study of some forty post–World War II conflicts, which concludes that countries within the same trading group are "more than twice as likely to fight than are nations which belong to different groups, or to none."[24] Similarly, another study revealed that among major powers of the nineteenth and twentieth centuries, trade partners were *more* likely than nontrade partners to engage in war against one another.[25]

Yet, Mill's argument that economic interdependence assists peace is supported by other research. A study based on trade statistics for thirty countries revealed that the states "with the greatest levels of economic trade engage in the least amounts of hostility. In fact, a doubling of trade on average leads to a 20 percent diminution of belligerence."[26] Additional—or rather, mixed—support for the conflict diminishing effect of trade is found in a study by Lois Sayrs, whose statistical analysis showed that trade lowered the overall level of conflict between states.[27] At the same time, however, trade did not seem to increase the incentives to cooperate, nor did it improve relations in those economic arenas where the norms of competition dominated or where the United States as a trading partner exerted extraordinary influence.

It seems clear that, at a minimum, international economic interdependence makes a nation-state more vulnerable to crises originating in other parts of the system. If a state is heavily invested abroad and dependent on trade or raw materials from beyond its borders, its incentive to intervene and to help resolve foreign conflicts will be very high when its economic interests are threatened.

Recent proposals to increase the interdependence of the North Atlantic region by creating a free trade area between North America and the European Union have

---

[24] Bruce M. Russett, *International Regions and the International System* (Chicago: Rand McNally, 1967), 198–199.

[25] Charles S. Gochman, "Military Confrontation and the Likelihood of War: The Major Powers, 1820–1970," *Peace Research (International) Papers,* 1975.

[26] Solomon William Polachak, "Conflict and Trade," *Journal of Conflict Resolution* 24 (1980): 55.

[27] Lois W. Sayrs, "Trade and Conflict Revisited: Do Politics Matter?" *International Interactions* 15 (1989): 155–175.

been motivated as much by political-military concerns as economic ones. Such action could serve to strengthen the cohesion among NATO countries—a cohesion that has been undermined as a result of the end of the cold war. Some Europeans have seen such economic linkages as a way of countering the U.S. trend toward isolationism now that the Soviet threat no longer exists. Similar arguments may be made with respect to broadening U.S. economic integration with the Far East and other regions of the world. By enhancing the interdependence of allies, the credibility of deterrence is strengthened. The logic is that would-be aggressors come to recognize that a united defensive reaction is all the more probable if vital economic interests are at stake.

# CONCLUSION

With the end of the cold war, economic statecraft is likely to play a more important role than military statecraft. Although economic imperialism has become a less significant feature of statecraft, some writers see foreign aid and trade taking the place of imperialism as developed states use these instruments to exploit less developed countries. Whether used for exploitative reasons or not, the fact remains that foreign aid can be a powerful part of the arsenal of reward and punishment in interstate relations. The desire to use foreign aid as an instrument of influence explains why the bulk of such aid remains bilateral rather than multilateral.

Economic sanctions have also become a preferred instrument for attempting to deal with all sorts of foreign policy objectives. Recent examples include efforts to quash violence in Bosnia by imposing sanctions on the Serbs, to generate economic privatization in the former Soviet bloc, and to improve human rights treatment in South Africa. The effectiveness of such sanctions is influenced by how dependent the targeted state is, by how many suppliers remain outside those applying the sanctions, and even by domestic political groups who may have an interest in seeing such sanctions end.

What the state does by way of its own economic statecraft, however, may have minimal impact when it comes to such issues as capital flows, currency valuation, and even inflation. That is because broader private economic forces increasingly determine the condition of the global economy. Particularly important in this regard is the role that the multinational corporation is able to play in the global economy, given its control of resources that make those of a number of smaller nation-states pale in comparison.

One thing is certain on the eve of the twenty-first century: The world is becoming more economically interdependent. This interdependence has increased the power of those able to reward or punish other actors in the system. Whether such interdependence will also improve the chances for peace remains uncertain, however, for economic interdependence increases the range of issues over which states will dispute—many of which were previously seen as the sovereign prerogative of the state.

CHAPTER 12

# North-South Economic Relations

By the 1990s, some forty countries were regarded as comparatively rich and highly developed, whereas the overwhelming majority, more than 140, were poorer and classified as in varying stages of underdevelopment. This division is based upon very crude distinctions, which typically measure economic output, the availability of technology, and average per capita income rather than the quality of people's lives. Just as crude are the terms **North** and **South** that have become the accepted shorthand to distinguish between the minority of highly developed states and the majority that are less prosperous. They refer to the fact that most of the richer countries happen to be located in the Northern Hemisphere (though Australia and New Zealand are obvious exceptions), whereas most of the less developed countries, when not actually south of the equator, lie at least to the south of most of the richer countries.[1]

We shall use the North-South shorthand to consider contemporary economic relationships between the planet's rich minority of states (often also called **DCs**, for *developed countries*) and its poorer majority (also known as **LDCs**, for *less developed countries*).

Relative wealth and poverty are political issues wherever the allocation of economic resources is seriously contested in the public arena. During the past half century, that contest has become global—or, more accurately, has come to be recognized as global by its entrance onto the agenda of world politics—in a way that it

---

[1]The term *third world* is still often used to refer to the less developed countries of the South. But it has lost whatever sense it made during the cold war period, when economic references were also to the *first world* of the rich, free market societies of the West and to the *second world* of the centrally planned economies of the Communist sector. While a handful of centrally planned economies still exist today, the former Communist bloc now consists of states that have adopted varying degrees of free market economics. *North* and *South* are therefore preferable to distinguish richer from poorer countries today.

was not in earlier periods. The emphasis of Chapters 10 and 11 was upon the dominant economic actors (mostly, the DCs), their behavior in the world's political economy, and the impact upon them of economic forces beyond the control of the state. Since our concern here is with economic development, we must consider the cooperation and conflict between those societies that have relatively more of the world's material goods and those that have less.

# THE NORTH-SOUTH GAP

Even though the world has paid increased attention in recent decades to the problems of economic development, many view the effort to close the gap between richer and poorer societies as having produced more failure than success. In general, those states that were already highly developed at the close of World War II have grown ever richer in the years since, while most of those that were economically underdeveloped fifty years ago have lagged behind, in relative if not in absolute terms.

A longer historical view depicts the widening gap even more starkly. Much of Europe was well on the road to industrialization by 1850 when, it is estimated, its most highly developed states were about twice as rich as the nonindustrialized rest of the world. By 1950, that gap between North and South had grown to about ten to one. Only ten years later, in 1960, the ratio had become nearly fifteen to one. Although the gap no longer is growing at the rate it did in the 1950s (which may reflect the very modest success of the development effort), it nonetheless continues to widen to the present day.

The Industrial Revolution evidently is at the heart of what accounts for this historical trend. Before it began in Western Europe late in the eighteenth century, what we today call the LDCs of the South accounted for nearly three-quarters of the world's manufactured goods, while Europe and North America produced less than one-quarter (see Table 12.1). But as industrialization took off on the latter two continents, those ratios changed dramatically. By 1900, Europe and the United States produced more than 85 percent of the world's manufactures, while the South's share had dropped to a mere 11 percent.

**TABLE 12.1** Relative Percentage Shares of World Manufacturing, 1750–1900

|              | 1750  | 1800  | 1830  | 1860  | 1880  | 1900  |
|--------------|-------|-------|-------|-------|-------|-------|
| Europe       | 23.2% | 28.1% | 34.2% | 53.2% | 61.3% | 62.0% |
| United States| 0.1   | 0.8   | 2.4   | 7.2   | 14.7  | 23.6  |
| Japan        | 3.8   | 3.5   | 2.8   | 2.6   | 2.4   | 2.4   |
| South (LDCs) | 73.0  | 67.7  | 60.5  | 36.6  | 20.9  | 11.0  |

**SOURCE:** Thomas D. Lairson and David Skidmore, *International Political Economy* (Fort Worth, Texas: Harcourt Brace, 1993), 183, adapted from Paul Bairoch, "International Industrialization Levels from 1750 to 1980," *Journal of European Economic History* 11 (1982): 296. Reprinted with permission of Cambridge University Press.

These data are clear and dramatic. Yet, determining their meaning is the subject of continuing debate about the nature of the development process. Two contending theories have dominated this discussion for several decades.

# MODERNIZATION THEORY

**Modernization theory** is essentially the product of the North's economic and political "establishment." Proceeding from a state or societal level of analysis, it is compatible with a liberal, or laissez-faire capitalist, outlook. It takes the view that the North has grown rich by learning how to sustain economic growth through capitalist modes of production. While the Industrial Revolution entailed harnessing the energy of, especially, fossil fuels so that production could be greatly increased, it was made possible by a confluence of social changes that transformed traditional societies into modern ones. Among those changes frequently cited are a rise in social mobility, a reward system based more upon merit than one's status at birth, the heightened value of innovation and intellectual diversity, the decline of authoritarian political and religious structures to the point that they are less intrusive into the increasingly "private" sphere of economic organization, urbanization, growth in the rule of law, and the development of greater economic specialization and a division of labor in society.[2]

According to this theory, traditional societies are hampered by characteristics no longer found in modern nation-states. Until traditional societies learn to value innovation and progress, creating reward systems reflective of those values, most of their people will remain mired in economies that remain largely at subsistence levels. Therefore, modernization theorists argue that the South should emulate the North. Sustained growth in the South requires that traditional societies undergo the same transition to modernity that the North has experienced over the past several centuries.

Modernization theorists see far more good than harm to countries in the South as the result of their extensive economic ties with the richer countries of the North. According to this view, multinational corporations, as we saw in Chapter 11, provide jobs, technical skills, and a broader understanding to people in the South of the workings of modern economic enterprises. Trade provides mutual gains to the trading partners. The dissemination of the North's modern popular culture through the entertainment and information media helps to undermine the traditional beliefs that stand in the way of development in the South.

Advocates of modernization theory argue that LDCs are most likely to better themselves economically on the coattails of the more highly developed countries. As

---

[2] Among the works influential in establishing the tradition of modernization theory are Gabriel Almond and James S. Coleman, *The Politics of Developing Areas* (Princeton, N.J.: Princeton University Press, 1960); Cyril Black, *The Dynamics of Modernization* (New York: Harper and Row, 1966); Alex Inkeles and David H. Smith, *Becoming Modern: Individual Change in Six Developing Countries* (Cambridge: Harvard University Press, 1974); Daniel Lerner, *The Passing of Traditional Society* (New York: Free Press of Glencoe, 1958); and Myron Weiner, ed., *Modernization: The Dynamics of Growth* (New York: Harper and Row, 1966).

the latter prosper, they will stimulate trade and investment with LDCs as well as with one another. They generally accept the **theory of comparative advantage,** which assumes that many LDCs are the producers of primary products because they are particularly well suited to produce them. Cocoa, coffee, bananas, and rubber, for example, are most cheaply produced in certain countries in the South. It may be to the comparative trade advantage for them to concentrate on producing these for export, rather than to try to build highly diversified economies of their own based on Northern models.

The proponents of modernization theory have various responses to the fact that the past half century of development efforts have failed to close the income gap between rich and poor countries. A pessimistic minority assert that traditionalism is frequently too deeply embedded in a culture to give modernization a foothold. A more optimistic majority claim that development is inevitably a slow and conflictual process, and that greater patience is necessary: Just because modernization has not yet produced results does not mean that success will not be realized in the future.[3]

Still, the fact remains that the continuing gap between North and South does not match what proponents of modernization theory predicted in the 1950s. They had supposed that limited economic assistance from the North to the South would soon permit the newly independent LDCs to experience economic "takeoff."[4]

Increasingly, a contending theory about the nature and problems of economic development came to be heard. Unlike modernization theory, which treats the state as its unit of analysis, **dependency theory** considers either the international system as a whole or its dominant economic system as the appropriate unit for analyzing contemporary development issues.

# DEPENDENCY THEORY

Proponents of dependency theory tend to be critical of the impact of the system of world capitalism on the South. By the early 1960s, several Latin American social scientists had advanced theories to explain the reasons for what they saw as the failure of the world's economic development policies.[5] They argued that the economic life of many LDCs was largely dependent upon economic decisions made in the North. Hence, their economic dependency was defined as "a situation in which a certain number of countries have their economy conditioned by the development and expansion of another" more highly developed society.[6]

---

[3]Richard N. Rosecrance, *The Rise of the Trading State* (New York: Basic Books, 1986).

[4]W. W. Rostow, *The Process of Economic Growth,* 2d ed. (New York: Norton, 1962).

[5]The first of these theorists, Raul Prebisch, had gained firsthand experience of development problems through his work on the United Nations Economic Commission on Latin America (ECLA). See Raul Prebisch, *Towards a Dynamic Development Policy for Latin America* (New York: United Nations, 1963).

[6]Theotonio dos Santos, "La crisis del desarrollo y las relaciones de dependencia en America Latina," in *La Dependencia politico-economica de America Latina,* ed. H. Jaguaribe et al. (Mexico, 1970), 180, as cited in J. Samuel Valenzuela and Arturo Valenzuela, "Modernization and Dependency: Alternative Perspectives in the Study of Latin American Underdevelopment," *Comparative Politics* 10, no. 4 (July 1978): 544.

For many dependency theorists, the key flaw in modernization theory's optimism is that today's LDCs face much more formidable competition from Northern states than that experienced by those first industrializing states some two hundred years ago. Whereas the Europeans had no rivals during the Industrial Revolution, today's developing countries must compete with the long established economic power of the North. But that merely hints at the central problem, which stems from the fact that capitalism is a global, and not merely a state-based, economic system. In today's world, the South is itself exploited as the resource base of the advanced capitalist sectors of the world economy.

One dependency theorist, Immanuel Wallerstein, is particularly clear in his view of the implications of capitalism as a global system.[7] Key terms in Wallerstein's analysis include *core, periphery,* and *semi-periphery* states. Core areas engage in the most advanced economic activities, such as banking, technology-based manufacturing and farming, and other aspects of production that require advanced technological skills. In contrast, the periphery provides raw materials, including minerals that fuel and feed the core's economic growth. Peripheral economies are quite deliberately suppressed by the core, so that they develop neither the skilled labor nor the technology that would help them compete with core states. The semi-periphery shares some characteristics of both the core and the periphery, serving also as an outlet for investment when labor in core regions becomes too costly. It follows that states in the periphery are "weak in that they are unable to control their fates, whereas states in the core are economically, politically, and militarily dominant."[8]

Analysts in this tradition see much evidence for their view of dependency as the central problem for LDCs. A common example occurs for exporters of raw materials to industrialized countries, which then turn them into finished manufactured goods. These typically are then sold not only at home, but back to the very country that supplied the primary product. The LDC is clearly dependent upon the manufacturer for the usable good, in such a case. Its raw material export may also be its chief source of foreign exchange. If its demand should suddenly plummet (as the result, say, of a severe recession in the North or the availability of a cheaper synthetic substitute), that country's foreign exchange earnings will likewise take a plunge. On the other hand, a bumper crop in such a primary product is no blessing either, for the greater supply of, say, coffee or latex will make prices for those products fall.

The argument is that, in contrast to manufactured products, primary products tend to suffer because the demand for them is relatively fixed. Since people can eat only so much and consume only a finite amount of energy, efforts to increase productivity of primary products will not increase income but, rather, tend to lower the price per unit. In contrast, many manufactured products have not yet saturated the world's markets, so that demand for them can expand substantially.

---

[7] Immanuel Wallerstein, *The Modern World-System I: Capitalist Agriculture and the Origins of the European World-Economy in the Sixteenth Century* (New York: Academic Press, 1974); and *The Modern World-System II: Mercantilism and the Consolidation of the European World-Economy, 1600–1750* (New York: Academic Press, 1980).

[8] Paul R. Viotti and Mark V. Kauppi, *International Relations Theory: Realism, Pluralism, Globalism,* 2d ed. (New York: Macmillan, 1993), 459.

A response attempted by some LDCs has been to try to produce more goods for export. While such strategies were sometimes successful, they often resulted in the increased influence of foreign multinational corporations (MNCs) whose assistance was needed to increase and diversify production. Again, the result was dependency on the more advanced capitalist states.

## THE DOMINANCE OF THE LIBERAL ECONOMIC ORDER

The debate between modernization and dependency theorists began to have an impact on the world's agenda of economic development issues by the 1960s. That was the period when Europe's empires were fast being dismantled, which gave most LDCs a voice in the development debate for the first time in recent history. Yet, that debate has continued within a world economic system that had been established some two decades earlier, at the close of World War II.

A comparative handful of Northern (and Western) governments created the Bretton Woods system at the close of World War II, a system largely inspired by the United States' vision of a liberal world economic order. As we saw in Chapter 10, one outgrowth of that vision has been the increased globalization of the world's present-day economy. A second outgrowth is an ongoing effort, first through GATT and now through the World Trade Organization, to keep reducing protectionist measures on a reciprocal basis. A third is the promotion of modernization theory as the way to overcome problems of poverty and underdevelopment in the South. Supporters of the system assume that economic development should proceed with infusions of private capital investment from the North, helped along by loans provided through the World Bank group.

### FINANCING DEVELOPMENT: THE WORLD BANK GROUP

The World Bank group includes the World Bank itself (officially, the International Bank for Reconstruction and Development), the International Finance Corporation, the International Development Association, and the International Monetary Fund.

---

### World Poverty

- **O**ne-fifth of the people of the world live in dire poverty;
- 70 percent of those people are women;
- 2 million children die every year from easily preventable infectious diseases;
- 1.3 billion people lack access to safe drinking water;
- 120 million people are unemployed;
- 80 million children do not attend primary school; and
- In 1995, poor countries owed $1.9 trillion in debt.

**THE WORLD BANK**  The institution informally known as the **World Bank (IBRD)** was created at the close of World War II to provide public international financing as a supplement to private loans. Its first priority was to make loans available to the war-torn countries of Europe to help them rebuild their industrial infrastructures. As that task was accomplished, the Bank gradually turned more of its attention to providing loans to governments in the South for various development projects. The amount of lending by the Bank has increased steadily, and in recent years has exceeded $20 billion annually.

The Bank's weighted-voting scheme grants voting power based upon the capital a country provides to the Bank's assets, which are determined by quotas. In accordance with this formula, a U.S. citizen always serves as the World Bank's president, and northern countries dominate its board of governors. Most of the Bank's assets come from borrowing in private capital markets. Both its structure and its loan policies mark it as strongly oriented toward the interests of its creditors, making it in that sense a "conservative" financial institution. It approves only *hard loans*—that is, those that protect the interests of the lender. For example, interest rates on loans must be about as high as those in the world's money markets and repayable only in convertible currency. Before loans are approved, recipients must first undergo analyses of their probable ability to repay them, typically over a period of from ten to thirty-five years.

**THE INTERNATIONAL FINANCE CORPORATION**  In 1956, the **International Finance Corporation (IFC)** was created as an affiliate of the World Bank to assist it in stimulating greater private investment in developing countries. It was meant to encourage private companies in the South by helping them secure investment capital from abroad, and by stimulating the formation of local investment. Compared to the World Bank, from which it can borrow, the IFC's resources are modest. But it may be seen as enlarging the effort to finance development through its focus on private capital formation.

**THE INTERNATIONAL DEVELOPMENT ASSOCIATION**  A second World Bank affiliate, the **International Development Association (IDA),** was created in 1960. Its primary aim was to attend to the special financial needs of the world's poorest countries. It was designed as a *soft-loan* agency—that is, one more favorable to the interests of the borrowing state. IDA permits up to fifty years for repayment of its loans, which are repayable on considerably easier terms than those demanded by the World Bank. No payments on loans are due for ten years. IDA then charges no interest, but does require a small annual service fee. In spite of these advantages for the borrowing state, however, loans must be repaid in a hard, or convertible, foreign exchange, which is often difficult for countries unable to earn much hard currency from their own exports.

**THE INTERNATIONAL MONETARY FUND**  One of the original postwar financial institutions, the **International Monetary Fund (IMF)** has become something of a crisis manager in development assistance. It provides temporary funds to help governments with balance of payments deficits, to compensate for sharp drops in the price of commodities, and the like. The IMF typically has demanded strong austerity measures from governments that turn to it for help when faced with serious economic downturns. These may include requirements that wages be frozen to stem inflation, and that local consumption otherwise be reined in to increase the probability of loan repayments.[9]

The activities of the World Bank group have assisted a very large number of LDCs. Yet, these institutions are the frequent object of criticism, on grounds that they are controlled by the richest states through their weighted voting provisions. The interests of those states are oriented, so it is charged, more toward protecting their capital resources than toward the real development needs of poor countries. For evidence, the critics note that the World Bank's loan terms are seldom much easier than those of private banks. They also point out that the IFC has never been financed by the North at a level that would have a dramatic impact on the development of the South. And the IMF's "belt-tightening" policies typically have harsh effects on the immediate economic life of the targeted LDC, which must encourage higher unemployment, even recession, if it is to comply with IMF terms for a loan.[10]

## OFFICIAL DEVELOPMENT ASSISTANCE

Between 1947 and 1951, the United States poured 2.5 percent of its GNP into the reconstruction of war-torn Europe through the Marshall Plan.[11] Europe's recovery was largely accomplished by the early 1950s. The United States then attempted to repeat this success story by contributing foreign assistance to stimulate the economic growth of many poorer states in the South. As with Marshall Plan aid, the United States and its allies, who also became donors, were strongly motivated by anticommunism. But as other Northern countries followed the U.S. lead, it became increasingly clear that the problems of underdevelopment were both different from and far more complex than those that were involved in Europe's recovery. That complexity is one of the themes of the North-South conflict. The other is the extent to which development assistance is to be regarded as a tool of political influence for the donor (as implied in the cold war foreign aid programs sponsored by both East and West) or an obligation owed by richer states to the poor.

As indicated in Chapter 11, **foreign aid** is a broad term that may include every kind of grant, such as military assistance, that one government bestows upon another. **Official development assistance (ODA)** is the preferred term for governmental grants or loans specifically intended to promote economic development. Foreign aid historically has been criticized as essentially an instrument of statecraft advancing the political and economic interests of donors more than the real development needs of recipients.[12] ODA is less controversial, at least to the extent that most decision makers (at least in principle) favor increasing the economic well-

---

[9]Richard Goode, *Economic Assistance to Developing Nations through the IMF* (Washington, D.C.: The Brookings Institution, 1985).

[10]Thomas D. Lairson and David Skidmore, *International Political Economy* (Fort Worth, Texas: Harcourt Brace, 1993), 65–66.

[11]World Bank, *World Development Report, 1985* (New York: Oxford University Press, 1985), 94.

[12]See the discussions of foreign aid as an instrument of statecraft (in Chapter 11) and dependency theory (earlier in this chapter).

being of the poor. The controversy over ODA centers on such questions as what development strategy is preferred, how much external financing of development is needed, and how it is to be provided.

**THE CHANGING PICTURE OF DEVELOPMENT AID**   ODA is awarded either bilaterally (directly from a donor to a recipient country) or multilaterally (through an IGO). Multilateral assistance presumably has the advantage of removing the donor's power to exercise direct political influence over the recipient. Nonetheless, in recent years, bilateral aid has accounted for 64 percent of the total of foreign assistance, while multilateral agencies dispensed some 22 percent.[13] For many years, the United States was the largest single donor of ODA, but Japan, with a population of less than half that of the United States, became the leader as of 1992. Among OECD members, three Scandinavian countries now lead the pack when aid contributions are calculated as a percentage of their GNPs, averaging about 1 percent. As Table 12.2 indicates, the much greater size of the U.S. economy still produces a larger aid package, although in percentage terms, the U.S. contribution was dead last in 1993: about 0.15 percent of its GNP. Note the sharp decline from the 2.5 percent of the American GNP given to Marshall Plan aid some forty-five years earlier.

What Table 12.2 does not report are the ODA contributions of several OPEC members of the Arab Middle East. Saudi Arabia's development aid approximates that

---

[13] World Bank, *World Development Report, 1990* (Oxford: Oxford University Press, 1990), 128. The remaining 14 percent comes through privately funded NGOs.

**TABLE 12.2**   Selected Official Development-Assistance Contributors, 1993

| COUNTRY | AMOUNT OF ASSISTANCE (IN U.S. BILLIONS OF DOLLARS) | PERCENTAGE OF GDP |
|---|---|---|
| Japan | $11.26 | 0.26% |
| United States | 9.72 | 0.15 |
| France | 7.92 | 0.63 |
| Germany | 6.95 | 0.37 |
| Italy | 3.04 | 0.31 |
| United Kingdom | 2.91 | 0.31 |
| Netherlands | 2.52 | 0.82 |
| Canada | 2.37 | 0.45 |
| Sweden | 1.77 | 0.96 |
| Denmark | 1.34 | 1.03 |
| Spain | 1.21 | 0.25 |
| Norway | 1.01 | 1.01 |

**SOURCE:** OECD, Development Assistance Committee (DAC), *Development Cooperation, 1994*, chart IV-1.

**TABLE 12.3** Geographical Distribution of ODA, 1982–1983 and 1992–1993

| PERIOD | SUB-SAHARA AFRICA | OCEANIA | ASIA | NORTH AFRICA AND THE MIDDLE EAST | LATIN AMERICA AND THE CARIBBEAN | SOUTHERN EUROPE |
|---|---|---|---|---|---|---|
| 1982–1983 | 27.9% | 3.5% | 33.3% | 21.4% | 11.7% | 2.2% |
| 1992–1993 | 37.3 | 3.0 | 29.4 | 12.3 | 11.4 | 6.3 |

SOURCE: OECD, Development Assistance Committee (DAC), *Development Cooperation, 1994,* table 39.

of Sweden, but constitutes an even larger share of its GNP—about 1.5 percent, which is down from a contribution in 1980 that equalled more than 5 percent of its GNP. The United Arab Emirates has made comparable contributions (1.66 percent of GNP in 1991). Those two countries, along with Kuwait, for some time have contributed a larger portion of their wealth to development assistance than any others.

Table 12.3 indicates the geographical distribution of ODA in 1982–1983 and 1992–1993. North Africa and the Middle East have seen a sharp decline in assistance over this period; aid has dropped off slightly for Asia and Oceania, and has increased substantially for sub-Saharan Africa and Southern Europe.

Finally, by examining the broad picture of foreign aid rather than the more narrow one of ODA, it becomes clear that nondevelopmental considerations often are at work on the part of the donor country. That is most pronounced in the case of the United States, whose aid typically is targeted at a handful of its strategic friends and allies; Israel, Egypt, Turkey, and Jordan are prime examples. A 1986 World Bank study found that only 8 percent of the U.S. aid budget in that year went to low-income countries.[14] As a result of such practices, Israel, a relatively rich country, has in recent years received nearly 300 times as much aid per capita as impoverished Nigeria.

## FOREIGN DIRECT INVESTMENT

Official development assistance does not approach the amounts of money involved in foreign investments made by individuals, banks, and multinational corporations. These activities of the private sector produce what is known as **foreign direct investment (FDI).** Not much of it is concerned with overcoming economic underdevelopment, as that term is used in this chapter. Investors, after all, wish to make a profit and so are unwilling to assume too much risk. Therefore, the majority of private foreign investments still go to the North, where the risk factor is likely to be comparatively low. For example, in the 1980s, more than one-half of all FDI

---

[14] *World Development Report, 1990,* 127–128.

**FIGURE 12.1** | Inflows of Foreign Direct Investment, 1988–1993

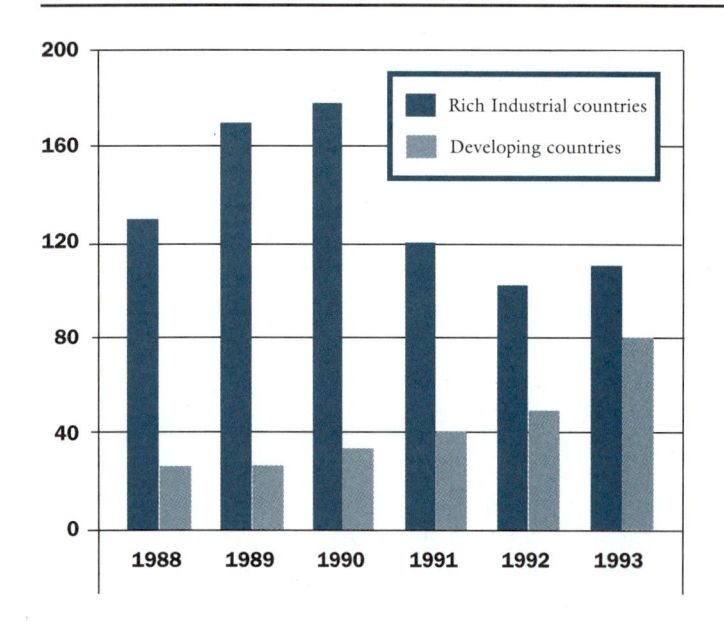

| Recipients of FDI |
| :--- |
| Ten largest developing countries, in billions of U.S. dollars |

| | 1988-1992 |
| --- | --- |
| China | $25.6 |
| Singapore | 21.7 |
| Mexico | 18.4 |
| Malaysia | 13.2 |
| Argentina | 10.6 |
| Thailand | 9.5 |
| Hong Kong | 7.9 |
| Brazil | 7.6 |
| Taiwan | 6.0 |
| Indonesia | 5.6 |

SOURCE: UNCTAD, "Trends in Foreign Direct Investment," TD/B/ITNC/2, February 18, 1994, Figure 1.

from investors in the United States was in five highly developed countries—the United Kingdom, Canada, Germany, Switzerland, and the Netherlands.[15]

Investment in the South has increased dramatically over the course of this century. In dollar amounts, such investment more than tripled from 1914 to 1960, then nearly doubled again by the late 1980s. But such figures can be misleading, because they are swallowed up by the enormous leap in total FDI during this period, most of which has been among the Northern states. Including those figures reveals that, in 1914, FDI in the LDCs accounted for 60 percent of the total worldwide investments; by the 1960s, that share had fallen to one-third; and by the mid-1980s, that proportion amounted to only about one-quarter of the total foreign investments.[16]

That pattern began to be reversed in the 1990s, however, as can be seen in Figure 12.1. Inflows of FDI into developing countries surged from $31 billion in 1990 to $80 billion in 1993.[17] That increase came at the same time that ODA was in a gradual decline, however—from $70.4 billion in 1991 to $68.5 billion in 1993.

---

[15]John R. O'Neal, "Foreign Investment in Less Developed Regions," *Political Science Quarterly* 103, no. 1 (1988): 137–138.

[16]Rhys Jenkins, *Transnational Corporations and Uneven Development: The Internationalization of Capital and the Third World* (New York: Methuen, 1987), 5, 13.

[17]"War of the Worlds," *The Economist,* October 1, 1994, 23.

This change in the balance between private and public sector sources of investment capital is reflected in where the investment actually goes. Figure 12.1 lists the ten states in the South that have received the most FDI in recent years. A majority of them are, by most measures, rapidly joining the rich countries as the result of their rapid growth rates and rising per capita income. In contrast, much of sub-Saharan Africa is impacted negatively by the shift, since its much poorer societies are less attractive to foreign investors and have been far more dependent upon assistance from governments and international organizations.

## IGOs in the North-South Debate

Most of the debate about development has taken place within the framework of IGOs over the past half century. That has influenced both the terms of the discourse and particular outcomes. IGOs typically provide forums where multilateral diplomacy takes place on the Westphalian principle of one state, one vote. Starting in the 1960s, that feature began to allow the South a greater voice in issues of development than ever before.

When the United Nations was created in 1945, a little more than one-half of its 51 members were what would today be described as LDCs; Northern states were dominant. Within twenty years, however, U.N. membership had more than doubled, and almost all the additional membership had come from the former colonies of European powers. The growth in the number of LDCs that were members of the United Nations led them increasingly to voice their views in the General Assembly, the one U.N. body where every member state is equally represented with a single vote. Newly independent states were particularly interested in using their membership to bring before the United Nations issues regarding their own economic development.

As a result of their voting strength, a number of U.N. agencies were created, particularly in the 1960s and 1970s, meant to advance the economic development of the South. Among these are UNCTAD, UNITAR, and UNDP (see Figure 12.2).

### United Nations Development Program

Donor states have tended to favor ODA in the form of **technical assistance,** for the obvious reason that it is the cheapest of all forms of aid. Technical assistance entails teaching people the skills and new technologies they need to modernize their economies. From its early years, the United Nations, as the world's principal multilateral forum for governmental interaction, has been the chief conduit of technical assistance channeled from member countries in the North to the South.

Since the mid-1960s, U.N. technical assistance programs have been unified under the **United Nations Development Program (UNDP).** Its activities fall broadly into two categories. The first is essentially a training function, which includes providing experts and demonstration equipment to train others in needed skills and technologies. Fellowships for training in industrialized countries are also included here. The second function focuses on "preinvestment" surveys of a country's natural resources and other potential for development.

**FIGURE 12.2** The United Nations System: Agencies with Roles in Economic Development Issues (in blue) and Other Organizations

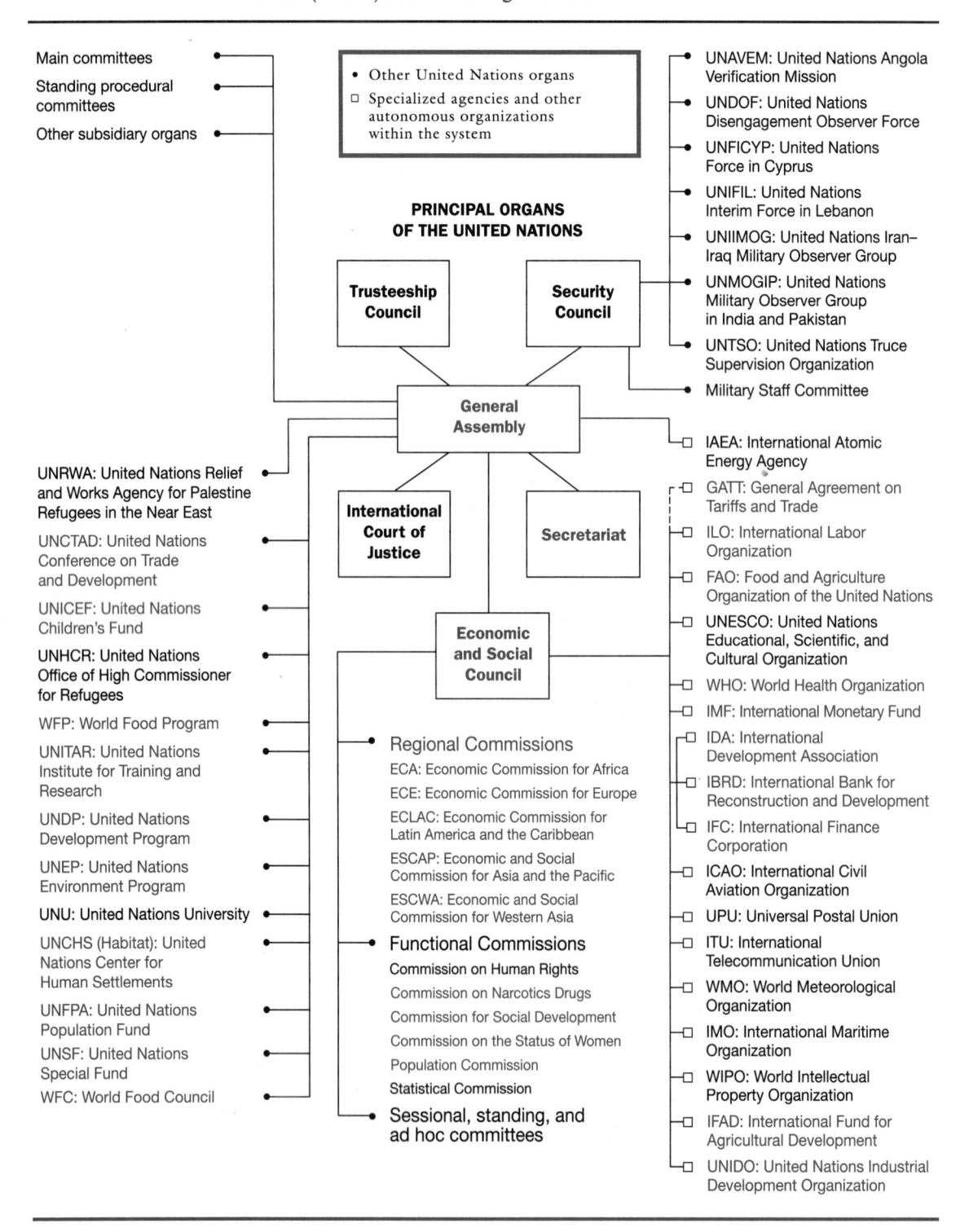

- Main committees
- Standing procedural committees
- Other subsidiary organs

- Other United Nations organs
- ☐ Specialized agencies and other autonomous organizations within the system

**PRINCIPAL ORGANS OF THE UNITED NATIONS**

- Trusteeship Council
- Security Council
- General Assembly
- International Court of Justice
- Secretariat
- Economic and Social Council

- UNAVEM: United Nations Angola Verification Mission
- UNDOF: United Nations Disengagement Observer Force
- UNFICYP: United Nations Force in Cyprus
- UNIFIL: United Nations Interim Force in Lebanon
- UNIIMOG: United Nations Iran–Iraq Military Observer Group
- UNMOGIP: United Nations Military Observer Group in India and Pakistan
- UNTSO: United Nations Truce Supervision Organization
- Military Staff Committee

- UNRWA: United Nations Relief and Works Agency for Palestine Refugees in the Near East
- UNCTAD: United Nations Conference on Trade and Development
- UNICEF: United Nations Children's Fund
- UNHCR: United Nations Office of High Commissioner for Refugees
- WFP: World Food Program
- UNITAR: United Nations Institute for Training and Research
- UNDP: United Nations Development Program
- UNEP: United Nations Environment Program
- UNU: United Nations University
- UNCHS (Habitat): United Nations Center for Human Settlements
- UNFPA: United Nations Population Fund
- UNSF: United Nations Special Fund
- WFC: World Food Council

**Regional Commissions**
- ECA: Economic Commission for Africa
- ECE: Economic Commission for Europe
- ECLAC: Economic Commission for Latin America and the Caribbean
- ESCAP: Economic and Social Commission for Asia and the Pacific
- ESCWA: Economic and Social Commission for Western Asia

**Functional Commissions**
- Commission on Human Rights
- Commission on Narcotics Drugs
- Commission for Social Development
- Commission on the Status of Women
- Population Commission
- Statistical Commission

Sessional, standing, and ad hoc committees

- ☐ IAEA: International Atomic Energy Agency
- ☐ GATT: General Agreement on Tariffs and Trade
- ☐ ILO: International Labor Organization
- ☐ FAO: Food and Agriculture Organization of the United Nations
- ☐ UNESCO: United Nations Educational, Scientific, and Cultural Organization
- ☐ WHO: World Health Organization
- ☐ IMF: International Monetary Fund
- ☐ IDA: International Development Association
- ☐ IBRD: International Bank for Reconstruction and Development
- ☐ IFC: International Finance Corporation
- ☐ ICAO: International Civil Aviation Organization
- ☐ UPU: Universal Postal Union
- ☐ ITU: International Telecommunication Union
- ☐ WMO: World Meteorological Organization
- ☐ IMO: International Maritime Organization
- ☐ WIPO: World Intellectual Property Organization
- ☐ IFAD: International Fund for Agricultural Development
- ☐ UNIDO: United Nations Industrial Development Organization

Contributions to the UNDP's programs from member states are entirely voluntary. As Southern states grew more vocal with the end of colonialism in the 1960s, they began to argue that voluntarism in the development effort was not good enough. Many insisted that Northern states should contribute 1 percent of their GNP to the development effort. The North refused, but eventually agreed to an ODA "target" of 0.7 percent. Table 12.1 reveals that, as of 1993, only four OECD countries—plus Saudia Arabia, Kuwait, and the United Arab Emirates—met or exceeded that target.

## UNCTAD AND THE NIEO

By 1964, the voting strength of the South was great enough in the U.N. General Assembly that LDCs could call for a special **United Nations Conference on Trade and Development (UNCTAD)**. The resulting conference is now known as UNCTAD I, since at its conclusion the majority voted to institutionalize UNCTAD as a permanent U.N. structure. The conference itself would meet at four-year intervals, while an ongoing secretariat would be charged with implementing the policies voted by the conferences.

UNCTAD, which should be viewed as largely the creature of the South, soon changed the nature of the development debate in world politics. In preparation for the first conference, the LDCs generally caucused as a unified bloc for purposes of presenting their policy positions, then voted on them at the conference. Because they were a numerical majority for the first time at UNCTAD I, this **Group of 77**—the name by which they still are known, even though their ranks are nearly double that number today—was able to see most of its positions adopted during the conference. In general, those positions were opposed by the minority of DCs from the North whose "establishment" proposals, built upon modernization theory, were voted down. This marked the emergence of a set of directives based largely upon dependency theory with the potential cures for economic underdevelopment.

Those directives were formulated, over the course of the next decade, into a call for what would be known as a **New International Economic Order (NIEO)**. The central argument of the NIEO agenda is that the dominant modernization approach to development perpetuates the dependency of the South.[18] The central issue for most LDCs remains that of their frequent dependence on a few primary products for trade.[19] The NIEO solution is to provide such countries with certain **trade preferences.** The argument is that in the initial stages of industrialization, it is impossible for the infant industries of the South to compete with the well-developed manufacturing industries of the North. Tariff protection and special preferences may be necessary to allow the South's industries to become competitive. One NIEO suggestion is to institute a system of indexation, so that the price the LDC receives for

---

[18]The NIEO agenda is most fully delineated in the U.N. Charter of Economic Rights and Duties of States, which was adopted by the U.N. General Assembly in 1974. U.N. General Assembly Resolution A/3281(29), 1974.

[19]For a fuller discussion of how arguments favoring a NIEO would alter the structure of the global economic system, see Stephen D. Krasner, *Structural Conflict: The Third World against Global Liberalism* (Berkeley: University of California Press, 1985).

its exports will not be out of line with the increasingly higher prices it must pay for importing manufactured products.

The NIEO critique also notes the problem of maintaining or restoring greater **economic self-sufficiency** to LDCs that have been increasingly pulled into the kinds of economic connections with the dominant world system cited above. Ironically, subsistence-level economies, poor as they are, nonetheless tend to be virtually self-sufficient, so that foreign trade is less relevant to their meager existence. Yet, the developing country may cease to be self-sufficient as it turns increasingly to trade to further its own growth and development. So, countries that traditionally fed their people adequately may become importers of food as they undergo industrialization. Peasant populations are drawn into the new factories in the burgeoning cities, perhaps to work for a multinational corporation that commands a major share of the nation's economic output. Even though the nation's GDP may grow, as a result, it almost certainly is entangling itself more thoroughly in the world economic system. It is moving from the preindustrial state's impoverished self-sufficiency into the dependency that accompanies development.[20]

According to the NIEO critique, another example of how dependence is perpetuated is seen in the so-called **brain drain:** the continuing tendency of intellectuals from developing countries to be drawn into the richer societies of the North—first for their higher education, then for their employment. Thus, the very valuable developmental resource they represent is lost to their native societies, often permanently.

Finally, the **transfer of technology problem** argues that most of the technology essential to the development of societies in the South remains in the hands of private experts and entrepreneurs in the North. These highly skilled professionals do not transfer their expertise to those who must become their counterparts in the developing world, any more than they themselves take up residence in developing societies, bringing their technological skills with them. Again, the result for LDCs is their continuing dependence on the essential skills of others outside their societies.[21]

**THE FAILURE OF THE NIEO** The call for an NIEO actually reached its peak during the 1970s. At the Sixth Special Session of the U.N. General Assembly in 1974, the Group of 77 advanced a broad set of proposals meant to restructure the international economic system to its greater favor. Although some limited reforms followed,

---

[20]The strategy usually advocated to restore greater self-sufficiency to countries that have become dependent upon world market factors is that of the "basic needs" approach. See Samir Amin, "Self-Reliance and the NIEO," *Monthly Review* 29, no. 3 (August 1977): 1–21; and Johan Galtung, "The NIEO and the Basic Needs Approach," *Alternatives* 4, no. 4 (March 1979): 455–476.

[21]"UNCTAD figures show that the distribution of scientists and engineers worldwide is overwhelmingly concentrated in the North. The rate per 10,000 inhabitants is 95 in the developing countries compared with 285.2 in the industrialized market economy. . . . The figure for technicians is even more dramatic, revealing a difference between North and South of an order of magnitude of 10.

"The number of scientists, engineers and technicians engaged in R&D in the developing countries is less than 1.5 per 10,000 inhabitants, compared with 16.6 in the market economies of the North. . . . In Africa and Latin America the figure is only 0.2 percent, and in Asia 0.5 percent. These figures have shown no significant increase in the last two decades." (Ivan L. Head, "South-North Dangers," *Foreign Affairs* LXVIII, no. 3 [summer 1989], 81.)

**"YOU'RE LIKE A BUNCH OF . . . OF . . . OF . . . CAPITALISTS!"**
Dennis Renault, *Sacramento* (Calif.) *Bee*, 1974.

the proposals associated with UNCTAD and the NIEO had not been implemented by the 1990s. The lesson learned by the Group of 77 was that their voting majority in an IGO framework did not persuade the minority with the effective power—in this case, economic resources—to carry out majority directives.

# NEWLY INDUSTRIALIZED COUNTRIES

Although the overall development record in recent decades has not been particularly encouraging, a few countries in the South have dramatically increased their economic output during this period. According to one assessment,

> From 1980 to 1993 . . . the average economic growth of the developing countries of South and East Asia (excluding China) averaged 5.8 percent a year, against about 2 per-

cent annual growth in the United States and the European Union, and 3.6 percent in Japan. Growth at two times the rate in the developed economies is likely to continue in many of these Asian economies for the rest of the decade.[22]

The **newly industrialized countries (NICs)** of Brazil, Hong Kong, Mexico, Singapore, South Korea, and Taiwan (not officially a "country" but a "province" of China) all achieved this status in the 1970s as the result of dramatic and quick growth. China, Indonesia, Malaysia, and Thailand have arguably joined their ranks in the 1990s.[23] Their developmental paths were varied, although all were aided by their ability to borrow from the experience and inventions of the older industrialized states. That allowed them to move up the development ladder quickly by avoiding many of the costly mistakes made by the older industrial states early in their own developmental stages.[24]

Partly because the economic takeoff of these countries is recent, opinion is divided as to the extent to which other LDCs will join their ranks. Developmental optimists are inclined to argue that their numbers will grow. They remind us that for the first industrialized states of the West, the developmental process actually took several centuries. From that standpoint, the development of the NICs is particularly impressive because it happened within a few decades. Optimists are inclined to see the growing gap in income within the developing economy as a natural part of the development process, and one that will lessen over time:

> Nineteenth-century Great Britain and 1930s Sweden did not have quite as much disparity in income before taxes as contemporary Brazil, but they had more than Argentina, India, and Mexico have today. It was taken for granted in the nineteenth century that the developmental process was one in which the income pyramid temporarily narrowed at the top, at least until income taxes were applied to equalize the difference.[25]

Pessimists are inclined to see the success of at least most of the NICs as having been driven by the repressive policies of less than democratic governments. They were able to stimulate industrial development through austerity policies that would not have been acceptable in countries where democratic competition was vigorous. Furthermore, although per capita income typically has grown as the result of industrial development, it has been accompanied—particularly in Brazil and Mexico—by massive migrations of poor farmers to overcrowded cities. Even if industrial jobs are available there, housing and other support services often are not. In social terms, one may view this development as a mixed blessing.

---

[22] United Nations, *World Economic and Social Survey* (New York: United Nations, 1994), 113.

[23] Nine of the ten top recipients of foreign direct investment from 1988 to 1992 are among these NICs (see Figure 12.1). That reinforces the conclusion that FDI is targeted at the wealthiest, rather than the poorest, states of the South. See Stephen Haggard, *Pathways from the Periphery: The Politics of Growth in the Newly Industrialized Countries* (Ithaca: Cornell University Press, 1990).

[24] Frederic C. Deyo, ed., *The Political Economy of the New Asian Industrialism* (Ithaca, NY: Cornell University Press, 1987).

[25] Rosecrance, *The Rise of the Trading State*, 52.

> **Global Changes**
>
> ## The Changing Dimensions of Economic Development
>
> In the absence of a significant new strategy for world development, the world economy is likely to become even more polarized and divided between the rich and the poor. Already today, about 1.3 billion people, more than 20 percent of world population, are seriously sick or malnourished, according to the World Health Organization.*
>
> The external debt of developing countries grew to an estimated $1.9 trillion in 1994, up more than 7 percent over 1993. . . . The worst situation today is that of sub-Saharan Africa, excluding South Africa. Collectively, the region's debt amounts to $180 billion—three times the 1980 total, and 10 percent higher than its annual output of goods and services. Debt-service payments come to $10 billion annually, about four times what the region spends on health and education combined.†
>
> - How is the debt problem of much of the South related to the development effort? What is its relationship to the continuing gap between rich and poor?
>
> In the rapid industrialization of the nineteenth century, one country, Japan, developed by exporting raw materials, mainly silk and tea, at steadily rising prices. Another, Germany, developed by leapfrogging into the "high-tech" industries of its time, mainly electricity, chemicals and optics. A third, the United States, did both. Both routes are blocked for today's rapidly industrializing countries—the first because of the deterioration of the terms of trade for primary products, the second because it requires an infrastructure of knowledge and education far beyond the reach of a poor country.‡
>
> - Does this assertion give greater support to the position of modernization or dependency theorists? How might the other theory—the one in contention with that you've just selected—be used to respond to this charge?
>
> ---
>
> *Alexander King and Bertrand Schneider, *The First Global Revolution* (New York: Pantheon Books, 1991), 90.
>
> †Gary Gardner, "Third World Debt Still Growing," in Lester R. Brown, Nicholas Lenssen, and Hal Kane (Worldwatch Institute), *Vital Signs: 1995* (New York: W. W. Norton, 1995), 72.
>
> ‡Peter F. Drucker, "The Changed World Economy," *Foreign Affairs*, vol. 64, no. 4 (spring 1986), 781.

Some who are skeptical about other countries entering the ranks of the NICs point to fundamental changes that are taking place in the global economy. Commodity prices have tended to fall steadily in recent years, thanks in part to synthetic substitutes.[26] Prices for nonfuel commodities, for example, fell in real terms by 30 percent from 1975 to 1989.[27] The decline in the price of oil in the 1980s seriously eroded Mexico's balance of payments situation, greatly increasing its external debt burden in the 1990s. Indeed, the debt burdens of both Brazil and Mexico in the 1980s stymied their continued growth and expansion for a number of years. This suggests at the least that their problems of dependency have not been entirely overcome.

---

[26]Robin Broad and John Cavanaugh, "No More NICs," *Foreign Policy* 72 (fall 1988), 81–103.

[27]Hal Kane, "Trade Continues Steep Rise," in *Vital Signs, 1993,* ed. Lester Brown et al. (New York: W. W. Norton, 1993), 74.

# Communism's Collapse and Economic Development

For many decades, authoritarian communism seemed to provide an alternative to the liberal economic model for modernization and development.[28] From 1917 through the 1960s, communist parties had their greatest success in attaining power in states that were at the early stages of economic modernization.

## ECONOMIC DEVELOPMENT UNDER COMMUNISM

The Soviet Union underwent remarkably rapid industrialization under the dictatorial rulership of Lenin and Stalin (1917–1953). At the beginning of Communist Party rule, Russia was still a predominantly agricultural country, while by the time of Stalin's death 36 years later, it had become the world's second-ranking industrial power. That had occurred thanks to a command system that forced industrial development at the expense of agriculture, the satisfaction of consumer wants, and environmental protection. In spite of the devastation the USSR suffered in World War II, recovery was rapid, and the economy continued to expand at an average rate of 10 percent annually through the 1950s.

Meanwhile, in 1949, communist rule was established over mainland China, the most populous nation in the world but also generally awash in the poverty of a traditional, preindustrialized society. Its leaders, under the direction of Mao Zedong, originally modeled their policies strongly on those of Stalinist Russia. Soon, however, China's distinctive problems—including its far larger peasant population and different culture—produced a growing animosity, and finally enmity, between the two governmental elites. Even though China began to try its own distinctive economic development strategies, these nonetheless were built, like those of the Soviets, on the ability of a single-party state to command its populace to engage in specified kinds of economic activity.

China had been considerably transformed by the time of Mao's death in 1976; but its development did not much resemble that of the Soviet Union or the other industrialized powers of the West. The country was also just beginning to recover from the turmoil of the Great Proletarian Cultural Revolution. That ten-year upheaval had disrupted much of the economic life of the nation by disgracing most of China's educated elite, including those responsible for economic planning, production and management.

**"REFORM" OF COMMUNIST SYSTEMS IN THE 1980s** Mao's successor as the effective leader of China, Deng Xiaoping, soon launched sweeping economic reforms that borrowed heavily from doctrines of free enterprise. Foreign investment was encouraged and the profit motive was reintroduced into many aspects of economic life. The result was that China prospered dramatically throughout the 1980s; some Chinese got rich, and China's economic ties with much of the rest of the world grew and

---

[28]Adam Ulam, *The Unfinished Revolution: Marxism and Communism in the Modern World*, rev. ed. (Boulder, Colo.: Westview Press, 1979).

flourished.[29] For a time, it appeared that Deng's policies marked the beginning of the end for the idea that an authoritarian Marxism could any longer serve as a doctrine of economic development.

That view was reinforced by events in the Soviet Union during the same period. By the beginning of the 1980s, the Soviets' remarkable period of economic growth had come to a halt. The period in which Leonid Brezhnev led the nation (1964–1983) came to be seen as a time when officially sanctioned corruption flourished while growth nearly ceased, and no one in a position of leadership had confronted the country's economic decline effectively. Mikhail Gorbachev's policy of perestroika, launched soon after he came to power in 1985, acknowledged how the authoritarian system of planning no longer was working for the state's economic advance. Gorbachev proposed dismantling the command system, along with the highly centralized control of the Communist Party over the society's political life.[30]

Gorbachev went far beyond any reformist initiatives the Chinese leaders had allowed. In general, during the period from about 1985 to 1989, the Chinese leadership focused on reform of the Chinese economic system while maintaining a rigidly authoritarian dictatorship over the nation's political life. In contrast, the Soviet leadership first launched far-reaching political reforms, with potentially earth-shaking implications for the stability of the system, and generally deferred tackling a substantial reform of the economic system. The priorities of the Chinese leadership were made clear in its murderous suppression of the pro-democracy movement at Tiananmen Square in June 1989. Having reversed those priorities, the Soviet leadership was soon faced with sweeping political challenges to the communist system. These eventually brought about the dismemberment of the Soviet Union itself at the end of 1991.

## THE FAILURE OF AUTHORITARIAN ECONOMIC PLANNING

These 1980s experiences in both China and the Soviet Union suggested that an authoritarian system built upon central planning and control was no longer able to assure continued economic growth and development under the changed economy of the late twentieth century. Both had been forced to introduce greater freedom through financial incentives to try to stimulate their economies. Well into the 1990s, it remained unclear whether economic freedom could be maintained under rigid political control, as in China—or whether, as in Russia, political freedom could induce economic reform without first bringing unacceptable disorder, rebellion, and collapse.

In 1990, the Marxist Sandinista government of Nicaragua was defeated at the polls. Within another year, authoritarian regimes once supported by Moscow had been driven from power in Mozambique and Ethiopia, at least in part because of their economic failures. With the crumbling of the Soviet Union as an authoritarian model of economic development, Moscow was no longer willing or able to continue economic assistance to what had been its client states.

---
[29] Benedict Stavis, *China's Political Reforms: An Interim Report* (New York: Praeger, 1988).
[30] Mikhail Gorbachev, *Perestroika* (New York: Harper and Row, 1987), chap. 2.

Latvians survey a statue of Lenin toppled in 1990.

## ECONOMIC LIBERALISM UNCONTESTED?

For the short run, at least, the end of communism left economic liberalism in an even more dominant position as the world's leading economic approach. It also left liberalism as the largely uncontested doctrine for addressing development issues.

In one respect, the dominance of liberalism was nothing new. Centrally planned economies never had been major participants in the world's trading system because planning requires governmental control of an order that cannot readily extend beyond the confines of the state. For forty years after the end of World War II, the most orthodox Marxist-Leninist regimes had promoted economic development on a state by state basis, each largely sealed off from the international economic system. There was considerable irony in this; Marx had viewed the nation-state as the historical instrument for the capitalists' domination of society, arguing that it would wither away once socialism had transformed economic and social relationships. Yet, Lenin and his followers had made the instruments of the state far more critical to the society's economic life than it had ever been in the capitalist world.

But during the existence of the Soviet bloc, its governments were at least sympathetic toward the arguments for statist protections that many LDCs advanced as tools in their own development. The USSR was also a major donor of foreign assistance. Today, the varied states of the former Soviet bloc are mostly preoccupied with a baffling array of economic difficulties at home as they attempt to adjust to economic life that is more or less responsive to the market. None of them currently has satisfied these domestic problems sufficiently to be able to address the needs of the developing countries of the Southern Hemisphere. It is not even clear in the mid-1990s how much of an economic "safety net" these societies now will retain for their disadvantaged citizens. Most of them are participating more fully than they have before in the world trading and investment system; yet, the economic and political challenges facing most of them are tending to push the development priorities of much of the rest of the world farther away from the international limelight.

# A 1990s Assessment

The end of the cold war did not result in a major rechanneling of resources from the West's military sector to development assistance. Some are convinced, nonetheless, that it helped bring about greater agreement between North and South on the central need for political and economic stability, popular participation, reliance on the market, and concern for the environment.[31] According to a spokesperson for the richest industrialized states of the North, the resulting *new paradigm* for development means that "distinctions between East and West, North and South, donor and recipient should become less significant."[32] The new development paradigm is overwhelmingly the product of liberal economic doctrines and modernization theory. It holds that LDCs will develop through privatization, relaxing trade and capital controls, and otherwise encouraging foreign investors into their markets.

The new paradigm received a jolt in the December 1994 foreign exchange crisis in Mexico. Foreign investors grew nervous over monetary policies of the new Mexican government and sold off their holdings in such a rush as to cause the near collapse of the peso. The crisis was ended after the United States came to the assistance of the Mexican government, though the experience left North and South in substantial disagreement as to what the crisis meant. "Many in the developing world felt that their worst suspicions of foreign investors had been confirmed. Many observers in developed countries, however, blamed the Mexican government."[33] Though steps were taken to try to prevent such a crisis elsewhere, support for the new development paradigm clearly had been shaken.

---

[31] George H. Mitchell Jr., "Economics and Development," *A Global Agenda: Issues before the 50th General Assembly of the United Nations* (Lanham, Md.: University Press of America, 1995), 113.

[32] James H. Michel, chairman of the OECD's Development Assistance Committee, quoted in Mitchell, "Economics and Development," 114.

[33] Mitchell, "Economics and Development," 114.

The debt owed by LDCs on loans from Northern banks doubled between 1965 and 1973, then doubled again by 1978. Throughout the 1980s, this debt crisis often overwhelmed development, as repayment of loans from South to North threatened to exceed new loans and grants from North to South. The South's total indebtedness stood at more than $1.7 trillion by 1992. One aspect of the new paradigm is the restructuring of much LDC debt on terms more favorable to the debtor countries. While that helps debtors, it also means that Northern banks are now assured of repayment of their loans. The fly in the ointment is that the resulting debt relief extends far more to the upper tier of debtor states than to those at the bottom. African debt-service payments still consume more than 30 percent of their export earnings.[34]

As a response to dependency, the OPEC cartel was extremely successful during the 1970s in increasing oil prices throughout the world and, therefore, oil revenues of its members. That marked the improved ability of some countries in the South to obtain a greater share of the profits from their raw materials. Similar cartels have been developed in conjunction with the production of bauxite and coffee, and have been partially successful in raising world prices on those products. Yet, by the 1990s, no other such effort had come close to OPEC's earlier success, and OPEC itself had long since seen its revenues decline with the falling price of oil.

The frustration of many in the South may stem less today from being exploited by capitalist states than from the fact that the LDC is so little needed as a supplier of raw materials, as a result of the development of synthetic materials and substitutes and the discovery of new sources of exploitable raw materials. The South has also become a less vital market for developed states. It accounted for more than 30 percent of the latter's exports in 1950 but less than 20 percent by the late 1960s. The one hopeful sign here was that by 1990, that percentage had increased slightly, to just over 20 percent. As noted earlier, Northern investment, as a percentage of total foreign investment, has also declined throughout the century.[35]

The North maintains its dominance in the world economic order largely because of the fact that it can produce the one thing, capital, that is in short supply in the South and essential to its development. That allows Northern governments largely to determine the conditions under which they will provide development assistance. It has mattered little that the Group of 77 can outvote the North where IGOs allow for equal state representation. Even though LDCs sometimes express their demands for development assistance as an entitlement owed their societies, the governments of the DCs, not surprisingly, tend to disagree. They are inclined to insist that such assistance is a "donation," not an obligation, and thus a matter left to the political determination of each donor state. They use the ideology of economic liberalism to support this stance, arguing that development is best encouraged through the "natural" workings of marketplace factors, rather than through substantial governmental interference in that process.[36]

---

[34] International Monetary Fund, *World Economic Outlook, October 1994* (Washington, D.C., 1994).

[35] See Jenkins, *Transnational Corporations and Uneven Development;* and the previous discussion of foreign direct investment.

[36] For a 1990s assessment and a call to action, see the report of the U.N. Secretary-General, Boutros Boutros-Ghali, *An Agenda for Development, 1995* (New York: United Nations, 1995).

> **Normative Dilemmas**
>
> ## Social Problems Accompany Development
>
> Brazil's mining of the vast mineral resources of the Amazon basin threatens primeval Indian communities living there. Logging and burning of the Amazon forest, which is one of the largest suppliers of oxygen for the planet, proceeds at a rate that could turn the basin into a desert by early in the twenty-first century.
>
> - Is there any way to accommodate the economic incentive to develop the Amazon basin with the cultural rights of the region's natives? How can exploitation of the Amazon forest be managed in a way that ensures it is sustained for centuries to come?
>
> A pair of Nike athletic shoes sells in the United States for $80 and up. Those shoes may have been manufactured in Indonesia, where minimum wages and labor rights generally are not enforced. In 1991, the typical worker in an Indonesian shoe factory was making just over $1 per day. In that same year, the Nike company reported net profits of $287 million.
>
> - What are the consequences of this situation for Indonesian social and economic life? If a substantially larger minimum wage were to be enforced in Indonesia, what would be Nike's likely reaction?
>
> "If, for example, an economy grows at an annual rate of 5 percent, it would, by the end of the next century, reach a level of 500 times greater (or 50,000 percent higher) than the current level."*A number of Asian economies grew at much higher rates than 5 percent throughout the early 1990s.
>
> - Based upon the above calculation, how long do you think it would be possible to sustain such growth rates? Are such rates indefinitely desirable? Why or why not?
>
> "[The] development gap poses moral, political, and economic challenges for the relatively wealthy countries of the North. . . . Third World debt could have far reaching consequences for Northern economies. Other issues of great concern to the North, such as illegal immigration, the destruction of the world's rain forests, and the drug trade, can be traced indirectly to continuing Third World poverty."†
>
> - What would be the consequences for the richer countries of the North if many Third World countries default on their debt repayments? Can you trace the social problems cited above to poverty in the Third World?
>
> ---
>
> *Eduard Pestel, former minister of culture, science and technology of Lower Saxony, as quoted in Alexander King and Bertrand Schneider, *The First Global Revolution* (New York: Pantheon Books, 1991), 6.
> †Thomas D. Lairson and David Skidmore, *International Political Economy* (Fort Worth, Texas: Harcourt Brace, 1993), 182.

# CONCLUSION

After nearly fifty years in which problems of economic development have been seriously addressed on the agenda of world politics, the results are at best mixed. Only a handful of countries have managed to improve their economic well-being dramatically. The overwhelming number of poorer countries have found themselves, like Alice in Wonderland, required to run ever faster to stay in the same place. Thanks to a host of factors, including those that stem from our rapid overcrowding of the planet, many of the poorest states of the South are actually seeing their living standards decline, while many others, better off than the former, are nonetheless unable to grow at the rate of the richest countries of the North.

As these trends first became visible, criticism grew of the world's liberal economic system, including its financial institutions, which were seen to favor loaners over borrowers. Theories of economic dependency gained a hearing in the South in the contest with the modernization theory generally embraced by the North. The call for a New International Economic Order, which was a thorough critique of the ideology of economic liberalism, received much attention in the 1970s. Along with UNCTAD, the NIEO marked the ability of LDCs, simply on the basis of their numbers, to make their voices heard in the formal institutions of IGOs.

But leading governments in the North generally spurned the NIEO program. In the 1980s, they tended to rely even more than before on the marketplace factors of capitalism as the sole engine of development. One result was a more pronounced disequilibrium in the economic growth of much of the North and the South during this period, as well as within those societies. A great many countries in the South staggered under an unexpected debt burden, soaring population growth, and economic stagnation.

Although a number of newly industrialized economies arose, starting in the 1960s, it was not clear in the mid-1990s whether others would join them. Structural changes in the world economy were altering the development discussion in new ways. Meanwhile, the collapse of the Soviet bloc as an authoritarian system meant also the decline of Marxism-Leninism as a viable doctrine of modernization and industrialization. In the immediate post–cold war world, the new development paradigm reflected a sense that the doctrine of economic liberalism seemed largely uncontested.

# CHAPTER 13

# Resource and Environmental Challenges

A book published more than two hundred years ago, Adam Smith's *The Wealth of Nations,* is the classic exposition of the theory of **laissez-faire** capitalism. Smith's central premise is that individuals should be left to seek their fortunes with a minimum of restrictions placed upon them by their governments, for it is assumed that the private quest for material wealth will ultimately work to the good of the whole society. Smith's argument also assumes that the material world is one of more or less unlimited potential bounty that only needs to be exploited by industrious human beings to yield up its riches. In capitalist theory, economic growth and development are the principal social goods of private enterprise.

Smith's premises remain those of contemporary capitalists. But the material world is vastly different as we approach the twenty-first century than it was in 1776, the year that *The Wealth of Nations* was published. The globe's human population has multiplied nearly sixfold, which itself would account for a sixfold increase in human demands for material goods. But the impact is actually far greater: The world's industrialization, which began soon after Smith's writings, has intensified that demand in an almost geometric fashion. That is, the well-off citizen of a highly industrialized society today—who employs or benefits from energy-consuming, labor-saving devices in great variety—thereby consumes far more resources or energy, much of which may be nonrenewable in the physical world, than his or her counterpart in a preindustrial society. On average, each one of the more than five billion persons alive today consumes more than *six times* the amount of energy used by each of the two billion humans on earth at the beginning of this century.

At the close of the twentieth century, we are at last being forced to confront the damage we are doing to the natural world as a consequence of our ongoing quest for material enrichment. However, this new awareness has by no means yet displaced the dominant values of material growth and development across most of the globe.

In this chapter, we shall explore how this conflict between the desire for greater economic well-being and the need for living within the limits imposed by nature itself are likely to shape world politics in the years ahead.

# SCARCITY VERSUS DEVELOPMENT DEMANDS

Economics has long been known as the dismal science because of its basic assumption that human demands for material goods will always exceed their inherently scarce supply. At one level, the attention of the past half century to the global problems of economic underdevelopment may be read as largely ignoring that assumption. That is, society has often proceeded as if all the world should reasonably aspire to the standards of material well-being now realized by a comparative handful of the richest states of the West. But, as we have come to recognize both the rapid depletion of nonrenewable resources and the damage development is doing to our natural environment, increasing numbers of critics have noted the flaws in such expectations.

## OIL CONSUMPTION AND OIL DEPLETION

Since the Industrial Revolution, "development" has been equated with the use of natural resources, most of which have been in the form of fossil fuels that can never be replaced. The most dramatic case in which such an energy source is being rapidly depleted is that of petroleum. A look at oil use in the United States, one of the most highly developed countries in the world today, tells much of the story.

Only about 6 percent of the world's population live in the United States. Yet, this country accounts for more than 30 percent of annual energy consumption, nearly one-half of which is from petroleum (16,640,000 barrels per day in 1991). Much of the gasoline and other petroleum products consumed in the United States annually go to fuel its more than one hundred million passenger cars—that is, one car for every 1.8 Americans by 1990. In that same year, approximately three billion barrels of oil were in production worldwide out of known reserves of approximately 708 billion barrels. At 1990's rates of worldwide production and consumption, supplies might last for another two centuries, although they will become more and more expensive to extract as the most easily tapped wells are exhausted. Yet, if oil consumption throughout the world were to approach American levels, these reserves would be thoroughly depleted in only about forty years. Therefore, to increase oil consumption (and car ownership) to those levels would be foolhardy, even if it were possible.

Nor can the current American rate of oil consumption be sustained much longer in the United States itself. By 1990, nearly one-half of all the oil discovered anywhere in the world had been consumed. Four-fifths of all the oil discovered in North America has already been burned. Proven oil reserves in the United States now total about 36 billion barrels, which is scarcely enough to sustain U.S. needs until the end of the century at current rates of consumption.[1] Clearly, the alternatives are for ever greater reliance on imported oil (as long as that is available) or the increased

---

[1] Lester R. Brown and Sandra Postel, "Thresholds of Change," in Lester R. Brown, et al., *State of the World: 1987* (New York: W. W. Norton Co., 1987), 11–12.

## TABLE 13.1 World Agricultural Energy Use and Grain Production, 1950–1985

| YEAR | ENERGY USE IN AGRICULTURE (MILLION BARRELS OF OIL EQUIVALENT) | GRAIN PRODUCTION (MILLION METRIC TONS) | ENERGY USED TO PRODUCE A TON OF GRAIN (BARRELS OF OIL EQUIVALENT) |
|---|---|---|---|
| 1950 | 276 | 624 | 0.44 |
| 1960 | 545 | 841 | 0.65 |
| 1970 | 970 | 1,093 | 0.89 |
| 1980 | 1,609 | 1,423 | 1.13 |
| 1985 | 1,903 | 1,667 | 1.14 |

SOURCE: Lester R. Brown and Sandra Postel, "Thresholds of Change," in Lester R. Brown, et al., *State of the World: 1987* (New York: W. W. Norton and Co., 1987), 11. Copyright by Worldwatch Institute.

use of other kinds of fuels. These alternatives carry new costs as well, whether in the increased price of oil itself (its price will inevitably rise along with those of production and supply, thanks to its growing scarcity), in greater pollution (coal, for example, is cheap but highly polluting), or in other social costs (since Chernobyl, nuclear power has come to be opposed by more than two-thirds of the populations of highly developed countries).

Since about 1950, worldwide agricultural production has increasingly been driven by oil as well. Before that date, most farmers were self-sufficient in energy, relying on draft animals for power and animal wastes as fertilizers. But with the greater mechanization of agriculture came large increases in productivity per acre. As a result, the agricultural use of energy has increased some sevenfold (see Table 13.1).[2] Modernization has demanded increases in agricultural productivity and the freeing up of labor to work in other sectors of the economy. Greater productivity is also demanded by swelling population growth with increased numbers of mouths to feed. This, too, has driven up the demand for oil. As the price of oil rises, the impact on food production will be to drive its costs up as well. People everywhere will have to pay a larger share of their income to satisfy basic needs. But the effect will be particularly harsh in the poorest societies, which may also increasingly have to resort to more primitive, subsistence farming methods to survive.

## ECONOMIC GROWTH VERSUS NATURE'S LIMITS

The situation with regard to oil consumption is echoed in the growing scarcity of other resources. Humanity increasingly must deal with the fact that its penchant for ever greater economic growth is running up against limits imposed by nature itself. The problem is a new one in historical terms, because the kind of economic

[2]Ibid., 10.

**TABLE 13.2** World Population, Economic Output, and Fossil Fuel Consumption, 1990–1995

| | POPULATION (IN BILLIONS) | GROSS WORLD PRODUCT (IN TRILLIONS OF U.S. 1980 DOLLARS) | FOSSIL FUEL CONSUMPTION (BILLION TONS OF COAL EQUIVALENT) |
|---|---|---|---|
| 1900 | 1.6 | 0.6 | 1 |
| 1950 | 2.5 | 2.9 | 3 |
| 1995 | 5.7 (est.) | 16.1 (est.) | 15 (est.) |

SOURCE: Adapted from Lester R. Brown et al., *State of the World: 1987* (New York: W. W. Norton and Co. 1987), 5. Copyright by Worldwatch Institute.

growth we tend to take for granted is itself a twentieth century phenomenon. Gross world product (the total amount of goods and services produced worldwide) increased from perhaps $640 million in 1900 to four and one-half times that amount in 1950 ($2.9 billion). But an even faster increase in wealth has come in the second half of the century. Per capita income roughly doubled over the next thirty to forty years, and at the same time the population was doubling as well. That actually meant more than a fivefold growth in the gross world product, to some $16 trillion by about 1995 (see Table 13.2).

Much of this huge increase in material production was made possible because of cheap energy costs. The fact that those costs generally continue to rise at the end of the twentieth century is one indication that, at least in some areas, we are now bumping up against the limits of what nature can provide. That, in turn, can only mean that the twentieth century's remarkable growth cannot be continued into the twenty-first. Writers for an important NGO in this area, Worldwatch Institute, put the issue this way:

> While the global economy has expanded continuously, the natural systems that support it unfortunately have not. Economist Herman Daly suggests that "as the economy grows beyond its present physical scale, it may increase costs faster than benefits and initiate an era of uneconomic growth which impoverishes rather than enriches." In essence, Daly points to an economic threshold with profound implications. As currently pursued, economic activity could be approaching a level where further growth in the gross world product costs more than it is worth.[3]

We have already seen evidence of such a trend in the increasing energy costs required to produce a ton of grain (Table 13.1).

[3] Lester R. Brown and Sandra Postel, "Thresholds of Change," *State of the World: 1987*, 6. (The Herman E. Daly quotation is from "Toward a New Economic Model," *Bulletin of the Atomic Scientists*, April 1986.)

There are many other indications that in the effort to satisfy our continually expanding appetites for growth, we have arrived at the thresholds of what nature can sustain. Some of these thresholds are obvious, as when the annual catch of fish exceeds the rate of replacement, or when the harvesting of lumber is greater than the rate of forest growth. But scientists are increasingly concerned about the more invisible thresholds that we may be crossing. Even though it may not always be possible to "prove" that a particular forest is dying because of damage from acid rain, scientists have documented acidification thresholds in soils. Once those thresholds have been crossed and forests die, it may take many years of expensive effort to restore soils to the point that they will again support healthy forest growth.[4]

## THE PRESSURE OF POPULATION GROWTH

Modern economic development has also been largely responsible for breakaway population growth. The poverty that most societies have endured since prehistory has always been accompanied by death, particularly for large numbers of newborn babies. That meant that for millennia before the start of the Industrial Revolution, the world's population grew very slowly, since so many children did not live to produce children of their own. Not until 1830 did the world's population reach one billion.

But by then, the Industrial Revolution was beginning in Western Europe and, with it, enough improvement in health care to begin to assure that more and more newborn babies would survive their childhood. In merely a century, by 1930, humankind numbered two billion. The third billion was added in only thirty years, by 1960; four billion people lived on the earth in another fifteen years. On July 11, 1987, world population reached five billion. At these rates of growth (which slowed slightly in the 1980s), the world's population could grow to 6.2 billion by the end of the century. That would amount to more than a threefold population increase during the lifetime of a man or woman who reaches the age of seventy in the year 2000. That same individual would have lived through more growth in population than occurred during the previous four million years of humanity's existence![5]

Every addition to the population obviously requires a subtraction from the resources necessary to sustain human life, whether those resources are a food supply, drinkable water, or the energy that humans draw from nature to give them mastery. For the first dozen or more years of their lives, human beings are especially dependent, requiring nurturing without providing much productivity of their own. Throughout many of the poorest regions of the world in recent decades, considerable increases in productivity have been eaten up (almost literally) by the need to sustain so many more young lives.

Here is one obvious explanation for the continuing increase in the income gap between North and South in spite of overall economic growth in the past half cen-

---

[4]T. C. Hutchinson and M. Havas, eds., *Effects of Acid Precipitation on Terrestrial Ecosystems* (New York: Plenum Press, 1980).

[5]Lester R. Brown, "Overview: The Acceleration of History," in Lester R. Brown, Nicholas Lenssen and Hal Kane, *Vital Signs 1995* (New York: W. W. Norton and Co., 1995), 15.

| **FIGURE 13.1** | World Population Growth (in billions) |

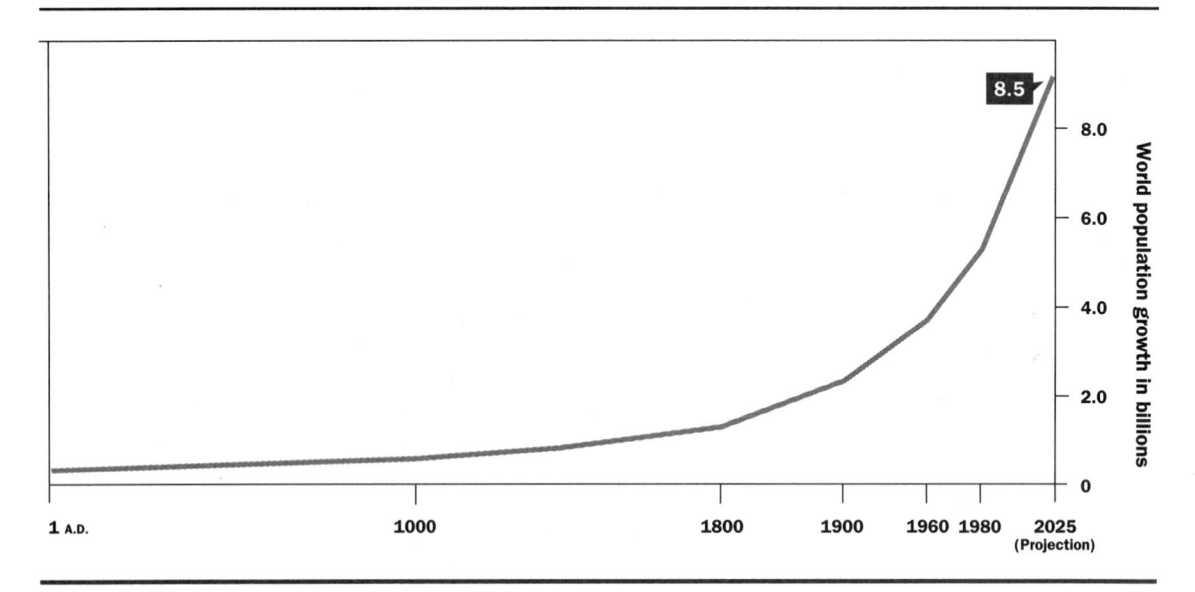

SOURCE: *1991 World Almanac: Atlas of the Environment.*

tury: In some of the world's richest societies, population growth is nearly nonexistent, while elsewhere—including some third world states that have more than doubled their productivity in recent years—a runaway growth in population is actually lowering living standards. A growth rate of only 2 percent per year doubles the population in thirty-five years. At 3.5 percent, that doubling takes a mere twenty years. Dozens of developing countries have been growing at those rates during the past three decades. We now add more than a hundred million people to the world's population annually. That is the equivalent of squeezing another Bangladesh onto the crowded surface of our planet every year (see Map 13.1).

For the short run, explosive population growth and development are mixing together to produce a volatile social cocktail in many third world nations. More and more of the increased population of the countryside is emigrating to cities in the search for the newly available forms of industrial work. The rate of growth of many cities in developing nations is far too rapid to be matched by governmental services. Mexico City has grown since World War II from fewer than two million to more than sixteen million today. As the president of the Worldwatch Institute has noted,

> in China alone, an estimated 120 million villagers have left their homes during the last few years, travelling to cities in search of employment. When they fail to find it, they move from city to city. The result is a huge floating population, one equal to the total population of Japan. Few trends concern officials in Beijing more than the potentially destabilizing effect of this mass movement of people.[6]

[6] Ibid., 20.

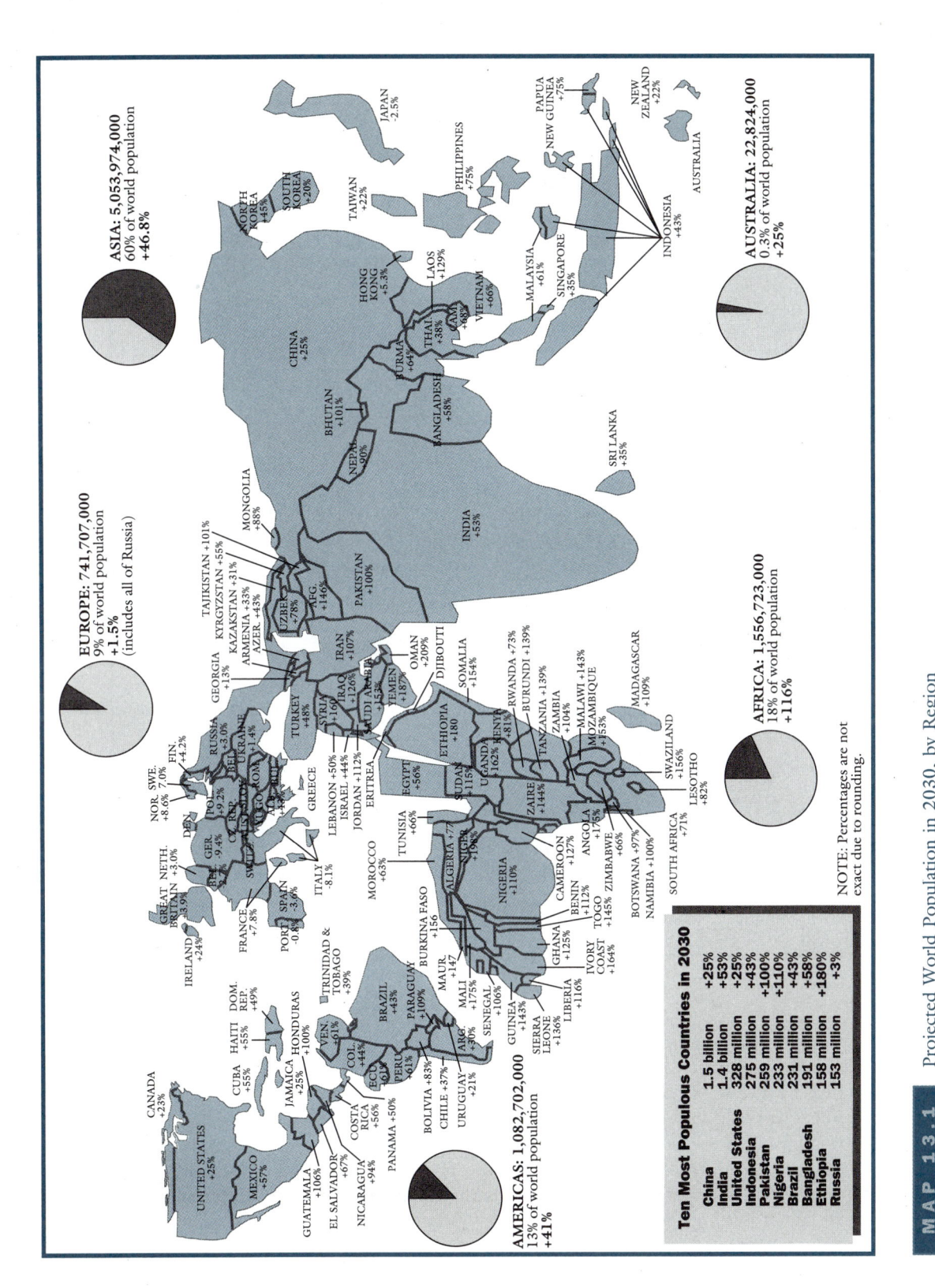

**MAP 13.1** Projected World Population in 2030, by Region

**EUROPE: 741,707,000**
9% of world population
+1.5%
(includes all of Russia)

**ASIA: 5,053,974,000**
60% of world population
+46.8%

**AUSTRALIA: 22,824,000**
0.3% of world population
+25%

**AMERICAS: 1,082,702,000**
13% of world population
+41%

**AFRICA: 1,556,723,000**
18% of world population
+116%

| Ten Most Populous Countries in 2030 | |
|---|---|
| China | 1.5 billion +25% |
| India | 1.4 billion +53% |
| United States | 328 million +25% |
| Indonesia | 275 million +43% |
| Pakistan | 259 million +100% |
| Nigeria | 233 million +110% |
| Brazil | 231 million +43% |
| Bangladesh | 191 million +58% |
| Ethiopia | 158 million +180% |
| Russia | 153 million +3% |

Country percentages shown on map:

CANADA +23%; UNITED STATES +25%; MEXICO +57%; GUATEMALA +106%; EL SALVADOR +67%; HONDURAS +100%; NICARAGUA +94%; COSTA RICA +56%; PANAMA +50%; CUBA +55%; HAITI +55%; DOM. REP. +49%; JAMAICA +25%; TRINIDAD & TOBAGO +39%; VEN. +61%; COL. +44%; ECU. +61%; PERU +83%; BOLIVIA +83%; CHILE +37%; URUGUAY +21%; ARG. +30%; PARAGUAY +109%; BRAZIL +43%;

IRELAND +24%; GREAT BRITAIN +3.9%; NETH. +3.0%; GER. −9.4%; DEN. +9.2%; NOR. +8.6%; SWE. +7.0%; FIN. +4.2%; RUSSIA +3.0%; UKRAINE −1.4%; FRANCE +7.8%; SWITZ.; PORT. −0.8%; SPAIN −3.6%; ITALY −8.1%; GREECE; TURKEY +48%; GEORGIA +13%; ARMENIA +33%; AZER. +43%; KAZAKSTAN +31%; KYRGYZSTAN +55%; TAJIKISTAN +101%; UZBEK. +78%; AFG. +146%; IRAN +107%; IRAQ +126%; SYRIA +169%; LEBANON +50%; ISRAEL +44%; JORDAN +112%; SAUDI ARABIA +153%; YEMEN +187%; OMAN +209%; DJIBOUTI; ERITREA;

MONGOLIA +88%; NORTH KOREA +45%; SOUTH KOREA +20%; JAPAN +2.5%; CHINA +25%; BHUTAN +101%; NEPAL +90%; BANGLADESH +58%; BURMA +64%; THAI. +38%; LAOS +129%; VIETNAM +66%; HONG KONG +5.3%; TAIWAN +22%; PHILIPPINES +75%; MALAYSIA +61%; SINGAPORE +35%; INDONESIA +43%; PAPUA NEW GUINEA +75%; NEW ZEALAND +22%; AUSTRALIA; SRI LANKA +35%; INDIA +53%; PAKISTAN +100%;

MOROCCO +63%; TUNISIA +77%; ALGERIA +90%; LIBYA; EGYPT +56%; SUDAN +115%; ETHIOPIA +180%; SOMALIA +154%; UGANDA +162%; KENYA +81%; RWANDA +73%; BURUNDI +139%; TANZANIA +139%; ZAIRE +144%; ANGOLA +170%; ZAMBIA +104%; MALAWI +143%; MOZAMBIQUE +153%; MADAGASCAR +109%; ZIMBABWE +66%; BOTSWANA +97%; NAMIBIA +100%; SOUTH AFRICA +71%; SWAZILAND +156%; LESOTHO +82%; CAMEROON +127%; NIGERIA +110%; BENIN +112%; TOGO; GHANA +125%; IVORY COAST +164%; LIBERIA +116%; SIERRA LEONE +136%; GUINEA +143%; SENEGAL +106%; MAUR. +175%; MALI +147%; BURKINA FASO +156%

NOTE: Percentages are not exact due to rounding.

SOURCE: Richard Furno, "2030: WOrld Population 8,474,017,000," *The Washington Post*, September 4, 1994, A41. ©1994, *The Washington Post*. Adapted with permission.

> **Global Changes**
>
> ## Humanity's Changing Impact on the Environment
>
> Population explosion: Every four or five days, an additional one million people are added to the world's net population. Every decade we add the equivalent of another India—one billion people—to the planet's crowded surface. Between 1995 and 2025, global population is expected to increase by some 60 percent—that is, from 5.3 billion to 8.5 billion.
>
> However, many of the richer countries of the North will experience almost no growth or even, in a few cases, population loss. Most of the anticipated increase, therefore, will be in less developed parts of the world. For example, India's population will nearly double, from 820 million to almost 1.5 billion. Mexico will also nearly double its population, from 85 to 150 million. Nigeria's population will almost triple, from 105 million to 301 million.
>
> - What are likely to be the chief social problems that the fastest-growing countries of the South will have to face over the next several decades? What will be the impact of those problems on the slow-growing countries of the North?
>
> Water: The demand for fresh water is thirty-five times what it was three hundred years ago. In many places, the cost of supplying water to burgeoning populations is becoming prohibitive.
>
> - Where in the world are demands for new fresh water supplies greatest? Where are those supplies most costly? What will be the consequences for human settlements where the cost of supplying fresh water becomes "prohibitive?"
>
> Greenhouse effect: Certain constituents of the earth's atmosphere trap the sun's heat, instead of reflecting it back into outer space. The Industrial Revolution started this process, increasing such heat-trapping gases as carbon dioxide in the atmosphere by 15 percent, nitrogen oxides by 19 percent, and methane by 100 percent. Carbon dioxide in the atmosphere has increased more in the past 150 years than in the previous 16,000 years, owing to the combustion of such fossil fuels as oil and coal. Also, the widespread elimination of tropical forests has reduced nature's capacity to absorb carbon dioxide through photosynthesis.*
>
> - If this warming trend continues, what effects on the world's climates might we expect? If coastal lowlands are eventually flooded as the result of the melting of polar ice, what will be the impact on human settlements? What about other kinds of economic activities that take place on coastal plains?
>
> ---
>
> *Alexander King and Bertrand Schneider, *The First Global Revolution* (New York: Pantheon Books, 1991), 33.

For the somewhat longer run, rapid population growth can only nullify whatever development gains have been made. First come the declining living standards that are already apparent in many developing countries. Next, as that decline continues, it could lead again to higher death rates as food and health supplies become insufficient. The result would slow population growth quite clearly, but at the cost of plunging much of humankind back into premodern economic and social conditions. The development efforts of this century then would have to be counted a total failure, setting the stage for a chillingly primitive struggle in the arena of world politics.

## DEVELOPMENT'S THREAT TO THE PLANET EARTH

Global politics in the coming years must increasingly reshape development strategies and goals if they are to prevent ecological collapse. The evidence for that conclusion has been mounting in recent years; yet by the 1990s, we had barely begun to reorient our thinking—let alone our policies and actions—away from traditional modes of industrialization and development. Many, perhaps most, statespersons throughout the world still were reluctant to find alternatives to the extraction and burning of fossil fuels to fire the engines of economic development and growth. Some even still seemed to think that those who warned of environmental dangers were simply finding a convenient excuse to deny development assistance to the LDCs.

But the dangers to our **biosphere** are real and unprecedented, thanks to the fast-growing number of our species and our ever greater exploitation of a fragile planet. Here we shall focus on two of the gravest threats to the health of the planet (which, of course, means to the health of all the species that inhabit it) in the closing decade of the twentieth century: global warming and the destruction of the ozone layer.

## GLOBAL WARMING

The so-called **greenhouse effect** of global warming is caused when carbon dioxide and other gases, produced mostly by the activities of humans, build up in the atmosphere and act like the glass of a greenhouse, letting in the sun's radiation but preventing much of it from escaping. This growing danger is a direct result of humanity's record of economic development. When the Industrial Revolution was still in its infancy, in 1860, perhaps 93 million tons of carbon were emitted worldwide. Carbon emissions jumped to 525 million tons by 1900, to 1.62 billion tons in 1950, and to about 5.92 billion tons by 1994 (see Figure 13.2).[7]

**BURNING FOSSIL FUELS** Fossil fuels are a principal source of the global warming problem. Not only are these resources finite and irreplaceable, but burning them releases carbon dioxide into the atmosphere in increasing amounts. The result is already potentially disastrous for the world's environment. Although it is difficult to measure a clear trend in global warming, ominous indications have appeared in recent years that it may be underway.[8]

For example, the ten warmest years since record keeping started have all occurred since 1980. Also since then, the temperature of the oceans has been rising at about twice the rate as had been anticipated in 1980 (0.2 degrees Fahrenheit). That has contributed to a rise in ocean levels of about one-twelfth of an inch per year. While that may seem a tiny amount, it could be enough to produce coastal flooding on a serious scale in coming decades, ravaging the world's great seaports and dislocating huge populations. More than that, rising temperatures could cause mass

---

[7]David Malin Roodman, "Carbon Emissions Resume Rise," in Lester R. Brown, Nicholas Lenssen and Hal Kane, *Vital Signs 1995* (New York: W. W. Norton and Co., 1995), 66.

[8]For a review of several studies investigating global warming, see Bill McKibben, "Is the World Getting Hotter?" *New York Review of Books,* December 8, 1988, 7–11.

**FIGURE 13.2** | Carbon Emissions from Fossil Fuel Burning by Economic Region, 1950–1993

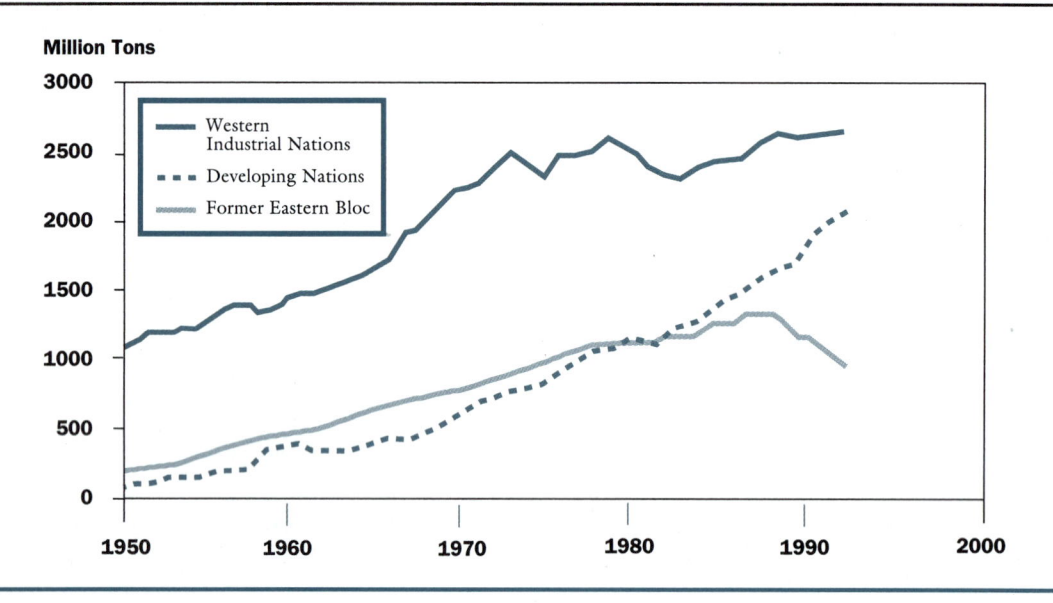

SOURCE: Lester R. Brown, Nicholas Lenssen, and Hal Kane, *Vital Signs 1995* (New York: W. W. Norton and Co., 1995), 67. Copyright by Worldwatch Institute.

starvation as food-producing regions are turned into deserts. Thousands of species could become extinct as their habitats are destroyed. Forty to 60 percent of the planet's vegetation could be disrupted and, with it, the planet's food chain.

Our motor vehicles are one of the principal culprits in the production of carbon dioxide, the chief contributor to global warming. As we have seen, the United States leads the world in automobile ownership. The United States now produces more than five times as much carbon dioxide—6.14 tons per person per year—as the world average. Here is a second reason why anything approaching American levels of automobile ownership worldwide would produce environmental catastrophe. But ironically, during the 1980s, just at the time when these dangers were becoming more widely understood, the U.S. government relaxed its fuel efficiency standards as gasoline prices fell. At the same time, investments in cleaner fuel technologies and mass transportation systems were cut back severely.

**DEFORESTATION** Humans are also increasing the amount of carbon dioxide in the atmosphere—and therefore global warming—through the rapid destruction of the world's tropical rain forests. The clearing and burning of these forests, which lie mainly in developing countries, is contributing between 1.1 and 3.6 billion tons of carbon to the air.[9] Since that figure is in addition to the nearly six billion tons of

---

[9] R. T. Watson et al., "Greenhouse Gases: Sources and Sinks," in *Climate Change 1992: The IPCC Supplementary Report,* Intergovernmental Panel on Climate Change (IPCC) (Cambridge: Cambridge University Press, 1992).

carbon dioxide released by the burning of fossil fuels, mainly in the industrialized North, it is clear that rich and poor alike now are responsible for global warming by raising carbon levels in the atmosphere. Of course, the clearing of tropical forests is itself a direct result of demands for these resources—including cleared land for farming—from an expanding population. And cleared land is subject to erosion, flood, and drought in a way it never was while forested. Turning good land into deserts and changing climates may be the long-term results.

By 1990, the clearing of the Amazonian rain forest of Brazil was proceeding at a rate that would transform the region into a desert within thirty years. From the Brazilian government's point of view, the first priority was to exploit the great mineral resources of the basin, which include iron ore, oil, gold, silver, and bauxite. Even as international opposition rose against Brazil's action, its government refused offers for outside agencies to contribute to the reduction of its huge foreign debt (well over $100 billion in 1990) in return for greater protection of its environment. In this, the Brazilian government resembled that of the United States as it reduced auto-emission standards in the interests of the automobile and highway lobby. In both cases, those we must rely upon as the principal instruments of global policy—the governments of states—are the least interested in promoting a global community good, and most attuned to the immediate selfish interests of powerful members of their own societies.

**GLOBAL ACTION**   These examples give some indication of the continuing resistance to concerted and effective international attempts to halt global warming. While in-

**FIGURE 13.3** | Global Average Temperature, 1881–1994

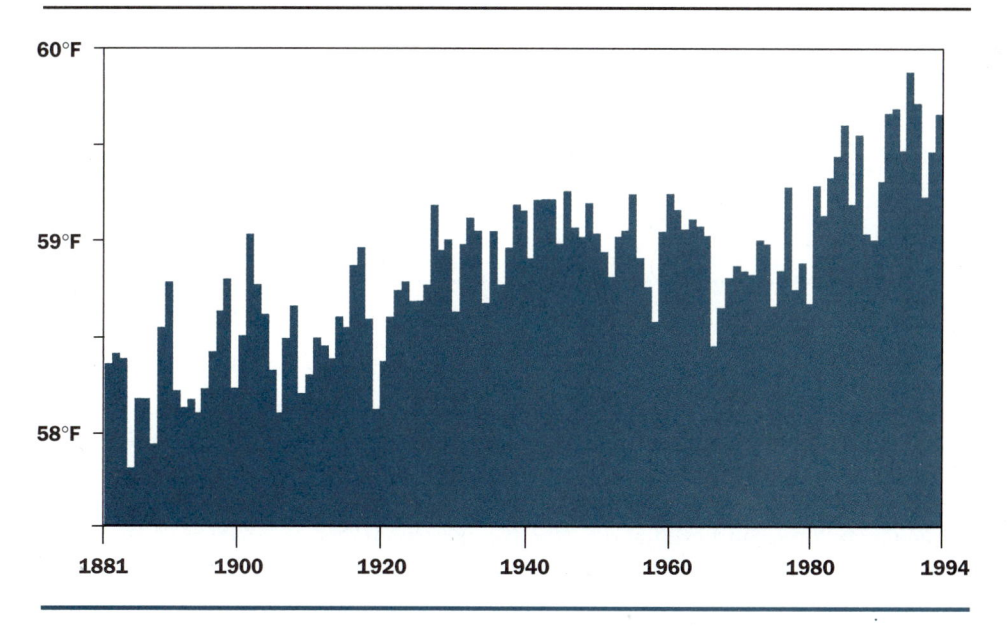

SOURCE: Vinnikov et al., *Trends '93* (Oak Ridge, Tenn.: Oak Ridge National Lab., 1994).

## FIGURE 13.4 — Tropical Deforestation, Fading Biodiversity

### TROPICAL DEFORESTATION

Deforestation has greatly increased in the tropics. Trees are key in recycling water, removing heat-trapping carbon dioxide from the atmosphere, and preventing soil erosion. Loss in thousands of hectares (1,000 hectares = about 2,500 acres).

| Region | Forest Area 1980 | Forest Area 1990 |
|---|---|---|
| Central America and Mexico | 77,000 | 63,500 |
| Caribbean Subregion | 48,800 | 47,100 |
| Tropical South America | 797,100 | 729,300 |
| South Asia | 70,600 | 66,200 |
| Continental Southeast Asia | 83,200 | 69,700 |
| Insular Southeast Asia | 157,000 | 138,900 |
| West Sahelian Africa | 41,900 | 38,000 |
| East Sahelian Africa | 92,300 | 85,300 |
| West Africa | 55,200 | 43,400 |
| Central Africa | 230,100 | 215,400 |
| Tropical Southern Africa | 217,700 | 206,300 |
| Madagascar | 13,200 | 11,700 |

### FADING BIODIVERSITY

Tropical deforestation is a threat to a disproportionately large share of the world's plant and animal species. Although tropical forests cover only about 7 percent of the globe, they are thought to contain more than half the world's species.

**The Utility of Biodiversity**

On the most profound level, the benefit biodiversity confers is a healthy biosphere, for species are the agents of essential biological processes. Varied species help scientists answer evolutionary questions. They also provide valuable products. New medicines are constantly being sought, as well as fibers, spices, oils, and lumber. Perhaps most immediately important to people, varied species insure vigorous crops and the resources for crop improvement. The strain of rice grown in Asia has been made immune to some diseases with genes from a single wild Indian species.

**Approaching Extinction?**

Populations of some animals have dwindled so rapidly that extinction within our lifetime seems inevitable. One in every three species of primate is in some danger, some conservationists believe, and one in every seven is highly endangered. The major cause is destruction of their habitats in tropical forests. They are also being hunted as a source of food or fur and to eliminate them from agricultural areas where they have become crop raiders.

SOURCE: *The New York Times,* May 5, 1992, C6. Copyright © 1992 by The New York Times Company. Reprinted by permission.

---

dividual states, such as the United States and Brazil, can do much to take corrective action through their own legislation (and both made some efforts to do so in the 1990s), the very nature of the issue demands a more global response as well. No government will undertake to reduce carbon dioxide emissions on a major scale while most others ignore and even add to the problem.

In 1972, the **United Nations Environmental Program (UNEP)** was created as an IGO with responsibility for encouraging environmental cooperation. As usual with IGOs, it was given only modest, coordinating powers and a tiny budget. But in 1988, another IGO, the **World Meteorological Organization (WMO)** created an Intergovernmental Panel on Climate Change (IPCC) to examine global warming. Together, these agencies played major roles in the direction of an international conference on global warming that met in November 1990. Then, in 1992, the most important environmental conference in history, the **U.N. Conference on the Environment and Development (UNCED)**, met in Rio de Janeiro. Among its many actions was the signing of a Convention on Climate Change (the "global warming treaty") intended to reduce greenhouse gases on the part of the signatories. At the insistence of the Bush administration, no deadlines for these reductions were included, although the members of the European Community pledged to stabilize their own emissions at 1990 levels by the year 2000. Soon after President Clinton came to office in 1993, he announced, in a reversal of Bush's policy, that the United States, too, was committed to reductions in emissions to their 1990 level.

Still, some three years after the global warming treaty was signed by 164 countries, few had actually begun to implement its provisions. Emissions of greenhouse gases continued to rise. Several oil-producing states of the Middle East had increased their opposition to stronger treaty provisions. And the United States, instead of stabilizing its emissions at 1990 levels as pledged by President Clinton, set a goal allowing a 3 percent growth in emissions annually. These developments probably did not mean that governments had turned their backs on the problems of global warming. Rather, when faced with pressures from their industrial sectors not to add to the costs of mineral production, they were inclined to drag their heels. In the words of a scientist with the Environmental Defense Fund (an important NGO in this field), "the treaty is lumbering along. . . . This problem is only going to be solved over a 20-, 30-, or 40-year period."[10]

A plan to prevent agricultural land from turning into desert also was initiated at UNCED, forming the basis of another treaty. The 1994 Convention to Combat Desertification is aimed at coordinating projects to protect and rehabilitate arid lands that have been overworked, and to encourage governmental contributions for that purpose. An estimated $10 to $22 billion will be needed annually to help rehabilitate land being lost to desertification. Without that kind of investment, the world can no doubt expect mass migrations from exhausted lands. That could result in even greater costs by creating the need for emergency resettlement aid.

## DESTRUCTION OF THE OZONE LAYER

In the 1980s, people throughout the world were suddenly awakened to the growing dangers to life on the planet as the result of the rapid destruction of the ozone layer. An international team of scientists reported that half the ozone over Antarctica was disappearing each spring; soon, severe disturbances were discovered in the ozone above the Arctic as well. The ozone layer in the outer atmosphere is the principal protection afforded higher forms of life from the deadly radiation of

---

[10]Mark Jaffe, "Disunity Undermining Treaty to Control Global Warming," quoting Michael Oppenheimer, *Philadelphia Inquirer*, March 26, 1995, A8.

the sun, so protection and restoration of the ozone layer is literally a matter of life and death.

This development, too, was mainly the result of industrial and technological practices of the late twentieth century. Scientists agreed that the chlorofluorocarbons (CFCs) produced for use as refrigeration coolants and in aerosol sprays were chiefly responsible for the hole in the ozone layer, although other chemicals, once released into the atmosphere, also contributed to the damage.

**ENDING CFC PRODUCTION**  Here, governments moved with relative speed to achieve formal agreement on controls on the use of some of these ozone-destroying agents. First came a treaty, the Vienna Convention for the Protection of the Ozone Layer, meant to foster cooperation, monitoring, and procedures for developing specific controls as needed. That treaty was signed in March 1985, by twenty, mostly highly developed, countries and the European Community. Those signatories then began deliberations on measures to limit CFC production. In September, 1987, twenty-four governments signed the Montreal Protocol on Substances that Deplete the Ozone Layer; its goal was to cut CFC production in half by 1998. Then, with more dramatic reports of ozone loss over the polar regions, parties to the protocol soon agreed to ratify, then expand, its provisions. The signatories next agreed to a phase-out of CFC production and consumption by the year 2000. At a second meeting, in June 1990, they were joined by senior ministers from many more countries not yet signatories, where it was agreed to ban all use of CFCs, halons, and carbon tetrachloride by the beginning of the twenty-first century.

Still the dangers grew. In 1992, the United States government measured ozone-damaging chemicals at the highest levels ever recorded over parts of North America. One of the most damaging chemicals, chlorine monoxide, had never before been found in such quantities any place above the earth. This report prompted President Bush to issue an executive order directing that the deadline for ending all CFC production in the United States be moved up to 1995. Soon, other CFC producers followed suit. As of December 31, 1995, all CFC production was ended by signatories to the Montreal Protocol. In the words of one observer, "through international cooperation, therefore, millions of skin cancer cases will be averted, and untold damage to agricultural productivity and ecosystems prevented."[11]

This most recent damage to the ozone layer reflected the lag between the time governments decided to halt their most injurious industrial practices and the years that would have to follow before the ozone layer would fully repair itself (it is estimated that it will not be fully restored until about 2055). There is a mixed lesson here. When scientific evidence is clear about a cause and effect relationship between a particular industrial practice and environmental damage, corrective action can be expected. But the threat to the ozone layer has been something of an exception in that regard. The impact of much other human behavior on the environment is sel-

---

[11]Hilary F. French, "Environmental Treaties Grow in Number," in Lester R. Brown, Nicholas Lenssen and Hal Kane, *Vital Signs 1995* (New York: W. W. Norton and Co., 1995), 90. Developing countries that have signed the treaty have an additional ten years to halt CFC production, although their output was a small fraction of that of the industrial countries in the mid-1980s.

dom so clear and direct. Therefore, particular governmental agreements may be harder to achieve (note the initial refusal of the United States to accept targeted cutbacks in carbon emissions, for example). Moreover, our experience with ozone-destroying chemical production, like that decades earlier with DDT, suggests that we will continue to create industrial products meant to serve useful purposes without understanding until afterwards that they may also create unacceptable environmental damage.

---

**Normative Dilemmas**

## When Environmental Threats Compete with Other Values

Assume that you are the minister of education in an extremely poor African country. You are asked to vote in cabinet on an offer from a leading European industrial nation to dispose of sizeable quantities of its toxic wastes, including radioactive isotopes, in your country for a very substantial fee. A twenty-year contract is proposed, with fees great enough to subsidize more than half the cost of basic education in your country for the life of the contract.

- What are the ethical factors you would need to consider before casting your vote?

China's economy grew faster than that of any other major country from the late 1970s into the 1990s, between 8 and 10 percent annually. At such rates, China would reach a developmental level more than 500 times greater than its 1990 GDP in less than a century.

- Can the environment sustain that kind of increase in energy consumption? A hundred Chinese cities were already "acutely short" of clean water by 1990. What would growth rates of the sort projected here do to China's water supply? Is the fostering of rapid growth the wrong policy for China? What alternatives should the Chinese consider that would allow for further growth without exhausting their resources?

The seabed and its resources have been designated the "common heritage" of humankind. Many argue that the concept means that the people of the planet have an equal right to share in the profits of whatever mineral resources are extracted from the ocean's floor, since the resources themselves are nonrenewable. Many mining companies, in contrast, argue that if they have the technology to engage in seabed mining, they have a "right" to the lion's share of the profits. Acting on that basis would allow some of the richest sectors of the world economy to gain the most from exploiting these seabed resources.

- What are the consequences for world order of each argument?

Note in Map 13.1 on page XXX that by 2030, India's population will grow by more than one-half, while China's will increase by less than one-quarter. Much of that difference is explained by the Chinese government's stringent population policy, encouraging no more than one child per family. Economic sanctions are imposed on couples who have more than two children. In contrast, India's official family planning program faltered in the 1970s amidst charges of coercive sterilizations.

- What are the moral implications resulting from these two cases? Is China's greater success at population control explained by the fact that, unlike India, it is not a democracy?

# THE CHALLENGE TO THE GLOBAL COMMONS

From their earliest times, most human societies have tended to appropriate patches of the earth (typically the territory where the society has made a permanent settlement) as their own. In modern times, as we have seen, the Westphalian arrangement was grounded on the principle that territory could be divided up on an exclusive basis—that is, what belonged to one sovereign could not be claimed by any other. But this principle generally has applied only to land, since that is where humans live and draw their sustenance. Throughout the modern era, the oceans have been regarded as belonging to all nations, and were thus considered the first **global commons.** The concept of a *commons* carries specific implications for ownership and use. It assumes joint ownership, with equal rights of access to whatever is held in common, and equal duties to protect and preserve it. In recent years, the concept has been applied to regions other than the oceans that have not been susceptible to sovereign ownership. As problems of global pollution have grown greater, we have seen an increasing tendency to insist that the earth's atmosphere should be viewed as the common property of all, and not simply as a free good needing neither protection nor management. Similarly, as humans have ventured into outer space, a body of law is developing that views outer space as a commons area.

## LAISSEZ-FAIRE AND FREEDOM OF THE SEAS

Hugo Grotius, the "father of international law," developed the doctrine of the freedom of the seas at the dawn of the Westphalian era.[12] That view came to dominate the modern world. In it, the concept of the world's oceans as a global commons is clear. Sovereign states may claim only a narrow band of water immediately offshore as sovereign territory. That allows them to protect their societies on land from sea-based attack, but prevents them from denying access to the peaceful merchant fleets of others, which presumably are engaging in mutually beneficial trade.

Viewing trade as mutually beneficial is at the heart of economic liberalism, and the greatest trading states have tended to become the leading states of the world in the modern period, as well.[13] Those are the same states that have made it a priority to insist upon a doctrine of free and open seas. They gained much from having assured access to distant ports; they conceded little, and upheld a principle of fairness, by granting the same right of free access that they claimed for themselves to all other states on the basis of reciprocity.[14] It is for these reasons that the great maritime states have tended to uphold the doctrine of freedom of the seas, rather than trying

---

[12] Grotius' book, *Mare Liberum* (Freedom of the Sea), was published in 1609.

[13] For two books—one by a historian, the other by a political scientist—that explore how states have achieved political power through trade, see Paul Kennedy, *The Rise and Fall of the Great Powers* (New York: Random House, 1987); and Richard Rosecrance, *The Rise of the Trading State* (New York: Basic Books, 1985).

[14] For a discussion of the importance of the principle of reciprocity in international law, see Chapter 15, "International Law and Global Order."

to appropriate parts of it for themselves, which they no doubt have had the capability to do.

But by the mid-twentieth century, a growing number of states began to claim an interest in closing large parts of the seas located just off their own shores. Their reasons had to do with growing demands for the resources of the sea (fish) or the seabed (oil or other minerals). The United States was the first important country to take a bite out of the doctrine of freedom of the seas. In 1945, President Truman issued several proclamations that claimed his country's exclusive jurisdiction over the living and nonliving resources of its continental shelf, while insisting that free navigation above the shelf would not be impaired.[15]

Yet, this action soon opened the door for Chile, Ecuador, and Peru—countries that had neither wide continental shelves nor substantial commercial shipping interests—simply to claim 200-mile swaths of the oceans off their shores as their exclusive "conservation zones." Their action was prompted mainly by the increased competition for fish beyond their three-mile territorial waters and by their desire to conserve certain other resources, as well. Population growth and increased technological efficiency were producing inexorable pressure on the sea's resources in ways that would keep increasing through the rest of the century. Other countries, finding themselves more frequently in conflict over scarce maritime resources, also began to assert exclusive claims to what had been free and open seas. The result was a growing tangle of jurisdictional claims that increasingly pointed to the need for radically revising the law of the sea.

**1982 LAW OF THE SEA TREATY** Those revisions took place in a series of meetings from 1972 to 1982 known as the Third U.N. Conference on the Law of the Sea (UNCLOS). The representatives of well over a hundred states wrestled with the conflicts arising from increased human demands upon the oceans. First, as we have seen, were those for the living resources of the sea, such as fisheries, which evidently required stronger conservation provisions than had been necessary in the past to prevent their destruction from overfishing. Second was a growing demand for the fixed mineral resources of the seabed, which include not only oil, but also manganese nodules on the ocean floor. Industrial technology had reached the point where these valuable minerals could be harvested economically, but the question of who owned them first had to be resolved. Third was the continuing need for open access to the seas for commercial and navigational purposes.

The **Law of the Sea Treaty** that was opened for ratification in 1982 is complex and unprecedented in providing a new regime for the seas. First, it permits all coastal states to claim an Exclusive Economic Zone (EEZ) of 200 miles offshore (where the coast of another state lies closer than that, the difference is split between them). The EEZ is not a territorial sea, inasmuch as the coastal state exercises only economic, not security, control over it. The rationale behind the EEZ provision is

---

[15]Truman Proclamation, 10 Fed. Reg. 12303 (1945). Note that, as a leading maritime state, the United States had an interest in insisting that these claims did not inhibit free travel and commerce on the traditional area of the high seas.

that the neighboring coastal state will have the primary interest in protecting that marine environment from pollution and depletion of fishing stocks. Yet it does largely remove approximately one-third of what had been high seas from the global commons.

Second, the treaty creates an International Seabed Authority (ISA) for the area beyond the EEZs of states whose purpose is to manage and license mining operations on the ocean floor. These provisions caused the most difficulty to achieve, inasmuch as negotiators were sharply divided as to the governing principle that ought to apply. In general, third world states strongly supported the novel principle that the seabed's resources constituted the **common heritage** of humankind, whose wealth should be shared, particularly among those states lacking the technology to engage in seabed mining themselves. From this point of view, the ISA should be empowered to control access to the deep seabed area on the part of state and private companies seeking mining permits. Most of the capitalist states of the North, on the other hand, argued for a weak ISA that would not be permitted to deny access to any company with the capital to invest in seabed mining.

In the end, a hybrid system was adopted. It allows the International Seabed Authority to engage in mining operations for the benefit of the poor only where it simultaneously grants a comparable license to a mining company from the North. In 1994, some twelve years after the treaty was opened for ratification, it was revised further in the direction of granting concessions to those states where the capability to engage in seabed mining resided. That was done at the behest of the United States, which had refused to ratify the treaty in its original form. By the end of 1994, the minimum number of states required for the treaty to enter into force among them (sixty) had at last been reached. It was assumed that, for the near future, the economic exploitation of EEZs would continue to overshadow mining of the deep seabed under the jurisdiction of the ISA.

## ANTARCTICA AS A GLOBAL COMMONS

Because of its inhospitable climate, Antarctica has not been an object of permanent human settlement like that of the world's other continents. By the mid-twentieth century, however, a number of states had begun to assert claims to various slices of Antarctica based upon sightings of the land, limited explorations, and even geographic proximity, in the cases of Argentina and Chile. To forestall conflict, the Antarctica Treaty was negotiated and ratified in 1961. It "froze" the various territorial claims that had been advanced and established a moratorium on mining as well as a framework for scientific cooperation on the part of the signatories.

But as the demand for oil grew during the 1970s, so did pressure on the treaty states to lift the ban on exploration and mineral exploitation in Antarctica. Huge oil reserves were thought to exist in Antarctica's continental shelf; other strategic minerals were discovered elsewhere on the continent. The Antarctica Treaty limited its signatories to those states able to engage in scientific research and exploration in the area. That meant that, by the early 1990s, treaty membership was limited to twenty-five states—those with the technology and interest to exploit Antarctica's resources for their own profit. By 1988, they had agreed to a second treaty that would permit regulated mining on this last natural frontier on earth.[16]

Throughout the 1980s, meanwhile, the **Group of 77,** largely excluded from treaty membership, argued that the common heritage principle should be applied to Antarctica. They saw the ISA as a useful precedent for a much more inclusive regime that would include all states in managing and protecting the region. Some NGOs called for making Antarctica the first "world park" in history. As the strength of antimining sentiment grew, a number of governments that had helped author the 1988 treaty allowing mining reconsidered their position. Their change of heart was reflected in a new replacement treaty, the **Madrid Protocol** of 1992. When ratified, the new treaty will prohibit any mining in Antarctica for a period of fifty years. The shift in the thinking of treaty states from 1988 to 1992 appears to have been a response to the rising tide of concern regarding the environmental risks associated with exploiting Antarctica's mineral resources.

## How to Manage the Environment?

The human creation we call the nation-state system does not seem ideally suited to ensure that human aspirations and development will support the health of the biosphere. The logic of Westphalia was and is that humanity can thrive in separate societies without the need for effective governance of the planet. The logic of nature, however, is holistic, in the sense that every aspect of the natural world is interconnected, linked in a chain of life—including the life of eons ago that produced the fossils we now use as fuel.

The political implications may not be altogether obvious. Is it our political fragmentation into separate states that is endangering our biosphere? That did not prevent the leaders of states from agreeing to measures that should effectively repair the hole in the ozone layer. But that fragmentation probably does make it more difficult to treat the much more diverse range of human activities that are contributing to global warming.

We are beginning to create international **regimes** (see Chapter 17) in the environmental area in response to the need for more effective governance of global resources and global commons. Yet, we know that wherever there is government—at local, state, or global levels—there is also resistance to many of its efforts from groups with a self-interest in what does not further environmental health. So, to look beyond the nation-state for ecologically sound treatment of these issues does not mean that the interests now in opposition will also disappear. Much of the problem was well put by Pogo when he said, "I have seen the enemy and he is us." Few of us

---

[16]Treaty-making requires, first, that negotiators—acting on behalf of their governments' executives—agree to a text, and second, that it then be ratified by the relevant governments' legislative authorities. When ratification involves many states, it may take a number of years. The 1988 Convention on the Regulation of Antarctic Mineral Resource Activities (CRAMRA) was effectively replaced by a substitute agreement (see the following discussion of the Madrid Protocol) before it was ratified and came into force.

have yet learned to behave so that we are not, at least some of the time, enemies of the environment and, therefore, of sound environmental policy.

This suggests a second respect in which our economic and political habits put us out of tune with the logic of nature. All human beings are dependent upon nature's health for their own well-being. While it is a tendency of many cultures (especially those we call "modern" or "developed") to regard nature as a force to be tamed and exploited, ecological logic reminds us that such thinking could ultimately doom us.

It is the attempt to move away from such dangerous patterns of thought that has led world leaders to rephrase the issue here as one of **sustainable development.** That term has been a catchphrase since the 1992 Earth Summit, although it clearly still suggests different priorities to political actors in the North and the South. Many leaders of poorer countries still fear that the North uses environmental arguments as an excuse not to help them grow economically, engaging in "a sort of 'green' colonialism that consigns them to a permanently unequal status."[17] And while the North argues that developing countries do not fully recognize the severity of environmental threats, neither do they invest the capital needed to encourage environmentally sound developmental practices throughout the South.

For both camps, the issue of state sovereignty is the excuse, at least, for sticking to their guns. The LDCs argue that limiting the exploitation of the natural resources that lie within their territory is an infringement on their sovereignty. When international taxes are proposed as an alternative financing mechanism for sustainable development, would-be donor states oppose the idea, also on sovereignty grounds.[18] Here it seems clear that the Westphalian political order is the enemy of what would be most beneficial to the planet. That also makes it the enemy of ourselves as a species.

# CONCLUSION

The production of material wealth that is characteristic of the modern world is revealing nature's limits as never before. Development demands have risen sharply with the human population explosion in recent centuries. As a result, the demand for certain of the nonrenewable resources that have fueled industrial growth is increasing their cost, and some—most notably oil—are rapidly running out. Automobile ownership and mechanized agriculture, which are hallmarks of the richest societies today, cannot be increased substantially elsewhere without ever mounting costs, both for the fuel itself and for the damage they do to the environment.

The kind of development that has been pursued in most of the world in this century is creating unprecedented threats to the balance of nature. In recent years, the prospect that we may be radically altering the world's climate through global

---

[17] Gail V. Karlsson, "Environment and Sustainable Development," *A Global Agenda: Issues Before the 50th General Assembly of the United Nations* (Lanham, Md.: UNA/USA and University Press of America, Inc., 1995), 131.

[18] Ibid.

warming is beginning to raise an alarm. The constant rise in the burning of fossil fuels (which warm the earth through carbon dioxide emission) and deforestation (to make way for expanding human populations) are the principal causes for the warming of the planet.

It has become apparent that some of our industrial activities are destroying the earth's ozone layer, the thin protective shield at the atmosphere's outer edge that protects us from the ultraviolet rays of the sun. The chief culprit is chlorofluorocarbons (CFCs), whose production is being curtailed as the result of international agreements negotiated in 1985, 1987, and 1990. In the 1990s, a number of other international meetings have produced a growing list of agreements and directives meant to move toward more environmentally sound policies. Issue areas include biodiversity, deforestation, and desertification.

The global commons are those arenas of human activity that have been excluded from the claims of sovereignty. Governance of the oldest of these, the oceans, has been based upon a laissez-faire principle allowing access and exploitation to all on a reciprocal basis. In recent decades, mounting pressures to exploit the ocean's resources and the mineral riches of the seabed have resulted in a radically different regime for the sea. The 1982 Law of the Sea Treaty created 200-mile exclusive economic zones for coastal states and an International Seabed Authority to govern mining activities on the remainder of the world's ocean floor. The emerging regime for the seas is built upon the principle that they constitute the "common heritage" of humankind. That places far more emphasis on managing and conserving the seas' resources than did the more traditional, laissez-faire principle that required almost nothing in the way of global management.

During the past half century, Antarctica has also become a global commons. Ratification of the Madrid Protocol of 1992 would prohibit any mining on the continent for a period of fifty years. Scientific activities, meanwhile, are permitted.

Resource and environmental concerns have found their way increasingly onto the agenda of world politics, primarily because of the rapid growth in world population. Given the still mounting pressures for energy consumption, resource use, and economic growth, environmental issues are certain to loom even larger in the future than they do today. The nation-state system is not always well suited to treat environmental issues effectively. Those issues are holistic and interconnected, while the Westphalian state system assumes political separation and a large measure of disconnection in social problem solving.

PART V

# The Construction of a Global Society

From its creation, the chief rationale for the Westphalian nation-state system was to allow separate sovereign communities to flourish without undue interference from one another.

Since the new arrangement was built upon the premise that no supranational political community could or should command our loyalty, it was believed that international order would most likely result when sovereigns scrupulously avoided interference with other states. When they allowed themselves to become entangled in each other's affairs, warfare would likely result. According to Westphalian theory, peace should be the product of sovereign detachment, self-help, and a willingness to live and let live. Those values ought to require little in the way of international rules and regulations, and even less by way of permanent institutions at the international level.

Yet, ironically, the nation-state system has had to develop both rules and institutions—and ever more rapidly in the twentieth century—to respond to the most pervasive trends of human development in modern times. These are, first, our conquest of territorial space and, second, the rapid increase of our species. The separate and largely detached communities that were the models of the 1648 Westphalian sovereign states have grown less and less able to satisfy human needs and desires in an age when we can fly over dozens of countries in a matter of hours and interact with individuals residing in many nations within seconds. Social interaction on more than an occasional basis requires rules and institutions, which are essential to governance. As interaction among social groups increases, it may at some point begin to change the attitudes and assumptions each has held about the other in a more positive direction.

The quest for world order is increasingly a matter of searching for modes of international governance without abandoning the Westphalian premise that nation-states should remain the only "sovereign" governments on earth. In Chapter 14 we will address the most recent—and also the most radical—dimension of these world order trends: the effort to integrate states transnationally. It is radical in the sense that it actually could result in abandoning the Westphalian premise for some different, but as yet unseen, principle for organizing the lives of societies across the globe. Yet, even while integration has occurred, it has been accompanied—perhaps paradoxically—by the disintegration of some multinational states, whose meaning we also need to assess.

The oldest effort in the search for world order has been the establishment of international law, which is the subject of Chapter 15. In Chapter 16, we explore one relatively new dimension of this quest for a world ruled by law that logically runs counter to the doctrine that the sovereign state is all-powerful: the effort to protect human rights as an appropriate concern of global politics. Along with the development of international law in the search for world order has come the creation of organizations and institutions that transcend the boundaries of states. The purpose of many of these IGOs is to assist states in resolving conflicts among them, sometimes by encouraging peaceful solutions to conflicts and sometimes, where a conflict has turned violent, by taking on policing roles. We consider this development in our concluding chapter.

Together, these trends in transnational integration, international law, human rights protection, and international organization reveal much of the structure of global society today. They suggest both its progress and continuing weaknesses, leaving us with a thousand challenges as we confront the future of global politics. ∎

*Overleaf* • Troops under the command of NATO's International Force (I-FOR) patrol the war-torn city of Tuzla, Bosnia, in 1996.

# CHAPTER 14

# Interdependence, Transnational Integration, and Separatism

*Interdependence* has become one of the catchwords of the age in which we live. Its connotations are nearly the opposite of those associated with independence, which suggests a sturdy self-sufficiency and the ability to stand alone in the world. But neither does interdependence imply simple dependence, in which one party to a relationship clearly relies more or less exclusively on another for its sustenance. A young child may be almost wholly dependent upon its parents; in more complex social groupings—such as clubs, clans, or communities—the relationships are typically those of mutual dependency, or interdependence.

Not until the late twentieth century has it become commonplace to describe nation-states as increasingly interdependent. Our purpose in this chapter is to explore why this trend has developed in recent times; to determine where it is the most—and the least—advanced, and why; and to assess where this trend is likely to lead the world in the near future, including especially its impact on the nation-state system.

Since the central premise of the Westphalian world is the independence of each sovereign actor, then the growing interdependence of sovereigns appears to contradict that concept. In other words, the normative ideal of the international system assumes the self-sufficiency of each independent state, while the socioeconomic reality of our time may increasingly deny it. So, our central question here is whether the organizing principle of the international system may eventually have to give way to a new social reality, which may then require a new normative arrangement. We shall also have to consider whether the recent upsurge in nationalistic separatism counters the trend toward interdependence.

# Interdependence, Integration, and Amalgamation

**Interdependence** can be defined as a condition of mutual dependence among several persons or groups. It describes a social relationship that may have been formed casually, in the sense that no one planned for it to come about; it is probably the product of unreasoned social and economic forces that no one controls. Moreover, those in large-scale interdependent relationships may be only dimly conscious of that fact, knowing little or nothing about the individuals or groups with whom they are connected. They may or may not benefit materially from this connection, and so they may or may not find it to their liking. Since interdependence may actually encourage conflict, it is not clearly a prerequisite condition for peace. When the price of oil skyrocketed in 1973–1974, for example, thanks to the OPEC cartel price agreement, millions of people in the industrialized North became suddenly aware of their interdependence with distant oil-producing countries. This interdependence is precisely what made the OPEC action so effective; yet the experience no doubt produced far more hostile than friendly feelings on the part of those affected.

**Integration** implies a more conscious mixing of individuals and groups that have been separated. It also may not be a planned, rational process, but those affected almost certainly are aware of it, either approving or opposing this new mix of peoples in accordance with their own perceived interests. While most of us are aware of governmental policies that seek the fuller integration of minority groups into the mainstream of national society (U.S. civil rights policies are a familiar example), we have typically had much less experience of comparable efforts among nation-states. Yet, that is the meaning of the ongoing integration of Europe, which we will examine shortly. Moreover, as has been demonstrated in the peaceful coexistence of Canada and the United States for more than two hundred years, this kind of "pluralistic integration," where separate nation-states are maintained but their disputes are resolved peaceably, may even be the result of only semiconscious policy choices.[1]

**Amalgamation** is used here to mean the formal joining of separate political groups through conscious governmental choice. It may or may not express the wishes of the majority of those joined together, but the end result is a new political entity that, in the terms of our subject matter, constitutes a "sovereign" state. Historical examples abound: Texas was separated from Mexico in the mid-nineteenth century and, shortly thereafter, amalgamated into the United States. A comparable process occurred in 1940 when the briefly independent states of Estonia, Latvia, and Lithuania were made republics of the Soviet Union. From 1958 to 1961, Egypt and Syria were amalgamated in a state called the United Arab Republic, but which, after Syria's withdrawal, became two independent states again. In each of these cases, and in many more like them, it does not much matter from the standpoint of the formal requirements of the Westphalian system whether true social integration of the

---

[1] The idea that two or more states can be integrated "pluralistically," in the sense that their citizens have "dependable expectations of peaceful change" in their mutual relationships, specifically does not assume that their separate political systems must first be joined. For a fuller exploration of this concept and the empirical evidence supporting it, see Karl Deutsch's pioneering study of international integration, *Political Community and the North Atlantic Area* (Princeton, N.J.: Princeton University Press, 1957).

amalgamated peoples follows. That is a social issue; the legal question of sovereignty has been resolved in favor of the newly amalgamated union.

As the above examples indicate, these three processes or conditions—interdependence, integration, and amalgamation of peoples—are not merely conceptually distinct; neither are they necessarily connected in a developmental way in the contemporary political and social life of the world. Interdependence describes a *condition* in world politics today, one that presents both positive opportunities for human beings to expand their lives and also negative prospects for overcrowding and ecological destruction. In contrast, integration and amalgamation are *processes* supported by at least some political actors.[2] One reaction to interdependence may be hostility and a desire to reestablish distance between interdependent groups, rather than a choice for integration. Integration may or may not lead to amalgamation of separate sovereign states. Conversely, amalgamation does not necessarily either reflect or help induce true social integration. If it did, there would be no secessionist movements among such peoples as the Basques in Spain, the Quebecois in Canada, or the Kurds in Iran, Iraq, and Turkey.

# THE EUROPEAN UNION

Europe has been the scene of much destructive warfare over the course of many centuries, none with more terrible global consequences than the two world wars of 1914–1918 and 1939–1945. As these wars and the Thirty Years' War of the seventeenth century proved, warfare that devastates whole nations can have revolutionary consequences in world politics.

## PROPOSED AMALGAMATION AFTER WORLD WAR I

After World War I, the European great powers were less interested in uniting Europe than they were in restoring a European power balance that would keep Germany in check. Nonetheless, a few leaders feared that a return to the balance of power system would only lead to war again, and argued the case for a more radical transformation of European politics through a plan of conscious integration. In 1929, the French foreign minister, Aristide Briand, embraced a proposal for European Union, and asked for the participation of all the other European governments in the League of Nations. In the next year, the League Assembly appointed a special commission to consider how such a scheme might be carried out. Following Briand's death in 1932 and the onset of a worldwide economic depression, there was little progress toward the union of Europe. Yet, these efforts not only foreshadowed far more serious moves in that direction after 1945; they also forced politicians to begin to examine many of the issues the integration movement would have to face after World War II.

One of the lessons of the failed European Union movement seemed to be that the quick amalgamation of many European states was an unrealistic goal. In spite of

---

[2] We are grateful to Professor Steven W. Hook for calling this distinction to our attention.

many values that they shared, the states of Europe had widely differing political interests, traditions, and institutions—all of which had produced an often distrustful nationalism that could not be overcome by a governmental fiat to unite. In 1954, the French National Assembly rejected an effort to create an integrated European Defense Community, which would have united under a single command the armies of what had so recently been bitter enemies. That action confirmed for many the lack of realism in continuing to seek political union through amalgamation.

## THE FUNCTIONAL APPROACH AFTER WORLD WAR II

After World War II, an alternative approach to integration proved to be far more successful. **Functionalism** refers to the theory, first developed by David Mitrany in the 1930s and 1940s, that progress toward international integration is most successful when state barriers are eliminated to make possible particular social "functions" that have been inhibited by national boundaries in the past.[3]

Unlike the effort to unite separate states through their formal amalgamation (an approach that addresses issues of what are called *high politics* because of the political passions and prejudices they may arouse), functionalism seeks out issues of *low politics* that are generally noncontroversial because they do not raise such overtly "political" questions. For example, as modern governments have created agreements for the worldwide delivery of mail, for telephone transmissions, for standardized health regulations, and for customs unions, or as some states have created regional airlines or worked jointly to harness the power of a river that flows through their nations, they have engaged in activities that, according to functionalist theory, may be moving them toward greater social and political integration.

In Western Europe, a number of the governments that came to power after World War II were more similar in their ideologies than at any time in the recent past. All had rejected the militarist and imperialist values that had led to war, and all were committed to pluralistic democracy and to capitalist development leavened by social welfare protections. Among their political elites were a number of individuals who were persuaded that economic functionalism could assist their states toward greater unity than was evidently possible through the kind of high politics effort that sought a quick European federation. Their ideas first generated a treaty signed by six Western European democracies in 1951 that sought to integrate two of their basic industries.

**THE EUROPEAN COAL AND STEEL COMMUNITY**  The authority this treaty created was the European Coal and Steel Community (ECSC), signifying an important effort to eliminate one of the most notable causes of nationalist rivalries in the modern period between, particularly, France and Germany. The richest coal deposits in Europe are located in the Saar and Ruhr regions of western Germany, close to the Rhine

---

[3] David A. Mitrany's path-breaking work is *The Progress of International Government* (New Haven, Conn.: Yale University Press, 1933). He expanded his ideas in *A Working Peace System* (Chicago: Quadrangle Books, 1966). For a brief and lucid interpretation of functionalism, see Inis L. Claude Jr., *Swords into Plowshares,* 6th ed. rev. (New York: Random House, 1986), chap. 17.

River border with France, while rich iron ore deposits are found nearby but to the west of the border in the French province of Lorraine. Among the causes of the Franco-Prussian War of the 1870s and the belligerencies of these countries in World Wars I and II were efforts by each to wrest control from the other of these vital resources. The Low Countries, too, of Belgium, Luxembourg, and the Netherlands, had been the victims of this quarrel and they, along with Italy, joined France and West Germany in the creation of the ECSC. By abolishing national economic barriers, a more efficient and profitable coal and steel industry was created. At the same time, the ECSC eliminated a major source of political friction among its members.

**THE EUROPEAN ECONOMIC COMMUNITY** The ECSC's success induced its member governments to go further. In 1957, they signed the Treaty of Rome that created the European Economic Community (EEC). That plan for a Common Market (as it is more familiarly known) was far more ambitious than its prototype, for it sought the integration, in stages, of virtually every aspect of the economic lives of the participating states. The formation of the EEC meant, first, the gradual reduction of all the tariff barriers that the member nations had erected among themselves and the simultaneous creation of a common trading policy with the outside world. Second, as agreements were reached by the separate governments on how to implement the EEC goals, they would increasingly hand over regulatory (and inevitably, more decision-making) power to the Community institutions created by the Treaty of Rome, thereby gradually ceding aspects of their sovereign capability to make economic decisions. Third, because a conscious choice was made by the original six members not to become an exclusionary economic actor, the process of integration has been deliberately slowed at times to adjust to the additions of new member states—to nine in 1973, when Denmark, Ireland, and the United Kingdom were added; to twelve by 1986, after Greece, Portugal, and Spain had been added; and to fifteen in 1995, with the entry of Austria, Finland, and Sweden. By the beginning of the twenty-first century, some half dozen additional members from Eastern Europe and the Mediterranean may have joined the European Union, as it is now called.

## THE INSTITUTIONS OF THE EUROPEAN UNION

In less than forty years, the European Union (EU) has transformed the economic and political life of Europe in many respects.[4] Whether it is also transcending the nation-state system by inventing post-Westphalian political structures remains an intriguing question. To put the issue in more familiar, if cruder terms, while the economies of member states are increasingly integrated into a single market, the amalgamation of their separate political systems is much less certain. That is largely because the integration effort has been pursued via the approach of functionalism, which deliberately avoids attacking the high political issues of amalgamation head-on.

---

[4]For further reading on the operation of the European Union, see Neill Nugent, *The Government and Politics of the European Community* (Durham, N.C.: Duke University Press, 1989); and Dale L. Smith and James Lee Ray, eds., *The 1992 Project and the Future of Integration in Europe* (Armonk, N.Y.: M. E. Sharpe, 1992).

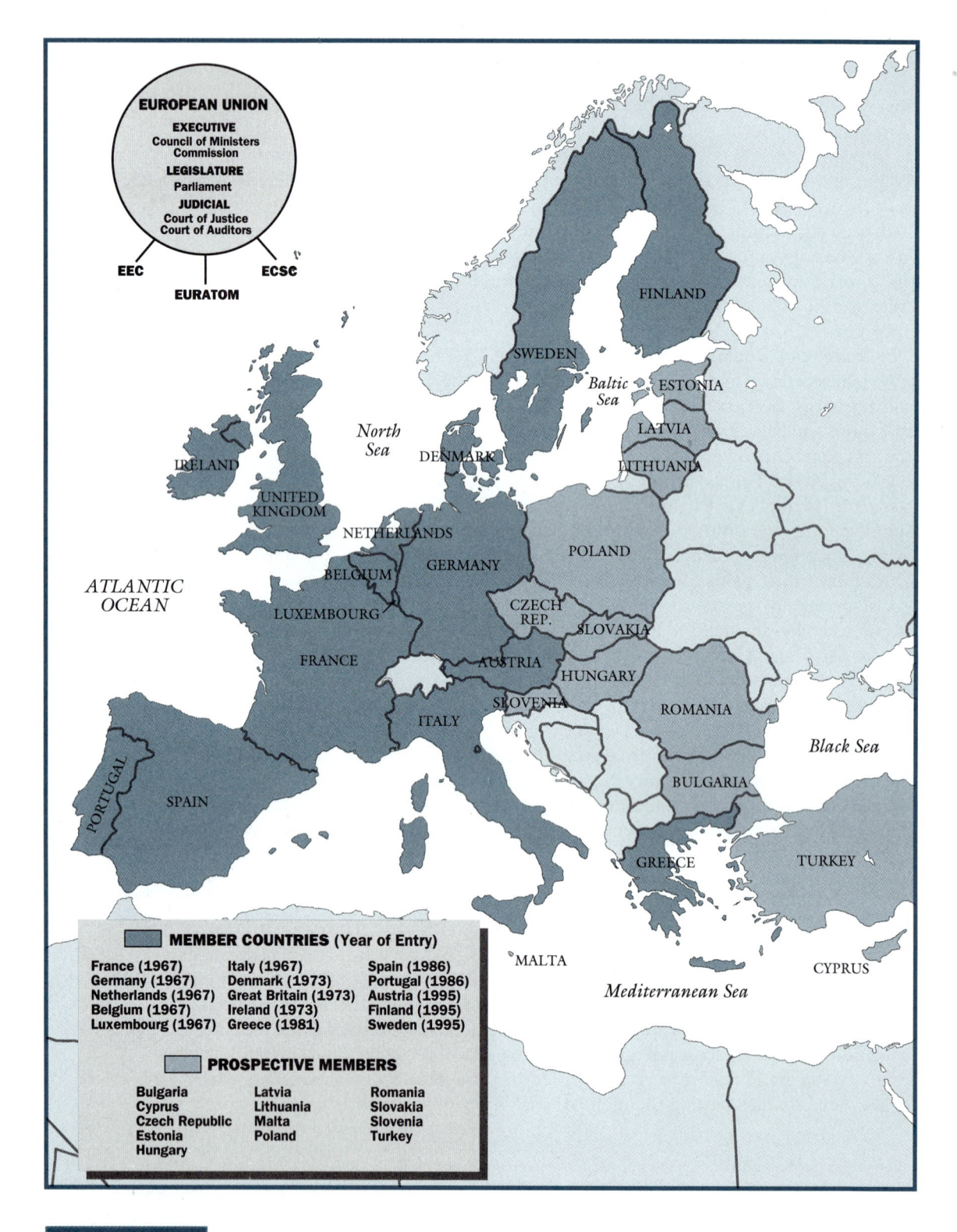

**EUROPEAN UNION**

EXECUTIVE
Council of Ministers
Commission

LEGISLATURE
Parliament

JUDICIAL
Court of Justice
Court of Auditors

EEC — EURATOM — ECSC

**MEMBER COUNTRIES (Year of Entry)**

| | | |
|---|---|---|
| France (1967) | Italy (1967) | Spain (1986) |
| Germany (1967) | Denmark (1973) | Portugal (1986) |
| Netherlands (1967) | Great Britain (1973) | Austria (1995) |
| Belgium (1967) | Ireland (1973) | Finland (1995) |
| Luxembourg (1967) | Greece (1981) | Sweden (1995) |

**PROSPECTIVE MEMBERS**

| | | |
|---|---|---|
| Bulgaria | Latvia | Romania |
| Cyprus | Lithuania | Slovakia |
| Czech Republic | Malta | Slovenia |
| Estonia | Poland | Turkey |
| Hungary | | |

**MAP 14.1**  European Union Member (and Prospective Member) Countries

Functionalist theory assumes that as patterns of cooperation grow and spread within the economic sphere, they increasingly will encourage affected populations to formalize their integration, perhaps by building new political institutions. This process, known to functionalists as the *spillover* effect of integration, may itself be gradual (like the economic integration from which it is spawned); if it works, issues of high politics eventually are marginalized, so to speak, by the habit of cooperation on low political matters. Surely in fifty or one hundred years, it will be possible to say with some confidence whether political spillover was irreversibly taking place in 1990s Western Europe. We may look for answers today, however, by examining the roles of EU institutions.

The **European Parliament** is the most democratic of these, since its members have been chosen through direct popular elections throughout the Union since 1979. Prior to that, members of the European Parliament were appointed by member governments and met only for a few weeks a year in Strasbourg, France, to discuss common problems and make recommendations to their governments. Now, however, the Parliament participates in the formulation of Union decisions and shares increasing power in policy making with the Council of Ministers.

The **European Commission** is composed of at least one private citizen from each member state appointed by mutual agreement of EU governments. Its members may not receive instructions from any national government, a requirement that makes this body, like the European Parliament, somewhat "non-Westphalian." The Commission has broad powers both to initiate proposals for the Council of Ministers and to ensure that the Union's treaty structure is respected. Typically, it refers hundreds of proposals annually to the Council for consideration.

The **Council of Ministers** makes the major policy decisions for the EU. Consisting of relevant cabinet ministers from member states (i.e., agriculture ministers meet to discuss farm prices, economic ministers deal with employment problems, and so on), the Council meets, on average, more than once a week to address proposals referred by the Commission. Its structure makes it the EC body that most clearly reflects the sovereign prerogatives of member states. It can alter proposals from the Commission only by unanimous agreement; unanimity is also required in the Council for certain important decisions.[5]

**THE TREATY OF MAASTRICHT** In 1992, an EC treaty signed in Maastricht, the Netherlands, committed all members except Britain to introduce a common currency and establish a European Central Bank before 1999. It also renamed the EC the European Union (EU). Yet the Maastricht agreement immediately opened a debate as to whether this kind of deeper integration should continue to be the goal now that the end of the cold war had removed the barriers between Eastern and Western Europe for the first time since World War II. Some argued that the EU now should open its doors to many more states of Europe, even though that would surely slow the pace of integration. Others predicted that if the European Union were to grow too fast, it would dilute the cohesiveness that has been acquired rather slowly among

---

[5]Other EU institutions include a *Court of Justice,* which ensures that all measures passed are compatible with Community treaty law; an *Economic and Social Committee,* which advises the Council of Ministers; and a *Court of Auditors,* to ensure budgetary and fiscal responsibility.

> **Global Changes**
>
> **The Changing Number of States:** The trend following the 1648 Peace of Westphalia was a consolidation of European states into larger units, most of which ultimately were identifiable as nation-states. From the time of Westphalia until 1900, the states of Europe were consolidated from several hundred to fewer than thirty. During the twentieth century, the Westphalian model was extended globally. A system still dominated by Europe at the start of World War II saw the addition of more than one hundred non-European states over the next fifty years. Near the end of the century, more than a dozen new states were born when the Soviet Union disintegrated. By 1995, there were nearly 190 sovereign states—the largest number in history.
>
> - Does the great increase in the number of sovereign states in recent decades make the conduct of diplomacy easier or more difficult? Does the idea of the legal equality of sovereign states seem more or less sensible today than it did in 1950? In 1900? In 1648? Why or why not?
>
> **The Growth of the European Union:** In 1957, six Western European countries agreed to form a European Common Market, the forerunner of today's EU. In 1973, three more members were added; in 1986, with another three, membership reached twelve. It grew to fifteen in 1995, when membership applications also were pending from several additional countries, including Eastern European states.
>
> - What are the main arguments in favor of a continued expansion in the membership of the EU? Why do some "Europeanists" argue against further expansion, at least in the near future?

the twelve members, even jettisoning its plans for integration in the direction of political amalgamation in order to expand the scope of its free trade area.

The Treaty of Maastricht became effective in November 1993, establishing new integrative goals for economic union and the creation of a common foreign and security policy for the EU. But that did not end the debate over how soon or even whether those goals would be met. In functionalist terms, Maastricht brought more high politics issues to the surface than had been present before.

## THE EU: ASSESSMENT AND PROJECTIONS

Looking back over the EU's history, several generalizations about the integrative process are clear. In the first place, the success to date of economic integration is perhaps best explained in terms of the competitive edge and profitability to be found in economies of scale for advanced industrial societies. That is, in the second half of the twentieth century, it was in the clear economic self-interest of the comparatively small states of Western Europe to combine their markets if they were to compete effectively with such economic superpowers as the United States (which has itself constituted a "common market" ever since the 1789 adoption of its Constitution prohibited its states from creating trade barriers among themselves). The dynamics of a free market system produce a continual demand for ever larger economic units in the modern world, with the fewest possible restraints on the wide-scale movement of capital, labor, and goods. The Common Market, therefore, has served the self-interest of its creators by adapting to the requirements of competitiveness in the modern world.

Second, the European Union has been built from the start upon a faith that gradualism could produce the organic growth of integration. The process started with what was simplest and most manageable, then provided an outline for beginning a far more complex process in the Treaty of Rome. That treaty amounts to a constitutional agreement on broad structural arrangements within which policy continues to evolve. It is assumed that not all problems have to be solved at once. What was an insoluble issue of high politics in the early years may not seem so by tomorrow, simply because a social process of cooperative interaction has extended its roots more deeply in the intervening period. The goal of achieving a fully integrated internal market by December 1992 only became realistic in the late 1980s, when it could be built upon three decades of gradual movement in that direction.

Third, European integration has also been based upon respect for political pluralism. Such respect is assumed in functionalist theory, which opposes the potentially coercive amalgamation of high politics traditionally associated with conquest and imperialism. Within the European Union, important political differences remain among the elites and among different ideological groups. Some leftist parties, for example, have opposed a loss of the "safety net" for disadvantaged members of their own societies as integration forces them to compete with a much larger population; some have viewed the EU's relationship with certain third world countries as increasingly one of economic imperialism. Others on the political right have feared that integration will centralize ever more economic decisions in a distant bureaucracy of "Eurocrats" unmoved by considerations of nationalism. At the moment, it is impossible to know whether these kinds of objections can ever gain the support necessary to pull the EU apart; but it should be noted that a diversity of political interests is always assumed to exist within a pluralistic, democratic society. The issue is whether they can be peacefully accommodated in ways that are at least minimally acceptable to all. If that does happen within the EU, then we can be sure that enough spillover has occurred to produce a measure of political, in addition to economic, integration.

Fourth, integration long since proceeded to the point that the members constitute what Deutsch and his associates defined as a pluralistic security-community.[6] Individuals and groups throughout the European Union now possess "dependable expectations of peaceful change" when political conflicts arise between them. No one supposes warfare will break out between them, as it has among many of them in the not-so-distant past. As the president of the EU commission noted in his inaugural address in January 1995, "it is something of a miracle that war between our peoples should have become unthinkable. To squander this legacy would be a crime against ourselves."[7]

Many would argue that the EU's dynamism, and particularly its goal of completing its integration as a common market by 1992, helped accelerate the sudden

---

[6]Deutsch, *Political Community and the North Atlantic Area*. For the reader interested in tracing the development of the EU as a security-community, see George Lichtheim, *The New Europe—Today and Tomorrow* (New York: Frederick A. Praeger, 1963); James Barber and Bruce Reed, eds. *European Community: Vision and Reality* (London: Croom Helm, 1973); and Ernest Wistrich, *After 1992: The United States of Europe* (London: Routledge, 1989).

[7]Jacques Santer, president of the commission, to the European Parliament, Strasbourg, January 17, 1995. *Bulletin of the European Union, Supplement 1/95*.

collapse of communism throughout Eastern Europe in 1989.[8] By the same token, these dramatic developments to the East instantly confronted the EU with unprecedented challenges, for it immediately became a magnet attracting most, if not all, the states of Europe into its orbit. Yet, because of the great differences in their recent economic histories, the two halves of Europe cannot merge economically without difficult, perhaps painful, periods of adjustment for which earlier enlargements of the EU are not appropriate models. While most, if not all, the reformed states of Eastern Europe seek closer ties with the EU, most observers believe that admission would first entail the creation of some kind of associate status for these nations until they are ready, economically and politically, for full membership.

With these rapid changes, the question of Europe's economic connections to the rest of the developed world has also loomed large. The United States and several of the former Soviet republics insist—and most officials of the EU seem to agree—that they not be shut out of the growing European market either. The United States needs the injection of dynamism that increased trade and investment with the rest of the developed world can bring. Russia now is opening its long-closed economic system to the outside world as it looks to Europe and the West for the increased trade that can further Russian development.

## OTHER TRENDS IN ECONOMIC COOPERATION

The European Union is unique in today's world in the extent to which the economies of separate, highly developed states have been systematically and thoroughly integrated into a single market. In the 1960s and 1970s, however, a number of regional treaties were created to further the economic cooperation—if not necessarily the integration—of neighboring states. Some, such as the **European Free Trade Association (EFTA)** and the **Latin American Integration Association (LAIA),** known as the Latin American Free Trade Association before 1980, were intended to reduce barriers to trade among the membership without literally creating a common market, in which labor, capital, and goods all are permitted to move freely within the market area. The **North American Free Trade Agreement (NAFTA)**, ratified in 1993, could be the most important regional treaty because it joins the economies of Canada and Mexico to that of the United States in one of the world's largest trading blocs. Other groupings, such as the **Association of South-East Asian Nations (ASEAN)** and the **Economic Community of West African States (ECOWAS)** are meant to encourage economic cooperation within an organizational context where various diplomatic and foreign policy initiatives may also be pursued in common.

While the European Community was thriving and expanding during the 1980s, many of the other associations of states failed to develop in a way that enhanced their economic interactions. Some, such as the Central American Common Market and the East African Community, made promising beginnings toward the economic in-

---

[8]See, for example, Reginald Dale, "Europe's New Arch," *Europe* (July/August, 1990), 7–8.

tegration of their members, then were disrupted by internal warfare that caused these efforts to collapse. Others simply did not materialize in any significant way once a formal agreement on their purposes and goals was announced. Still others (ASEAN and EFTA are good examples) generally succeeded in the goals they set for themselves, but these were more modest than the complete integration of member-state economies. What can we conclude from the dramatic success of the EC over several decades and the much more modest record of other regional agreements to create more cooperative economic relationships? (See Table 14.1).

The success of regional agreements seems to be related to the level of a society's economic development. In general, the impulse toward economic integration is strongest among the most fully developed countries (DCs). Today, their economies exhibit a strong tendency toward ever greater expansion and specialization. Ever since the Peace of Westphalia created what for the time were the large territorial units of nation-states in Western Europe, their leading economies have tended first to fill, then push outward from, the state's political boundaries. Those boundaries are increasingly anachronistic to the demands of further economic development, expansion, and growth. In contrast, the economic life of a less developed country tends either to be much more localized—perhaps not yet having integrated or "filled" the entire nation-state—or more dependent upon trade with DCs than with its less developed neighbors.

Today, the governments of the world's leading market economies find that they must cooperate as leaders of what is increasingly a single, world economy. This **Group of Seven (G-7)** meets regularly to set the economic agenda and respond to issues that arise from both their own economic interdependence and their competition.[9] For example, as interdependent partners, the G-7 met in Detroit in 1994 to consider how they might mutually stimulate the growth of jobs in their countries. That concern for mutual productivity reflected their awareness of how more jobs would lead to more consumption of each member's products. As competitors, they are divided into at least three different economic groups: the four European states are members of the EU, and are becoming increasingly integrated; the United States, Canada, and Mexico have formed their own free trade arrangement in NAFTA; and Japan's continued economic dynamism makes it the dominant economic power in the Pacific. Still, the G-7 will almost certainly continue as a rich nations' club with interdependent interests even as these other cleavages divide them.

## MNCS IN THE GLOBAL VILLAGE

Humankind's economic activity is increasingly breaking through the boundaries of individual nation-states; while this process is creating ever larger markets, it is also producing varying degrees of economic and social dependence, interdependence, and integration. Much of this transnational development is being fostered by multinational corporate activity. That is, the growth of transnational commercial exchange may be producing conditions of dependence in some relationships, interdependence in others, and integration in others still. The image of the global village may remind us how the planet has shrunk thanks to the speed of communication

---

[9]Since 1993, the president of Russia has been invited to participate in sessions of the G-7.

**TABLE 14.1** Regional IGOs Designed for Economic Cooperation

| IGO | FORMED | AIM | MEMBERS |
|---|---|---|---|
| African, Caribbean, and Pacific Countries (ACP) | April 1, 1976 | Preferential economic and aid relationship with EC | 69 |
| African Development Bank (AfDB) | Aug. 4, 1963 | Economic and social development | 76 |
| Andean Group (AG) | May 26, 1969 | Harmonious development through economic integration | 5 |
| Arab Bank for Economic Development in Africa (ABEDA) | Feb. 18, 1974 | Economic development | 16 |
| Arab Cooperation Council (ACC) | Feb. 16, 1989 | Economic cooperation and integration, possibly leading to an Arab Common Market | 4 |
| Arab Fund for Economic and Social Development (AFESD) | May 16, 1968 | Economic and social development | 20 |
| Arab League (AL) | Mar. 22, 1945 | Economic, social, political, and military cooperation | 20 |
| Arab Maghreb Union (AMU) | Feb. 17, 1989 | Cooperation and integration among the Arab states of northern Africa | 5 |
| Arab Monetary Fund (AMF) | April 27, 1976 | Arab cooperation, development, and integration of monetary and economic affairs | 19 |
| Asia Pacific Economic Cooperation (APEC) | Nov. 1989 | Trade and investment in the Pacific basin | 15 |
| Asian Development Bank (AsDB) | Dec. 19, 1966 | Regional economic cooperation | Regional: 35; Nonregional: 15 |
| Association of Southeast Asian Nations (ASEAN) | Aug. 6, 1967 | Regional economic, social, and cultural cooperation among the non-Communist countries of Southeast Asia | 6 |
| Benelux Economic Union (Benelux) | Feb. 3, 1958 | Closer economic cooperation and integration | 3 |
| Caribbean Community and Common Market (CARICOM) | July 4, 1973 | Economic integration and development, especially among the less developed countries | 13 |
| Caribbean Development Bank (CDB) | Oct. 18, 1969 | Economic development and cooperation | Regional: 20; Nonregional: 5 |
| Central African Customs and Economic Union (UDEAC) | Dec. 8, 1964 | Establishment of a Central African Common Market | 6 |
| Central African States Development Bank (BDEAD) | Dec. 3, 1975 | Loans for economic development | 9 |

## TABLE 14.1 Continued

| IGO | FORMED | AIM | MEMBERS |
|---|---|---|---|
| Central American Bank for Economic Integration (BCIE) | Dec. 13, 1960 | Economic integration and development | 5 |
| Central American Common Market (CACM) | Dec. 13, 1960 | Establishment of a Central American Common Market | 5 |
| Colombo Plan (CP) | July 1, 1951 | Economic and social development in Asia and the Pacific | 26 |
| Council of Arab Economic Unity (CAEU) | June 3, 1957 | Economic integration among Arab nations | 11 |
| East African Development Bank (EADB) | June 6, 1967 | Economic development | 3 |
| Economic Community of Central African States (CEEAC) | Oct. 18, 1983 | Regional economic cooperation and establish a Central African Common Market | 10 |
| Economic Community of the Great Lakes Countries (CEPGL) | Sept. 26, 1976 | Regional economic cooperation and integration | 3 |
| Economic Community of West African States (ECOWAS) | May 28, 1975 | Regional economic cooperation | 16 |
| European Bank for Reconstruction and Development (EBRD) | April 15, 1991 | Transition of seven centrally planned economies in Europe | 35 |
| European Union (EU) | April 8, 1965 | A fusing of Euratom, ESC, and the EEC, or Common Market | 15 |
| European Free Trade Association (EFTA) | Jan. 4, 1960 | Expansion of free trade | 6 |
| European Investment Bank (EIB) | Mar. 25, 1957 | Economic development of the EC | 12 |
| Gulf Cooperation Council (GCC) | May 25–26, 1981 | Regional cooperation in economic, social, political, and military affairs | 6 |
| Inter-American Development Bank (IADB) | April 8, 1959 | Economic and social development in Latin America | 44 |
| Islamic Development Bank (IDB) | Dec. 15, 1973 | Islamic economic aid and social development | 43 |
| Latin American Economic System (LAES) | Oct. 17, 1975 | Economic and social development through regional cooperation | 26 |
| Latin American Integration Association (LAIA) | Aug. 12, 1980 | Freer regional trade | 11 |
| Nordic Council (NC) | Mar. 16, 1952 | Regional economic, cultural, and environmental cooperation | 5 |

## TABLE 14.1 Continued

| IGO | FORMED | AIM | MEMBERS |
|---|---|---|---|
| Nordic Investment Bank (NIB) | Dec. 1975 | Economic cooperation and development | 5 |
| Organization for Economic Cooperation and Development (OECD) | Dec. 14, 1960 | Economic cooperation and development | 24 |
| Organization of African Unity (OAU) | May 25, 1963 | Unity and cooperation among African states | 50 |
| Organization of American States (OAS) | April 30, 1948 | Peace and security as well as economic and social development | 35 |
| Organization of Eastern Caribbean States (OECS) | June 18, 1981 | Political, economic, and defense cooperation | 8 |
| Organization of the Islamic Conference (OIC) | Sept. 22–25, 1969 | Islamic solidarity and cooperation in economic, social, cultural, and political affairs | 47 |
| South Asian Association for Regional Cooperation (SSARC) | Dec. 8, 1985 | Economic, social, and cultural cooperation | 7 |
| South Pacific Commission (SPC) | Feb. 6, 1947 | Regional cooperation in economic and social matters | 27 |
| Southern Cone Common Market (MERCOSUR) | Mar. 26, 1991 | Regional economic cooperation | 4 |
| West African Development Bank (WADB) | Nov. 14, 1973 | Economic development and integration | 7 |
| West African Economic Community (CEAO) | June 3, 1972 | Regional economic development | 7 |

**SOURCE:** Adapted from *The World Factbook* (Washington: Central Intelligence Agency, 1992).

and travel today. But it also suggests how the whole world is fast becoming the kind of economic system that once scarcely extended beyond a single village.

Multinational corporations (MNCs) are clearly a major factor in this development. Chapter 11 asked whether MNCs are creating conditions of dependence or interdependence. Outside of the European Union, it is by no means clear whether multinational corporate activity is increasing the **integration** of societies within which it operates today. That is because integration stems from a conscious effort to mix separate groups together in the hope of securing more dependable expectations of peaceful change among them. Yet, what most characterizes economic initiatives in the private sector is a profit motive, not some social or political purpose such as the creation of a pluralistic or amalgamated security-community.

Therefore, if integration is advanced as the result of transnational economic activity, it will generally be the by-product of the effort to achieve private gain. In this

> **Normative Dilemmas**
>
> ## The Challenge of the Information Society
>
> In 1995, officials of G-7 countries agreed to cooperate in a number of projects meant to advance the global information society. They included developing or advancing the following: international links between high-speed networks; cross-cultural training and education, particularly for language learning; a global network interconnecting electronic libraries, accessible to the general public; electronic museums and galleries; environment and natural resources management; global emergency management, including emergency response; global healthcare applications; procedures for conducting electronic administrative affairs between governments, companies, and citizens; the interoperability of electronic services and information on trade on a global scale; and maritime information systems through information and communication technologies.
>
> - Which of these projects would you expect to be most significant in furthering social and political integration cross-nationally? How?
> - Will any of these projects be susceptible to increasing conflict, separatism, and the like? Why or why not?
> - Will any of these projects risk increasing social divisions between the educated and the uneducated? What would be the likely impact of such a development on world politics?
> - Can the intergovernmental effort to advance information technologies increase the ability of governments to control people? Why or why not?

respect, the European Union is an exception only in the sense that its authorities are consciously promoting the creation of a common market in part through multinational economic activity as an aspect of the goal of integration.

## THE DISINTEGRATION OF MULTINATIONAL STATES

While transnational integration was advancing rapidly in Western Europe, threats to the continued unity of multinational states erupted suddenly in the 1990s. Unrest within many of the non-Russian republics of the Soviet Union and the fragmentation of Yugoslavia constituted the leading examples, but the trend extended to a renewal of separatist sentiment among French-speaking Canadians in Quebec, Kurds in Iraq, Basques in Spain, and to continuing unrest over British rule in Northern Ireland, to name several of the most prominent cases. Some of these movements implied that they might be satisfied with greater autonomy within a federated framework, while others demanded their sovereign independence.

But an examination of these movements across the globe suggests that this potent force can still loosely be called nationalism—that is, the determination of distinct cultural communities to rule their affairs autonomously. So it was that in recent years Tibetans have sought greater autonomy from China, Sikhs from the Hindu majority of India, and Tamils from the Sinhalese of Sri Lanka. In some cases, and most notably in the Middle East, the increase of separatism came in the aftermath of warfare and its resulting turmoil. Kurdish separatists attempted to break away from

Iraq at the close of the Iran-Iraq war in 1988, and again after Iraq had been driven from Kuwait in 1991. Lebanon had largely disintegrated as a viable nation-state in the 1970s as the result of intercommunal conflict, rather than from a separatist uprising by a particular minority group. But Lebanon's troubles, too, had been compounded by military action—in that case, civil war followed by Syrian intervention, then Israel's forced occupation of the southern part of the country.

The principal lesson in all of these cases was that nationalism and statism still maintain an uneasy marriage at the end of the twentieth century. After the decolonization of most of Africa in the 1960s, it became apparent that the newly independent African governments were determined not to let their countries fragment along ethnic lines. But at the time, the older states of the North generally seemed to be impervious to secessionist threats of that kind. Thirty years later, secessionist nationalism is more salient in many "older" states than it appears to be in Africa.

## SEPARATISM WITHIN THE FORMER SOVIET EMPIRE

The region spanned by the Soviet bloc seemed for years to demonstrate stability and cohesiveness among its multinational states. It was clear that the explanation for revived nationalist separatism there during the 1990s was related to the decline of communism and the end of the cold war.

**IMPOSED UNITY AND ITS AFTERMATH** The communist system had for some forty years imposed unity and order on a number of multi-ethnic states in Eastern Europe and the Soviet Union, in effect freezing those divisions under authoritarian rule. Some, such as Czechoslovakia and Yugoslavia, were created after World War I from the ruins of the polyglot Austrian and Ottoman empires. Their instability as newly independent states, however, had been one of the factors that led to World War II, and in the aftermath of that war, greater cohesion was bought at the cost of single party rule. Their communist dictatorships were supported by that of the Soviet Union, which threatened to intervene if communist rule in Eastern Europe were seriously challenged. As those authoritarian systems began to collapse in 1989 and 1990, some of those communities over whom they had ruled began to demand a greater voice in their own affairs. Political thaw brought the revival of intrastate ethnic conflicts.

**THE EMERGENCE OF PLURALISM** The replacement of single-party rule by greater pluralism and democracy throughout most of the region was the second factor producing a revival of ethnic conflicts in Eastern Europe. With the end of police states, dissident groups were free again to protest their positions in societies they shared with others; new political leaders likewise arose to advance their interests. Since democratic theory generally supports the right of "nations" to govern themselves, reformers found themselves caught in a dilemma, as Mikhail Gorbachev discovered: He could either stand by and preside over the disintegration of his country as a product of the democratization he had himself encouraged, or try to maintain the unity of the state through police or military action. By the end of 1991, all of the republics had become independent and the Union of Soviet Socialist Republics was officially abolished (see Table 14.2).

### TABLE 14.2 — The Disintegration of the Soviet Bloc

| MEMBERS OF SOVIET BLOC: 1988 | INDEPENDENT STATES FROM SOVIET BLOC: 1995 |
|---|---|
| Union of Soviet Socialist Republics (USSR) | Russia |
| | Belarus |
| | Ukraine |
| | Moldova |
| | Armenia |
| | Georgia |
| | Azerbaijan |
| | Kazakhstan |
| | Uzbekistan |
| | Tajikistan |
| | Turkmenistan |
| | Kyrgyzstan |
| **Soviet Bloc** | |
| East Germany | Germany (united with W. Germany) |
| Bulgaria | Bulgaria |
| Czechoslovakia | Czech Republic |
| | Slovakia |
| Hungary | Hungary |
| Romania | Romania |
| Poland | Poland |

The new government of Russia accepted the disintegration of the empire that had made up the Soviet bloc—that is, of non-Russian states. But it demonstrated its unwillingness to allow the splitting off from the Russian republic of ethnic communities residing within it. In December 1994, the Russian army attacked separatist Chechnya, in Russia's Caucasus region, to make the government's point in a bloody and destructive show of force that it would not tolerate the breakup of Russia itself.

**THE EXAMPLE OF WESTERN ECONOMIC PROGRESS** The third explanation for Soviet ethnic and national dissolution may have been the very success of the process of economic integration in Western Europe. By the end of the 1980s, the growth of the European Community (now the EU) had brought unparalleled economic success to its member countries within a multinational framework. Some viewed this accomplishment as both a model and a magnet for many of those to the East who wanted to duplicate those successes.

Within a few years, the EU was committed to the eventual admission of Bulgaria, the Czech Republic, Hungary, Poland, Romania and Slovakia. It was widely assumed that another several Eastern European countries might eventually join the EU as well.

## THE RESULT OF SEPARATISM

In sum, loosening the repressive reins of government—whether through reform or warfare—has frequently provided discontented ethnic or other minority groups a window of opportunity to press for greater autonomy. Where a commitment to democratic pluralism can be sustained, such movements may meet with some success. In the immediate post–cold war period, that may have accounted for the "velvet divorce" that in 1993 split Czechoslovakia into separate Czech and Slovak republics. As of 1995, after a referendum on independence was only narrowly defeated, it seemed that Quebec might also leave its union with English-speaking Canada. Northern Ireland's long-simmering "troubles"—marked for years by many acts of violence between militant members of Catholic and Protestant communities—took a step toward resolution in 1995 through an agreement that could lead to greater autonomy for the North from British rule, although progress toward a peaceful solution was marred by renewed violence in 1996. The dilemma for the established, or pro-union, government in every case is whether it can grant greater autonomy to disaffected communities without their complete secession. But where democratic values are strong, we may expect that even complete separation may be negotiated, however painfully, in relative peace.

On the other hand, where a central government is still willing and able to exercise sufficient repressive force (as was the case in 1991 with China in Tibet and Iraq in Kurdistan), then separatist movements are likely to be suppressed indefinitely. The outcome of Russia's armed intervention in Chechnya, though badly bungled, was never in doubt because of the overwhelming strength of the Russian army. This show of force was no doubt meant as an object lesson to would-be separatist communities elsewhere within Russia as well.

**CONSTRAINTS AGAINST INTERVENTION IN SEPARATIST CONFLICTS** In keeping with the logic of the nation-state system, the governments of other countries have usually not given more than rhetorical support to the desire for self-determination of minority groups in other states. This is especially true when the government in question seeks friendly relations with the state experiencing separatist friction. No friend of Canada, for example, would call publicly for the independence of Quebec; nor would any Western nation in 1990 risk demanding the independence of the Baltic republics at a time when the improvement of their own relations with the Soviet government remained a top priority. Nor, for that matter, did the governments of other multi-ethnic states, most notably in Africa, call for the breakup of similarly constructed states elsewhere in the world.

The constraints supporting nonintervention in the "domestic" affairs of others even led the United States, following the Gulf War in 1991, to refuse to lend material support to Kurds and Shiite Muslim communities in Iraq. That refusal was justified on grounds that it likely would have produced large numbers of American casualties; nonetheless, it was particularly striking in light of President Bush's repeated call for the Iraqi people to overthrow the rule of Saddam Hussein. Eventually, as the plight of Kurdish refugees fleeing the Iraqi army grew to near catastrophic proportions, the Bush administration did reverse its "hands-off" policy toward the Kurds,

**"ETHNIC CLEANSING"**
Auth © 1994 *The Philadelphia Inquirer*. Reprinted with permission of UNIVERSAL PRESS SYNDICATE. All Rights Reserved.

and U.S. troops were deployed to hold off Iraqi forces from parts of Kurdistan so as to encourage the refugees to return to their homes. Still, it was the special circumstances of the Kurdish problem, brought into play by the U.S. defeat of Saddam Hussein, that made this latter case the obvious exception to the rule against intervention in the minority problems of other states.

**THE LESSONS OF BOSNIA** The bloody civil war in Bosnia-Herzegovina has been the most dramatic testament to nationalist separatism run amok. The conflict was triggered by the break-up of Yugoslavia, a federated state of several "nations."[10] With Yugoslavia's dissolution, rival groups were given free rein to gain what they could through force of arms. It was the sharpest reminder in the 1990s of how a violent quest for ethnic autonomy could produce human suffering on a massive scale. The

---
[10]Croats, Serbs, and Bosnian Muslims are all ethnically Slavic. They are more distinguishable by their different religious affiliations than by genuine ethnic, or genetic, differences. Therefore, the term *ethnic cleansing*, used to describe the forcible elimination by (largely) Serbian militias of Bosnian Muslims from territories they conquered, is something of a misnomer.

refusal of other governments to intervene forcibly enough to stop the fighting was also a stark illustration of Westphalian principles of state self-interest still at work. In spite of the suffering, in spite of the obvious case for collective action against Serbian aggression, other governments were unwilling to go beyond a timid, "neutral" intervention through the United Nations and NATO. That remained the case until such time, late in 1995, when the military forces among the rival communities in Bosnia came sufficiently into balance to end any prospect of a sweeping victory for any of them. That at last permitted a precarious peace settlement dividing the territory between rival communities. The settlement was brokered by the United States and enforced by a substantial NATO deployment of peacekeepers.

As usual, our worldviews predetermine a fair amount of what we find when we attempt to analyze such events as these in Bosnia. Realists and neorealists will tend to see the bloody intercommunal strife as evidence that nationalism is not a spent force and will likely continue as a major cause of warfare in the future. Idealists and neo-idealists may note that the bloodshed there produced an international war crimes tribunal to reinforce the normative view that criminal behavior in time of war cannot and will not be tolerated—a view that assumes that meting out such punishment is a way to discourage at least the worst excesses of war in the future (see Chapter 16 for a discussion of this and other war crimes tribunals).

## WHO'S IN CHARGE OF THE WORLD?

The developments discussed in this chapter are beginning to "open up the system" of world politics in unprecedented ways. Yet, in most respects, it is still only the representatives of states who have the formal authority to make policy for the world community (which is another reason why statehood is such an important goal for some social groups). Our planetary society today may be a bit like that of France on the eve of the French Revolution. In a sense, we are living in a time when the "ancien regime" of nation-states maintains its long monopoly over the authoritative allocation of values, even though a much broader array of social groups now clamor for a share in global governance. This disjunction helps explain the sense of ferment and turbulence in efforts to analyze world politics today.[11]

States remain reluctant to share global decision-making power with other groups for an obvious reason: The Westphalian order serves the interests of sovereign states, which have been empowered by those arrangements. Intergovernmental organizations, for the most part, remain more the tools of states—which created them—than agents capable of reforming the interactions of states. As the history of national revolutions has shown, institutional structures and norms tend to lag in meeting the needs of a new social reality, precisely because they serve the interests of an old social order instead.

---

[11] See the important exploration of these themes by James N. Rosenau (a leading international relations scholar): *Turbulence in World Politics: A Theory of Change and Continuity* (Princeton, N.J.: Princeton University Press, 1990).

> ## The Impact of the Internet
>
> Now that people all over the world can communicate via the Internet, does that make war less likely? ... Those who tout the Internet as global benefactor believe that networking will tighten social ties across borders. ... The positive impact on political trendlines is easy to imagine. If tens of thousands of Russian and American scientists chat regularly in cyberspace, they might build ties so strong that rulers couldn't revive old tensions. ...
>
> The political downside is that the human outreach can come not just from those who want to communicate but also from those who want to create networks of evil. When you open the door to cyberspace, you have no idea who will come in. ... The Net can also inspire ethnic nationalists—like small tribes in the Caucasus—to try to promote dreams of secession that otherwise might die in obscurity.
>
> SOURCE: Trudy Rubin, "Cruise Missiles or Just Cruising?" *The Philadelphia Inquirer*, February 2, 1996, A15. Reprinted with permission from *The Philadelphia Inquirer*.

## FROM GEOPOLITICS TO INFOPOLITICS?

In contrast, our economic and social life generally is lived further away, so to speak, from the constraints of Westphalian norms. Technology usually originates in such a place, largely removed from the rules that govern the interactions of states. Yet, unprecedented technological developments have boosted human communication (among other things) during the twentieth century, with enormous consequences for our social and economic lives. Speeding up how and what we communicate is creating new demands in the realm of politics, which in turn create new pressures on the formal worldwide political structure. Some go so far as to argue that our ancient fixation with *geopolitics* is giving way to *infopolitics*. In other words, we are moving from an obsession with territory and security to a world in which our social connections, enterprises, and even our loyalties increasingly spring from our ability to communicate instantly with virtually any other group of individuals on the planet.[12]

**EUROPE'S NEW ECONOMIC MOTORS** Our discussion of the world political economy (Part IV) showed a trend in which economic exchanges of all kinds are breaking down the political barriers erected by states. One aspect of that trend reveals how the formal nation-state arrangement no longer fully controls some dynamic economic developments. The lowering of barriers to commerce, combined with electronic communication, are transforming certain cross-national regions into semi-autonomous economic actors. This is particularly apparent for several booming "city-states" within the European Union, where internal economic barriers are now nearly nonexistent (see Map 14.2). It is also likely to create powerful economic centers wherever else in the world political boundaries are weakened or erased to encourage economic

---

[12]See, for example, Mihaly Simai, *The Future of Global Governance: Managing Risk and Change in the International System* (Washington, D.C.: United States Institute of Peace Press, 1994).

352   PART V   The Construction of a Global Society

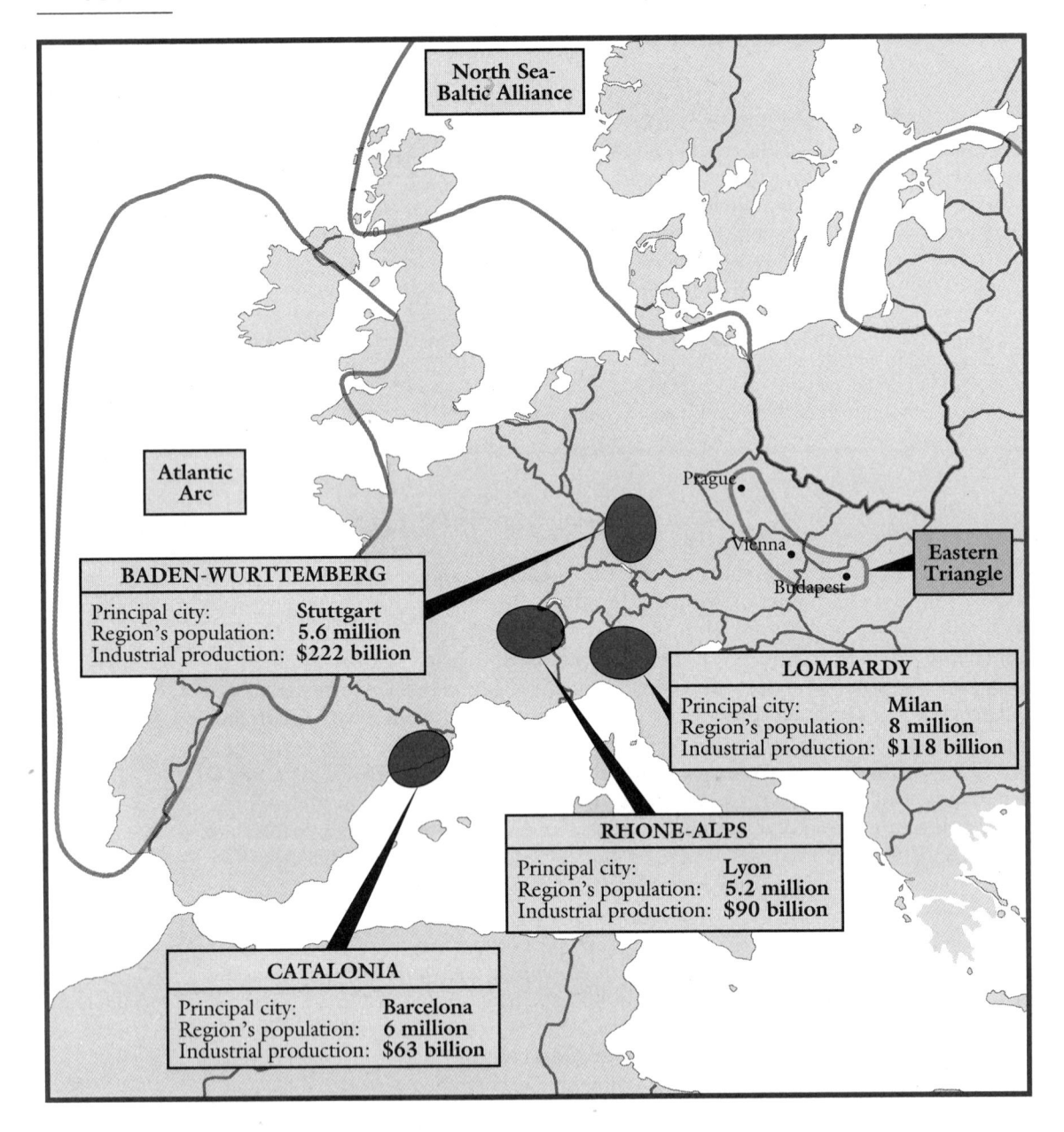

**North Sea-Baltic Alliance**

**Atlantic Arc**

**Eastern Triangle**

Prague
Vienna
Budapest

**BADEN-WURTTEMBERG**
Principal city: Stuttgart
Region's population: 5.6 million
Industrial production: $222 billion

**LOMBARDY**
Principal city: Milan
Region's population: 8 million
Industrial production: $118 billion

**RHONE-ALPS**
Principal city: Lyon
Region's population: 5.2 million
Industrial production: $90 billion

**CATALONIA**
Principal city: Barcelona
Region's population: 6 million
Industrial production: $63 billion

**MAP 14.2**   Europe's New Economic Motors

Across Europe, economic interests are starting to overpower national ones. Road, rail, and communications networks have begun to supersede national borders. Broad alliances are forming along the Atlantic; the Baltic and North seas; and in Eastern Europe. In particular, four "poles of prosperity" have emerged: the regions around Stuttgart, Lyon, Milan, and Barcelona.

SOURCE: Dave Cook, *The Washington Post*, March 27, 1994, C3. © 1994, *The Washington Post*. Reprinted with permission.

growth. That could lead to the further disintegration of some long-standing political units, while at the same time creating new cross-national units. Obviously, contemporary infopolitics have become a strong influence upon geopolitics.

**LEADING THE INFOPOLITICAL WAY** The contemporary period is also characterized by a greater mingling of peoples than ever before in history. Unquestionably, this is inciting ethnic conflict in many places, but it also has the potential to stimulate more cosmopolitan development of human societies than the world has had the capacity to evolve before now. One recent writer, Joel Kotkin, claims that certain transnational ethnic groups—or *tribes*—dispersed across the planet, have taken the enterprising lead in building a truly global village. He argues that Jewish merchants and financiers in the Middle Ages were followed by migrations of Anglo-American settlers of the New World, whose heritage spread a common system of financial accounting and commercial law, making English the global language of business today. They, in turn, were succeeded by the Japanese, who have built world-spanning enterprises out of their resource-poor homeland. Now, according to Kotkin, we are seeing overseas Chinese and Indians beginning to emerge as powerful economic actors in global networks.[13] Are geopolitics or infopolitics dominant here?

# CONCLUSION

We have seen that political and social interdependence is on the rise in the modern world as the result of a largely unplanned and, in some sense, nearly inevitable product of the conditions of human life today. Integration, in contrast, implies a conscious effort by political authorities to produce the conditions providing for "dependable expectations of peaceful change" on the part of social groups that had been previously independent of one another. Amalgamation, finally, brings the formal joining of what have been separate nation-states into a single (although perhaps federated) political system.

Functionalism constitutes the effort to integrate political communities "from the ground up"—that is, by encouraging economic interests to increase the wealth of members of society by building common markets and the like. The antecedents of the European Union were largely developed in keeping with functionalist theory. The EU has now evolved to the point that it will soon be as much a single economic system as is that of the fifty United States of America. Whether a more formal amalgamation of its member states produces a common foreign policy or centralized political institutions remains to be seen.

Although functionalism has inspired the creation of arrangements for greater economic cooperation in other parts of the world, nowhere have the results been comparable to that of Western Europe. That is largely explained by two factors: (1) in many parts of the non-European world, nation-building has come first, pushing regional economic integration aside; and (2) European-style economic integration has evidently been encouraged by the requirements for specialization and larger

---

[13]Joel Kotkin, *Tribes* (New York: Random House, 1993).

markets that are characteristic of the most advanced industrial and postindustrial societies.

Multinational corporations have been important agents in the creation of economic and social dependence, interdependence and—less certainly, outside the European Union—integration.

One phenomenon at first glance appears to contradict that of transnational integration: In many parts of the world, minority communities are demanding greater autonomy or independence from multi-ethnic states. In what used to be the Soviet empire, in particular, such separatist movements clearly have been the product, first, of the loosening of highly centralized and repressive governmental control and, second, of the attractions of Western pluralistic democracy and economic growth. Much turbulence and, in the worst cases, interethnic warfare, have been an immediate result. Yet, there and elsewhere, the democratic ideal of self-determination may increasingly be communicated by our modern technology of mass communication.

Nation-states remain largely (and "officially") in charge of the world system. But they cannot account for all of the influences on world politics, or on the economic and social ties of people. The rapid availability of information that empowers people in new ways is giving rise to a transnational infopolitics that is beginning to challenge some of the fundamental assumptions of traditional geopolitics.

CHAPTER 15

# International Law and Global Order

The prefixes *intra* (within) and *inter* (between) reveal some fundamental distinctions between political life within the nation-state and between or among states. Clearly, law plays a stronger role within most individual states than it does in governing the relations among states. In fact, it seems that the very logic of the Westphalian system requires the law that governs the interactions of sovereign governments to be fundamentally different from the domestic law that governs the behavior of individuals within the state. After all, according to its strictest definition, sovereignty indicates that no higher authority—and so no "law"—exists above and beyond the government of the sovereign.

However, is it enough to say that international law today is, strictly speaking, merely the intergovernmental law of sovereigns? That is what is emphasized especially in the **positive law** tradition, which asserts that all law is made by human beings authorized to do so—that is, governments. Without doubt, the body of rules and regulations that "govern" the interrelationships of states has always been at the heart of what it means to speak of international law.

But the international legal tradition has also insisted upon regulating and protecting the lives of human beings. Much of that tradition since the Peace of Westphalia has been posited on the assumption (or the hope) that international law could protect individuals *through* the medium of their own governments, which would have the first responsibility to protect the human beings over whom they rule.[1] But what

---

[1] The issues discussed above reveal two long-standing views of the way in which international law is meant to act on individuals. The **dualist** outlook views international law as commanding sovereign governments, which in turn incorporate international laws into their own domestic legal systems where they impact upon individuals. **Monism** views the world legal system as capable of acting directly upon individuals, whether government officials or private citizens. The dualist view is most deferential to the sovereignty of states, whereas monists are supported when international laws are enacted requiring states to protect individuals under their jurisdictions. Note too that the dualist outlook likely will dominate realist and neorealist analysis of the place of law in world politics, while idealists and neo-idealists are more likely to accept a monist view.

if governments act unjustly against their subjects? Do human beings have legal "rights" apart from those guaranteed by their own states? Do they have recourse through international law to protections from abuse by their rulers? The international legal tradition has always insisted that they do. This is most evident in the doctrines of **natural law,** whose authority derives from the very laws that govern the universe. In the social sphere, natural law is viewed as providing moral imperatives that transcend human-made rules. The importance of the natural legal tradition is its assumption that justice for humanity must on occasion override the sovereignty of states. Intraspecies justice therefore should supersede international law.

# DOMESTIC VERSUS INTERNATIONAL LAW

Law seems a more powerful force within states than it does among them for the following reason: a sovereign is theoretically able to command its subjects, whereas sovereigns, by definition, cannot command one another. Commanding and obeying commands suggests an arrangement in which authority is hierarchical. That is the characteristic arrangement of domestic legal systems.

## THE HIERARCHY OF DOMESTIC LAW

In any well-developed domestic legal system, a hierarchy of rules and obligations is clearly established. While justices of the peace or local magistrates possess the authority to carry out certain legal functions for the society—performing marriages and judging traffic violations are typical examples—that authority is clearly limited. Moreover, the decisions of low-level authorities are typically subject to review by those above them. This is perfectly clear, for example, in the federal system of the United States, where the power of judicial review has required that the highest court in the land be the institution with the final authority to determine the law. That sometimes results in its overruling lower courts. Once the Supreme Court makes its decision, the executive branch is required to enforce it. But that, in turn, rests upon the executive's capability to enforce the law, for enforcement power comes from the state's monopoly of police power.

In well-run states, most subjects of the law submit to it most of the time without reflecting a great deal about whether or not they might get away with violating it on this or that occasion. We know that sanctions exist if we engage in many kinds of illegal acts; therefore, we find it more convenient generally to obey the law than to run the risks involved in violating it.

## THE LACK OF HIERARCHY IN INTERNATIONAL LAW

When we turn to the international legal system, we find that it is much less hierarchical. States are its principal subjects, although not the only ones; yet states (or their governments) possess almost all the lawmaking, enforcement, and interpretation capability that exists in the world. The array of power and authority here is far more horizontal—that is, based more on cooperation than on a "vertical" system of command—than would be imaginable within any state.[2] This is still true in spite of the fact that states have created some institutions that seem to represent a more vertical authority in the world. For example, the United Nations contains a Security Council and General Assembly that somewhat resemble the upper and lower houses of a national legislature, a Secretariat that looks a bit like the executive arm of a government, and an International Court of Justice that seems to be a kind of supreme court for the world.

**THE ABSENCE OF WORLD GOVERNMENT** But the fact is that the United Nations was not created to be a world government where authority for maintaining law and order across the globe would be combined with powers to permit law enforcement, even against state governments. Instead, the U.N. system was meant to be the instrument of sovereign states, assisting them in a variety of ways to administer an international legal system based upon coordination of state agreements. Even in an age in which intergovernmental organizations (IGOs) and nongovernmental organizations (NGOs) have proliferated at a rapid pace, most of the capability for creating, enforcing, and judging international law remains in the hands of its most important subjects, the nation-states. Therefore, some have concluded that international "law" amounts to little more than unenforceable pieties about how states ought to behave but seldom do.

Yet, in the real world, such a view grants too much to the theoretical claims of sovereignty. Even though a state may exercise its power to disrupt or violate international legal standards, that same power may be used to uphold or enforce international law as well. Only the discredited ideologies of totalitarianism have ever argued that sovereignty could be used to crush the legal rights of human beings under its sway. So, even though the independent power of states remains the central fact of global political life today, a state's power is nonetheless constrained in several ways by legal rules, norms, and ideas of justice.

**THREE PURPOSES OF INTERNATIONAL LAW** The result is a system of world order that cannot be understood simply by synthesizing all of the various international legal rules, even though those rules may explain much about the nature of the international system. Instead, we must consider the three principal purposes or features of the contemporary international legal order. First, it is designed to maintain the system of sovereign equality of states. Second, it is intended to rationalize and justify the alignment of state capability with international norms. Third, it is presumably intended to extend justice to human beings throughout the world. It is obvious at the outset that these three functions can lead to conflicting demands on behavior, since they reflect the contradictory impulses between the need for individual self-assertion and group harmony.

---

[2]For a clear discussion of the implications of this difference, see Richard A. Falk, "International Jurisdiction: Horizontal and Vertical Conceptions of Legal Order," *Temple Law Quarterly* 32, 295–320.

> ## States of War, States of Peace
>
> Nation-states can settle their disputes by using force, and do so lawfully, whereas forcible dispute resolution is not allowed within a state's domestic legal system. This is often viewed as the most fundamental difference between domestic and international order. But the difference may be more of degree than of kind. First, civil wars can and do break out within states, so warfare is not just an international phenomenon. Second, warfare is by no means a frequent occurrence among states, but a last, extreme resort. Third, between many (even neighboring) states, warfare doesn't occur at all.
>
> - Why have the United States and Canada, with the world's longest undefended border, not fought a war in more than two hundred years?
>
> - How did France and Germany move from being bitter enemies and antagonists in three major wars within eighty years to friends soon after the end of World War II?
>
> - Because the difference between domestic and international order is *not* based on a clear-cut distinction about how force may be used to solve disputes, why is warfare lawful? Can it still serve any positive functions for human betterment?
>
> - What are the factors that have discouraged or even ended the prospects of warfare between some states?

# THE RIGHTS AND DUTIES OF STATES

Most of what we think of as international law is intended to serve the interests of all states, at least most of the time. Sovereign governments themselves frequently want and need to create rules and regulations to guide them in their interactions for a simple reason: It is easier to live in a world in which one can anticipate fairly dependably how others will behave than to try to coexist amid erratic and unpredictable behavior. The more highly developed the legal system, the more predictable are at least the patterns of behavior one can expect from its members. But even in the relatively primitive international legal system, some "ground rules" for interaction are required.

## THE GROUND RULES FOR STATES

The Peace of Westphalia provided some of the ground rules for European political actors in the seventeenth century that have now been enlarged upon and extended worldwide. At the heart of those rules is the body of law that defines the rights and duties of sovereign governments. This includes law relevant to diplomatic exchange (which is what facilitates governmental interaction), to the jurisdiction of states over persons and territory (which defines the limits of sovereignty), and to the regulation of activities on the high seas or other global commons (which are areas, they have mutually agreed, that are beyond the reach of sovereignty).

Since the law that defines the rights and duties of states—the ground rules for the system's operation—tends to express the clear mutual self-interest of all states, enforcing this kind of international law is seldom a problem; it tends to be self-enforcing by the very state actors that mutually benefit from it. Governments defer to the norms they have created because it serves their interests to do so. Occasionally, of course, some governments do violate diplomatic immunity or fail to regulate ships that fly their flag in accordance with international requirements. But, by the same token, even in a well-ordered domestic system, some citizens violate state laws from time to time. The test in both cases should be whether the general expectation of subjects and rulers alike is that certain legal standards will prevail.[3]

## CODIFIED AND CUSTOMARY LAW

Much of the law that clearly serves the mutual self-interest of sovereigns is now codified in **treaties.** Codification frequently was facilitated in recent times by the fact that it was built upon the **customary practice** of states, often over centuries. The existence of clear customary practice is itself a good indication that the mutual self-interest of states is being served, since state governments, like the actors in any social system, customarily interact in ways that serve their mutual interests.

One example is the law surrounding diplomatic immunity, that, by the end of the twentieth century, has been fully codified in treaties to which virtually all states are parties. But the customs regulating the rights and duties of diplomats were already well established by the time the Westphalian system was created. Why? The obvious answer is that international actors have always understood their mutual need to communicate with one another in certain dependable ways—to explain their intentions, suggest how mutual cooperation would be helpful, even to issue clear threats. None of these interests would be served if their envoys were to be forever at risk of their own lives in foreign capitals, or if they could not speak freely on their government's behalf to another without having their work disrupted by seizure and arrest.

**RECIPROCITY** The concept of **reciprocity** is essential to the development of customary practices that in turn may eventually become treaty law. Reciprocity can be defined as the exchange of mutual, comparable favors between actors to maintain an equality of opportunity between or among them. More familiarly put, you, the sovereign, do unto other sovereigns as you would have them do unto you. In the largely horizontal legal order, reciprocity may be exemplified in a number of ways. One is when a sovereign grants the same diplomatic favor—or disfavor—to another sovereign's ambassador that it receives in return. Thus, the exchange of ambassadors is entered into reciprocally, and if one should be declared *persona non grata* by the government to which it is accredited, his or her government will reciprocate by expelling the representative of the government that acted first. A second example of reciprocity is revealed in the mutual acceptance by many sovereigns of their rights and duties in navigating the high seas. All ships are expected to observe uniform safety and communication standards, and to display the flag of the licensing state, which is responsible for enforcing the standards on board that vessel. A third example is seen in a government's protection of foreign nationals temporarily residing

---

[3]For a fuller discussion on this theme, see Roger Fisher, "Bringing Law to Bear on Governments," *Harvard Law Review* 74 (1961), 1130.

within its territory, a protection that is extended at least in part on the assumption that its own citizens will receive similar treatment on another sovereign's soil.

Wherever the exercise of reciprocity is effective in upholding the international order, it is because the police power readily available to the state is deployed to enforce a right or duty mutually agreed upon by the sovereigns. So far, our discussion has emphasized those areas of state relations where the theoretical equality of sovereigns is little challenged by their great inequality in capability. The smallest state must have roughly the same ability to protect the lives of those within its territory as the largest state; the greatest maritime state has at least as much interest in observing the rules of safety on the high seas as the state with only a tiny seagoing capability. Reciprocity helps to reinforce the equality of states by granting them identical rights and duties as they interact on a day-to-day basis. Yet it is clear that the very great differences in the power of states also has an impact on the nature of the international legal order.

# THE IMPACT OF STATE POWER ON GLOBAL ORDER

Because the international legal system is largely nonhierarchical, most of the determinations about what is lawful and unlawful rest with the very states that are likely to have an interest in the outcome. That gives a political coloration to what our experience in domestic law tells us should be more objective determinations by third parties. For example, in the name of **domestic jurisdiction,** states may assert that they cannot permit an external authority to determine the legality of their actions that presumably do not affect the outside world. That argument for years allowed the government of South Africa to assert that its apartheid policy, disenfranchising its black citizens, was not subject to review under international law. Also, many of the foreign policy disputes that arise between nations may be regarded as **nonjusticiable,** which means simply that they are likely to raise issues of a state's foreign policy, rather than of accepted international rules. Finally, even though sovereign governments agree to be bound by the treaties they have ratified, according to the legal dictum of *pacta sunt servanda,* they may reasonably insist that they should not be bound any longer if the conditions assumed when the treaty was ratified have changed fundamentally. Yet, it is the very signatories to the treaty who must determine whether this principle of fundamental change, known as the doctrine of *rebus sic stantibus,* should apply in a given circumstance.[4] In these and other cases, the judges (governments) are not disinterested, but act as judges in their own case.[5]

---

[4] *Pacta sunt servanda* requires the good faith performance of treaty obligations—that is, the expectation that parties to a treaty will attempt to abide by its purpose and terms. *Rebus sic stantibus* asserts, in contrast, that a change in the governing circumstances since the signing of the treaty may allow its parties to avoid or renegotiate their obligations on grounds that the treaty's objective is now either very difficult or impossible.

[5] John Locke's well-known assessment of what made the state of nature inadequate as a justice system was its lack of impartial judges. If each individual could judge his or her own case, then the rights of the stronger would typically prevail over those of the weaker. Thus, the first chapter of Locke's *Second Treatise of Government,* "Of the State of Nature," can be read as a still largely accurate depiction of the international system.

The Peace Palace in the Hague, the Netherlands, is the home of the International Court of Justice.

## THE SPECIAL ROLES OF GREAT POWERS

Moreover, the international system remains one in which the mighty have more freedom to do as they please than do the weak. The Concert of Europe offers a clear example of how the nineteenth-century world order system produced a group of great powers that exercised their right—which they did not share with small powers—to police Europe and much of the rest of the world. The Security Council of the United Nations, and its veto-wielding Permanent Members in particular, institutionalized a comparable principle in the twentieth century. These examples suggest that one of the clear special roles likely to accrue to great powers in the Westphalian system is this collective right to enforce the peace.[6]

---

[6]For some contemporary examples, see Thomas Ehrlich and Mary Ellen O'Connell, *International Law and the Use of Force* (Boston: Little, Brown, 1993).

**GREAT POWER INTERVENTION** The veto granted to U.N. Permanent Members is also an example of the larger measure of freedom great powers insist upon for themselves than they are willing to grant smaller states. So is the much greater inclination of the relatively more powerful to intervene in the affairs of others. They have the capability to do so, and almost always justify such intervention on grounds that it upholds some particular legal principle. For instance, the United States argued that its intervention in Vietnam in the 1960s was lawful on the grounds that it was helping South Vietnam to defend itself against the unlawful aggression of the Viet Cong under the direction of North Vietnam. When the Soviet government intervened in Afghanistan in 1979, it argued that it did so also at the invitation of the Afghans' Marxist government, which was facing an increasingly difficult civil war with rebel Muslim groups. And when Iraq invaded Kuwait in August 1990, Saddam Hussein justified his takeover on grounds that the Kuwaiti government had refused to negotiate Iraq's grievance over disputed oil rights along their border, as well as the claim that Kuwait had historically been an integral part of Iraq.

In a hierarchical legal system of the kind we are used to domestically, these claims might be tested in courts of law, with judgments enforced by the policing power of the state. But in the much weaker international legal system, these claims were advanced on the battlefield. Private citizens, as well as other governments, had to judge them, and enforcement of that judgment was left to the contending adversaries themselves. In the case of Vietnam, the final success of the North reunited the country by force of arms. Its claim—that the real legal principle at stake was the right to end its colonial domination and to achieve national unity—prevailed. In Afghanistan, the Soviet justification for its intervention seemed even more spurious, because the very Afghan president who had allegedly invited the Soviets to enter his country, Hafizullah Amin, was promptly executed in a coup engineered by the invaders. Another Marxist government was installed as the client of the Soviets; yet the war ground on, inflicting many casualties. Finally, Soviet troops were withdrawn more than eight years later with the admission by the new Soviet foreign minister, Eduard Shevardnadze, that the intervention had contravened international law.

In both these cases, the stronger, external actor claimed to be exercising a kind of police power through its intervention, in spite of the fact that among the well-established duties of states in the modern world is that they refrain from intervention in the civil conflicts of others. This suggests again that the main reason for the international order's weak system of justice is that the principal subjects of the law (the states) are also usually their own judges and enforcers. Yet, as the Vietnam and Afghanistan cases show, the "judgment" of the stronger party might not prevail if its weaker opponent can mobilize action around a more convincing cause of its own.

Note how different was the end result of Iraq's invasion of Kuwait.[7] Iraq's ability to impose its will on its much less powerful neighbor was unacceptable to greater powers than Iraq, and they chose to use their own military capability to force an Iraqi withdrawal. On these terms, the coalition with the overwhelming military power carried the day. But that power was more clearly employed to strengthen world order by halting an unlawful aggression. The collective might of the coalition that de-

---

[7]Ibid., 6–74.

feated Saddam Hussein in 1991 was a much clearer expression of the will of most of the world community (and was viewed as an exercise of the police power of the United Nations) than was the great power involvement in either Vietnam or Afghanistan.

**WHY FREEDOM OF THE SEAS?** Where the international legal order is strongest, one can usually see how the impact of state power over time has served the interests of all. That is especially clear in the history of the freedom of the seas. For several centuries, states have recognized their mutual right to use the seas for navigation and various commercial purposes, and their mutual duty to refrain from claiming sovereign jurisdiction over any of the fixed resources of the high seas—that is, those of the seabed.[8] In contrast, territorial seas, which today may extend up to twelve miles from a state's coast, are the exclusive territory of the adjacent state, unlike the high seas beyond them. All states, whether or not they possess great navies or commercial fleets, have accepted these standards and have generally abided by them since the beginning of the modern era. Why should this be, when powerful maritime states obviously have had the capability to enforce any claims they might have made to jurisdiction over wide stretches of the open seas?

The answer lies in the commercial interests of the greatest maritime states themselves. With the rise of capitalism, they recognized that their continued economic growth (a measure of their power in the world) largely depended upon trade with distant nations. It has therefore always been in their interest to treat as much of the oceans as possible as an open highway, allowing their traders free access to the rest of the world. Whatever economic advantage they might find in laying claim to the seas nearest them would be far less than what they would gain if their ships could rove freely to ports across the globe. Since the greatest seagoing states are, for that reason, those with the greatest ability to police their own claims upon the high seas, they are obviously able to prevent smaller coastal states from making good on any claims they might assert to restricting access to waters near their shores. The result has been a firmly established doctrine of free access to the high seas.

All these examples show that the relative power of a state obviously has an impact on the shape of international order, but that power is to a greater or lesser extent directed and constrained by legal norms. Such norms characteristically are advanced as protections for those who, although equal to the mighty in theory, are comparatively weaker and, therefore, in danger of losing their equal rights to the dominance of unequal power.

# INTERNATIONAL LAW AND FOREIGN POLICY

Our discussion thus far has proceeded mainly from the level of analysis of the global system. That is, it has assumed the perspective of one who observes the world as a single political system in which nation-states are merely its most important sub-

---

[8] A good summary of the law of the sea can be found in Michael Akehurst, *A Modern Introduction to International Law,* 6th ed. (London: Unwin Hyman, 1987), 168–196.

> ### Global Changes
>
> ## The Changing Face of International Law
>
> Women's rights: A number of multilateral treaties now exist with the potential to promote the improved treatment of women. It is hoped that women, roughly one-half the world's population, will now have access to international legal standards where the protections of their own state governments have failed them.
>
> Outer space: The sovereignty of the state extends to the air above it—but how far up? That question did not require an answer until after Sputnik was launched in 1957. Soon, the two original space powers, the United States and the U.S.S.R.—along with all other governments—agreed that sovereignty did not extend to outer space. Space thus became a global common. Today, more than a dozen treaties express the world community's interest in governing humanity's activities in outer space.
>
> Territorial sea: In the eighteenth century, states generally agreed that their sovereignty extended only three miles out to sea—that distance being the approximate range of a cannon shot at the time. In the 1990s, a new treaty became effective, allowing territorial seas to be expanded to a distance of twelve miles. More far-reaching was the fact that the treaty also gave all coastal states huge new exclusive economic zones, extending 200 miles offshore.
>
> Subjects of international law: In the nineteenth century, only states were the subjects, or "legal persons," of international law. Today, in addition to states, individuals, corporations, and IGOs all possess some degree of international legal personality. Some legal scholars now think it is time to empower individual citizens to carry out international obligations.
>
> - Which of the above developments reflect a change in thinking about what is appropriate subject matter for international law? Which reflect the need to treat "old" subject matter differently as the result of new human capabilities? Which reveal an evolution in our idea of the nature of international law and how it functions?

jects, with rights and duties when it comes to making, judging, and enforcing world law. Now we need to consider the standpoint of the individual state. The very fact that sovereignty is the basis of the international system means that sovereign states have wide latitude to intermix their roles as legal authorities with the foreign policy choices they make. Their policy presumably reflects their national interests, as they perceive them; yet their obligations under international law are meant to produce behavior that is less self-interested and more directed toward the common good. It may often seem that governments will always favor their selfish, political interests over their more selfless, legal obligations.

## INTERNATIONAL LAW'S UTILITY IN FOREIGN POLICY

This perspective undoubtedly induces cynicism about the ability of international law to advance justice. Even when state officials take action that we may agree upholds international law, that action is almost surely self-serving to some degree. That is particularly true if no obvious power or group of powers is present to challenge a state's interpretation of the law or enforce a different view. If it did not serve the interest of the state to uphold a particular law, its officials could quite likely find a different principle or doctrine of the law against which to justify its policy. Law therefore should be viewed as a tool of foreign policy from this point of view, rather than, in the words of Stanley Hoffmann, "as a set of rigid commands somehow independent from and superior to the political processes of statecraft."[9]

**WHEN THE STATE'S INTEREST AND THE LAW COINCIDE** First, in many kinds of foreign policy situations, states draw upon and support international normative standards because their national interests and global order interests are harmonious rather than conflicting. This is typically the case in the routine conduct of foreign policy. Bureaucracies must be run on the basis of rules and regulations, and foreign ministries routinely behave in accordance with accepted standards of international conduct; otherwise, they could not operate in the kind of predictable milieu that is essential to "business as usual" in any social setting. If some clear standard of international law is to be violated because a government has concluded it is in its interest to do so, the responsibility rests upon the highest authority, not upon the bureaucrats whose duty it is to adhere to recognized norms unless otherwise directed from above. Thus, it is no surprise that international legal rules tend to be clearest and most widely accepted as they relate to the routine conduct of foreign policy.

**LAW AS AN INSTRUMENT OF COMMUNICATION** Second, governments use international law as an instrument of communication. When they present their policy positions in legal terms, they thereby clarify to others what they regard as essential or important in their claim, and simultaneously indicate how they intend to proceed to press that claim. A dramatic example came at the outset of the Cuban Missile Crisis of October 1962.[10] When President Kennedy explained the unacceptability to the United States of allowing Soviet missiles to be emplaced on Cuban soil, he concluded his speech by ordering a naval "quarantine" of Cuba to prevent Soviet vessels to deliver their cargo of missiles to that island nation. To the layperson, the quarantine no doubt looked like a naval blockade; yet the president carefully refrained from using the term that, in the language of international law, is legal action only for a state that is at war with another. Since the United States sought to avoid allowing the missile crisis to become the opening act of war with the Soviet Union, it was essential that the president coin another term for his action.

**LAW'S USE IN CONFLICT MANAGEMENT** Third, governments rely on international law to channel conflict with other states in ways they hope will make the conflict more manageable. This may take a variety of forms, but it is often useful to states to be able to argue that they have avoided using force because international law requires it. Throughout the cold war, it was typical in time of crisis between the superpowers for each to frame its position in such a way as to put full responsibility

---

[9]Stanley Hoffmann, "Introduction," in *International Law and Political Crisis,* ed. Lawrence Scheinman and David Wilkinson (Boston: Little, Brown, 1968), xi.

[10]For a full account of the Cuban Missile Crisis, see Henry M. Pachter, *Collision Course: The Cuban Missile Crisis and Coexistence* (New York: Frederick A. Praeger, 1963).

for the initial use of force on its adversary. The Cuban Missile Crisis is a clear example, for President Kennedy's "quarantine" speech made clear that the United States would not use force if the Soviets complied with the demand to remove their missiles from Cuba. When a crisis has erupted into violence that threatens to escalate out of control, with likely unintended consequences for one or both antagonists, neutral great powers will often attempt to exercise influence to try to bring hostilities to a halt before the scope of the conflict is enlarged. This was essentially what characterized the unsuccessful Soviet effort to produce a cease-fire in the Gulf War in February 1991, at the moment when coalition forces were poised to begin their ground assault against Iraq.

**LAW'S USE IN MOBILIZING OPINION** Fourth, the state may use international law as a tool of its policy rather than because its application clearly contributes to world order. Foreign policy officials may, for example, cite international law to protect or advance their position in relation to others, as in the case of Western access to West Berlin during the cold war. The Western powers referred to the occupation agreement entered into by the Allies at the close of World War II as justifying their right of access to what became West Berlin. Clearly, that legal right was their strongest card, since West Berlin was physically surrounded by a pro-Soviet regime where thousands of Red Army troops were stationed, giving the Soviets the military advantage.

Similarly, in the 1980s, when Britain's Prime Minister Thatcher dispatched troops to the distant Falkland Islands to fend off the effort of Argentina to annex the islands, she cited Britain's right to the territory based upon its long settlement and governance of the Falklands. Argentina, in contrast, justified its attempted seizure of what it called the "Malvinas" by appealing to other legal doctrines, including the unacceptability of European colonialism in South America, Argentina's own earlier claims to the islands, and its close proximity to the islands. One can be fairly certain that when the foreign policies of two states are in conflict over a particular matter, each will seek to justify its own position at least in part by appeal to an international legal standard. Seldom, of course, is that standard identical for states with disputing claims.

Governments also seek to mobilize international support for their policies by invoking international law. When the Palestine Liberation Organization argues that Palestine has a "right" to self-determination, it is primarily appealing for public support by invoking a fundamental doctrine of the Westphalian era. Conversely, when Israel opposes that argument by insisting that the PLO's real goal is the unlawful elimination by force of the "nation" of Israel, its appeal to the same doctrine is apparent. In ordering U.S. troops to invade Panama in December 1989, President George Bush sought public support both by giving the operation the value-laden name "Just Cause" and by emphasizing the illegal activities of the country's leader, General Noriega, whom he accused of drug trafficking.[11] The president was obviously interested in countering the fact that an invasion by one country of another is likely to be regarded by many as one of the most serious wrongs possible under international law. However, Bush was able to use the unlawful aggression argument to make his own case against Saddam Hussein's invasion of Kuwait in 1990.

---

[11]See Ehrlich and O'Connell, *International Law and the Use of Force*, 75–107.

## THE CONFLICT BETWEEN FOREIGN POLICY AND INTERNATIONAL LAW

It may also be said that legal rules and standards play a negative role in foreign policy. States sometimes ignore or violate international law, either because it does not suit them in a particular instance or because they wish to change the law. Such behavior is closely related to taking advantage of conflicting legal doctrines to advance one's cause, and occurs when governments wish to make the point either that applying the legal doctrine will have negative effects, or that old rules are outmoded.

**WHEN GOVERNMENTS IGNORE LEGAL RULES** Governments sometimes find it mutually useful to ignore legal rules in order to achieve a political settlement of a controversy between them. Through the first half of the twentieth century, when socialist revolutions resulted in the expropriation of property held by foreign investors, the United States and other Western nations typically insisted that international law demanded compensation in accordance with standards established by the capitalist world. Revolutionary governments disagreed—frequently on grounds that the very revolutions that had brought them to power were motivated by popular opposition to foreign ownership of much of the country's wealth, and that they should be compensated for past economic exploitation. When an accommodation eventually was reached (for both parties had an interest in resuming normal diplomatic relations), they usually agreed to settle for limited compensation. This amounted to a political compromise in practice, while it allowed the capitalist government to cling to its insistence that the legal principle of prompt, adequate, and effective compensation still was in place, presumably for future cases.

**WHEN GOVERNMENTS VIOLATE LEGAL RULES** Governments sometimes deliberately violate a traditional rule of law to induce world order change. For example, anticolonial and other revolutionary governments in the twentieth century have made a point of opposing the rules created by imperialist and capitalist powers with a monopoly on the Westphalian legal system. When, for example, President Castro came to power in Cuba, he offered to compensate U.S. nationals with sugar holdings in his country in accordance with the very low assessments (favorable to them for tax purposes) made of those properties under the old regime his revolutionary forces had overthrown. That was unacceptable to the property owners and to the government of the United States, and a rupture of diplomatic relations followed. Nonetheless, Castro's refusal to accept the traditional capitalist rule in such matters was consistent with his view that the old rule had worked an injustice on Cuban society.[12]

This example reminds us that much of what the older states insist is binding in international law amounts to rules they alone created in an earlier period. As many more states have entered the international system in the twentieth century, they naturally have objected to the argument that they are bound by rules that not only were

---

[12] See commentary on the case of *Banco Nacional de Cuba v. Sabbatino, American Journal of International Law* 55 (1961), 822–824.

designed without their consent but were often intended to prevent them from participating as independent actors in world politics.

**RETALIATION AND REPRISAL** This discussion of the varied reasons why states heed international law has not included what may sometimes be the most obvious reason of all—the threat of retaliation or reprisal by other sovereigns against a state that violates a legal standard. Governments may argue that such retaliation is itself unwarranted or unlawful, but if they do so—as did Saddam Hussein in 1990–1991—they may have to face the consequences of police action by others to rectify the alleged wrong.

International law, like all law everywhere, is a conservative social force, which helps explain why those most desirous of social and political change may resent and resist submitting to its authority. But there is another truth about the law, which is that in conserving social order, it provides needed predictability and peaceful competition for the allocation of values. Even in the relatively anarchic international system, constraints probably have been strengthened in the years following the cold war

---

### Normative Dilemmas

## Making Law Conform to Social Norms

Ending colonialism: In the 1950s, the then-new state of India wished to annex the tiny Portuguese colony of Goa, because it was part of the Indian "nation" and entirely surrounded by Indian territory. Portugal refused, citing a 300-year-old treaty for its right to remain in Goa. The World Court agreed with Portugal. India then seized Goa forcibly, arguing that European colonialism was no longer an accepted norm in contemporary international practice. Since 1961, Goa has been a part of India.

- What are the factors that allowed the Indian position to prevail in this case? To what extent was India's normative claim a plausible one? How does the outcome of this case impact upon India's argument that colonialism had become unlawful by the mid-twentieth century? Can you think of similar conflicts in world politics today between an "outmoded" international rule and changing social norms?

Policing fishing: In 1995, Canadian patrol vessels seized a Spanish trawler that was sailing in international waters on the grounds that it was catching too much turbot. The European Union, speaking for Spain, called Canada's actions "international piracy," because the seizure was beyond Canada's 200-mile economic zone. But Canada's action was directed by a Canadian law giving its government the right to seize ships that overfish on the high seas off Canada.

- Canada's seizure of the Spanish trawler was lawful according to Canadian law. Should it also be considered lawful on the basis of emerging international norms? Why or why not? This "fishing war" was resolved in April 1995, by a Canada–EU agreement that included (1) smaller quotas for turbot than those previously agreed to, and (2) tougher inspection and enforcement of fishing on the high seas. Does this outcome support the view that a tougher norm against overfishing is emerging? If so, what are the reasons for this development? Does this case suggest that the patrol boats of any country have a right to police international waters, and, if so, for what purposes?

against the reign of brute force and in favor of peaceful change.[13] President Gorbachev put the matter clearly at the Soviet Union's 28th Communist Party Congress in July 1990. When attacked by Party conservatives for having allowed the countries of Eastern Europe to leave the Soviet orbit during the dramatic period six to eight months earlier, he asked, "What would you have had me do? Direct an invasion to hold them by force?" Gorbachev's predecessors had done just that, violating perhaps the most important norm of international law when they invaded Hungary in 1956 and Czechoslovakia in 1968. Yet even Gorbachev's critics did not suggest that such action should be repeated in 1990.

# JUDICIAL INSTITUTIONS

Given the nonhierarchical, largely horizontal nature of the international legal system, it should not be surprising that its judicial institutions are weaker than are those within states, where a more vertical, or hierarchical, authority prevails.

## ARBITRATION

Arbitration is one technique for settling disputes under international law that has been used for more than a century. Yet, no one would claim that arbitration is often utilized for the major peace-threatening issues that confront the world.

**Arbitration** is essentially a judicial proceeding, for the arbitrators are required to apply specified legal rules to address an issue.[14] In the twentieth century, a number of standing arrangements have been created to provide arbitration when disputes arise over particular kinds of issues between countries. An example is the Arbitral Tribunal created in 1935 by the United States and Canada to determine responsibility for such matters as transboundary pollution along their mutual border.

Frequently, however, an arbitration panel is created by the parties to a particular dispute for only that particular situation, and the judges may be restricted to addressing only one narrow set of issues, rather than to all of those that might underlie the conflict. While the decision produced by arbitration is meant to be binding on the parties, it has resulted from the sovereign consent of the disputants to try to reach a settlement. The same parties must then be presumed to want a settlement, since only they have the ability to enforce it. For example, consider an arbitration decision in 1968 that helped end armed hostilities between India and Pakistan.[15] A three-judge tribunal divided territory between the two countries that had been in dispute ever since the birth of an independent India and Pakistan some twenty years

---

[13]See Bruce Russett, *Grasping the Democratic Peace: Principles for a Post–Cold War World* (Princeton, N.J.: Princeton University Press, 1993).

[14]Adam Boleslaw Boczek, *Historical Dictionary of International Tribunals* (Metuchen, N.J.: Scarecrow Press, 1994).

[15]*India v. Pakistan*, Case Concerning the Indo-Pakistan Western Boundary (Rann of Kutch) between India and Pakistan, Award of February 19, 1968, *United Nations Rep. Intl. Arbitral Awards* 1 (1980).

before. Even though Pakistan won more territory than India, the arbitrators' decision was respected and new boundary markers were put in place.

# THE WORLD COURT

An understanding of the role of the **International Court of Justice (ICJ),** or World Court, also reveals the basic lack of hierarchy in the structure of the international system. Dating from the creation of the United Nations in 1945, the ICJ replaced the Permanent Court of International Justice established with the League of Nations in 1920. The World Court is located in the Hague in the Netherlands. Its fifteen members are distinguished jurists, each of whom must be of a different nationality. They are elected by the United Nations so that "the representation of the main forms of civilization and . . . the principal legal systems of the world" are included.[16] (It is ironic that, in spite of such inclusive language, not one of the sixty judges elected to the Court since 1947 has been a woman.)[17]

Only states may be parties in cases brought before the Court. Given the doctrine of sovereign equality, the impact the Court has been able to have upon the global legal order has been severely limited. Since states are viewed as sovereign equals when they come before the ICJ, their appearance is essentially voluntary, for they cannot be coerced into submitting to the Court's jurisdiction.[18] That, in turn, assures that cases the Court will be asked to hear are usually those that each party to the dispute assumes it has a good chance of winning. This means that highly charged political conflicts, such as those typical of the U.S.–Soviet rivalry during the cold war period, are not likely to be submitted to the Court's adjudication. Consequently, the ICJ never experiences the crowded docket of cases typical for domestic courts in the United States. There have been periods when it had no business before it at all, although its caseload has increased in recent years. Early in 1996, nine contentious cases were pending before the ICJ.

Whatever its jurisdictional limitations, the World Court's decisions do carry weight. They may influence even those governments that assert they won't be bound by a particular decision of the Court. In 1984, the goal of the Reagan administration was to help bring Sandinista rule to an end in Nicaragua. When it became known that the United States had mined Nicaragua's harbors, the government of Nicaragua went to the ICJ to charge that the U.S. action violated international law, as did U.S. support for those who were attempting to overthrow the Sandinistas. Three days before the complaint was brought, the Reagan administration announced that for the next two years it would not submit to the jurisdiction of the ICJ in any Central American dispute. Even so, the Court agreed to hear the case, denying the right of the United States to excuse itself. So the case went forward without the participa-

---

[16] Article 9, statute of the International Court of Justice (executed at San Francisco, June 26, 1945).

[17] Dorinda G. Dallmeyer, ed., *Reconceiving Reality: Women and International Law* (Washington, D.C.: The American Society of International Law, 1993). More remarkably, out of the 128 persons elected to the United Nations' International Law Commission (ILC) since 1949, not one of them has been a woman!

[18] Note the paradox: International law upholds the principle of the equality of states, but that same equality actually weakens the jurisdiction of the ICJ!

**"EVENTUALLY? WHY NOT NOW?"**
Jay N. Darling. Reprinted with permission. *The Des Moines Register* and Tribune Company, 1945.

tion of the "defendant."[19] The Court's decision largely upheld the Nicaraguan contention and found to be illegal a number of the U.S. actions that had prompted Nicaragua to file charges.[20]

Yet, the U.S. government actually complied with much of the Court's ruling, even while it insisted that it was not really bound. Even before the case was brought, the United States had stopped mining Nicaragua's harbors—largely because of the outcry that action had caused when it leaked out within the United States. When the Court ordered the United States not to hinder access to Nicaragua's ports, the

---

[19]As a mark of the fact that states, and not individuals, are the only parties who may engage in litigation before the ICJ, they are termed *applicants* and *respondents* rather than plaintiffs and defendants in World Court actions.

[20]*Nicaragua v. United States of America*, Case Concerning Military and Paramilitary Activities in and against Nicaragua. For a summary of the Court's judgment, see *American Journal of International Law* 80, 3 (July 1986): 785–807.

State Department announced that the United States would comply. While the case was in court, the U.S. Congress cut off additional funding to the CIA because of the mining operation. The end result was that, by the time the ICJ decision was announced in 1986, the U.S. government's actions against the Nicaraguan government had largely ended.

The end of the cold war may generate a somewhat greater use of the ICJ than in the past. Still, it is doubtful that the Court will soon assume a place of importance in the global legal order remotely comparable to that of, say, the Supreme Court of the United States within the American legal system. The nonhierarchical principles of Westphalia remain much too firmly in place for that.

# CONCLUSION

We have considered the principal reasons why international law is generally weaker than intranational, or domestic, law, noting that the very concept of sovereignty in the modern world militates against the idea that sovereigns should themselves be bound by a higher authority. Yet, because sovereign governments also have an interest in the orderly, predictable behavior that law provides, they have created an international legal order in which they are not only the law's principal subjects, but also its chief legislators, enforcers, and interpreters. The international legal order, in contrast to the more familiar domestic legal systems, is essentially nonhierarchical; it coordinates state behavior more than it subordinates it, and the structure of the legal system is largely horizontal rather than vertical.

The role of law in world politics can be conceived of in three different ways, each of which has an essential, but somewhat contradictory, purpose when considered in conjunction with the others. The first is to maintain the system of sovereign equality as the foundation upon which the international system rests. This describes international law in its narrowest, most literal sense as the body of treaties and customs by which sovereign governments have consented to be bound. The second is to rationalize the alignment of political capability with legal norms, which is what joins power ("real" capability) with authority (defined as the "ideal" or formal grant of competence to govern) sufficiently to constrain behavior in various predictable ways. This describes the study of international law much more broadly than is possible by focusing solely on the positive laws consented to by sovereigns. The third is to extend justice to human beings either with or without the assistance of sovereign governments themselves. It describes the precept that any concept of justice (i.e, moral rightness) must focus on human beings, rather than such abstractions as states, as its subjects of concern (see the next chapter for a fuller discussion of human rights).

In terms of the law's role in foreign policy, (1) governments may abide by international law out of a self-interest that in fact is in harmony with the general interests of world order; (2) they may use international law to try to advance their particular foreign policy interests at the expense of another state's claim; or (3) they may ignore or violate an international law either because they agree to a political rather than a legally binding solution to their conflict, or because they wish to advance the argument that the relevant legal rule is unjust.

Finally, we noted how the principles of sovereign equality and the essentially nonhierarchical nature of the international system impacted on judicial authority for the world. Arbitration, a special kind of judicial technique, is only occasionally useful in resolving some kinds of disputes. The International Court of Justice can only adjudge cases that states have voluntarily brought before it, which typically prevents it from having a major impact on the most volatile and war-threatening issues of world politics.

CHAPTER 16

# Human Rights

By the end of the twentieth century, human rights issues had become more prominent on the agenda of world politics than ever before in history. In strict Westphalian terms, this development was something of a paradox. Sovereigns presumably cannot judge how other sovereigns treat the individuals over whom they rule. So why are such judgments a growing feature of world politics today?

## BRINGING JUSTICE TO HUMAN BEINGS

Two characteristic trends and features of the contemporary world have figured prominently in this text, both of which largely explain the new prominence of human rights in global politics. The first is humanity's ever growing interdependence. Today it is possible for groups, and even individuals, to penetrate the once nearly impermeable walls of nation-states to help bring human rights protection to those who are abused by their own political elites. The fact that we increasingly live in a global village means that we can immediately know, in a way our ancestors could not, about the mistreatment of human beings in distant parts of the globe. It also allows us to use that same power of communication and action to bring pressure to bear to correct such situations. For centuries, we necessarily supposed that whatever improvements were to be made in advancing human dignity had to come from within particular states and largely be confined to the life of a single polity. Today we have begun to grasp that human rights and aspirations are not only common to all of us in theory; perhaps they now can—even must—be advanced globally in practice if the rights of any of us are to be maintained.

The second trend is the development of technologies that can be used to abuse, as well as serve, humankind. When these technologies are exercised by political groups intent on dominating others, they may result in an abuse of human rights on a scale unimaginable to the tyrants of the past. Genghis Khan is remembered as one of the cruelest conquerors in history; yet his Golden Horde mostly could only butcher hapless victims one at a time, with pikes and swords.[1] Today, some in positions of power are capable of wiping out whole populations at a single command, while almost any of us has the capability to commit individual acts of terrorism that can result in the loss of hundreds, even thousands, of innocent lives in an instant.

## CREATING INTERNATIONAL STANDARDS FOR HUMAN RIGHTS

The twentieth century has produced some of the most murderous governments in history. Adolf Hitler's orchestration of a genocidal campaign against Jews, Slavs, and other minorities may remain the most obscene example. For the distinction of being history's greatest mass murderer, however, he has had such rivals as Joseph Stalin and Cambodia's Pol Pot. Nor should we forget that millions of innocent civilians—far more than in the previous two or three hundred years—have also become the victims of this century's use of instruments of mass destruction in time of war. While concepts of "military necessity" are generally used by apologists for warring governments to justify behavior that would be unacceptable in time of peace, from the point of view of the victims, the violation of their right to live is clear.

At the close of World War II, when the dimensions of the Nazi Holocaust became clear, the architects of the post–World War II order began to create international standards for the conduct of governments. The Nuremberg and Tokyo war crimes trials—carried out on the premise that universal standards of acceptable conduct had been violated by the alleged war criminals—showed that treaties were needed to clarify what those standards were. The organizational meeting for the United Nations, which took place in San Francisco in the spring of 1945, considered creating an international bill of rights as an adjunct to the U.N. Charter. But it was concluded that the effort would delay completion of the Charter itself, and so should be undertaken as soon as the new international organization had begun its work, in January 1946.

**THE UNIVERSAL DECLARATION OF HUMAN RIGHTS** The first result of the effort to define worldwide standards of human rights came with the adoption by the U.N. General Assembly of the **Universal Declaration of Human Rights** in December 1948. The Declaration was *not* a treaty requiring governmental ratification that would then bind governments to its standards. That was because the whole nature of this standard-setting enterprise seemed so radical to many of those representing governments at the United Nations that they preferred a nonbinding declaration of intent as the easier, first step. Thus, the Universal Declaration proclaims itself "a common standard of achievement for all peoples and all nations, to the end that every indi-

---

[1] Despite primitive instruments of warfare in the thirteenth century, Genghis Khan's armies were responsible for 1,600,000 deaths within a week of the fall of Herat, a city in today's Afghanistan.

vidual and every organ of society . . . shall strive . . . to promote respect for these rights and freedoms."[2]

The Universal Declaration is a relatively brief document. Many of its thirty articles set out the kinds of civil and political protections from arbitrary government that were familiar in Western constitutional systems.[3] Others address economic, social, or cultural standards of the sort that in most Western countries have more typically been addressed through legislation than by constitutional mandates. Examples are expressed as rights to join trade unions, to periodic holidays with pay, and to free primary education.[4]

**INTERNATIONAL COVENANTS ON CIVIL AND POLITICAL RIGHTS AND ON ECONOMIC, SOCIAL AND CULTURAL RIGHTS** It would be many years after promulgation of the Universal Declaration before not one, but two, treaties on the subject finally were completed and open to ratification by member states. These are the **International Covenants on Civil and Political Rights,** and on **Economic, Social and Cultural Rights,** both of which became operational in 1976. By the early 1990s, some one hundred states had signed or ratified the two Covenants, which spelled out in somewhat greater detail the standards of the Universal Declaration. Since they were treaties rather than just declarations of intent, the Covenants were binding for every state that ratified them.

The obligations assumed under the two Covenants were not identical. Political and civil rights were viewed as essentially restraints against governmental intervention in the lives of people. As such, they could theoretically be protected by any conscientious government that adhered to the relevant set of standards. Economic, social, and cultural rights, on the other hand, more frequently require governmental activism to achieve them. When, for example, governments agree to the right of every adult to work at meaningful employment or to adequate housing for all, they commit themselves to undertake and sustain social policies that require governmental intervention in the lives of people.[5]

**OTHER U.N. TREATIES ON HUMAN RIGHTS** Meanwhile, the world community has undertaken other efforts to protect human rights. In addition to the general human rights standards set down in the two Covenants discussed above, the United Nations has generated some twenty treaties in the past half century that address a range of

---

[2]Preamble, *Universal Declaration of Human Rights,* Dec. 10, 1948, U.N.G.A. Res. 217 A (III), U.N. Doc. A/810, at 71 (1948).

[3]Typical examples are these: "Everyone has the right to life, liberty and the security of person" (Article 3); "No one shall be subjected to arbitrary arrest, detention or exile" (Article 9).

[4]"Everyone has the right to form and to join trade unions for the protection of his interests" (Article 23[4]); "Everyone has the right to rest and leisure, including . . . periodic holidays with pay" (Article 24); "Everyone has the right to education. Education shall be free, at least in the elementary and fundamental stages . . ." (Article 26).

[5]For a discussion of the Universal Declaration, the two Covenants, and other human rights treaties produced by the United Nations, see David Forsythe, "The United Nations and Human Rights, 1945–1985," *Political Science Quarterly* 100 (summer 1985): 249–269.

more specific issues. These include such problems as racial discrimination, the rights of women, of refugees and stateless persons, the suppression of terrorism, and the prohibition of torture (see Table 16.1).

The United Nations has occasionally convened special treaty-making conferences, some of which have dealt with human rights standards. More typically, however, these treaties have been generated within the standing bodies of the U.N. system. Here, the process has usually been lengthy. It is customary for a draft treaty to originate in the Commission on Human Rights, then be reconsidered by the Commission's parent body, the Economic and Social Council (ECOSOC). After that, it will likely undergo further scrutiny and revision in the General Assembly's Third (Social, Humanitarian, and Cultural) Committee before it is approved, or "adopted," by the General Assembly in a plenary session. Only then is the document open for ratification by member governments in accordance with their various constitutional requirements. That, too, is often a slow process, since a government's vote to approve a treaty in the General Assembly is no assurance that it will then ratify the treaty speedily, if at all.[6]

A glance at Table 16.1 shows that human rights treaties typically enter into force some years after they are opened for ratification. That delay reflects the time it takes for individual states to ratify the treaty but also the usual requirement that some minimum number of states must ratify them. For example, both of the general Covenants required thirty-five ratifications before they could begin to operate, a process that took about ten years.

## THE U.N. FRAMEWORK FOR ENFORCING HUMAN RIGHTS

A few treaties have required the creation of new international institutions to implement their provisions. Thus, the Covenant on Civil and Political Rights provides for the election of an eighteen-member **Human Rights Committee** whose members take no instruction from governments. The committee reviews annual reports from states and listens to petitions from individuals whose states have accepted an optional protocol that permits such petitions. Within a few years of the committee's creation, national laws in both Sweden and Senegal reportedly were changed as the result of questioning by the committee. In the words of one authority, "the committee has been energetic and assertive, seeking to make the review process as rigorous as possible, but staying within the bounds of a generally cooperative attitude toward states."[7] Its work has provided an interesting example of how an inter-

---

[6]The United States has been particularly slow to ratify a number of human rights treaties, partly because of Constitutional jealousies over the division of legislative and treaty-making powers between Congress and the Executive. See William Korey, "Human Rights Treaties: Why Is the U.S. Stalling?" *Foreign Affairs* 45, no. 3 (April 1967): 414–424; and Vernon Van Dyke, *Human Rights, the United States, and World Community* (New York: Oxford University Press, 1970). For example, the United States ratified the U.N.'s Genocide Convention only in 1988, some forty years after President Truman first submitted it to the Senate. In 1992, the Senate consented to ratification of the *Covenant on Civil and Political Rights*. That came twenty-six years after the treaty was completed and fifteen years after President Carter had submitted it to the Senate for ratification.

[7]Forsythe, "The United Nations and Human Rights, 1945–1985," 254.

**TABLE 16.1** Major Human-Rights Treaties, with Ratifications

UNITED NATIONS
**The International Bill of Rights***
1. International Covenant on Economic, Social, and Cultural Rights (1966) [1976]†
2. International Covenant on Civil and Political Rights (1966) [1976]

**Slavery, Servitude, Forced Labor, and Similar Practices**
1. Slavery Convention (1926) [1927]
2. Protocol Amending the Slavery Convention (1953) [1955]
3. Supplementary Convention on the Abolition of Slavery, the Slave Trade, and Institutions and Practices Similar to Slavery (1956) [1957]

**War Crimes and Crimes against Humanity**
1. Convention on the Prevention and Punishment of the Crime of Genocide (1948) [1951]
2. Convention on the Non-Applicability of Statutory Limitations to War Crimes and Crimes against Humanity (1968) [1970]
3. Convention on the Taking of Hostages (1979) [1983]

**Prevention of Discrimination**
1. International Convention on the Elimination of All Forms of Racial Discrimination (1965) [1969]
2. International Convention on the Suppression and Punishment of the Crime of Apartheid (1973) [1976]
3. Convention on the Political Rights of Women (1952) [1954]
4. Convention on the Elimination of All Forms of Discrimination against Women (1979) [1981]
5. International Convention against Apartheid in Sports (1985)

**Protection of Persons Subjected to Detention or Imprisonment**
1. Convention against Torture and Other Cruel, Inhuman, or Degrading Treatment or Punishment (1984)

**Freedom of Information**
1. Convention on the International Right of Correction (1952) [1962]

**Nationality, Statelessness, Asylum, and Refugees**
1. Convention on the Nationality of Married Women (1957) [1958]
2. Convention on the Reduction of Statelessness (1954) [1975]
3. Convention Relating to the Status of Refugees (1967) [1967]

**Marriage and the Family, Childhood, and Youth**
1. Convention on Consent to Marriage, Minimum Age for Marriage, and Registration of Marriages (1962) [1964]
2. Convention for the Suppression of the Traffic in Persons and of the Exploitation of the Prostitution of Others (1949) [1951]

THE INTERNATIONAL LABOUR ORGANIZATION
1. Forced Labour Convention (1930) [1932]
2. Abolition of Forced Labour Convention (1957) [1959]
3. Freedom of Association and Protection of the Right to Organize Convention (1948) [1950]
4. Right to Organize and Collective Bargaining Convention (1949) [1951]
5. Workers' Representatives Convention (1971) [1973]

**TABLE 16.1** *(continued)*

6. Equal Remuneration Convention (1951) [1953]
7. Discrimination (Employment and Occupation) Convention (1958) [1960]
8. Employment Policy Convention (1964) [1966]

UNITED NATIONS EDUCATIONAL, SCIENTIFIC, AND CULTURAL ORGANIZATION (UNESCO)
1. Convention against Discrimination in Education (1960) [1962]

REGIONAL CONVENTIONS
Council of Europe: European Convention on Human Rights (1950) [1953]
Organization of American States: American Convention on Human Rights (1969) [1978]
Organization of African Unity: Banjul Charter on Human and People's Rights (1981) [1987]

*The Universal Declaration of Human Rights (1948), while not a treaty, is also considered an integral part of the "international bill of rights."
†The year in which the treaty was adopted appears in parentheses (); the year in which the treaty became effective is bracketed [].

governmental body can interact with governments to persuade them to change practices that do not conform to treaty standards they are obligated to uphold.

A second uninstructed body (that is, whose members do not take instructions from their governments) is the U.N.'s **Subcommission on Prevention of Discrimination and Protection of Minorities.** Over time, the Subcommission has proved willing to use public pressure against governments to which it objects. It employs working groups to tackle problems that occur outside its regular sessions. It also reviews communications from individuals and groups that consider themselves discriminated against under the terms of relevant treaty law.

The **Human Rights Commission,** elected by the United Nations Economic and Social Council (ECOSOC), currently consists of forty-three U.N. member states, whose views they are expected to reflect. Since the 1980s, the Commission has become increasingly active in protecting civil and political rights. With the end of the cold war, members of the Commission from the former Soviet bloc were no longer opposed to such activities, and the Human Rights Commission seemed to assume an even bolder role. It is now the norm, rather than the exception, for Commission members to support initiatives to protect the human rights of citizens against violations by their own governments.

While these are some of the most important structures in the U.N. system for advancing human rights, there are many others. Some act in an uninstructed capacity as experts in the human rights field. They include the *Committee of Experts of the Racial Discrimination Convention,* the *Commission on Women,* the *Commission on Discrimination against Women,* the *High Commissioner for Refugees,* and the *High Commissioner for National Minorities.* There are also U.N. administrators in the

Secretariat, from the Secretary-General on down, with responsibilities for implementing human rights policies.

U.N. human rights activities have taken off in recent years in large part because of the increased presence of nongovernmental organizations (NGOs). *Amnesty International, Human Rights Watch,* the *International League for Human Rights,* and the *International Commission of Jurists* are a few of these. They work as powerful pressure groups because information is more readily available than ever before about much inhumane treatment of individuals by governments. The age of communication in which we live also gives these NGOs the ability to publicize their concerns, both directly to offending governments and to U.N. bodies responsible for their correction.

Finally, this complex framework for addressing human rights issues is strengthened from time to time by relevant international conferences, which themselves assess progress and provide new directives and goals. The U.N.'s 1993 *World Conference on Human Rights* in Vienna is an important recent example.[8]

The declaration produced at the Vienna Conference asserted the "universality" of human rights in spite of attempts by some to insist that they might vary in strength and salience in different cultural contexts. It linked democracy, development, and respect for human rights as "interdependent and mutually reinforcing." It declared its support for the cultural identity of indigenous peoples (calling for completion of a declaration on the rights of such peoples), included a section on the "Equal Status and Human Rights of Women" (anticipating the *World Conference on Women* held in Beijing in 1995), and called for the United Nations to create a **High Commissioner for Human Rights** to coordinate all the human rights activities within the U.N. system.

## REGIONAL EFFORTS TO ADVANCE HUMAN RIGHTS

Human rights protections have also been advanced regionally, most notably in Europe, beginning soon after World War II. The most important European approach grows out of the **European Convention for the Protection of Human Rights and Fundamental Freedoms,** which became effective in 1953. Drafted largely in the effort to ensure that a recurrence of the Nazis' crimes against humanity would never be repeated, it has created a strong institutional system in Europe—consisting of a Council of Ministers, a Commission, and a Court of Human Rights—with a genuinely supranational authority. More than once, for example, the British government has had to answer charges that certain laws it had enacted to aid in suppressing Irish separatism in Northern Ireland were violations of the rights of individuals under the terms of the convention.

With the end of the cold war, the Czech Republic, Slovakia, Hungary, and other Eastern European states formerly within the Soviet bloc came under the ju-

---

[8]A full account of the work of the Vienna Conference can be found in Charles H. Norchi, "Human Rights," *Global Agenda: Issues before the 48th General Assembly of the United Nations* (Lanham, Md.: University Press of America, 1993), 213–239.

risdiction of the European Court when they joined the **Council of Europe,** the Court's parent body. Another trans-European body, the **Organization for Security and Cooperation in Europe (OSCE),** joins fifty-four nations—almost all the former NATO and Warsaw Pact antagonists—that have promised to protect human rights in the region.[9] The organization's *Charter of Paris,* signed in November 1990, affirms that protecting human rights is a valid international concern of the members, obligating their governments to respect such rights at home.

In 1991, the OSCE states authorized investigations of member countries' human rights practices without their consent. It was an ironic comment on how rapidly the world had changed that the government of the USSR (then still clinging to life) and the infant democracies of Eastern Europe supported sanctions against member states violating human rights, while many Western governments opposed such action. The final document clearly abandoned the long-held principle prohibiting interference in the affairs of sovereign states, declaring human rights to be "the legitimate concern" of all countries. Under the policy adopted, three-member fact-finding teams would investigate human rights allegations in a member state without its consent if ten countries called for an inquiry, or if it were authorized by a committee of senior OSCE officials.

In the Americas, the American Declaration of the Rights and Duties of Man was adopted by the members of the **Organization of American States** in 1948. It formed the basis for a treaty, the **American Convention on Human Rights,** which became effective in 1978 and established an inter-American Court of Human Rights. Some two-thirds of the thirty-five OAS members had ratified the treaty by the 1990s, although the United States was not among them. Proponents of the convention argue that it is essential to overcome the history of authoritarianism, rule by military dictatorships, and the like in much of Latin America. When the Court was established in 1978, most OAS members were under authoritarian rule. But by the time a democratically elected government was restored in Haiti in 1994, only Cuba's government remained classified as authoritarian. The Court's caseload has increased significantly in recent years.

In Africa, too, human rights issues are beginning to be addressed on a regional basis. In 1981, the **Organization of African Unity (OAU)** adopted the **African Charter on Human and People's Rights,** known as the Banjul Charter. It created a Commission patterned after that of the U.N. Human Rights Commission, to which member states were to make detailed reports on their human rights activities at home. The Commission's work has been hampered both by lack of adequate financial support and a failure by some states to comply fully with the reporting obligation.[10] Nonetheless, one of the observable trends of the early 1990s was the growth in the number of competitive political systems in Africa—a process that could stimulate the development of human rights machinery as well.

---

[9]The original name of the OSCE was the Conference on Security and Cooperation in Europe (CSCE). The name was changed in 1994 to reflect its increasingly formalized existence as an intergovernmental organization. The OSCE headquarters are located in Vienna.

[10]See Claude E. Welch Jr., "The African Commission on Human and Peoples' Rights: A Five-Year Report and Assessment," *Human Rights Quarterly* 14, no. 1 (February 1992): 43–61.

> **Global Changes**

> # The Changing World of Human Rights
>
> ## Rights vs. Government
>
> The idea of human rights should intimidate governments or it is worth nothing. If the idea of human rights reassures governments, it is worse than nothing.*
>
> - Do you agree that governments necessarily threaten human rights? What techniques have some societies devised to try to assure that governments protect the rights of all their citizens?
>
> ## Migration
>
> International migration is moving to the top of the foreign policy agenda. . . . In global terms, international migration movements have been large and on the rise over the past decade, and the numbers designated as refugees and displaced persons have increased sharply.†
>
> The right of people to migrate from their native country is strongly suggested in the Universal Declaration of Human Rights (Article 13). High migration scenarios are forecast through the year 2030. By that year, a net number of 2,000,000 migrants may have been moved to North America; 1,000,000 to Western Europe; and 350,000 to Japan, Australia, and New Zealand.‡
>
> - Why are large-scale migrations a problem for governments? How are human rights issues raised by migration?
>
> ## Women's Rights
>
> Women have achieved political and civil rights in much of the world (not until the twentieth century did they begin to acquire the right to vote). Yet, this does not mean that women have achieved genuine equality with men.
>
> > The major forms of oppression of women operate within the economic, social, and cultural realms.
>
> Economic, social, and cultural rights are traditionally regarded as a lesser form of international right and as much more difficult to implement.§
>
> - Can you think of examples of how women are oppressed in "economic, social, and cultural realms"? Which of the treaties in Table 16.1 attempt to address the human rights issues specifically of women?
>
> ## War Crimes
>
> In 1993, the United Nations established an international tribunal to hear cases stemming from violations of international humanitarian law in the former Yugoslavia. That was the first such "war crimes" tribunal established since Nuremberg, in 1945. It was followed by the 1994 U.N. tribunal to try people accused of genocide and other crimes in Rwanda.
>
> - What kinds of crimes are most prominent among those being prosecuted in these two cases? Which of the treaties listed in Table 16.1 designates such conduct as an international crime?
>
> ---
>
> *Philip Allott, *Eunomia: New Order for a New World* (New York: Oxford University Press, 1990), 288.
>
> †*Threatened Peoples, Threatened Borders: World Migration and U.S. Policy*, Final Report of the Eighty-Sixth American Assembly, November 10–13, 1994 (New York: W. W. Norton, 1994), 3–4.
>
> ‡Wolfgang Lutz (Director, Population Project, IIASA, Austria), "The Future of World Population," *Population Bulletin* 49, no. 1 (June 1994).
>
> §Hilary Charlesworth, Christine Chinkin, and Shelley Wright, "Feminist Approaches to International Law," *American Journal of International Law* 85 (1991), 635.

# A Human Rights Consensus?

Human rights can never be secured at all times in all places. The very nature of our evolving life on the planet makes that impossible, for changes threaten some rights while advancing others. The standard-setting and implementation efforts sketched above are similar at the international level to those in the United States that began with the creation of a Bill of Rights in 1792 and continue in the nation's legal system to the present. The struggle for human rights is at least partly a matter of creating a consensus as to what is and is not acceptable treatment of human beings—not just within a particular country, but everywhere. Only with such a consensus, so the reasoning goes, can one expect predictable and effective enforcement of those standards to follow.

The kind of consensus that may advance human rights seems, paradoxically, to grow out of domestic political systems where vigorous competition over values is the rule. That is because sustained political competition requires a respect for the rights of minorities if they are to be able to continue to compete for majority status. Such respect is at the heart of all human rights. When the world is divided between popularly governed states with vigorous political competition and autocratically ruled states where participation is curtailed, the human rights agenda may be highly politicized, restricted to little more than name-calling and posturing. So it was through much of the cold war. Even now, the post–cold war shakedown of the world is not complete. Therefore, it is no doubt too soon to be certain that a liberal democratic consensus can be sustained across enough of the globe to strengthen the tentative agreements concerning human rights issues.[11]

At the moment, the most intriguing—and difficult—questions are these: Has the growing web of declarations, organizations, conferences, and treaties dealing with human rights actually pushed authoritarian governments toward democratic reform? Have the precedents for transnational attention to human rights grown strong enough that we may expect more vigorous and effective action to correct flagrant human rights abuses by governments?

It is impossible to determine a cause-and-effect relationship between the world's attention to human rights and the apparent trend toward greater democratization in the world. In fact, the durability of that trend is itself not clear from today's perspective. The evidence sought depends upon the time frame chosen. For example, a study published in 1981 argued that in the two decades from 1960 to 1980, militarized and antidemocratic governments had increased nearly threefold throughout the world, from about sixteen to more than forty. There had also been an additional doubling, to nearly ninety, in the number of more or less repressive regimes.[12]

But during the 1980s, a dozen or more of these regimes were replaced with more progressive ones (Argentina and the Philippines are examples), or were made more accountable to their own citizens (South Korea and Chile). Then in the 1990s, these and other countries generally saw democratic processes strengthened, as more competitive political systems also grew in parts of Africa and the former Soviet bloc of states. Yet, these latter developments are so recent and so fragile that it is uncertain whether they will be sustained into the twenty-first century. If they do continue, then we may be much more confident in another ten or twenty years that a democratizing trend has truly taken root (see Figures 16.1 and 16.2).

The global attention to human rights issues may indeed mean that more action will be taken to correct abuses in the future. That is, the world now has in place

---

[11]See Francis Fukuyama, *The End of History and the Last Man* (New York: Free Press, 1992).

[12]Richard Falk, *Human Rights and State Sovereignty* (New York: Holmes and Meier, 1981), 63–124.

**FIGURE 16.1** Percentage of Countries with Democratic Systems, 1985–1986 and 1995–1996

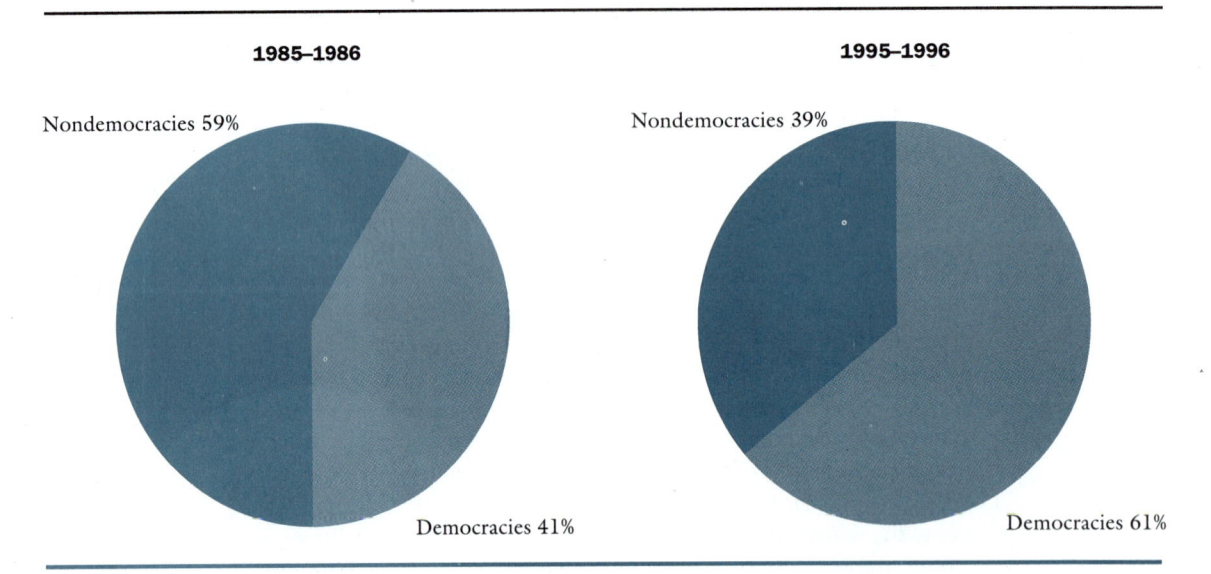

1985–1986

Nondemocracies 59%

Democracies 41%

1995–1996

Nondemocracies 39%

Democracies 61%

SOURCE: Freedom House's December 19, 1995 press release of the *Comparative Survey of Freedom*.

**FIGURE 16.2** Percentage of Free, Partly Free, and Not-Free Countries, 1985–1986 and 1995–1996

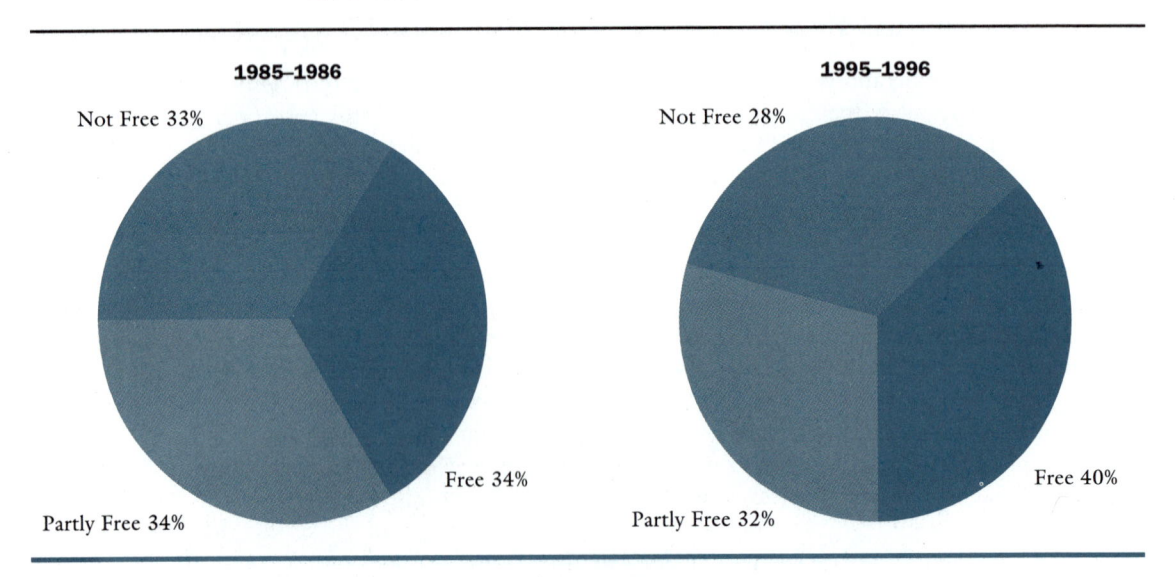

1985–1986

Not Free 33%

Partly Free 34%

Free 34%

1995–1996

Not Free 28%

Partly Free 32%

Free 40%

SOURCE: Freedom House's December 19, 1995 press release of the *Comparative Survey of Freedom*.

quite a lot of "machinery" for treating human rights issues. Yet, the nature of the state system is such that it is still comparatively easy for determined governments to avoid using that machinery or submitting their practices to its inspection. No government is eager to be criticized for its treatment of people under its jurisdiction. The greater the perceived importance to the government of the issue in question, the less willing it is likely to be to submit it to the impartial scrutiny of others. So, it would be naive to suppose that every time a government engages in flagrant violations of human rights it will be called to account by the world community.

But, if it means anything at all, the growing importance of human rights issues in global politics means that governmental leaders cannot be nearly so confident as they once were that they can "get away with murder" without paying some price for it. This price may not always be to answer directly to some international human rights authority. Rather, the cost may be in loss of diplomatic or economic support from other sovereign states and of domestic support from disaffected groups within the society. Such losses can inflict damage, sometimes seriously so, on the rights violator, even without the involvement of an international authority.

Some such consequences may result from the Russian army's attack, at the end of 1994, on the capital of the would-be independent state of Chechnya in the Caucasus. From the start, Russian President Yeltsin's campaign was unpopular with large numbers of the Russian public. Other governments were initially reluctant to criticize it on grounds that it was a purely Russian matter. But when, in January 1995, the Russian offensive produced substantial civilian casualties in and around Grozny, some Russian opponents of the attack called on Western governments to pressure Yeltsin to stop the fighting. Some Western leaders then urged mediation and a peaceful settlement of the conflict through the "collective and European intervention" of the OSCE.[13] Although no effective intervention was possible for some time, the OCSE succeeded in bringing Chechnyan and Russian negotiators together for peace talks in May 1995. Meanwhile, both the war and the world reaction to it damaged Yeltsin's effort to portray himself as a democratic champion of people's rights. His stature was further damaged when, in January 1996, Russian troops killed many Chechnyan separatists along with most of the hostages they had seized, rather than allowing them to have safe passage back into Chechnya.

## THE HUMAN RIGHTS CHALLENGE TO STATES

As much of the discussion to this point has shown, the human rights focus is a serious challenge to a doctrine that is deeply embedded in the implications of Westphalia—namely, that a state's sovereignty prevents it from being accountable to any higher authority for its treatment of its people. Until quite recently, that lack of accountability was a reality that no amount of idealism could overcome. We spoke earlier of the sovereign state's traditional "impermeability," given the military technology of the prenuclear age. The term might also describe the sovereign's traditional ability to withstand efforts to enforce a single universal standard for the treatment of human beings, as some idealists might have proposed.

---

[13] *New York Times,* January 5, 1995.

Moir/*Sydney Morning Herald*, Sydney, Australia/Cartoonists & Writers Syndicate.

Unfortunately, throughout most of modern history, if human beings were maltreated by their own governments, there was very little that the outside world could do to assist them. Before the twentieth century, when the Westphalian order was largely confined to Europe, one could hope that sovereigns were generally bound by their own common ethical and legal traditions—at least when they interacted among themselves. But in recent times, that hope has been shattered. The rise of totalitarianism in the heart of "civilized" Europe was a horrifying demonstration of the brutalities governments could inflict on individuals and groups within their care. Again, we should remember that totalitarianism arose when it did in the first half of the twentieth century because new communications and military technologies gave governments an ability to penetrate the lives of their subjects, inflicting them with brutalities beyond the capabilities of the tyrants of the past.

# POLITICAL CONFLICT AND THE CHALLENGE TO HUMAN RIGHTS

Still another factor makes it difficult to be confident that we are securing human rights throughout the world more effectively than in the past. Such progressive developments as the end of authoritarian communism in most of the former Soviet bloc give rise to new problems and, therefore, new threats to people's rights. The end of Soviet rule from central Europe to Siberia has produced much economic hardship and social disorder, at least for the short run. Such conditions can easily undermine the development of pluralistic democracy and perhaps strengthen the prospect for the reemergence of harsh, authoritarian rule. Moreover, political disintegration was accompanied by a revival of ethnic conflict that quickly led, most tragically in Yugoslavia, to the final deprivations of human rights that come in bloodshed, the destruction of people's livelihoods, and death. Likewise, a year and more after Saddam Hussein's defeat, the suffering of the innocent Iraqi population was far greater than before the Gulf War, thanks to a lack of adequate food, clean water, and health care.

Further complications can arise when seeking to evaluate the role of elections in guaranteeing human rights and democracy. For example, when elections were held in Algeria in 1992, after thirty years of dictatorial rule, about one-half the potential workforce was unemployed and a fundamentalist and backward-looking Islamic movement dominated the voting. This Islamic Salvation Front was poised to take over the National Assembly in a run-off election when the long-time president of Algeria resigned and the election was called off by a new governing council, one-half of whose members were army generals. Ironically, this cancellation of the election was supported by many democratically minded Algerians who had opposed both the single-party state and the repressive ideology of the Islamic Salvation Front. Yet, one result was an increase in the incidence of terrorism—especially against non-Islamic foreigners residing in Algeria—that seemed inspired by the frustration of the Islamic Salvation Front.

The Algerian case is a dramatic example of the contradictory tendencies that may accompany human rights efforts, when progress in one area may produce new problems in another. The freedoms that come with democracy will not seem so important to a family that cannot find food in the stores, thanks to the breakdown of a distribution system. The end of security arrangements that threatened unprecedented death and destruction will not constitute so great a human rights gain for the world if it leads to a wide dispersal of nuclear weapons into the hands of terrorist groups and outlaw governments. Nor will the ability of minority groups to govern themselves be an advance in human rights if the result is warfare that turns peace-loving individuals into refugees or corpses.

In many respects, the situation in South Africa in the early 1990s served as a microcosm illustrating many of these dilemmas. It also offered a remarkable example of how the world's attention to human rights can produce positive outcomes.

For more than four decades, beginning soon after World War II, the government of South Africa was made a pariah in the international community because of its apartheid policy, which segregated the races and gave the comparatively tiny white minority nearly all the land, wealth, and exclusive voting rights for governing the country. International pressure against the white minority regime was expressed in pronouncements of the United Nations. These took the form of increasingly strong arms embargoes and trade and investment sanctions as the years passed, including the U.N. treaty, *On the Suppression and Punishment of the Crime of "Apartheid,"* which became effective in 1976. For many years, these actions had little impact on the South African situation. Then, late in 1989, several leaders of the long-outlawed African National Congress were released after many years of imprisonment by the government of the new prime minister, F. W. De Klerk. Negotiations soon began between representatives of the government and the major opposition factions for the creation of a new constitution based upon universal suffrage and the end to apartheid.

Early in 1992, Prime Minister De Klerk called for a national referendum (restricted to white voters only) asking for approval of the liberalizing direction in which he was asking the society to move. To the surprise of many, when the vote was taken, an overwhelming majority of the white electorate approved of the prime minister's initiative. Years of pressure, as well as the increasing economic and social hardships of South Africa's political isolation, apparently had led a large number of the white community to conclude that it was time to "get in step" with the values of the world community on the matter of equal rights for all.

Then came a difficult transition period as the new constitutional arrangement was negotiated and elections were planned that for the first time were based upon universal suffrage. Outbreaks of violence between rival political factions—which caused the deaths of several thousand people—sometimes threatened to derail the entire process. But in April 1994, massive numbers of new voters turned out to elect

---

**Normative Dilemmas**

## Peace or Human Rights?

Throughout the twentieth century, a number of new states have achieved their independence through rebellions led by individuals and groups viewed as terrorists and criminals by the state against whom they fought. For example, consider the independence struggles of Ireland (and Northern Ireland), Israel, Kenya, Algeria, and Palestine. Yet, today, the leaders of the Bosnian Serbs are under indictment by a U.N. tribunal for crimes against humanity and they are forbidden to hold office in an independent Bosnia.

- How do you account for this difference in the treatment of "freedom fighters" in the latter case as opposed to many others? Is it based upon a greater criminality in the case of Bosnian Serb leaders? Does it show that the international community is for the first time trying to counter the cynical view that "anything goes" when a society is fighting for its freedom? Do such prosecutions help or hamper the effort to achieve a lasting peace in Bosnia? Is it preferable to prosecute individuals for war crimes rather than to blame the entire community in whose name they acted?

## A Right to Emigrate or to Control Immigration?

The Universal Declaration of Human Rights strongly implies the right of people to emigrate from their places of birth to other countries (Article 13, paragraph 2). Yet, virtually all states have enacted legislation regulating immigration into their own countries, which suggests that they regard immigration as a privilege, rather than a right.

- What justification do governments use to block or limit immigration of various nationality groups into their countries? If "everyone has the right to leave any country, including his [sic] own," as stated in the Universal Declaration, how can other countries lawfully deny that person entry into their own? To what extent does the rapid increase in emigration/immigration of the current period threaten the national basis of nation-states?

Nelson Mandela as South Africa's new president. He quickly set about to reassure opponents and incorporate political rivals into his governing coalition. Within a few months, the signs for a hopeful future for South Africa seemed much strengthened. However, enfranchisement of the nonwhite majority did not eliminate its underprivileged economic, educational, and social status: Greater distributive justice would no doubt constitute the next generation of struggle in South Africa. Meanwhile, however, the developments of the early 1990s seemed proof positive that in the contemporary world, no state, not even South Africa, is impervious to the persistently stated values of the world community.

# INTERNATIONAL CRIMINAL LAW

States, even governments, are abstractions—abstractions that are composed of real, living human beings. So it is the human government officials who must, in the final analysis, be made accountable for their treatment of others. This radical insight inspired the prosecution of alleged war criminals at the close of World War II. More generally, what has furthered the development of international criminal law in recent decades is the truism that no person, public or private, holds rights without corresponding obligations to other members of society.

## NUREMBERG AND TOKYO TRIALS

The greatest landmark in this respect is the International Military Tribunals held at Nuremberg and Tokyo soon after the end of World War II.[14] These tribunals were created by the victorious Allied powers to bring certain officials of the defeated governments of Germany and Japan to trial for alleged war crimes. They illustrate many of the underlying issues and provide a foundation for ongoing concerns in the human rights field. These Tribunals defined three categories of behavior for which criminal charges could be brought: crimes against the peace, war crimes, and crimes against humanity.

**CRIMES AGAINST THE PEACE** Crimes against the peace were defined by the Tribunals as the "planning, preparation, initiation or waging of a war of aggression, or a war in violation of international treaties, agreements or assurances, or participation in a common plan or conspiracy for the accomplishment of any of the foregoing." By the time of the Nazi attack on Poland in 1939, several treaties had outlawed the initiation of warfare through aggression. The most important of these was the Covenant of the League of Nations; also relevant were the Locarno Treaties of 1925, in which Germany had agreed to peaceful settlement of disputes with several of the states it

---

[14]The judgment of the Nuremberg Tribunal can be found in the *American Journal of International Law* 41, no. 1 (January 1947): 172ff. In the same issue, see an analysis of the tribunal's significance by Quincy Wright, "The Law of the Nuremberg Trials," 38–72. The language detailing the three categories of crimes is from the Agreement for the Prosecution and Punishment of the Major War Criminals of the European Axis, art. 6, 59 stat. 1544, 1547–1548.

subsequently attacked, and the Kellogg-Briand Treaty of 1928, which outlawed war as an instrument of national policy. Defendants held either that no international law existed at the outbreak of World War II that defined crimes against the peace or that, if it did, the accused were not bound by it, since Hitler had withdrawn Germany from the League several years before the attack on Poland. Similarly, the 1928 Kellogg-Briand Treaty outlawing war could not be taken as a serious legal obligation since it had included no supporting enforcement provisions. These arguments stemmed from a strictly traditional interpretation of international legal obligation, which assumes that states are almost never bound to a legal standard contained in a treaty to which they have not voluntarily submitted.

But a broader interpretation of international obligation argues that some international standards—including the criminalization of a violation of the peace—were the result of an international consensus by the time World War II broke out that bound even those sovereigns that had not given their formal consent to a particular treaty. This viewpoint, which was important to the prosecution at Nuremberg, has since gained greater prominence. Its proponents note that all customary law has developed in a similar fashion, although typically over many years. They insist that the existence of IGOs, in particular, permits the majority of governments to express their views in a way that simply speeds up the determination of what is the customary practice of most states. At Nuremberg, it was relevant that the Assembly of the League of Nations had passed a resolution in 1927 that prohibited all wars of aggression. Even though such resolutions did not have the force of treaties, this action marked an early example of the way IGO recommendations today may express the emerging consensus regarding world community standards.

**WAR CRIMES** The second category, *war crimes*, raised less controversial issues in terms of accepted theories about the legal obligation of states. It dealt with "violations of the laws or customs of war"—that is, with the conduct in wartime of governmental officials, including armed forces, that had been recognized as unlawful prior to World War II. In the words of the London Charter, "such violations shall include, but not be limited to, murder, ill-treatment or deportation to slave labor or for any other purpose of civilian population of or in occupied territory, murder or ill-treatment of prisoners of war or persons on the seas, killing of hostages, plunder of public or private property, wanton destruction of cities, towns or villages, or devastation not justified by military necessity."

Because the modern state system had long accepted warfare as legal under certain conditions, it followed that the conduct of war should be held to certain agreed-upon standards.[15] In the nineteenth century, treaties were adopted that dealt with standards for treatment of the wounded in the field, and with prohibiting some kinds of projectiles. The two **Hague Peace Conferences** held in 1899 and 1907 produced

---

[15]These standards had been codified by the twentieth century. See *The Rules of Land Warfare* (Washington, D.C.: U.S. Government Printing Office, 1917). For interpretations of the laws of war, see, for example, Terry Nardin, "The Laws of War and Moral Judgment," *British Journal of International Studies* 3 (1977): 121–136; Telford Taylor, *Nuremberg and Vietnam: An American Tragedy* (Chicago: Quadrangle Books, 1970); and Quincy Wright, "The Outlawry of War and the Law of War," *American Journal of International Law* 47 (1953): 374–376.

a general Convention with Respect to the Laws and Customs of War on Land; after World War I, additional treaties were drafted and adopted regarding, for example, the use of poison gas and the treatment of prisoners of war. Since such war crimes were fully established prior to World War II, certain Axis officials, including officers in the field, could legitimately be tried for allegedly having violated them. Because these standards served the interests of states in their claim that warfare could remain lawful if its conduct were bound by these rules, they could even be regarded as intended primarily to maintain the Westphalian international system, although they obviously served secondarily as human rights protections.

**CRIMES AGAINST HUMANITY** The Tribunals defined *crimes against humanity* as "murder, extermination, enslavement, deportation, and other inhumane acts committed against any civilian population, before or during the war, or persecutions on political, racial or religious grounds in execution of or in connection with any crime within the jurisdiction of the Tribunal, whether or not in violation of the domestic law of the country where perpetrated." Here was the most controversial category of charges brought, for some argued that traditional international law had not yet recognized such offenses. To bring individuals to trial on such charges in 1945 therefore would amount to imposing *ex post facto* punishment. The definition of this category of crimes made explicit the prosecutors' view that certain universal standards for protecting human rights not only existed at the time these acts allegedly took place, but that they were superior to domestic law. According to this view, such standards must be upheld even if it meant overturning Nazi legislation that had been enacted to make "lawful" mass deportations, slave labor, and genocide. Here was a dramatic challenge to the doctrines of **positive law,** which views the institutions of government as the ultimate source of law. That challenge was supported by the **natural law** tradition, which asserts that human life is or ought to be ruled by universal values thought to be inherent in our nature.[16]

Proceedings at the International Military Tribunals also raised the question of whether the accused could claim "superior orders" by the state in defense of their actions. Since officials are expected to carry out loyally the commands of their own authorities—and in fact are subject to charges of treason if they disobey—they are in obvious jeopardy if they are then to be held to a higher authority. The London Charter recognized this dilemma by noting that officials who justified their actions on the grounds of superior orders might have their punishments mitigated, as a result, although such a defense would not exonerate them from their obligation to adhere to universal standards regarding the treatment of human beings. In fact, at both Nuremberg and Tokyo, the severest punishments were reserved for the very highest officials, those with the fewest claims to receiving orders from someone superior to themselves.

These international military tribunals have been criticized as constituting "victors' justice," since they were carried out by the victorious allies of World War II against their defeated adversaries, who were in no position to resist. That contro-

---

[16] Legal positivism is part of the philosophical tradition of realism, and the natural law school starts from an idealist position. See Chapter 1 for a further discussion of these terms.

versy remained for several decades during which the world failed to establish similar tribunals to deal with the alleged war crimes committed by U.S. officers in Vietnam; by Cambodians against their own people; by Soviet soldiers in Afghanistan; and by Iraqis against Iranians, Kuwaitis, and Kurds.

It is obvious that sovereign governments are most reluctant to concede wrongdoing by those who are carrying out their own policies. Loyalty and patriotism are such strong forces in the nation-state system as to make action by governments against their own officials an extremely rare occurrence. For those who contend that no higher allegiance exists than to one's nation-state, Nuremberg and Tokyo were a wrong turn in the road of world politics. But the idea that human beings have universal rights argues powerfully that certain universal norms must be regarded as superior to the commands of one's own sovereign. If that were not the case, it would not be possible even to imagine trying to advance human rights beyond the nation-state.

## WAR CRIMES TRIBUNALS IN THE 1990S

In 1993, for the first time since the Nuremberg and Tokyo trials, an international panel of jurists was constituted to bring alleged war criminals to trial. The *International Tribunal for Violations of International Humanitarian Law in the Former Yugoslavia* was established by the United Nations after media reports of atrocities accompanying the breakup of Yugoslavia produced growing public outrage. The statute creating the tribunal provided for a prosecutor to conduct investigations and prepare indictments, for two trial chambers consisting of three judges each, and for five judges to sit in an appellate chamber. The tribunal's jurisdiction addresses war crimes and crimes against humanity, including genocide, committed after January 1, 1991, when the Yugoslav federation began to come apart.[17] The harshest sentence it may impose is life imprisonment.

The tribunal began its work with an obstacle that had been unknown to the Nuremberg and Tokyo prosecutors. That is, most of those suspected of war crimes in the Bosnian conflict were protected by the warring parties as the civil war continued, and were not in the custody of the tribunal itself. The tribunal is not permitted to hold trials in absentia. It therefore seemed likely that at least some of the more than fifty Serbs and Croats indicted for crimes against humanity would escape arrest and the tribunal's judgment. Yet, when a Bosnian peace agreement finally was achieved late in 1995, it required the Croatian and Serbian governments to cooperate with the tribunal in return for economic assistance.

The two leaders of the Bosnian Serbs were among those charged with directing many of these crimes. The peace agreement stripped them of their political power. These two, like others still outside the tribunal's custody, would likely find themselves virtual prisoners at home, risking arrest if they traveled. In the words of one tribunal official, "it's the duty of every authority that respects itself and that wants

---

[17]For an examination of the tribunal's origins, powers, and authority, see James C. O'Brien, "The International Tribunal for Violations of International Humanitarian Law in the Former Yugoslavia," *American Journal of International Law* 87, no. 4 (October 1993): 639–659.

to be respected in the international community to cooperate with the tribunal."[18] Meanwhile, trials were underway in several European states, including Austria and Denmark, where alleged war criminals from the former Yugoslavia had been residing when charged. It would be some time before it would be possible to judge the full impact of the tribunal. Yet, its early actions suggested that its impact might be greater than a summary judgment about its jurisdictional problems would suggest.[19]

Toward the end of 1994, the U.N. Security Council authorized establishment of a similar war crimes tribunal for Rwanda. That central African country had been devastated by a civil war earlier in the year, in which an estimated 500,000 to 1,000,000 people had been massacred. Most of those killed had been members of the Tutsi minority or those opposed to the former Hutu-led government that gave rise to vengeance killings as it was being overthrown.

Thirteen of the fifteen members of the Security Council voted to establish the tribunal: The Chinese delegate abstained and Rwanda voted against it. Interestingly, the negative vote from the new Tutsi-dominated government of Rwanda evidently did not imply its opposition to bringing those to trial who had engaged in genocidal massacres. Instead, it seems to have reflected some question that the tribunal might undermine its own sovereign prerogatives to conduct such trials itself, combined with a concern that the tribunal's justice might not be harsh enough. The Rwandan government objected that the tribunal would likely hear some cases outside Rwanda. It opposed provisions under which those convicted would serve their sentences in the jails of countries offering the tribunal prison facilities, rather than in Rwanda. It also objected to the fact that the international tribunal would not be permitted to impose the death penalty.

## AN INTERNATIONAL CRIMINAL COURT?

The past half century has greatly heightened the effort to make individuals responsible for complying with universal standards in their treatment of other human beings. The war crimes trials at the close of World War II were particularly insistent that even a state's highest officials must comply with standards to which the rest of humanity is held. The war crimes trials of the 1990s may prove to be more remembered for insisting that ordinary soldiers—including those acting largely on their own—are accountable to the human rights standards that have been established at the global level.

We should probably not expect very frequent prosecutions of heads of state and others for authorizing mass violence—not as long as the massive use of force, both domestically and in foreign affairs, continues to be regarded as an acceptable instrument of state policy. But it may not be long before the world decides to create an international criminal court to try individuals for those crimes now fully rec-

---

[18]Theo van Boven, as quoted in the *Philadelphia Inquirer*, November 8, 1994, A3.

[19]In January 1996, it was reported that most of the data on human rights violations in Croatia (although not those in Bosnia) had been stolen from the U.N. Center for Human Rights in Zagreb, Croatia. That was a serious blow to the work of the U.N. Tribunal, but also a mark of how seriously it was taken by those who feared its prosecutorial power.

ognized as outlawed by the world community. The idea of such a court is not new, but it has begun to receive greater support since the end of the cold war.

Such an international criminal court would no doubt be created by a U.N.-sponsored treaty. One proposal foresees two tracks for the court's jurisdiction. On the first, the Court would deal with *international crimes,* usually the product of state policy and action, that affect the peace and security of humanity or are particularly offensive to fundamental human values. Crimes included here are wars of aggression, war crimes, crimes against humanity, genocide, unlawful use of weapons, apartheid, torture, unlawful human experimentation, slavery, and slave-related practices.[20] On the second track, the court would prosecute individuals for misdemeanors "that offend human values but are not usually the product of state action or policy and do not threaten the general peace and security."[21] These include piracy, aircraft hijacking, the threat and use of force against diplomats and other internationally protected persons, attacks upon maritime navigation, taking civilian hostages, drug offenses, environmental damage, theft of nuclear weapons, and the like.

The fact that virtually all the above offenses are now recognized as contrary to international human rights standards is a remarkable commentary on how much work has been done in this area since World War II. The task of earlier decades was to determine those standards and set them out in documents for the world to accept. Now increasing attention is going to enforcement of those standards. The creation of an international criminal court may be a logical next step.

# CONCLUSION

Agreement on human rights standards and how to enforce them have constituted increasingly important concerns on the agenda of world politics since World War II. The increased interdependence of peoples across state lines, our greater ability to "penetrate" the behavior of other states through information technology, and the corresponding new threats to human rights that result from modern weapons and transportation technologies all combine to make human rights problems ever more a global matter.

Starting in the late 1940s, the United Nations first worked to create an agreed-upon statement expressing the fundamental human rights values that the community of states could accept (the Universal Declaration of Human Rights). Then it created treaties, both general in their coverage and related to particular rights issues, which bind ratifying states to uphold them. With a variety of institutional arrangements to assist in the effort, attention increasingly has turned to assuring greater compliance with these standards. Meanwhile, human rights treaties and other agreements have emerged from several regions of the world.

---

[20]This proposal is the work of the United States Commission on Improving the Effectiveness of the United Nations, created by act of Congress, and composed of prominent American officials and former officials. See the Commission's final report, *Defining Purpose: The U.N. and the Health of Nations* (Washington, D.C.: Author, September 1993), 26–27.

[21]Ibid., 27.

Wherever human rights are enforced against recalcitrant governments, the old doctrine of absolute state sovereignty is challenged. Therefore, states continue to resist accountability. Nonetheless, human rights activity has proceeded to the point that some have discerned a growing global consensus on the values that need to be protected internationally. Yet, progress on some fronts can sometimes raise new challenges on others, as is clear in the greater disorder that has followed new freedom within much of the former Soviet bloc.

Where human rights standards are enforced, individuals must be made accountable for their actions. The Nuremberg and Tokyo war crimes trials following World War II were an attempt to make that case to the world community. In 1993, two new tribunals were established to investigate war crimes and other human rights violations in the former Yugoslavia and Rwanda. A call is growing to create a new international criminal court, which would have jurisdiction over the kinds of crimes typically initiated by state officials (war crimes, genocide, and so on) as well as for reprehensible actions less generally disruptive of international order (such as airplane hijacking, hostage-taking, and drug running).

# CHAPTER 17

# The World's Structural Framework for Peace

Much of the rationale for the nation-state when it first emerged nearly four hundred years ago was that it effectively "contained" almost all the common human interactions of that time. But, by the nineteenth century, cross-state interactions had increased to the point that organizations were created to link governments for specified purposes. Such organizations multiplied rapidly after the middle of the twentieth century, when other kinds of cross-national associations, in addition to those created by governments, also grew and thrived. The development of all these associations reflects the fact that modern life requires them. As our intergovernmental and transnational connections grow ever more extensive, they are creating an organizational framework for the conduct of world politics.

## INTERNATIONAL INSTITUTIONS FOR A SHRINKING PLANET

Intergovernmental organizations (IGOs) express the traditional premise of Westphalia, that the governments of states are the only governing agencies for the world. Since they are *sovereign*—that is, no authority exists above them—they may agree to take joint action for specified regulatory purposes. But they presumably cannot consent to be perpetually bound by the will of even all the rest of their associated governments, for that would mean true sovereignty had passed to their associational arrangement, which should then be called a supranational organization, perhaps even a supranational government.[1]

---

[1] See Chapter 14 for the more overtly supranational development of the European Union.

## IGOS: SUPPORTING SOVEREIGNTY

IGOs may be regarded as having fundamentally conservative purposes, for they are primarily designed to shore up and build upon the Westphalian premise that the sovereign governments of nation-states are the sole effective actors in the international political system. In this sense, they are the tools that governments have decided to create in the effort to improve the quality of their own interactions. Such tools are designed for various purposes; these may range from the somewhat modest goal of developing commerce on an international waterway that crosses their territories to the much more ambitious effort to prevent or reduce the incidence of war among them. But if they work as they are intended, they should strengthen the traditional nation-state system by making it serve human needs while minimizing human suffering.

Yet, the record of IGOs makes clear that they do not always work as they were intended. First, they sometimes fail in their regulatory purposes, thanks almost entirely to the very nature of sovereignty itself. When they sign the treaty that establishes the IGO, governments may sincerely mean to commit themselves to its terms. These terms might include, for example, the military defense of fellow signatories in the face of aggression from some other sovereign. Yet, when that time arrives, those same governments may claim to see extenuating circumstances that allow them to judge this particular case as not requiring the kind of action called for by the treaty. It is their sovereignty that, above all, allows states to judge their own cause. So agreements among sovereigns are clearly more susceptible to being disrupted by the very prerogatives of sovereignty than is the case where nonsovereigns are subject to the law.

## IGOS: CHALLENGING SOVEREIGNTY

But there is a second unintended result that is less often evident in the work of IGOs, though it may be at least as significant for understanding the evolution of the global system. Intergovernmental organizations, even when they seem to fail, may produce new expectations about what is appropriate international interaction. This in turn may result in the gradual erosion of the very premise that state sovereignty is an absolute and that the sovereign government is unaccountable to others. For example, when the government of South Africa released Nelson Mandela from prison in 1990, that marked a dramatic response to the many years in which South Africa had been the target of censure and economic sanctions by the United Nations for its construction of a repressive apartheid system. In this case, the sovereignty of the government of South Africa could not prevent the United Nations from taking action against it, which increasingly made South Africa a pariah within the international community; nor could it evidently delay indefinitely an end to apartheid.

It is true in general terms that IGOs have both conservative intentions and potentially more radical consequences for the state system; but that is too sweeping a generalization to explain very much about the various kinds of institutions that are identifiable as intergovernmental organizations today. These range from groupings of a few states with very specific and limited aims (the Danube River Commission is an example) to worldwide bodies, such as the United Nations, with a multitude of

> ## China Distances Women's Groups from U.N. Women's Conference
>
> The U.N. World Conference on Women was convened in Beijing in September 1995. Several months earlier, China's hard-line prime minister, Li Peng, had directed that the parallel NGO Forum on Women must be housed at a site some ninety minutes away from the main conference. The Chinese government evidently was unwilling to see the more than 35,000 members of advocacy groups attending the NGO forum in such close proximity to governmental delegates. Women's groups were outraged, some angrily demanding that the conference be postponed or moved to another country. But in June, it was decided to hold the conference as scheduled. "We women are fighters, and . . . the wisdom of a fighter is to see what you die for," said Gertrude Mongella, the secretary-general of the conference, in announcing the decision. "Do we die for the venue, or do we die for the issues?"
>
> - What might the Chinese government have feared would result if the NGO Forum on Women had been allowed easy access to government delegations? What does this conflict reveal about the tensions in contemporary world politics between the work of many NGOs and the governments of (at least some) states?

purposes and programs. They do not always resemble one another in their component parts, any more than do the components of any complex physical structures. But, considered as a whole, they constitute a framework for peace in this sense: Either they are intended to harmonize conflicting policy goals before divisions escalate to the point of violence or, alternatively, they are designed to restore peace where it has been broken.

Since 1920, most governments have interacted within a single international organization—first, the League of Nations, followed after World War II by the United Nations—meant to serve governments as a kind of macro-institution for multilateral diplomacy throughout the world. While both the League and, especially, the United Nations have had multiple goals and have spawned many programs and agencies to promote them, both have been intended first and foremost to assist in keeping or restoring world peace. Both, therefore, have been meant to provide the world with a **collective security** system. The core idea behind collective security is that an act of military aggression against any state amounts to an attack upon every state that has sworn to maintain the peace. Such an event, therefore, should prompt nearly universal action against the offending state.

# COLLECTIVE SECURITY AND THE LEAGUE OF NATIONS

The Covenant of the League of Nations, drafted at the Versailles Peace Conference at the close of World War I, created the first standing international organization whose members were pledged to take collective action against any state that resorted to war. This novel effort was spawned by the lesson of how World War I had begun. War apparently had broken out in 1914 because the European great powers had attempted to counterbalance the threat of aggression from their adversaries through military commitments to their allies. The disastrous result had been that a single act of violence in the Balkans (the assassination of the heir to the throne of Austria-Hungary) had produced a chain reaction of threats, counterthreats, and preparations for war. Within a matter of weeks, all Europe, and soon thereafter, much of the rest of the world, was engaged in the most destructive war yet known.

President Woodrow Wilson of the United States and other supporters of a League of Nations were convinced that it was the prewar effort to achieve security by balancing power in rival alliances that had to be ended.[2] That could be accomplished only if something resembling a global alliance committed to keeping the peace could replace balance of power politics. The answer was to be found in the collective security provisions of the League of Nations Covenant.

## THE REQUIREMENTS FOR COLLECTIVE SECURITY

Collective security of the sort envisioned by the League's founders was built upon certain assumptions and conditions, some of which are clearer with hindsight than they were at the time. The first is that the collective security organization must strive for the nearly universal membership of the states of the world. Most, or at least the most powerful, governments need to be committed to the collective security system, for if important actors are not bound by it, they might feel free to create a security arrangement of their own. The result would merely be a new variation on the kinds of alliances that had produced World War I, rather than an alternative to them.

A second assumption is that the leading members must have an essentially status quo orientation toward the rest of the world. That is, they must be satisfied enough with their own position in world politics that they have no desire to try to improve their lot through the spoils of war. They must also be willing to thwart any upstart state that tries to increase its standing through the use of force, since that would be a threat to the status quo in which great powers hold a stake. Not coincidentally, the League of Nations was created by the victorious allies of World War I, who presumably were satisfied with what they had achieved and acquired in that victory. The Covenant they wrote reveals their pledge to uphold the status quo. Article 10 asserts that "the Members of the League undertake to respect and preserve as against external aggression the territorial integrity and existing political independence of all Members of the League." If no territory would be permitted to change hands through the exercise of force, then war might be effectively outlawed; but that assumed a perpetual commitment to defend the territorial status quo against all comers.

A third requirement for collective security is that the sanctions threatened against a would-be violator of the peace must amount to effective police power, cer-

---

[2]During the war, Wilson had often criticized the alliance policies of the European powers as responsible for the conflict. In one example, he attacked "that old and evil order which prevailed before this war began" as the "ugly plan of armed nations, of alliances, of watchful jealousies, of rabid antagonisms, of purposes concealed, running by the subtle channels of intrigue through the veins of people who do not dream what poison is being injected into their systems." Ray S. Baker and William E. Dodd, eds., *The Public Papers of Woodrow Wilson, War and Peace*, vol. 2 (New York: Harper, 1927), 234, 235.

tain and severe enough to do the job. If collective action is not effective and the aggressive state achieves its aim, then collective security is weakened and would-be aggressors are emboldened to seek their foreign policy goals by force.

Finally, it is assumed that collective security will occur almost automatically in the event that some state undertakes to use force against another in violation of the terms of the collective security agreement. In the case of the Covenant, the commitment to participate in collective action was presumably a legal obligation by all member states of the League, leaving them no choice to decide whether they were in or out once the peace had been disturbed.

## THE FAILURE OF COLLECTIVE SECURITY UNDER THE LEAGUE

None of these expectations was fully realized in the world in which the League of Nations had to operate. The goal of a nearly universal membership was crippled at the outset when the United States refused to join, in spite of the fact that the Covenant was chiefly the work of its own president.[3] That refusal delayed for another twenty years the full participation of the United States as a great power on the world stage, but it also deprived the League of a leading voice committed to the rule of law in the world. Meanwhile, Russia, in the aftermath of its Communist Revolution, and the defeated Germany were deliberately excluded from membership at the beginning (both countries later were admitted, although Adolf Hitler took his country out of the League in 1934, and the Soviet Union was expelled for its attack on Finland in 1939). By the mid-1930s, a general exodus from the League began; at one time or another a total of sixty-three states had been members, but of these seventeen withdrew.

**THE ETHIOPIA CASE** The fact that several important states were outside the League by 1935 would soon have a crippling effect on the collective economic sanctions that the League enacted against Italy for its attack on Ethiopia. Since nonmembers were under no obligation to engage in sanctions against Mussolini's government, they generally felt free to continue trade with Italy. Worse than that, states not held to the League's sanctions could actually be rewarded for their noncompliance by increasing their trade with Italy. This in turn became an excuse for certain League members finally to vote for an end to sanctions.

By the time of the Ethiopian engagement, too many of the world's important states were led by governments committed not to maintaining the territorial status quo, but to revising it to suit themselves—through the use of force, if necessary. Starting in 1931, Japan had made a puppet state of the Chinese province of Manchuria. This action produced criticism from the League, but no sanctions against Japan, which nonetheless quit the League in 1933 over the Manchurian affair. Hitler had pulled Germany out of the League because of his own territorial ambitions, and once sanctions were voted against Italy, the Nazi government made common cause with Mussolini. That not only helped keep the Italians from feeling the full force of the sanctions, it also raised the fear in the capitals of the Western democracies that

---

[3]The vote for ratification fell short of the Constitutional requirement of a two-thirds majority in the Senate.

the League action was only forcing Mussolini into Hitler's embrace. One argument for ending the sanctions in July 1936 was that it was harmful to continue to ostracize Italy even after its military conquest of Ethiopia was an accomplished fact. Ending sanctions, some hoped, might allow Italy to play a constructive role again in world politics, perhaps even to act in association with the League in an effort to contain Hitler. Yet that was not to happen, for Mussolini became ever more disdainful of the League, and withdrew his country from membership at the end of 1937.

Finally, the League's collective action against Italy failed because the sanctions at its disposal—the severance of diplomatic and trade relations with the target state—were not strong enough to produce a clear and tangible result. Instead of being brought to its knees, Mussolini's government stepped up its campaign against Ethiopia, annexing that ancient African nation as a colony before the League's sanctions could cripple the Italian economy. Clearly, if economic sanctions are to work at all, they take time, whereas the military conquest of a country outmatched by the forces of the invader may be accomplished in weeks or even days. The League's action ironically may have strengthened both Italian resolve to back Mussolini and certain sectors of the Italian economy by increasing demand for local production of goods that could no longer be imported from abroad.[4]

**SUCCESSES OF THE LEAGUE OF NATIONS**   Because of its inability to reverse the Italian aggression in Ethiopia, the League tends to be remembered as a failure, even though it produced a record of considerable success in its earlier years. Quincy Wright, a distinguished professor from the University of Chicago, found that of the sixty-nine important disputes that came before the League, fifty-five were successfully resolved (although some by agencies other than the League). Seventeen of these involved the use of force; of those, only seven were settled. In the other ten, the aggressor was not stopped until World War II.[5] Almost all the failures occurred once the Ethiopian case began. After 1936, all the world knew that the League's collective security potential would not be invoked again, and those with radical demands to make on the status quo—Germany, Italy, and Japan—pressed their claims until, less than three years later, they led directly to the outbreak of World War II.

## THE U.N. FRAMEWORK FOR PEACE

As they fought their way to victory, the Allied Powers of World War II agreed that they would recreate the League, this time with sufficient "teeth" to do the job. The creators of the United Nations attempted to learn from the League's mistakes as they built the new organization.[6] They agreed to a much longer and more elaborate constitutional document, the U.N. Charter, which provided a framework in

---

[4]F. P. Walters, *A History of the League of Nations* (Oxford: Oxford University Press, 1964), 53.

[5]Quincy Wright, *A Study of War*, abridged ed. (Chicago: University of Chicago Press, 1965), 408.

[6]For discussions of the ways in which the League's experience influenced the founders of the United Nations, see Inis L. Claude Jr., *Swords into Plowshares,* 4th ed. (New York: Random House, 1971), 57–80; and Peter R. Baehr and Leon Gordenker, *The United Nations in the 1990s* (New York: St. Martin's Press, 1992), 20–43.

> **Global Changes**
>
> ## IGOs on the Global Scene
>
> In the first forty-two years of the United Nations' existence—that is, from 1945 through 1987—it authorized a total of thirteen peacekeeping operations. By 1988, the cold war was ending. Then, in just six years, from 1988 to 1994, the United Nations fielded twenty-two new peacekeeping operations, including several of the largest and most expensive operations ever undertaken.
>
> - How did the end of the cold war contribute to the huge upsurge in the number of U.N. peacekeeping operations after 1988? Apart from the increased costs of such efforts, what other problems has the United Nations faced as the result of this heightened level of peacekeeping?
>
> The International Telegraph Union (ITU) and the Universal Postal Union (UPU) were created in 1865 and 1874 respectively. They were needed to supervise the rising amount of mail and telegraph communication that was beginning to crisscross national boundaries in the period. They are still thriving agencies today (the ITU is now the International Telecommunication Union), and became the models for the world's current huge array of "functional" IGOs. These include everything from the Agency for Cultural and Technical Cooperation, which promotes such cooperation among French-speaking countries, to the World Meteorological Organization, which ensures cooperation for global weather forecasting (with hundreds more in an alphabetical listing between). Consider the following four IGOs, all of which have been created in the past half century, to determine how they may be said to constitute planks in a world "structure for peace":
>
> 1. INTERPOL (International Criminal Police Organization), whose aim is to promote international cooperation between criminal police authorities among its more than 150 member states;
> 2. INTELSAT (International Telecommunications Satellite Organization), which was created to develop and operate a global communications satellite system;
> 3. MTCR (Missile Technology Control Regime), whose twenty Western members seek to arrest missile proliferation by controlling the export of key missile technologies and equipment; and
> 4. WIPO (World Intellectual Property Organization), a U.N. agency concerned with protecting literary, artistic, and scientific works.

which collective security might constitute only one of several techniques for securing peace.

## ALTERNATIVE APPROACHES TO PEACE

In general, the varied approaches embodied in the U.N. effort to strengthen the fabric of peace in the world can be described as, first, treating the circumstances that underlie disputes; second, attempting to achieve peaceful settlement of disputes that nonetheless arise; and third, when peaceful settlement fails, dealing with the threat or outbreak of war. The prominent role the United Nations has played in advancing the economic development of poor countries is relevant to the first of these approaches, in the sense that these activities were conceived by the Charter's authors as one of those alternative approaches to peace. Both the Economic and Social Council (ECOSOC) and many of the specialized or associated agencies of the United

Nations are meant to advance the thesis that peace can be the product of greater economic well-being and social integration, which today is broadly termed as functionalism (see Figure 17.1).[7]

An array of provisions in Chapter VI of the Charter lays out procedures and expectations for the pacific settlement of disputes. Parties to a dispute are urged to negotiate their differences, and resort to enquiry, mediation, conciliation, arbitration, judicial settlement, the involvement of regional agencies, or "other peaceful means of their own choice" (Article 33). Additional provisions would involve the Security Council at any stage of the dispute, allowing it to make recommendations for the terms of a settlement. All of these directives presume a desire on the part of disputants to reach an amicable and peaceful solution. The objectivity and resulting moral authority of these third-party agencies were intended to help them help the disputants maintain the peace.

## COLLECTIVE ENFORCEMENT UNDER THE SECURITY COUNCIL

Should these efforts fail, however, the Charter framers provided for much more forcible action by the Security Council to keep the peace. The Council was created in the expectation that it would have nearly exclusive responsibility for peace and security functions in the new organization, while the General Assembly would deal with all other issues. Although matters have not worked out quite that clearly in practice, it should first be noted that the Security Council was conceived as a great-power directorate in which the interests of other groups of nations would be represented on a rotating basis. The five permanent members were, not coincidentally, the five major Allies of World War II: China, France, the Soviet Union (now Russia), the United Kingdom, and the United States.[8] Chapter VII of the Charter created a blueprint for a collective security system under the control of the Security Council that was meant to be much stronger than anything suggested for the League. The permanent members were expected to agree to a plan to be worked out by their joint chiefs of staff for a combined military force capable of collective action.

Although all member states were committed to make available whatever armed forces or other assistance the Security Council might ask them for, it was the Security Council as such that would take the leading role in any collective action on behalf of the organization. That in turn assumed that the Council, or at least its five permanent members with the power each was given to veto any action approved of by a majority, would find themselves in agreement on action they might take collec-

---

[7]The principal author of functionalist theory was David A. Mitrany, who drew on the experience of the League of Nations to conclude that states could develop more peaceable relationships by cooperating where "functional" or social needs for cooperating were noncontroversial and, in that sense, not "political." See Mitrany's *Progress of International Government* (New Haven, Conn.: Yale University Press, 1933); and *A Working Peace System* (Chicago: Quadrangle Books, 1966). See Claude, *Swords into Plowshares*, chap. 17, for a concise interpretation of functionalism.

[8]Because these five are named as Permanent Members, a Charter amendment would be required to expand this group. Although various candidates were proposed for Permanent Member status in the 1990s, no agreement had been reached by 1996 on an enlarged Security Council.

**FIGURE 17.1** United Nations System

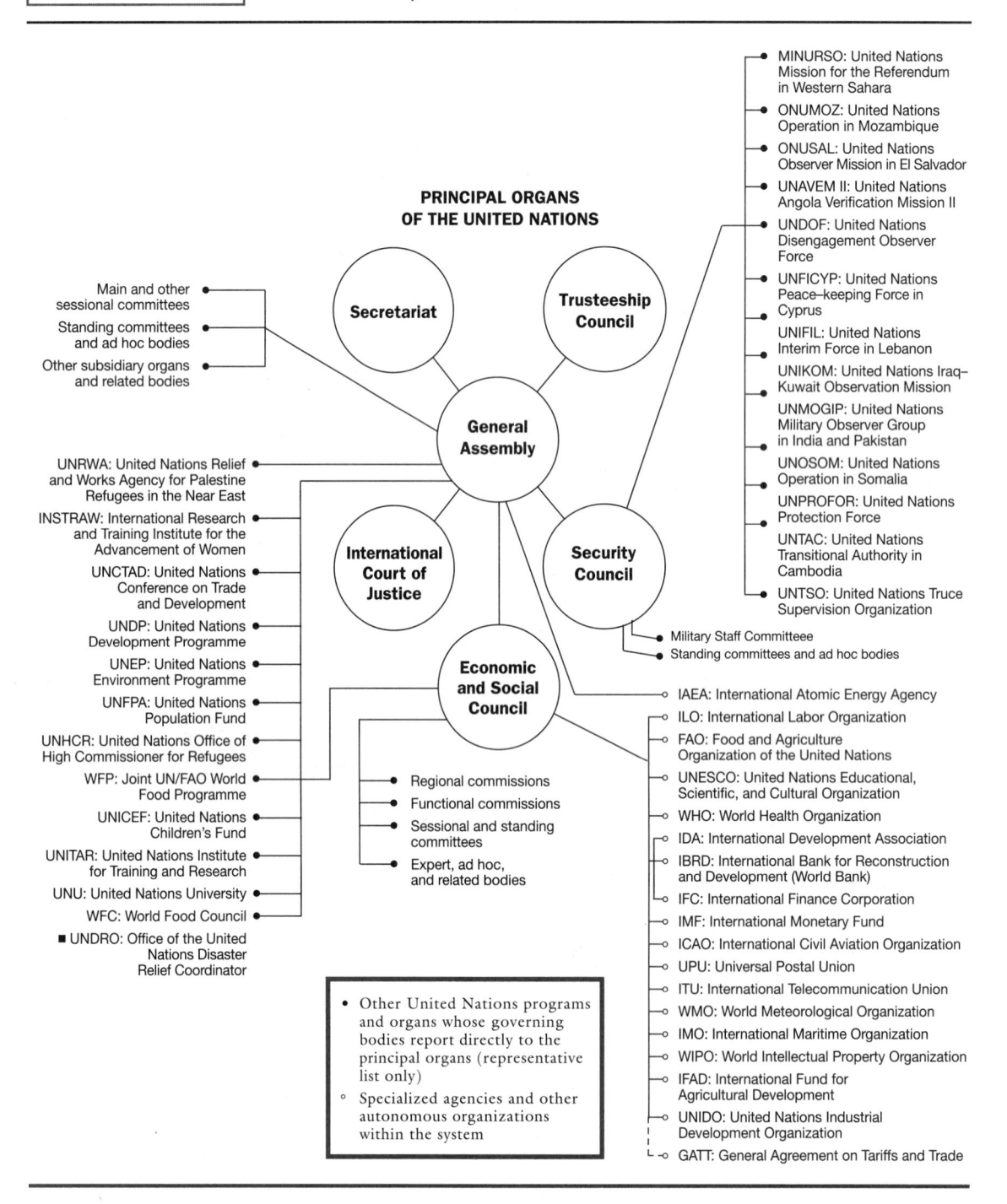

### PRINCIPAL ORGANS OF THE UNITED NATIONS

**Principal Organs:**
- Secretariat
- Trusteeship Council
- General Assembly
- International Court of Justice
- Security Council
- Economic and Social Council

**Connected to Secretariat / General Assembly:**
- Main and other sessional committees
- Standing committees and ad hoc bodies
- Other subsidiary organs and related bodies

**Connected to General Assembly:**
- UNRWA: United Nations Relief and Works Agency for Palestine Refugees in the Near East
- INSTRAW: International Research and Training Institute for the Advancement of Women
- UNCTAD: United Nations Conference on Trade and Development
- UNDP: United Nations Development Programme
- UNEP: United Nations Environment Programme
- UNFPA: United Nations Population Fund
- UNHCR: United Nations Office of High Commissioner for Refugees
- WFP: Joint UN/FAO World Food Programme
- UNICEF: United Nations Children's Fund
- UNITAR: United Nations Institute for Training and Research
- UNU: United Nations University
- WFC: World Food Council
- ■ UNDRO: Office of the United Nations Disaster Relief Coordinator

**Connected to Security Council:**
- MINURSO: United Nations Mission for the Referendum in Western Sahara
- ONUMOZ: United Nations Operation in Mozambique
- ONUSAL: United Nations Observer Mission in El Salvador
- UNAVEM II: United Nations Angola Verification Mission II
- UNDOF: United Nations Disengagement Observer Force
- UNFICYP: United Nations Peace-keeping Force in Cyprus
- UNIFIL: United Nations Interim Force in Lebanon
- UNIKOM: United Nations Iraq-Kuwait Observation Mission
- UNMOGIP: United Nations Military Observer Group in India and Pakistan
- UNOSOM: United Nations Operation in Somalia
- UNPROFOR: United Nations Protection Force
- UNTAC: United Nations Transitional Authority in Cambodia
- UNTSO: United Nations Truce Supervision Organization
- Military Staff Committeee
- Standing committees and ad hoc bodies

**Connected to Economic and Social Council:**
- Regional commissions
- Functional commissions
- Sessional and standing committees
- Expert, ad hoc, and related bodies

**Specialized agencies:**
- IAEA: International Atomic Energy Agency
- ILO: International Labor Organization
- FAO: Food and Agriculture Organization of the United Nations
- UNESCO: United Nations Educational, Scientific, and Cultural Organization
- WHO: World Health Organization
- IDA: International Development Association
- IBRD: International Bank for Reconstruction and Development (World Bank)
- IFC: International Finance Corporation
- IMF: International Monetary Fund
- ICAO: International Civil Aviation Organization
- UPU: Universal Postal Union
- ITU: International Telecommunication Union
- WMO: World Meteorological Organization
- IMO: International Maritime Organization
- WIPO: World Intellectual Property Organization
- IFAD: International Fund for Agricultural Development
- UNIDO: United Nations Industrial Development Organization
- GATT: General Agreement on Tariffs and Trade

- • Other United Nations programs and organs whose governing bodies report directly to the principal organs (representative list only)
- ○ Specialized agencies and other autonomous organizations within the system

tively to maintain or restore the peace. If all five had the status quo orientation of satisfied great powers, then it might be reasonable to expect they could frequently agree on collective action in the face of peace-threatening behavior from lesser states.

## BIPOLARITY IN THE COLD WAR

But almost immediately after the United Nations was created, the cold war began. Among the permanent members, deep divisions began to surface. With China in the throes of a civil war that would lead to a communist government in control of its mainland by 1949, and with France and Britain both reduced in stature by World War II and the loss of their overseas empires, the United States remained as the leading antagonist to confront its opposite superpower, the Soviet Union. In such a situation, the great power cooperation assumed by the U.N. Charter quickly proved impossible.

As a result, none of the plans for an international armed force under the Security Council's direction could be implemented. Instead, the United Nations soon became a forum where the Soviet Union and the West squared off against each other rhetorically. In the early days, before the end of colonialism added dozens of non- (and frequently anti-) Western countries to U.N. membership, the United States was the most influential member, typically able to get much of what it wanted in resolutions passed by the General Assembly and other U.N. bodies. Yet the United States and the West were just as often prevented from getting pro-Western resolutions passed in the Security Council, since there the veto power was operative and the Soviet government was not reluctant to use that power to prevent actions it regarded as harmful to its own interests. Under these conditions, genuine collective security action seemed even more an impossibility than under the League of Nations.

**THE KOREAN CONFLICT** An event in June 1950, however, unexpectedly pulled the United Nations into what was alleged at the time to be a collective security action even though it violated the spirit, if not the actual letter, of the Charter. Armed troops of the Korean communist regime, based north of the thirty-eighth parallel, crossed into pro-Western South Korea, thereby instigating the Korean War. The Soviet government was boycotting the United Nations at the time to show its displeasure at the refusal to seat the representatives of the Chinese communists, who had come to power the previous October, in the place reserved for "China." The United States introduced several resolutions in the Security Council that condemned the North Korean action and committed the United Nations to back the U.S. decision to assist the beleaguered South Korean regime. By the time Stalin directed his U.N. delegation to return to the Council so that it could exercise its right of veto, it was too late to undo the "collective" security action—which remained almost exclusively a joint South Korean and American effort—already authorized in the name of the United Nations.

Meanwhile, flush with the possibilities for engaging the United Nations in enforcement action on its terms, the United States sponsored a resolution in the General Assembly meant to circumvent the veto problem in the Council. In November 1950, the Assembly adopted the Uniting for Peace resolution, which permitted the General

Assembly to consider an issue that had been vetoed in the Council, and to recommend action that might restore or save the peace. Specifically, the resolution asked member states to designate military contingents that might be called into action to carry out such a recommendation by the Assembly. It marked the initiation of a trend that would give the all-member Assembly somewhat greater stature than the Charter had indicated, although the Assembly would never prove to be an effective substitute for the enforcement capabilities that had been entrusted to the Security Council.[9]

**THE UNIQUENESS OF THE KOREAN OPERATION** The Korean action was to remain an aberration during the cold war period, for it marked the only time in which it was possible to mobilize the collective security authority of the United Nations so clearly on behalf of one of the major antagonists. By the autumn of 1950, even some of America's allies began to have second thoughts after U.S. and South Korean forces went on the offensive, pushed deep into North Korea and, nearing the Yalu River border of China, saw their action bring thousands of Chinese "volunteers" into the conflict on the side of North Korea. Although ultimately the United States settled for an armistice that redrew the partition of Korea almost exactly where the line had been before June 1950, the experience suggested to many that in the future the United Nations would be more wary of serving, or seeming to serve, the cold war interests of one of the superpowers.[10] To enlist the United Nations in so partisan a way again would undermine the stability of bipolarity and might even encourage a nuclear response.

# THE EVOLUTION OF U.N. PEACEKEEPING

As the organization's record after Korea makes clear, the United Nations soon would show it had an ability to play a different kind of peacekeeping role in a bipolar world than anyone had foreseen in 1945 (see Figure 17.2). Beginning in the mid-1950s with its involvement in the Suez crisis, the United Nations was adapted to help stabilize the bipolar structure of international politics. In 1956, when the United Kingdom, France, and Israel intervened in Egypt to try to force the Nasser government to return the Suez Canal to international control, the Soviet Union threatened to come to the support of Egypt. That in turn raised the prospect that the United States would be drawn in on the side of its allies, with a general East-West war the possible outcome. Instead, once a Security Council resolution calling for the withdrawal of the intervening forces was vetoed by Britain and France, the Uniting for Peace resolution was invoked. That permitted the General Assembly to call for the creation of an international force that would police a cease-fire and withdrawal of the intervening armies.

This was the first of a number of occasions in which cold war rivalries in some part of the world threatened to turn a local conflict into a superpower confrontation, with all the dangers this held for a U.S.–Soviet engagement spiraling upward

---

[9]For a brief discussion of what made the Korean case an aberration for the United Nations, see Thomas G. Weiss, David Forsythe, and Roger A. Coate, *The United Nations and Changing World Politics* (Boulder, Colo.: Westview Press, 1994), 43–45.

[10]Leland M. Goodrich, *Korea: A Study of U.S. Policy in the United Nations* (New York: Council on Foreign Relations, 1956).

**FIGURE 17.2** The United Nations at Age 50

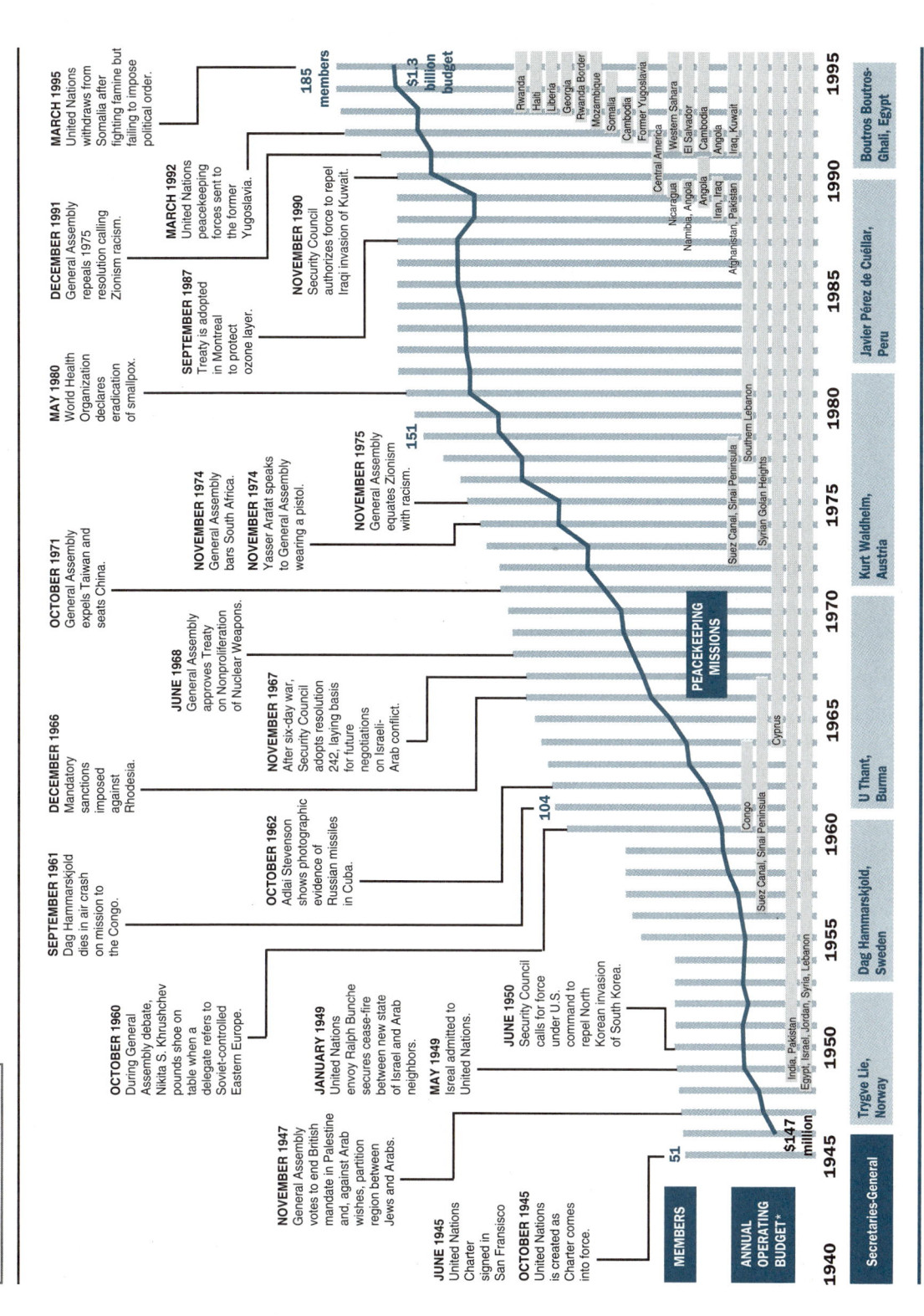

*Adjusted for inflation, in 1994 dollars. Starting in 1974, the United Nations adopted two-year budgeting. Data after 1973 are the annual average for the corresponding two-year period.

**SOURCE:** *The New York Times*, October 22, 1995, 10. Copyright © 1995 by The New York Times Company. Reprinted with permission.

toward a thermonuclear exchange. Fear of such an outcome, even when it seemed a fairly distant possibility, encouraged the cold war antagonists to engage the United Nations in an effort to stop a local war or oversee a cease-fire so that neither of the superpowers would be tempted to intervene unilaterally.

**THE PURPOSES OF PEACEKEEPING**  U.N. interventions of this type have come to be known as **peacekeeping operations**[11] because they differ in the following specific ways from the theory of collective security:

**No Act of Aggression Is Determined**  Such operations are not meant to punish aggression or another act of war that the rest of the international community regards as illegal (as was the case with the collective security actions against Italy in 1935, against North Korea in 1950, and against Iraq in 1990). Peacekeeping reverses the first premise of collective security; that is, it starts by recognizing that the Security Council may not always be faced with an obvious act of aggression that would require it to engage in collective security to punish a state violating the peace. Instead, the critical issue is to contain, if it cannot stop, the conflict before it spreads to others. In such cases, the United Nations can play a useful role only as a neutral party, interposing its forces between foes—usually after a cease-fire is reached—then supervising the truce without taking sides in the conflict.

**Use of Interpositionary Force**  It follows that the nature of the operation in the field is interpositionary rather than coercive. It resembles the use in a city of police officers to ensure crowd control at a public demonstration; it is not like a police raid on a site where criminals are known to be hiding. Typically, U.N. peacekeeping operations have needed only comparatively few soldiers—a few hundred to a few thousand—to perform their tasks. Starting in the 1960s, a few U.N. member states (Canada and the Netherlands are examples) began training some of their armed forces in the special requirements of interpositionary forces, thus preparing them for U.N. peacekeeping assignments when called upon.

**No Settlement Is Enforced**  As already implied, peacekeeping operations are aimed at halting wars rather than settling the conflicts that gave rise to them. While it may be hoped that a peace conference will follow where a settlement can be arranged, that is not necessarily included in the peacekeeping operation itself. Indeed, in a number of cases, peacekeeping operations have been in place for many years while the diplomatic effort to achieve a peace treaty remains unsuccessful. The leading example is the U.N. Force in Cyprus (UNFICYP), which, in one form or other, has been stationed between rival Greek and Turkish Cypriot communities ever since 1964 while peace negotiations have been stymied.

**Operations Are Voluntary**  In sharp contrast to the obligatory nature of collective security in theory, peacekeeping operations are voluntary in almost all respects. The

---

[11]For a discussion of the general principles guiding the development of peacekeeping under the United Nations (although he calls it "preventive diplomacy"), see Claude, *Swords into Plowshares,* chap. 14. See also Paul F. Diehl, "The Conditions for Success in Peacekeeping Operations," in *The Politics of International Organizations,* ed. Paul H. Diehl (Pacific Grove, Calif.: Brooks/Cole, 1989), 173–188.

noncoercive quality of peacekeeping is marked by the fact that the very parties to the conflict must agree to the operation. Other member states then must volunteer to provide troops so that a multinational force can be created. Ever since the mid-1960s, even payment for peacekeeping operations has been essentially voluntary on the part of U.N. members. Thus, if any of the participants turn sour on the operation, they may decide to withdraw their approval, perhaps effectively ending the operation. That was the case with the first United Nations Emergency Force (UNEF I) in the Middle East, which helped keep the peace between Israel and Egypt from 1956 to 1967. When, in the summer of 1967, President Nasser of Egypt demanded that U.N. troops be withdrawn from his nation's territory, the U.N. secretary-general, U Thant, felt he had no choice but to comply, although he thus set the stage for the Israeli attack that launched the Six-Day War. Those who contribute peacekeeping forces also may decide to withdraw them. When U.S. peacekeepers in Somalia came under increasing fire in 1993 and 1994, the Clinton administration pulled them out of the U.N. operation.

**Administration by the Secretary-General**  The administration of peacekeeping operations has been largely the responsibility of the U.N. secretary-general, who has usually been given specific authority to carry out the Security Council's mandate. The Charter provisions for a standing international force under the direction of the permanent members' military chiefs of staff have never been realized; without it, supervision for the more ad hoc military operations characteristic of peacekeeping had to rest somewhere. Since the secretariat is the neutral administrative branch of the United Nations, it is the logical administrator of peacekeeping operations.

**PEACEKEEPING DURING THE COLD WAR**  Several studies have attempted to assess the record of the United Nations at keeping or restoring the peace during the cold war period. This research generally does support the view that U.N. efforts were more successful at stopping hostilities than at resolving the issues that produced the conflict. One study of the 1946–1964 period examined forty-four cases involving the use or threat of force that came before the United Nations. In seventeen of those where war had broken out, the United Nations or some other agency brought about a cease-fire, although only after serious hostilities in five of the wars. In fourteen cases, hostilities ended without a formal cease-fire. In twenty-five, the dispute leading to the hostilities was not settled by 1964, leaving seven territories divided by armistice lines and fifteen quiescent.[12]

Another study of the same period showed that almost one-half of the fifty-five disputes considered by the United Nations were left unsettled, while about one-third were settled in part or wholly through United Nations action.[13] A third study considered more than one hundred international wars and crises from the end of World War II through 1977. It showed that less than 20 percent of these outbreaks elicited a U.N. resolution calling for a halt or threat to the use of force. Where such action was called for, in only about one-half of the cases did the parties comply with the

---

[12]Wright, *A Study of War*, 408.

[13]Ernst Haas, "Collective Security and the Future International System," *University of Denver Monograph Series in World Affairs* 5 (1967–1968), 41, 51.

**FIGURE 17.3** United Nations: Referrals and Success Rates

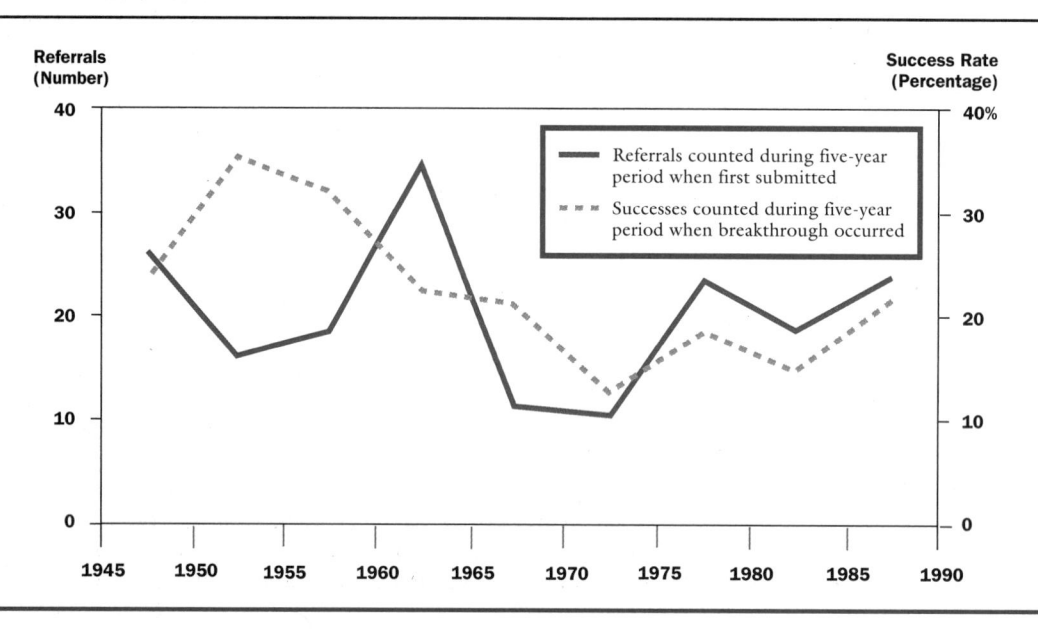

[Graph showing referrals (number) and success rate (percentage) from 1945 to 1990. Legend: solid line = Referrals counted during five-year period when first submitted; dashed line = Successes counted during five-year period when breakthrough occurred.]

SOURCE: Ernst B. Haas, "Collective Conflict Management: Evidence for a New World Order?" in *Collective Security in a Changing World*, ed. Thomas G. Weiss (Boulder, Colo.: Lynne Rienner Publishers, 1993), 66. Copyright © 1993 by Lynne Rienner Publishers, Inc. Reprinted with permission of the publisher.

U.N. directive soon after it was issued.[14] This last, more pessimistic, assessment no doubt reflects the decline in the United Nations' ability to discourage use of military options in the dozen years following 1965.

Indeed, that downward trend apparently continued until late in the 1980s, when the United Nations suddenly became the focus once more for important efforts to prevent and control warfare. A 1993 study showed that in the five-year period from January 1, 1986, through December 31, 1990, there was a concurrent rise in the rate of the number of cases referred to the United Nations and its rate of success in resolving them (see Figure 17.3). That was an unprecedented record for the organization, suggesting increased effectiveness in dispute settlement.

**PEACEKEEPING AFTER THE COLD WAR** Once the cold war ended, so did the predictable divisions within the Security Council that had so often kept it from taking action in areas where either of the superpowers had a primary interest. The cold war's end also "let the lid off" conflicts that long had been repressed by the very rivalry of the superpowers and their fear of being drawn into a direct shooting war with each other. The result was a dramatic increase in the number and scope of peacekeeping operations authorized by the United Nations. As Map 17.1 indicates, nearly

---

[14]Jock A. Finlayson and Mark W. Zacher, "The United Nations and Collective Security" (Los Angeles: Annual Convention of the International Studies Association, 1980), 3.

two-thirds of the total number of such operations in the history of the United Nations have been authorized since 1988–1989, years that mark the end of the cold war period. Moreover, the recent operations tend to be large and complex. During the first six months of 1992, the number of U.N. peacekeeping soldiers in the field quadrupled, to 44,000, with an annual price tag of some $2.8 billion attached. Such a huge increase in peacekeeping activities was raising serious questions as to whether the United Nations' modest size and very limited means would not be overtaxed.[15]

Those questions grew sharper over developments involving the U.N. Protection Force in Bosnia (UNPROFOR) during 1995. UNPROFOR had been created three years earlier on the premise that both parties to the Bosnian conflict—Serbs and the Muslim-dominated government—in fact desired a peaceful settlement. Yet, each of the warring parties increasingly viewed UNPROFOR as, at best, irrelevant to the outcome it preferred, if not an actual hindrance to it. Bosnian Serb militias began to treat UNPROFOR troops as their enemy, rather than as a neutral or peacekeeping force. In May they seized several hundred of the out-manned peacekeeping troops, holding them hostage to thwart any further U.N.-directed, NATO air strikes at Serb positions.

The situation was a dramatic illustration of what can go wrong when peacekeeping collides with aggressive military action. Throughout most of UNPROFOR's engagement in Bosnia, the actions of the Serbian faction convinced most of the world that it expected to achieve its goals through force of arms, not negotiation. Yet, the Security Council continually insisted that the UNPROFOR they had created must act neutrally toward the warring parties. The peacekeeping force should continue to behave, in other words, as if both sides welcomed its presence, which was blatantly not the case by 1995. What accounted for this apparent blindness on the part of the leading U.N. members? The answer lay, not so much in their impaired vision as in their unwillingness—largely for domestic political reasons—to commit the much larger, more assertive force that might have overpowered the Bosnian Serbs in their war aims.

Many of these conditions changed later in that same year. Bosnian Serb forces began to be defeated on the battlefield. Meanwhile, NATO largely shook off U.N. control and for the first time launched effective attacks on Serb targets. The United States, meanwhile, increased its pressure on the weary belligerents for a cease-fire. That led, in November, to a successful peace conference in Dayton, Ohio. The agreement hammered out there preserved Bosnia, at least theoretically, as a single nation, although it was divided into two parts, a Muslim-Croat federation controlling 51 percent of the land and a Serb republic holding the remainder.

The Bosnian peace agreement heralded the end of the much-maligned UNPROFOR operation and the creation of a NATO peacekeeping force of up to 60,000 soldiers. One significant difference apparently was that the NATO peacekeepers—named I-FOR, for *International Force*—were much more heavily armed than UN troops had been, and would operate under more "robust" rules of en-

---

[15] Put somewhat differently, the peacekeeping budget of the United Nations increased from $230 million in 1988 to $3.6 billion in 1994. See Barbara Crosette, "U.N. Chief Chides Security Council on Military Missions," the *New York Times,* January 6, 1995, A3.

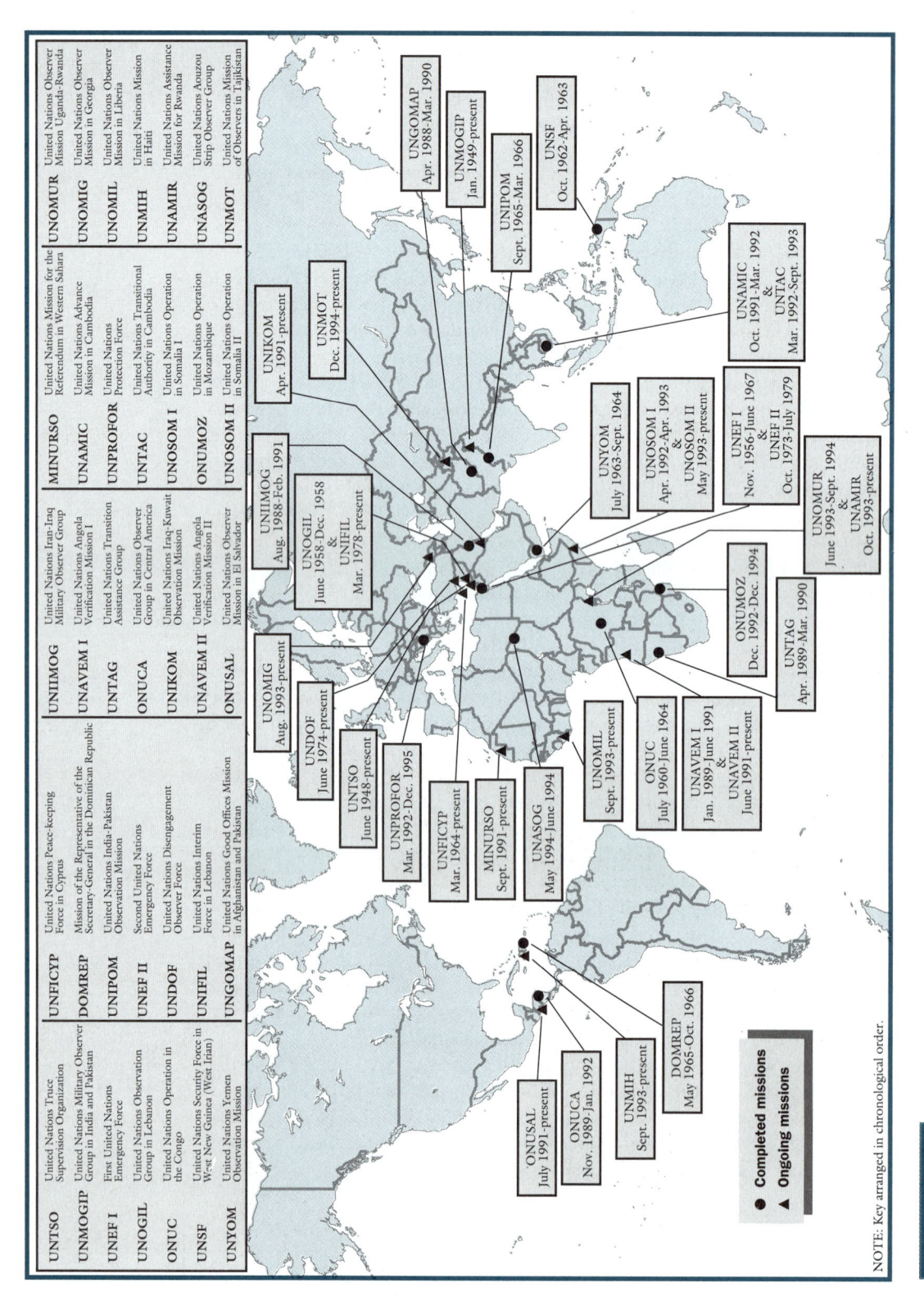

**MAP 17.1** U.N. Peacekeeping Operations, Past and Present

SOURCE: United Nations DPI/1306/Rev.4, February 1995, 7.5M.

gagement. In addition to large troop contingents from France, the United Kingdom, and the United States, I-FOR included representation from other NATO members and such non-NATO countries as Russia.

## THE UNITED NATIONS, COLLECTIVE SECURITY, AND THE GULF WAR

World leaders used the United Nations to respond very differently when the Iraqi government of Saddam Hussein invaded Kuwait in August 1990: Saddam's action was treated from the beginning as an act of aggression. As a result, the Security Council soon brought the sanctions provisions of the Charter into play in a way that had never been equalled during the cold war period.

**THE IMPOSITION OF SANCTIONS** The Council's action apparently constituted a return to collective security of a sort that had been intended by the United Nations' creators, forty-five years earlier. Over the course of the next four months, the Council passed a dozen resolutions, all based on the agreement of the majority (including the permanent members) that Iraq's attack on Kuwait was an act of aggression, that aggression was specifically outlawed by contemporary legal standards, and that it, therefore, had to be reversed—by the collective action of the world community, if necessary. The early resolutions put in place a general trade embargo of Iraq, calling on all U.N. members to comply. The resolution adopted on November 19, 1990, authorized the use of force against Iraq if it did not pull back from Kuwait by mid-January 1991. By the end of 1990, twenty-seven nations had posted armed forces to neighboring Saudi Arabia, although more than 80 percent of the several hundred thousand troops stationed there were from the United States.

**THE ARMED ATTACK AGAINST IRAQ** The United States then launched an attack against Iraq on January 16, 1991, almost immediately after Iraq had failed to meet the U.N. deadline for withdrawal from Kuwait. The war that followed was remarkably brief and decisive. After forty-two days, Kuwait was liberated and the Iraqi army nearly decimated. Deaths within the U.S.-led coalition forces numbered in the low hundreds, while Iraqi battlefield fatalities reached well beyond 50,000. At the war's conclusion, both the military coalition and the U.N. Security Council mandate that had helped produce it were still intact. Yet there had been signs of strain, suggesting a marked uneasiness with the way in which the United States had used the Council's resolutions to conduct a war largely of its own design.

**LESSONS ABOUT COLLECTIVE SECURITY** The Gulf War holds several lessons for collective police action in a post–cold war world. First, it was clear that one reason the Security Council was able to present a united front against Iraq was because the Iraqi invasion had been a classic—even an "old-fashioned"—example of what was expressly forbidden under the international law of the late twentieth century—that is, the seizure by force of territory recognized as sovereign. Much of the world's experience with violent conflict in the second half of the twentieth century was far less clear-cut in revealing an act of aggression. The Gulf War was a reminder that, now as in the past, collective security action would likely be limited to cases where blame could be assigned as unambiguously as when Saddam ordered his troops to seize Kuwait. (Contrast this situation with the outbreak of such civil conflicts as those that began later in 1991 in the former Yugoslavia.)

Second, the case quickly revealed the problems and frustrations of using economic sanctions to force an opponent to submit to the United Nations' authority. It is no doubt easier to achieve consensus to invoke economic—as opposed to military—sanctions, particularly as a first stage of collective action. But economic sanctions clearly can have little bite for weeks, months, or—depending on the nature of the economy of the target state—even years. The target government can use that time, as Saddam Hussein did, to try to pry apart the United Nations coalition through appeals to self-interest, regional or ethnic solidarity, or the willingness of many over time to accept a new political status quo. Moreover, as the close of the Gulf War revealed, during the months when economic sanctions were in effect, Iraqi forces were free to plunder Kuwait and to torture and murder large numbers of its citizens.

Third, achieving authorization for collective military action is much more troublesome to many precisely because its dangers are so obvious. This may be particularly true given the fact that the United Nations today still does not have the kind of structure for a unified command of military forces that was anticipated in the Charter. Such an arrangement was a victim of the cold war. In the early months of the Gulf crisis, the Soviets sought to make the Military Staff Committee the agent of any U.N.-sponsored military action that might be authorized. Perhaps agreement on how to provide the Security Council with its own standby forces will be possible in the future; nonetheless, its absence in 1990 led some to regard Security Council involvement in the Kuwait affair as little more than a fig leaf covering the U.S. determination to wage its own war against Iraq once sanctions failed.

Fourth, the U.N. sanctions against Iraq were a sharp reminder that collective security and compromise do not easily go hand in hand. President Bush's refusal to compromise in any way with Saddam Hussein's demands against Kuwait was connected to his view that the seizure of Kuwait was an illegal act that had to be reversed and the aggressor punished. Yet, one of the chief characteristics of most international conflicts is that they are not so unambiguous to most of the actors who confront them. The very basis of the state system supposes that sovereigns are free to interpret world events in light of their own interests, and since their interests are seldom identical, neither will be their views of right and wrong in their foreign policies. In such a world, there will be frequent disagreement on what kind of behavior must be stopped through the military enforcement capability of the international community. As the Gulf War showed, it is always troubling in ethical terms to try to justify using massive amounts of violence in the name of peace and justice.

## IMPROVING THE UNITED NATIONS' ABILITY TO KEEP THE PEACE

The Gulf War revealed a potential for building a U.N. structure for keeping the peace, while it also showed us that, in the words of a long-time U.N. official, "the United Nations has so far not provided a *system* for peace and security so much as a last resort, or safety net."[16] Because of the forty-year stalemate of the cold war, the United Nations' system for building peace and security remains rudimentary.

Yet, even with the heightened expectations for an improved security system after the cold war, the Security Council's permanent members have not yet been willing to engage in real reform. In the heady days immediately after the end of the cold war, the members of the Security Council directed the U.N. secretary-general to prepare recommendations for improving the United Nations' ability to treat violent conflicts.[17] The secretary-general's report was presented in June 1992. It contained both an analysis of the new dimensions of peace and security issues in the post–cold war world and new ideas for treating them.[18] Although the report was much discussed, its recommendations have not been implemented.

Meanwhile, a number of commentators and public officials have presented analyses and proposals of their own, both for reforming the Security Council to make it more nearly reflect current political realities and for improving the United Nations' ability to treat violent conflicts more effectively.[19]

## SECURING PEACE THROUGH REGIONAL FRAMEWORKS

As we have seen, states have never been willing to entrust their security entirely to the collective police powers of the United Nations or its predecessor, nor is there any reason to suppose they will do so in the foreseeable future. Instead, leading military powers and their friends have frequently sought security through their own arrangements.

### EUROPE: THE COLD WAR AND AFTER

Two such security arrangements—the **North Atlantic Treaty Organization (NATO)** and the **Warsaw Pact (WP)**—have been most closely identified with the cold war. The sudden and radical changes in the capability of these two organizations that occurred in 1989–1990 were themselves a significant indicator of the cold war's end. That is because world organizational structures always reflect characteristic patterns of interstate interaction, changing along with those patterns.

---

[16]Brian Urquhart, "Learning from the Gulf," *The New York Review of Books*, March 7, 1991, 34.

[17]That was the first time in history, January 31, 1992, that the Council had met at the level of heads of state. They asked the secretary-general for an "analysis and recommendations on ways of strengthening and making more efficient . . . the capacity of the United Nations for preventive diplomacy, for peace-making and for peace-keeping" (U.N. document S/23500, 115).

[18]Boutros Boutros-Ghali, *An Agenda for Peace*, 2d ed. (New York: United Nations, 1995).

[19]See, for example, Ingvar Carlsson, "The U.N. at 50: A Time to Reform," *Foreign Policy* 100 (fall 1995): 3–18; Ruben Mendez, "Paying for Peace and Development," Ibid., 19–32; James S. Sutterlin, *The United Nations and the Maintenance of International Security: A Challenge to Be Met* (London: Praeger, 1995); and Brian Urquhart, "Proposal for a U.N. Volunteer Military Force," *The New York Review of Books*, June 10, 1993; and "Who Can Police the World?" *The New York Review of Books*, May 12, 1994, 29–33.

> **Normative Dilemmas**

# Self-Interested Judges, Collective Goods, and Police Power

## Self-interested Judges

John Locke noted more than 300 years ago that "inconveniences" result where no government exists and each individual must judge his or her own case. Such a state of affairs is "unreasonable," said Locke, because

> self-love will make men [sic] partial to themselves and their friends and . . . ill-nature, passion, and revenge will carry them too far in punishing others, and hence nothing but confusion and disorder will follow.

- In the absence of world government, do states in a conflict act as judges of their own cases? What are some of the familiar ways in which they assert the justness of their own case when they go to war? What historical examples can you find where states have been "carried too far in punishing others" when they won their case on the battlefield?

## The Hunters and the Stag

A group of primitive hunters agree to form a circle in the forest and slowly close in to trap a stag. Then all will join in the spoils of the hunt. But as they move toward the trapped animal, a hare suddenly hops in front of the path of one of the hunters. He makes a quick calculation that catching the hare will satisfy his own hunger, and leaps after it. The stag, meanwhile, discovers the break in the circle and dashes through it to safety over the next hill.

- In what respects does this tale by Jean-Jacques Rousseau shed light on the central problem for collective security? Given the autonomy of individuals (or governments) to act in keeping with perceptions of their own interests, how can they be induced to choose the collective good of the whole (of which they are a part) over their immediate, selfish aims? What should the other hunters do about their selfish colleague to ensure that he (and all others) will support the collective good in their next stag hunt? Is it always ethically preferable to act on behalf of the collective good rather that one's immediate self-interest? Why or why not?

## Policing the World

The 1990s saw a great increase in the effort to police conflicts through multilateral action—by way of the United Nations, NATO, and other IGOs. Supporters of *multilateralism* saw it as a way to overcome the Lockeian problem of promoting, or seeming to promote, one's own self-interest when intervening to try to stop a conflict. Opponents argued that such "policing" intervention *should* be unilateral—that it should only take place when a state has a strong national interest in a particular outcome.

- On which side of this argument are you likely to find idealists, and on which side realists? Why does police action conducted by an IGO appear more legitimate as an attempt to act justly in a conflict than intervention by a particular great power? Must intervention that satisfies justice criteria always be neutral in its treatment of the parties to the conflict? Why or why not? What are the main differences between *neutral* and *objective* behavior when one is exercising a police power?

**THE NORTH ATLANTIC TREATY ORGANIZATION** NATO was created in 1949 by the United States and its allies bordering the North Atlantic.[20] In political terms, it marked the European cold war standoff, committing the members of the alliance to defend one another against any threat of attack from the Soviet Union. In this sense,

---

[20]The original members of NATO, in addition to the United States, were Belgium, Canada, Denmark, France, Iceland, Italy, Luxembourg, the Netherlands, Norway, the United Kingdom, and Portugal. Greece and Turkey joined in 1952, West Germany in 1955, and Spain in 1982. France withdrew its forces from the NATO command in 1966 without withdrawing from the alliance.

it was an alliance, pure and simple. But, given the novel requirements for deterrence as opposed to the mere defense of territory in the nuclear age, it was not enough to wait until an attack was imminent to activate NATO's treaty provisions. All members recognized the need for consultation and military planning if the enemy were to be deterred from initiating a conflict; over time, that meant the considerable integration of troops and weapons systems as well as the planning of strategy, which required complex organizational structures unlike any known for earlier alliances.

**THE WARSAW PACT**  In 1955, the Soviet Union entered into a similar treaty arrangement with the states of Eastern Europe that had fallen into the Soviet orbit in the years just after the end of World War II.[21] As in NATO, the Warsaw Pact engaged in joint military planning while its leading member, the USSR, stationed many thousands of its own troops throughout the territory of its smaller allies. Unlike the experience of NATO, however, the WP was used by the Soviet government to "police" its own allies and assure the orthodoxy of their governments. In 1968, troops from several WP countries—East Germany, Poland, Hungary, and Bulgaria—joined the Soviets in an invasion of Czechoslovakia to crush the liberal communist regime of Alexander Dubcek. That action was justified by the WP allies as insuring that Czechoslovakia not be wrested away from its "fraternal" ties through subversion from the NATO countries.

The invasion was actually an ingenious use of the alliance by the Soviet leaders to engage in a harsh enforcement action against a presumed friend. (The Soviets had not been so careful in 1956, when their troops invaded Hungary for a similar purpose, but without any participation from other WP governments.) It made a mockery of the idea that this treaty arrangement was anything other than the instrument of hegemony; it also foreshadowed the revolutionary reforms that would sweep most of the WP members more than twenty years later. Indeed, it was apparent, especially after the invasion of Czechoslovakia, that the only solidarity in the alliance was on the part of orthodox communist regimes that lacked the popular support of their own people.

**THE COLD WAR PEACE STRUCTURE**  Nonetheless, the two alliance systems of NATO and the Warsaw Pact together did constitute, in a primitive sense, the cold war framework that kept the peace, essentially by neutralizing each other. The iron curtain between them defined the cold war's frontier, over which neither side dared step without risking nuclear holocaust. And the alliance structures defined the regions within which threats to the political status quo would be resolved, whether by force or diplomacy, by the parties (or dominant party) to the relevant treaty. The existence of these organizations helped underscore an unwritten rule of cold war behavior, that neither superpower would intervene in the affairs of the other's alliance. That contributed far more to the stability of the bipolar period than was often apparent at the time, since what dominated our awareness was the cold war antagonists' rhetoric of mutual antagonism and threat.

[21]Warsaw Pact members, in addition to the USSR, included Bulgaria, Czechoslovakia, East Germany, Hungary, Poland, and Romania.

It is also revealing that both the superpowers stumbled into their gravest military defeats in areas where they had been much less successful at establishing their claim, through construction of a regional IGO, to an exclusive interest in the region in question. For the United States, that was the Vietnam War of the 1960s and 1970s, and for the Soviets, the Afghanistan conflict that began in 1979 and lasted nearly a decade. In both cases, the superpower governments mistook the strong ideological support of a client regime for a legitimate claim to the allegiance of that regime's people. Both conflicts seriously undermined the superpowers' presumed military superiority, partly because these wars were regarded increasingly as unnecessary and misguided by their own citizens—those who had to sacrifice their lives in combat.

**EUROPEAN SECURITY STRUCTURES AFTER THE COLD WAR** By 1990, the Warsaw Pact was a dying institution, since some of its members, by then led by non-communist governments, had demanded and received agreement from the Soviet Union to remove its troops from their territory. In November of that year, the WP was officially abolished. Although no comparable demands had been raised by any of the European NATO members against the United States, NATO's future was in question for the obvious reason that few in the West any longer expected or feared an attack by the Warsaw Pact on NATO territory. If these security plans were no longer urgent, why should Western Europeans submit to having foreign troops indefinitely garrisoned on their soil?

These questions were further complicated by the rapid moves toward the integration of the two Germanies after more than forty years in which they had been on opposite sides of the front line. When the East Germany reforms late in 1989 set the stage for reunification, President Gorbachev first argued that a united Germany should be made neutral and, therefore, detached from any military alliance. But in October 1989, the reunified Germany remained as a NATO member. Soon, NATO membership became a goal for several of the Eastern European states of the former Soviet bloc, for whom the memory of domination from the East made NATO's protection appealing. In 1994, in its Partnership for Peace plan, the United States proposed creating a new associate membership category for those states, and for Russia and other newly independent states that had been carved out of the former Soviet Union.

Meanwhile, another evolving organization seemed more suitable to some as a European security arrangement. The **Organization for Security and Cooperation in Europe (OSCE)** grew out of a 1975 effort at East-West detente, the so-called **Helsinki process.** It unites more than fifty countries of Europe (and Asia, since nearly all the former Soviet republics joined in 1992), including the neutrals that never were tied to either NATO or the WP, as well as the United States and Canada. Although its broad membership could make it an appropriate forum for treating issues of a post–cold war Europe, it permits action only by consensus; that is, any member can block an action that might be acceptable to the rest. This requirement has prevented it from taking a strong role in the conflicts of Central and Eastern Europe that have broken out since the cold war's end. The OSCE now seems likely to be more effective on issues that require economic, social, and environmental cooperation than toward the questions of military capability about which governments tend to feel most protective and adversarial.

The civil war in Bosnia raised fundamental questions about the use of European security structures after the cold war. From 1992 through 1995, NATO was largely the agent of the United Nations. But the lightly armed UNPROFOR—which was meant to be neutral in its relations with the combatants—could not function effectively where there was no peace to keep. Neither was NATO allowed to do much effective policing as the conflict raged. That situation changed with the Bosnian peace agreement at the end of 1995, in which NATO was charged with securing the peace with a much more substantial deployment of forces. This would be easily the most ambitious effort to date to adapt a regional security structure to the needs of post–cold war Europe. If successful, it would no doubt make a "new" NATO the foundation of a multilateral security system for Europe.

Whatever the outcome in Bosnia, cross-national influences on foreign policy decision making will remain and, no doubt, grow. For example, whether or not the OSCE develops a more elaborate framework for European security, it has spawned work in the nongovernmental sector in various countries to support the Helsinki process. One such important NGO in the United States is the Helsinki Watch Committee, founded in 1979 to monitor domestic and international compliance with the 1975 agreements that launched the OSCE. Its *News from Helsinki Watch* provides investigative reporting on relevant developments in OSCE countries as a way of mobilizing opinion and action on behalf of Helsinki's goals. It is linked to similar organizations that attend to conditions in Africa, Asia, Latin America, and the Middle East. This example demonstrates how networks of transnational responsibility, both governmental and nongovernmental, are increasingly being constructed to add to the global framework for peace.

## OTHER REGIONAL STRUCTURES FOR PEACE

Europe became the focus of efforts to build regional security structures because it was where the cold war originated at the close of World War II. Security needs elsewhere in the world, however, have made other IGOs salient in regional peacemaking, as well.

**THE ORGANIZATION OF AMERICAN STATES** The early cold war period saw the United States take the lead in transforming its historic relationship with Latin America into an IGO with conflict-management potential. Specifically, the United States attempted to use the **Organization of American States (OAS),** which was created in 1948 out of the older Pan-American Union, as an instrument of its own security interests in the hemisphere, and one meant to exclude Soviet or communist influence.

The United States had some success with this into the 1960s, while the cold war was at its height. In 1954, U.S. covert action removed a leftist government from Guatemala while Washington argued successfully at the United Nations that the Security Council need not take up the matter on grounds that the OAS Council of Ministers was considering the situation. In 1965, a unilateral intervention by U.S. Marines in the Dominican Republic brought much criticism to the administration of President Johnson both at home and abroad. The United States then persuaded reluctant Latin American governments to turn its unilateral action into an OAS-sponsored Inter-American Peace Force that helped prepare for democratic elections. Even so, a number of these governments were not inclined to see the threat of communism behind every sign of unrest or change in the hemisphere and refused to contribute troops to the Dominican effort.

The 1960s marked the high point of Washington's concern about Fidel Castro's capacity to foment anti-U.S. unrest south of its border. But, as Castro's revolution aged, the only other government with Marxist leanings to come to power in the hemisphere in the 1970s, that of Salvador Allende in Chile, was brutally ended in assassination. Allende was succeeded by a military dictatorship, and the threat of "communist subversion" in Latin America receded. By the time President Reagan came to office and made support of the anti-Sandinista contras in Nicaragua a major pillar of his foreign policy, no one seriously supposed that he could make an OAS-based coalition serve his purposes.

The fact is that the OAS was never conceived as an anticommunist alliance by most Latin American states, which explains its very limited, early success in that role. It has remained a "structure for peace" after the cold war to the extent that it encourages regional economic and social development, while taking on relatively few cases of direct military conflict.

**THE MIDDLE EAST, AFRICA, AND ASIA** Much the same can be said of several other regional IGOs formed with the intent of improving political relationships among neighboring states by facilitating communication and economic and social exchange. That is, they have survived and, in some cases, thrived as instruments of cooperation in areas of activity *outside* the immediate security sphere.

This contradicts the expectation by some authorities in the 1950s and 1960s that these arrangements would develop a regional police capability that would complement, or compete with, what was being somewhat erratically exercised by the Security Council at the time. Among the candidates, in addition to the OAS, were the **Arab League,** which was created at the end of World War II to join all the Arab states of the Middle East and North Africa in a cooperative framework; the **Organization of African Unity (OAU),** which was formed in 1964 when much of Africa was achieving independent statehood for the first time in the modern age; and the **Association of South-East Asian Nations (ASEAN),** which was created during the cold war to unite the most pro-Western states of that region. It remained for a nominal economic union to marshal the largest regional peacekeeping force in Africa: In 1990, the seventeen-nation **Economic Community of West African States (ECOWAS)** sent a force of 7,000 troops from five states to quell violence in Liberia and to help administer the country.

Today, hundreds of IGOs unite several or many nation-states. Of these, very few—other than those just previously discussed—either have had, or are soon likely to develop, any serious roles as regional peacekeepers. Almost all of them have overt purposes that do not seem to relate directly to the security of states at all. Rather, they provide investment capital (e.g., the African Development Bank), cooperation for space exploration (the European Space Agency), regulations for an international waterway (the Mekong Committee), or the promotion of regional trade (the Latin

American Economic System). As reflections of humanity's interdependence today, they may be contributing to our perception of the growing unity of humankind.

## THE CONSTRUCTION OF INTERNATIONAL REGIMES

It is not the visible institution, whether of an IGO or an NGO, that alone explains where or why particular patterns of international interaction have occurred and persisted to the point that they constitute actual governance of certain intersocietal relationships today. The structure of a building does not tell us what life is like inside it, even though we may be sure that a building meant to house a family will likely be designed to enhance family life to some degree, while a structure meant as a school or factory will be designed in a comparably different way. Similarly, the institutional structure of the United Nations was created to make possible and encourage certain kinds of interstate behavior, but, as the evolution of peacekeeping shows, it provided wide leeway for adaptations to a new and unexpected world political environment. After about forty years of evolution, U.N. peacekeeping is now an international regime that tells us something of how life is lived within the U.N. framework. As such, it is one of countless such regimes in the world today.

**International regimes** constitute those distinctive patterns of behavior engaged in by global actors that are produced by their particular structural or organizational arrangements.[22] In the words of Stephen D. Krasner, "international regimes can be defined as sets of implicit or explicit principles, norms, rules, and decision-making procedures around which actor expectations converge in a given area of international relations" that may help to coordinate their actions.[23] It is the existence of all of these factors together that produces what we have called patterns of international behavior. As their definition suggests, regimes are seldom the one-time creation of a single act of international negotiation, but are the organic product of repeated common policies, agreements on incremental goals, and reinforced perceptions of the need for "sovereign" cooperation to treat the real demands of social life.

International organizations with their treaty structures can have a major role in producing such regimes, but their connection to them is more complex than that. An IGO may serve to regulate (and, in the process, further refine) a desired regime, as does the International Civil Aviation Organization for commercial air travel. Or it may mark the culmination of a prior evolution of a regime, as occurred with the creation of the U.N. Development Program (UNDP) in 1965. Regimes may also arise without real assistance from international organizations. It is in this sense that the U.S.-Soviet relationship of mutual nuclear deterrence during the cold war is sometimes viewed as a regime.

---

[22]Oran R. Young, "International Regimes: Problems of Concept Formation," *World Politics* XXXII, no. 3 (April 1980): 331–356.

[23]Stephen D. Krasner, "Structural Causes and Regime Consequences: Regimes as Intervening Variables," in *International Organization: A Reader,* ed. Friedrich Kratochwil and Edward D. Mansfield (New York: Harper Collins, 1994), 97.

The U.N. General Assembly has frequently had an important, instrumental hand in regime creation. Take the development of a regime for outer space as an example. The space age began with the launching of *Sputnik* by the Soviet Union in 1957. Soon thereafter, the United States followed suit. Almost immediately, the General Assembly passed a nonbinding resolution (as is the case, theoretically, of all Assembly resolutions) calling upon the space powers to engage in space exploration for peaceful purposes. Over the next several years, as the U.S. and Soviet space programs developed, the Assembly adopted several additional resolutions expressing views of the members on principles to govern their space activities. Of critical importance was the fact that the two states with the capability to obey or disobey these directives—the space powers themselves—supported these resolutions. Meanwhile, the Assembly had created a standing Committee on the Peaceful Uses of Outer Space, whose legal subcommittee eventually was charged with drafting a treaty to govern the exploration of outer space and extraterrestrial bodies. That drafting group's central members were U.S. and Soviet negotiating teams, which, with the other negotiators, agreed to a text in 1966. The Outer Space Treaty has served as the basis of the regime that has remained in place, with refinements and expansion, ever since.

We are so habituated to associate *government* with what the sovereign state alone does, and *anarchy* with a description of relationships between states, that we are not used to noticing the growing skein of *governance* without the existence of a formal government.[24] The development of international regimes is simply a recent indicator that human societies, including their interrelationships, continue to evolve. It may be helpful to remember that the sovereign nation-state system that tends to look to us today like an eternal feature on the historical landscape is in fact a very recent invention for a species that has lived in highly organized communities for many thousands of years. It may be that, scarcely without noticing it, we—our governments and ourselves, acting in private groups—are remolding that system in unimagined ways as we work to strengthen the framework and the fabric of peace.

# Conclusion

Intergovernmental organizations have grown in number and intent throughout the twentieth century, and their effect on world politics may differ in various ways from the expectations of those who created them. In general, IGOs have contradictory impulses. They are conservative to the extent that they preserve and strengthen the nation-state system, and radical when, through collective action, they may transform that system.

The League of Nations was the first IGO to attempt to implement collective security, in which a large number of states agree to repel an attack upon any one of them. The League's effort to turn back the Italian aggression against Ethiopia in

---

[24]A new scholarly journal is entirely devoted to such questions. See, in its first issue, James N. Rosenau, "Governance in the Twenty-first Century," *Global Governance: A Review of Multilateralism and International Organizations* I, no. 1 (winter 1995): 13–43. See also Harlan Cleveland, "The Future of International Governance," *The Futurist*, May/June 1988, 9–12.

1935 by the use of economic and diplomatic sanctions was a failure. In the attempt to learn from the lessons of that failure—which included the slowness of economic sanctions to work, avoidance of sanctions by those outside the League, and willingness to accept a new status quo built upon force—the founders of the United Nations modified and tried to strengthen collective security in the new organization. They also built a more complex framework for peace, encouraging economic and social development as a means to reduce the incidence of violent conflict by improving human well-being.

With the onset of the cold war, the Korean conflict ostensibly complied with the security provisions of the U.N. Charter, but did so over the objections of one of its permanent members, the Soviet Union. Thereafter, peacekeeping evolved as a way of stabilizing the bipolarity of the cold war by preventing large-scale conflict in regions where the superpowers competed for influence. With the end of the cold war, the collective enforcement provisions of the Charter were dramatically revived in the Security Council's opposition to the Iraqi seizure of Kuwait. But the U.N./NATO intervention in Bosnia was no success because it was predicated on peacekeeping in a situation where armed factions sought a military solution instead. Once a peace accord was reached, NATO had a more plausible role to play as its enforcer.

At regional levels, the cold war created rival European security arrangements in NATO and the Warsaw Pact. With the end of the cold war, the WP was abolished and a reunited Germany remained as a NATO member. Other security structures in Europe—for example, the OSCE—have been slower to achieve their promise. Early in the cold war period, the OAS had sometimes been used as an anticommunist security structure in the Americas; after the late 1960s, however, that regional organization increasingly became an instrument of intraregional economic and social cooperation rather than of U.S. security interests. In this, it resembled the great majority of other world regional organizations of the late twentieth century.

Along with IGOs, international regimes have been created in increasing numbers since the end of World War II. These are less the conscious products of efforts to build institutions than they are the patterns of behavior and relationships that result from particular structural or organizational arrangements made by global actors. The world operates in many such regimes today, including those for civil aviation and the exploration of space. An understanding of international regimes is a reminder that governance can and does exist in a world where formal government frequently is lacking.

# Glossary

**Aberrant personality.** A personality that diverges from what is considered as normal.

**Accidental war.** War resulting from some sort of accidental firing or unauthorized use of weapons by subordinates.

**African Charter on Human and People's Rights.** A human rights treaty completed in 1981 by the Organization of African Unity. Also known as the Banjul Charter.

**Amalgamation.** In international politics, the formal merger of what have been distinct political units, whether by mutual agreement of their governments or by conquest and annexation of one unit by another.

**Ambassador plenipotentiary.** Literally an ambassador with complete power.

**American Convention on Human Rights.** A human rights treaty for the states of the American hemisphere establishing an inter-American Court of Human Rights. Some two-thirds of the members of the Organization of American States are signatories; the United States has not yet ratified the treaty, which became effective in 1978.

**Antiballistic Missile Treaty.** Part of the SALT I agreements restricting antiballistic missile development in the United States and the Soviet Union.

**Arab League.** Organization formed in 1945 among Arab states to promote economic, social, political, and military cooperation.

**Arbitration.** A dispute-settling technique, in which the arbitrators act as judges to apply specified legal rules to an issue. Often, each of the two parties to the dispute select an arbitrator, who in turn select a third. The judgment is typically decided by a majority.

**Asia-Pacific Economic Cooperation (APEC).** Organization designed to promote trade and investment in the Pacific basin with the goal of establishing a free trade area by 2020. Its eighteen members include the United States, Canada, and both Far Eastern and Southeast Asian countries.

**Associations of governments.** Alliances or other international organizations designed to coordinate policy among two or more governments.

**Association of South-East Asian Nations (ASEAN).** A regional IGO established in 1967 to link six noncommunist countries of Southeast Asia to further their economic, social, and cultural cooperation. Vietnam became a member in 1995.

**Autarky.** National economic self-sufficiency.

**Authoritarian personality.** A personality favoring complete obedience or subjection to authority in contrast to individual freedom.

**Ayatollah.** Title in the religious hierarchy of Shiite Muslims. In 1979, the Ayatollah Khomeini became the de facto head of Iran.

**Balance of payments.** Difference between a nation's total payments to foreign countries for goods, services, investments, and so on and its total receipts from foreign countries.

**Balance of power.** Distribution of power among states such that no single nation or group is able to dominate.

**Balance of trade.** Difference between the value of the import and export of goods between countries. Said to be favorable if exports exceed imports.

**Banjul Charter.** See *African Charter on Human and People's Rights*.

**Beggar-thy-neighbor.** Policies designed to change the trade balance by imposing import restrictions or devaluing domestic currency.

**Beliefs.** Tenets that are regarded as true and therefore usually accepted on faith.

**Biological Convention.** A 1972 treaty that required the signatories to destroy any biological weapons then in their possession, and committed them never to develop such weapons in the future.

**Biosphere.** Every aspect of the natural world, including the earth's resources, atmosphere, and all the living creatures of the planet.

**Bipolarity.** When power in the system is divided between two major groups.

**Blacklisting.** The use of trade as a political instrument to pressure states and corporations to behave in a desired way: Anyone involved in trade with a stated enemy is forbidden from business activities with the country issuing the threat. Used by Arab states against corporations that conducted business with Israel.

**Bourgeoisie.** In Marxist terminology, the middle and capitalist classes of industrial society, whose beliefs and values are shaped mainly by concern for property values and conventional respectability.

**Boycott.** Refusal to buy goods from a given country or seller.

**Brain drain.** The term describing a tendency in the modern world for intellectuals from less developed countries to emigrate to highly developed countries for their education and employment.

**Bretton Woods system.** The rules and institutions that were established at the close of World War II to govern economic relations in the postwar period; some of these institutions, such as the World Bank and the International Monetary Fund, continue to play central roles in the world economic system.

**Bureaucracy.** In every modern society, the state employees charged with administering governmental policies.

**Bureaucratic politics.** The role played by various bureaucracies in shaping policy and influencing the foreign-policy choices of political leaders.

**Camp David Accords.** A 1978 agreement in which peace was established between Egypt and Israel, and the Sinai was returned to Egypt.

**Capability.** The economic and military resources of a state. It provides the basis for exerting power, but is not power itself.

**Capitalism.** An economic system in which investment and the ownership of the means of production and distribution are in the hands of individuals or corporations rather than the state.

**Chlorofluorocarbons (CFCs).** Chemicals once widely used in refrigerants and aerosol sprays, now being phased out of production. They are responsible for most of the destruction to the earth's ozone layer.

**Codification.** In international law, those rules of behavior that have been expressly written down and agreed to, usually in treaty form, by states.

**Cognitive dissonance.** Anxiety produced by simultaneously trying to maintain contradictory or incompatible beliefs or attitudes.

**Collective security.** The concept that an attack upon one state committed to collective defense constitutes an attack upon all, who are pledged to respond unitedly to end the aggression. The concept formed the basis for the central security arrangement for the League of Nations and, with modifications, for the United Nations.

**Command and control.** Issues related to decisions regarding the use of weapons, usually nuclear ones.

**Common heritage.** A principle advanced in U.N. Law of the Sea negotiations in the 1970s claiming that the resources of the deep seas constitute the common heritage of humanity and, therefore, cannot lawfully be appropriated for the profit of some without compensation to other states.

**Common market.** An organization in which there is free exchange of goods, capital, and labor. The prime example is the European Common Market.

**Comparative advantage.** Theory that every country can benefit from international trade if it concentrates on goods that it can produce less expensively than its trading partners.

**Conventional Forces in Europe (CFE) Agreement.** A 1990 agreement between NATO and Warsaw Pact countries that required deep cuts in the numbers of conventional weapons the Soviet Union had deployed in Europe, and far fewer cuts in NATO's conventional weapons.

**Council of Europe.** Created in 1949 by eighteen Western European states to promote greater unity on a wide variety of issues. It produced the European Convention on Human Rights (in force from 1953), and now includes most of the countries of Eastern Europe among its members.

**Council of Ministers (European Union).** The principal decision-making arm of the European Union, the Council consists of ministers from member governments, whose participants change with the agenda (agriculture ministers determine farms prices, economics ministers deal with employment issues, and so on). However composed, the Council meets every few days, usually in Brussels, the EU headquarters. When heads of state of the members meet—as they do two or three times annually—this body becomes the European Council.

**Cruise missile.** A pilotless, low-flying vehicle with precision guidance that can be launched from the ground, from the sea, or from the air and can carry either nuclear or conventional warheads.

**Convertible currency.** Currency that can be interchanged for another national currency or into gold.

**Customary practice.** In international law, standards that have evolved out of repeated, or customary, behavior of states in their interactions. In the contemporary period, much customary practice has in fact been codified (see Codification).

**Customs union.** A free trade area in which barriers to the free flow of goods are abolished and a common external tariff is established.

**Damage limiting strategy.** Strategies designed to limit nuclear damage to oneself by either offensive or defensive means. The latter would include civil defense programs and antiballistic missile systems.

**Demographic profile.** Statistical information of a given population, especially as related to age, income, education, and so on.

**Dependency theory.** The theory that in the modern world economy, most less developed countries are in varying degrees dependent upon decisions made by the highly developed states for their livelihood. Dependency theory is typically a critique of capitalism for the economic inequalities it allegedly spawns throughout the world.

**Detente.** The relaxation of tension between adversaries.

**Deterrence.** The use of threats to discourage undesired behavior (such as military attack).

**Developed countries (DCs).** The term generally used for modern, highly industrialized—or postindustrial—states.

**Diplomatic immunity.** Immunity from the jurisdiction of the state to which the diplomat is accredited. Diplomats are typically immune from such local jurisdiction (with some reservations) to ensure their ability to act as effective agents of their governments in their host country.

**Diplomatic recognition.** Recognition given to another state either *de facto* (noting the fact of its existence) or *de jure* (accepting it as a legitimate and sovereign entity).

**Dollar diplomacy.** A U.S. policy of the Taft Administration (1909–1913) that asserted North America's right to intervene in the domestic affairs of Latin American states to protect or advance the economic interests of American companies.

**Domestic jurisdiction.** An international legal concept asserting that issues and state policies whose impact is essentially internal to a state are off-limits for action by external, or international, actors.

**Dove's dilemma.** The fact that restricting arms trade to a state may lead it to develop its own arms production facilities and become an exporter of arms, thus adding to the global supply of arms.

**Dualism.** The concept that views international law as authoritative in a dual process: first, it binds sovereign governments that assent to it; second, it is thereby incorporated into that sovereign's domestic law, where it binds its citizens. In this view, individuals are thus only indirectly the subjects of international law.

**Dumping products.** Price discrimination between exports and domestic sales of a given product. Viewed as violating GATT rules by providing an opportunity to gain market share, thus resulting in unfair competition.

**Economic Community of West African States (ECOWAS).** Organization formed in 1975 involving sixteen members with the aim of promoting regional economic cooperation, with some attention given to security issues in West Africa.

**Economic determinism.** The idea that the economic situation or condition of a state or persons predisposes it to act in certain ways. The term often implies that those who hold this view regard such economic factors as controlling, determining political and other developments in a society's life.

**Economic union.** The highest stage of economic integration, which not only includes the free flow of capital, goods, and labor, but also involves harmonizing economic policies among its membership, particularly with regard to macroeconomic and regulatory policy.

**Embargo.** Prohibition placed upon trading with another state in the hope of influencing its policies.

**Ethnocentrism.** Belief in the inherent superiority of one's own group, according to which all other groups are evaluated.

**European Commission.** The executive office of the European Union. It is composed of at least one citizen from each member state, appointed for four-year terms by mutual agreement of member governments. Members may not receive instructions from any government and are supervised by the European Parliament. The Commission is charged with ensuring that rules and principles of the EU's treaties are respected. It may also propose measures for the Council of Ministers to enact.

**European Community (EC).** See *European Union*.

**European Convention for the Protection of Human Rights and Fundamental Freedoms.** The first treaty to give specific legal content to human rights, the convention established a commission, a committee of ministers, and the European Court of Human Rights to address issues raised under the treaty. Twenty-seven European states are members.

**European Economic Community (EEC).** The 1957 treaty that launched a common market for members of what today is called the European Union. It is the most important of three Community treaties—the European Coal and Steel Community and the European Atomic Energy Community are the others—managed by the common institutions of the EU.

**European Free Trade Association (EFTA).** An organization of states not originally members of the EEC, EFTA was created in the late 1950s to eliminate trade barriers among its members, thereby making it easier for their possible entry into the Common Market. To date, EFTA members that have joined the EU are the

United Kingdom, Ireland, Portugal, Austria, Finland, and Sweden. Iceland, Liechtenstein, Norway, and Switzerland are EFTA states that remain outside the EU.

**European Parliament.** The quasi-legislative arm of the European Union. Its members have been elected by universal suffrage since 1979, for five-year terms. Deputies take their parliamentary seats on the basis of political groups rather than nationality. Meeting in Strasbourg, France, the Parliament comments on policy proposals made by the Commission, which may amend them as a result. It also has budgetary and other supervisory functions. The Parliament is expected to develop stronger legislative powers as European integration proceeds.

**European Union (EU).** The name (since 1995) of what had been the European Community. With fifteen members in 1995, the EU was built upon a number of treaties that originally bound only six countries in 1948. Their purpose is the gradual integration of member states. Members have created a common market (see European Economic Community) and are working toward a common financial system, social policies, and ultimate political union.

**Export restrictions.** State limits to the export of certain kinds of goods to specific countries.

**Extended deterrence.** Extending the deterrent threat to protect another state or ally.

**Fascism.** A governmental system headed by a dictator that suppresses all opposition, emphasizing an aggressive nationalism and often racism.

**Feminism.** A theoretical approach critical of much traditional thinking about world politics on grounds that it is biased in favor of maleness, reflecting male domination of world political practice.

**Feudalism.** The social and economic system of the Middle Ages, especially in Europe, based upon holding lands in fief or fee. Social roles are hierarchical, linking a superior (lord) with an inferior (vassal). The lowest class, serfs, were attached to the land.

**Finite deterrence.** Notion that one does not need unlimited numbers of weapons to deter another or even to match the numbers of the other side. All that is needed is enough to assure the other side that it will be subject to unacceptable damage.

**Fixed exchange rate.** Currency exchange rates established by a government for various foreign currencies. It may or may not be related to the real market value of the currency.

**Flexible response.** U.S. and NATO military doctrine developed in the 1960s providing for an adjustment in response based upon the level of provocation of the adversary. A conventional strike would not initially be met with a nuclear response.

**Floating exchange rate.** Letting the market determine the value of a state's currency, which in turn would be dependent upon the confidence of others in a given economic and political system.

**Foreign aid.** A broad term including every kind of grant or loan that one government bestows upon another.

**Foreign direct investment (FDI).** Investments made by the private sector—individuals, banks, and multinational corporations—in economies other than their own.

**Foreign office.** Any government's ministry or executive department responsible for the conduct of its foreign policy.

**Freedom fighter.** One who fights to free a particular social group from rule by another.

**Fractionation.** Notion developed by Roger Fisher, suggesting the utility of breaking issues into smaller ones and developing agreements on those where common interests can be found, rather than waiting until more comprehensive solutions can be found.

**Free trade area.** Area in which parties agree to allow the free flow of goods without tariffs and other restrictions. Each state, however, may retain its own external restrictions upon trade with outsiders.

**Freezing of foreign assets.** A state economic sanction prohibiting foreign citizens from withdrawing capital goods or profits from the country.

**Frustration-aggression hypothesis.** Notion that people who are frustrated will respond aggressively. Although initially viewed as an innate tendency, subsequent research suggests that there are other ways to respond to frustration than simply being aggressive.

**Functionalism.** A theory of international integration that asserts that states should first develop cooperative ventures in "functional" areas of *low politics*, where political passions are not involved. As those ventures produce benefits to citizens in the cooperating countries, they will support additional such actions. Eventually, the habit of cooperation may produce a "spillover" effect so that integration may proceed at the level of *high politics*.

**General Agreement on Tariffs and Trade (GATT).** A U.N.-affiliated organization established in 1947–1948 as a forum for promoting international trade. Its functions have basically been assumed by the World Trade Organization, which was created as a result of the Uruguay Round of negotiations that concluded in 1994.

**Geneva Protocol.** A 1925 agreement banning the use of chemical and biological weapons.

***Glasnost.*** Russian term used by Mikhail Gorbachev in the late 1980s calling for political openness and freedom in the Soviet Union.

**Global commons.** The principle in international law that some areas of the planet are not subject to territorial claims by states, but constitute a shared commons of the international community. The high seas are defined as a global commons and, in recent decades, so are space and extraterrestrial bodies.

**Global warming.** The hypothesis that human activities, particularly industrial practices, are heating up the earth's atmosphere to the point that climate changes may be underway. See *Greenhouse effect.*

**Globalism.** A neo-idealist approach to the study of world politics. This perspective views states as increasingly interdependent component parts of a single world social system.

**Government.** The legal authority of a state, typically manifested in established institutions through which the state's public policy is administered and controlled.

**Greenhouse effect.** Caused by the release of carbon dioxide and other gases into the atmosphere, which trap the sun's short waves from being reflected back into space. The effect is much like the principle whereby the glass walls of a greenhouse retain the sun's heat.

**Gross domestic product (GDP).** Preferred measure of the total value of economic goods and services produced and marketed, since it excludes profits on foreign investments that may skew the results without really measuring productivity.

**Group of Seven (G-7).** Regularized summits among leaders from Britain, Canada, France, Germany, Italy, Japan, and the United States that deal primarily with economic issues. Although not a full member of the group, Russia has also become a participant.

**Group of 77.** Initially formed by 77 less developed states in 1967, it now includes more than 120 countries concerned about promoting economic cooperation and development.

**Groupthink.** The tendency of cohesive, insulated groups to suffer from a deterioration of mental efficiency, reality testing, and moral judgment. Used to explain why bureaucratic groups often make less desirable decisions than an individual might.

**The Hague Peace Conferences.** Meetings at the Hague, in the Netherlands, in 1899 and 1907, when representatives of most of the sovereign governments then in existence convened to consider ways to improve international cooperation. They in some respects served as models for the creation of the League of Nations Assembly in 1919, and for the United Nations General Assembly in 1945.

**Hard currency.** Refers to those currencies that are readily convertible and accepted as currency in countries other than those that produce them.

**Hegemonic stability theory.** The theory that a dominant state is necessary to assure cooperation and to enforce the rules. It is applied primarily to the operation of the international economy.

**Helsinki process.** The process whereby human rights issues are considered in periodic meetings of the fifty-four–nation Organization for Security and Cooperation in Europe (OSCE). The process was launched by the 1975 Helsinki Agreements, signed in a period of cold war detente by thirty-five states in Europe and North America that at the time formed the core of the Soviet and Western blocs. The Agreements expressed the intention of all signatories to respect human rights.

**High Commissioner for Human Rights.** A United Nations post, created in 1993, responsible for promoting and protecting human rights throughout the world. The Commissioner is responsible for overseeing country-specific and thematic mechanisms throughout the U.N. system.

**Historical determinism.** The notion that free choice is limited and that a basic inevitability underlies the unfolding of events.

**Hot line.** Direct communication link between Moscow and Washington established in 1963 to minimize the danger of accidental war in the light of the Cuban missile crisis. It has been upgraded several times since then and similar hot lines have been established between other adversaries.

**Human Rights Commission, U.N.** A standing body of the U.N. Economic and Social Council, its 53 members are selected to assure equitable representation from member states. The commission has drafted many of the human-rights treaties that appear in Table 16.1.

**Human Rights Committee.** Created to help implement the Covenant on Civil and Political Rights, this eighteen-member committee of experts takes no instruction from governments signatory to the treaty. It reviews annual reports from states and listens to petitions from individuals whose states have accepted an optional protocol permitting such petitions.

**Idealism.** In world politics, the philosophical outlook that emphasizes the fundamental unity of political and social behavior from the local to the world level, and typically seeks to apply the lessons of stable domestic political systems to the global system. It tends to view the world system as undergoing continuous transformation, rather than stasis.

**Ideology.** A belief system that is accepted as truth by some group and provides the believer with a fairly thorough picture of the world.

**Idiosyncratic variables.** Factors used to explain foreign-policy choice based upon individual characteristics such as personality traits, motivation, perception, and the like.

**Imperialism.** The practice of extending a state's authority by territorial acquisition or through establishing economic or political hegemony over other nations.

**Import quota.** Limit placed upon the number of items of a particular product that can be brought into a country.

**Innate aggression.** The notion that aggression is caused by conditions inherent in the biological makeup of human beings.

**Intangible factors.** Variables that cannot be measured. For example, such qualities as leadership, morale, educational level, and technological capability all affect the power of a state, but are very difficult to measure.

**Integration.** The usually purposeful process whereby two or more states merge their economic, social, and political systems to the point that they constitute a single community. An integrated community may be achieved, in the sense that its members generally can expect conflicts to be resolved peaceably among them, without the formal amalgamation of unified institutions of government.

**Intercontinental ballistic missile (ICBM).** A ballistic missile with a range in excess of 5,500 kilometers.

**Interdependence.** Mutual dependence, often today of societies and states.

**Intergovernmental organization (IGO).** A standing body created by treaty among several or many sovereign governments to assist them in advancing certain kinds of policies.

**Interim Offensive Arms Agreement.** Part of the 1972 SALT I agreements that froze ICBMs and SLBMs at existing levels for five years.

**Intermediate Nuclear Force Treaty (INF).** 1987 treaty that eliminated U.S. and Soviet intermediate- and shorter range weapons from European soil.

**Intermediate-range ballistic missile (IRBM).** A missile with a range of between 1,000 and 5,500 kilometers.

**International Bank for Reconstruction and Development (IBRD).** See *World Bank*.

**International Court of Justice (ICJ).** A "principal organ" of the United Nations, its headquarters are in the Hague. All U.N. members are automatically members of the ICJ, or World Court. It is composed of fifteen judges elected by concurrent vote of the General Assembly and the Security Council, no two of whom may be of the same nationality. Judges serve nine-year terms. Additional judges may be appointed on an ad hoc basis to hear a case in which no sitting judge is a national of one or more of the parties to the case. Only states may bring cases to the ICJ, although no state is required to do so. The ICJ is also authorized to give advisory opinions on legal questions submitted by the General Assembly or the Security Council.

**International Covenant on Civil and Political Rights.** Completed for state ratification in 1966, it entered into force in 1976, and now has about one hundred signatories. This treaty spells out the relevant standards expressed in the Universal Declaration of Human Rights.

**International Covenant on Economic, Social and Cultural Rights.** Like the above treaty, it was completed in 1966 and entered into force in 1976, and now has about one hundred signatories.

**International Development Association (IDA).** Created in 1960 as an affiliate of the World Bank, it is intended to serve the special financial needs of the world's poorest countries as a "soft loan" agency.

**International Finance Corporation (IFC).** Created in 1956 as an affiliate of the World Bank, it is intended to stimulate private investment in developing countries.

**International law.** Literally, law "between nations;" states have been its principal subjects as well as the chief makers, enforcers, and interpreters of the law. Increasingly, nongovernmental groups and individuals are viewed as participants in the international legal order.

**International Monetary Fund (IMF).** A specialized U.N. agency concerned with world monetary stability and economic development.

**International politics.** Literally, politics "between nations"; the relations between sovereign nation-states or their governments.

**International regime.** See *regime*.

**Interventionism.** Usually refers to the tendency to undertake military intervention in the affairs of another state.

*Intifada.* Violent uprising of Palestinians in the Israeli West Bank and Gaza strip between 1987 and 1993.

**Intranational law.** Literally, law "within the nation"; all law made by states for effect domestically, as distinguished from international law.

**Invulnerable retaliatory capability.** Reducing the vulnerability of nuclear retaliatory capability by putting it underground, dispersing it, or making it mobile.

**Isolationism.** Policy of looking out for domestic security interests rather than cooperating with others to provide a more general global security. The United States had a long tradition of isolationism during the nineteenth century and between the two world wars.

**Jihad.** A holy war undertaken as a sacred duty by Muslims.

***Keiretsu* system.** The Japanese system of interlocking manufacturers, suppliers, and distributors that discriminates against outside competitors.

**Kennedy Round of Trade Negotiations.** Multilateral trade negotiations under GATT, 1963-1967, that produced tariff reductions of up to 35 percent on a wide variety of goods in trade.

**Laissez-faire.** From the French, "allow to do," the phrase describes the central principle of capitalism—that governments should not regulate or interfere with economic affairs.

**Latin American Integration Association (LAIA).** Known as the Latin American Free Trade Association from its founding in 1960 until 1980, its aim is to promote freer regional trade among member states Argentina, Bolivia, Brazil, Chile, Colombia, Ecuador, Mexico, Paraguay, Uruguay, and Venezuela.

**Law of the Sea Treaty.** Opened for ratification in 1982, it came into force in 1994. The treaty grants a 200-mile exclusive economic zone to coastal states and creates an International Seabed Authority to regulate commercial activities on the seabed beyond those limits.

**League of Nations.** The multipurpose intergovernmental organization created in the Treaty of Versailles (1919), with headquarters in Geneva. It effectively ceased to function with the outbreak of World War II in 1939, but formed the basic model for creation of the United Nations in 1945.

***Lebensraum*.** German for "living space"; the central concept in the geopolitical belief held by Karl Haushofer, and adopted by Hitler, that held that Germany would have to expand (gain living space) or it would wither.

**Left.** In politics, leftist individuals, groups, or parties seek varying degrees of change, gradual or revolutionary. Leftists may be described accordingly as liberal, progressive, radical, and so on.

**Less developed country (LDC).** The term generally used today to describe countries either at preindustrial levels of development or that are early in the industrializing process. As a large class of states, they are by most measures substantially poorer than the developed countries.

**Liberalism.** (a) Especially in the United States, a doctrine of progressive, left-of-center politics, in contrast to conservatism. (b) In economic terms, laissez-faire liberalism is the doctrine supporting capitalism—that is, the view that government should avoid economic regulation.

**Limited Test Ban Treaty.** Treaty ratified in 1963 banning nuclear testing in the atmosphere, in outer space, and under water. Testing was still allowed underground.

**Long postwar peace.** Reference to the long period of peace in Europe, from 1945 to the end of the cold war in 1989, which was alleged to have resulted from the stability brought by two antagonistic alliance systems—including their nuclear threat.

**Maastricht Treaty.** A 1992 agreement of the members of the European Community, it changed the name of the EC to the European Union, and committed members (except the United Kingdom) to establish a common currency and central bank by 1999.

**Managed trade.** The notion that government should intervene to shape trade relations to favor the national interest or at least to avoid unequal tariff trade-offs.

**Manifest destiny.** Nineteenth-century belief espoused by some political groups in the United States that it was the destiny of the United States to expand its territory over the whole of North America.

**Martyrdom.** Extreme suffering or death of a martyr-that is, one who chooses death rather than abandon religious principles.

**Massive retaliation doctrine.** U.S. defense policy in the 1950s threatening the Soviet Union with a massive nuclear strike if it sought to expand its territorial control further.

**Mediator.** A neutral third party invited by two conflicting parties to help them reach an agreement.

**Mercantilism.** Philosophy that seeks to secure economic and political supremacy for a state. Emphasis is placed upon encouraging exports over imports in order to accumulate money.

**Military-industrial complex.** Notion that business and military interests collude in order to increase military spending.

**Missile Technology Control Regime (MTCR).** Agreement of suppliers of sophisticated missile technology prohibiting the transfer of such technology and supplies to states that lack them.

**Modernization theory.** The dominant theory since World War II regarding the origins of industrialization and economic modernization; it is compatible with a liberal, or capitalist, economic outlook.

**Monism.** The concept that views international law as binding states and individuals in a single skein of authority.

**Monroe Doctrine.** A U.S. foreign policy doctrine introduced by President James Monroe in 1823, which opposed further European colonization or interference with independent nations in the Western Hemisphere.

**Most-favored nation clause.** Requirement in a commercial treaty that an accommodation reached between two states should be extended to all others that are parties to the treaty.

**Multiculturalism.** A doctrine that promotes ways to increase shared understanding as the result of the greater mixing of people, cultures, and ideas that is characteristic of the contemporary period.

**Multinational Corporation (MNC).** A business enterprise organized in one society, but with such substantial investments and operations in other countries that the MNC is beyond the regulatory reach of a single state. Also called a transnational corporation (TNC).

**Multiple independently targetable reentry vehicle (MIRV).** Reentry vehicles, carried by one nuclear-tipped missile, that can be directed toward other targets.

**Multipolarity.** When power is distributed among more than two states in the international system.

**Mutual assured destruction (MAD).** A system of mutual deterrence where both sides have the capability to survive a first strike and are able to launch a retaliatory strike.

**Nation.** A collection of people who, because of ethnic, linguistic, or cultural affinity, perceive themselves to be members of the same group.

**National interest.** Concept that the state should take into consideration whether or not a policy is supportive of its basic interest, often defined in terms of security.

**Nationalism.** Belief in and identification with the nation-state or one's ethnic group.

**Natural law.** The law that governs the workings of nature, as opposed to human-made law. The natural law tradition asserts that international legal norms should be derived by discerning the laws of nature relevant to how humans ought to behave.

**Neo-imperialism.** Following their formal imperial rule, the continued economic domination of these rich and developed countries over (usually) their former colonial states.

**Neomercantilism.** Twentieth century version of mercantilism that emphasizes high tariffs to discourage imports while pursuing policies to increase exports.

**Neorealism.** Like realism, neorealism also views state interests as the most significant forces in world politics, although it accepts that the conduct of interstate politics is now much modified by the growth in intergovernmental organizations and by nongovernmental and private participants in the process.

**Newly industrialized countries (NICs).** The name for a number of countries whose economies grew rapidly starting in the 1970s. Most are in Asia, and two are in Latin America. (Also called NIEs, or newly industrialized economies.)

**New International Economic Order (NIEO).** An alternative agenda to that of traditional modernization theorists for assisting the South in its economic development; it was most prominent in the 1970s.

**Nongovernmental organization (NGO).** The creation of a private group or groups whose interests or activities cross nation-state boundaries. Many NGOs act as pressure groups toward governments regarding policy choices. Increasingly, some NGOs are themselves participants in international negotiations and organizational bodies.

**Nonjusticiable dispute.** A dispute not appropriate for or subject to a court trial.

**Nonstate actors.** Actors other than the state itself that have an impact upon the international system. Such actors include international organizations, terrorist groups, multinational corporations, other private associations, and so on.

**Nontariff barrier.** Any device, other than taxes on imports and duties, that restrains the free flow of goods and services between states.

**Norm.** A standard, model, or pattern regarded as typical for a particular group. Especially in international law, an appropriate standard of behavior.

**Normative behavior.** Behavior relating or conforming to a norm or social standard.

**North.** A term designating the group of highly developed, relatively rich contemporary states, most of which happen to lie north of the equator.

**North American Free Trade Agreement (NAFTA).** Agreement among the United States, Canada, and Mexico that eliminates barriers to trade among the three states.

**North Atlantic Treaty Organization (NATO).** The 1949 alliance among fifteen North Atlantic countries pledging to defend one another in case of an outside attack.

**Nuclear proliferation.** The spread of nuclear weapons capability to additional states.

**Nuclear Nonproliferation Treaty (NPT).** A global treaty negotiated in 1968 and extended in 1995 that prohibits nonnuclear-weapon states from producing or receiving nuclear weapons.

**Official development assistance (ODA).** A grant or loan made by one government to another that is specifically intended to promote the latter country's economic development.

**Organization for Economic Cooperation and Development (OECD).** Its twenty-four members include the most highly developed states of Europe, in addition to Australia, Canada, Japan, New Zealand, and the United States. Its purpose is to promote economic cooperation and the development of poorer countries.

**Organization for Security and Cooperation in Europe (OSCE).** An organization comprised of over fifty Western and former Soviet bloc states that deals with economic, security, and human rights issues. (Formerly called the Conference on Security and Cooperation in Europe.)

**Organization of African Unity (OAU).** Established in 1963, the OAU is meant to promote unity and cooperation among African states, all of which except Morocco are members.

**Organization of American States (OAS).** Created in its present form in 1948, the OAS includes all thirty-five of the independent states of North and South America (Cuba has been excluded from formal participation since 1962). Its aims are to promote peace, security, and economic and social development in the hemisphere of the Americas.

**Organization of Petroleum Exporting Countries (OPEC).** Established in 1960 by thirteen oil-producing countries in Africa, Asia, and South America, its aim is to coordinate petroleum policies among the members.

**Paradigm.** A model, pattern, or example. When the principles, presuppositions, and expectations about human behavior are fully articulated, the rationale they provide for studying the phenomena of world politics in a certain way constitutes a paradigm.

**Parliamentary diplomacy.** Open diplomacy as practiced through an international body such as the General Assembly of the United Nations.

**Peace of Westphalia.** The peace agreements of 1648, consisting of the Treaty of Münster and the Treaty of Osnabruck, that formally ended Europe's Thirty Years' War (1618–1648). The agreements set the stage for the modern period in world politics, in which participating states are recognized as equal sovereigns, each possessing exclusive jurisdiction over its territory and people.

**Peacekeeping operations.** The term used for most United Nations military and supervisory operations, in which a comparatively small international force is provided to assist warring parties to achieve peace, or to help keep the peace in the aftermath of war.

**Perception.** Reality as seen by an observer or a decision maker.

*Perestroika.* The economic restructuring of Soviet society, as proposed by Mikhail Gorbachev in the late 1980s.

**Persona non grata.** Declaring a foreign diplomat as unacceptable to a recipient state.

**Pluralism.** In world politics, this theoretical approach views states as abstractions containing multiplicities of conflicting forces that interact with other groups, both internal and external to particular states.

**Positive law.** Human-made law, as opposed to natural law. In international law, that which is (almost always) written and formally assented to by sovereign governments.

**Power transition model.** A theory, suggested by A.F.K. Organski, for predicting the likelihood of war. It assumes that war is most likely when differences in power among rivals have narrowed.

**Preponderance of power.** Situation believed by some to be the optimum condition for peace, when one power is able to dominate the weaker parties in the system.

**Preemptive war.** A military strike by a party that believes itself about to be attacked.

**Private market forces.** Allowing the market to operate freely on the basis of supply and demand; it assumes private rather than state ownership.

**Proletariat.** Marxist terminology for the working class in industrial society.

**Public diplomacy.** Propaganda and informational activities in which governments engage in an effort to influence public attitudes abroad.

**Qualitative arms race.** One that focuses upon the development of more powerful and sophisticated weapons.

**Quantitative arms race.** Reciprocal effort to increase the number of weapons by increasing production of specified weapons systems.

**Rational actor model.** An approach that assumes that decision makers select the best alternatives as they attempt to maximize their goals.

**Realism.** In world politics, the philosophical outlook that attributes great significance to the durability and interests of nation-states, whose goals are frequently in conflict with one another. Those goals typically require accommodation, rather than transformation.

**Reciprocal Trade Agreements Act.** Legislation authorizing the president of the United States to negotiate reciprocal reductions on trade barriers with other states. It has been renewed many times since it was passed in 1934, but with varying titles.

**Reciprocity.** Responding in kind to the moves of the other party. As a principle of justice, reciprocity assumes that what one actor in the system claims for itself may also be claimed by fellow actors.

**Regime.** An international regime consists of those distinctive patterns of international behavior that are the product of principles, norms, rules, and procedures that then shape relevant decisions on particular issues.

**Right.** In politics, rightist individuals, groups, or parties seek stability and continuity with the past, abhorring change. Rightists may be described accordingly as conservative, reactionary, and so on.

**Risk-averse behavior.** Situation in which the primary concern is to avoid risk in making decisions.

**Self-Actualizer.** One who has his or her basic needs of food, safety, belongingness, and self-esteem satisfied and is thus able to make more mature and rational choices.

**Self-determination.** The principle that an ethnic or other community should be free to determine for itself whether to be independent.

**Shiite.** Minority group in Islam, many of whom are found in Iran.

**Short-range attack missiles (SRAM).** Missiles placed on the wings of a bomber, allowing it to fire at a greater distance from the target.

**Shuttle diplomacy.** The practice whereby a high-level diplomat "shuttles" between the capitals of parties to a conflict to try to assist them in achieving a peaceful solution to their dispute.

**Soft currency.** A currency that is largely nonconvertible and in which outside states have limited confidence.

**South.** A term designating the very large group of less developed, comparatively poor states of the contemporary period, most of which happen to lie south of the equator.

**Sovereignty.** Supremacy of authority or rule as exercised by a sovereign state.

**Special Drawing Rights (SDRs).** Reserves created and held by the International Monetary Fund from which member states can draw to help stabilize the value of their currencies.

**State.** A sovereign political unit whose government exercises exclusive jurisdiction over a particular territory and its people.

**Strategic Arms Limitation Treaty (SALT).** Treaties between the United States and the Soviet Union signed in 1972 and 1979 that included restrictions on antiballistic missiles and ceilings on strategic delivery systems. The SALT II Treaty was never ratified.

**Strategic Arms Reduction Treaty (START).** Includes two treaties signed by the United States and the Soviet Union in the early 1990s that provide for substantial reductions of their strategic nuclear capabilities.

**Subcommission on Prevention of Discrimination and Protection of Minorities, U.N.** This body is composed of individual experts rather than officials representing states. It screens private petitions—from individuals and NGOs—before sending them to the Human Rights Commission.

**Submarine-launched ballistic missile (SLBM).** A ballistic missile carried in and launched from a submarine.

**Subsidy.** Government assistance to the development, production, or export of specific goods.

**Subsystem.** In international politics, a system smaller than the global system as in the case of a regional subsystem such as the Middle East.

**Summit diplomacy.** Diplomacy carried out by officials at the highest levels of government—that is, the heads of state.

**Sunni.** The majority sect of Islam.

**Super 301.** U.S. legislation that authorizes the president to retaliate against perceived unfair trade practices by other states.

**Sustainable development.** Strategies for industrialization and economic development that do not exhaust nonrenewable resources and thus can be sustained without risking environmental collapse.

**Tangible factors.** Measurable factors of power such as numbers of people, weapons, industrial capability, and so on.

**Tariff.** A tax placed upon the importation of goods, making them less competitive in one's own domestic market.

**Technical assistance.** In economic development programs, that form of assistance that entails teaching people the skills and technologies necessary for their society's development.

**Technosphere.** The human-made environment of the planet.

**Terrorism.** Violence against citizens and property designed to intimidate others for a specific political cause.
**Theory.** A set of interrelated propositions that purports to explain or predict.
**Threshold Test Ban Treaty (TTBT).** Treaty signed in 1974 placing a 150-kiloton limit on underground nuclear tests.
**Tokyo Round of Trade Negotiations.** Multilateral trade negotiations completed in 1979 under the GATT.
**Trade preferences.** Protectionist measures other than tariffs to assist, usually, LDCs in their development by favoring certain of their products over similar ones produced elsewhere.
**Transfer of technology problem.** The tendency of much multinational corporate activity to avoid transferring technological skills from highly advanced countries to LDCs, leaving the latter dependent on the former for critical economic decisions and activities.
**Treaty.** A formal agreement between sovereign states that, when ratified, binds them.
**United Nations.** The general-purpose intergovernmental organization created at the close of World War II in 1945. Headquartered in New York City, the U.N. system includes a wide array of specialized agencies and other bodies designed to assist states in their interactions.
**United Nations Conference on the Environment and Development (UNCED).** The 1992 "Earth Summit" conference in Rio de Janeiro that produced a number of directives and treaties dealing with environmental issues.
**United Nations Conference on Trade and Development (UNCTAD).** Created in 1964 when the first conference was made a permanent agency of the United Nations, it has been dominated by the developmental views of many less developed countries.
**United Nations Development Program (UNDP).** The principal U.N. agency for administering the contributions of member states for technical assistance to LDCs.
**Universal Declaration of Human Rights.** Adopted by the U.N. General Assembly in 1948, this was the first comprehensive set of human rights standards drafted in the aftermath of World War II. Although not a treaty, it has guided the drafting of later treaties on human rights.
**United Nations Environmental Program (UNEP).** Created in 1972 to oversee U.N. activities dealing with environmental matters, UNEP has a permanent headquarters in Nairobi, Kenya.

**United Nations Human Rights Commission.** See Human Rights Commission.
**United Nations Subcommission on Prevention of Discrimination and Protection of Minorities.** See Subcommission on . . .
**Universal Postal Union (UPU).** Created as an IGO in 1874, it is now a specialized agency of the United Nations. Its purpose is to assure that the postage of one country is honored in others, so that mail will be reliably delivered internationally.
**Uruguay Round of Trade Negotiations.** Multilateral GATT negotiations concluded in 1994 that reduced a number of impediments to global trade and created the World Trade Organization.
**Voluntary export restraints (VERs).** Agreements to restrict one's exports, often in response to another country's threat of even greater trade restrictions if the state does not "voluntarily" agree to such limits.
**Voluntary import expansion.** Situation in which a state agrees to import a given amount of goods, often in response to pressures of a state experiencing balance of trade difficulties.
**War.** (a) **International war.** The organized use of large-scale violence by the governments of two or more states, typically in the pursuit of conflicting governmental policies. (b) **Civil war.** The organized use of large-scale violence by groups within a state. Frequently, one group is led by a challenged government and its challenger is an insurgent group seeking to become the government.
**War by miscalculation.** War that arises as a result of decision makers' misreading the intentions of others and initiating a strike that is neither necessary nor desirable.
**War Crimes tribunals.** The first of these were created at the close of World War II to try German (the Nuremberg trials) and Japanese (the Tokyo trials) war criminals. In 1993, a third tribunal was launched to investigate violations of international humanitarian law in the former Yugoslavia. A fourth tribunal, authorized in 1994, had a similar purpose for Rwanda, following large-scale massacres there.
**Warsaw Pact (WP).** Military alliance between the Soviet Union and its Eastern European satellites, which collapsed as a result of the breakup of the Soviet bloc in 1989.
**Westphalia.** The shorthand reference to the modern period or world system ushered in by the Peace of Westphalia (1648). It began to make sovereign nation-states the principal actors in world politics. At first, the

Westphalian system was limited to Europe, but as European nations began to dominate the world, Westphalia's organizing principle—the sovereign equality of nation-states—was extended across the globe.

**World Bank.** (Formally, the International Bank for Reconstruction and Development, or IBRD.) Created at the close of World War II to provide public international financing to supplement private loans. Initially, its funds went mainly for the reconstruction of war-torn economies, but since the 1950s, the World Bank's effort has been directed largely toward the economic development of LDCs.

**World politics.** Political interactions that are manifested at global levels, including the relations among sovereign governments, intergovernmental organizations, nongovernmental organizations, and many private groups, organizations, and individuals.

**World Trade Organization (WTO).** Organization created as a result of the GATT Uruguay Round of negotiations that is empowered to rule on unfair trade practices.

**Xenophobia.** An unreasonable hatred or fear of foreigners or strangers.

**Zero sum.** A "game" situation, in which everything gained by one side results in a comparable loss to the other—as distinguished from a nonzero situation, in which both sides have something to gain in a given solution.

# Author Index

Abravanel, Martin, 134
Adorno, T. W., 114
Akehurst, Michael, 363
Albin, Cecilia, 176
Allon, Philip, 382
Almond, Gabriel, 284
Amin, Samir, 296
Anderson, M. S., 215, 218
Ardrey, Robert, 206
Armstrong, J. D., 120
Aron, Raymond, 124
Axelrod, Robert, 21

Baehr, Peter R., 401
Bagdikian, Ben, 73
Bailey, Thomas A., 229
Bairoch, Paul, 283
Baker, Ray S., 399
Balaan, David N., 242
Baldwin, Robert E., 252
Bandura, Albert, 207
Barber, James, 339
Bauer, Raymond A., 135
Beard, Charles A., 204
Beer, Francis A., 188
Bell, Coral, 227
Bergesen, Albert, 20
Berkowitz, Leonard, 207
Berman, Maureen R., 230
Biderman, Albert, 204
Black, Cyril, 284
Blainey, Geoffrey, 161, 202
Blaker, Michael, 233
Blechman, Barry M., 169
Bloch, Marc, 6
Boczek, Adam Boleslaw, 369
Bodin, Jean, 32
Boone, Louis E., 259
Boutros-Ghali, Boutros, 304, 415
Brass, Paul, 37
Breuilly, John, 48
Broad, Robin, 299
Brown, Lester R., 253, 299, 308, 309, 310, 311, 315, 320
Buchan, Alastair, 200
Bull, Hedley, 20
Burns, James MacGregor, 117
Butwell, Richard, 203

Caldwell, Dan, 115
Caldwell, George, 116
Camilleri, Joseph A., 61
Campbell, A. C., 16
Carlsson, Ingvar, 415
Carr, E. H., 22
Cashman, Greg, 195, 196
Cavanuagh, John, 299
Champion, Michael, 195
Chan, Steve, 200
Charlesworth, Hilary, 382
Chazan, Naomi, 37
Checkel, Jeff, 92
Chinkin, Christine, 382
Clarke, Michael, 127
Claude, Inis L., Jr., 334, 401, 403, 408
Clausewitz, Karl von, 4
Cleveland, Harlan, 422
Cline, Ray S., 159
Clutterbuck, Richard, 175
Coate, Roger A., 406
Codevilla, Angelo, 191
Cohen, Raymond, 232
Coleman, James S., 284
Collins, Peter, 93
Cook, Dave, 353
Coplin, William D., 201
Corbin, Jane, 228
Crabb, Cecil V., Jr., 95
Crosette, Barbara, 411

Dahl, Robert A., 159
Dale, Reginald, 340
Dallmeyer, Dorinda G., 370
Daly, Herman, 310
Daniel, Donald C., 153
Davis, E. D., 232
De Magalhaes, Jose C., 213
De Tocqueville, Alexis, 124
Dell, Edmunds, 272, 278
Deudney, Daniel, 21
Deutsch, Karl W., 159, 194, 208, 332, 339
Dexter, Lewis, 135
Deyo, Frederic C., 298
Diehl, Paul F., 159, 408
Diesing, Paul, 209

Dinan, Desmond, 66
Ding, X. L., 93
Dixon, William J., 201
Dodd, William E., 399
Dollard, John, 206
Dorpalen, Andreas, 147
Dos Santos, Theotonio, 285
Doxey, Margaret, 273
Draper, Theodore, 168
Drodziak, William, 241
Drucker, Peter F., 299
Druckman, Daniel, 223, 230, 232

East, Maurice A., 39
Ehrlich, Thomas, 361, 366
Elliott, Kimberley A., 247, 258, 273
Enloe, Cynthia, 22
Espito, John L., 99
Etheredge, Lloyd S., 115, 208
Evans, Alfred B., Jr., 92
Evans, Peter B., 226

Falk, Richard A., 21, 357
Fanon, Frantz, 72
Feierabend, Ivo K., 203
Feierabend, Rosalind L., 203
Fieleke, Norman S., 247
Fingleton, Eamonn, 263
Finlayson, Jock A., 410
Fisher, Roger, 228, 231, 359
Forsythe, David, 376, 406
Fouraker, L. E., 230
Fox, Annette Baker, 124
Frankel, Jeffrey A., 254
Frederick, Howard H., 73, 157
French, Hilary, 320
Freud, Sigmund, 206
Frieden, Jeffrey A., 242
Fukuyama, Francis, 104
Fulbright, William J., 97
Furedi, Frank, 50

Galbraith, John Kenneth, 127
Galtung, Johan, 296
Gandhi, Mohandas K., 17
Gardner, Gary, 299

George, Alexander L., 243
Gerber, David J., 34
Gills, Barry, 198
Gilpin, Robert, 20, 68
Glad, Betty, 117
Gochman, Charles S., 280
Goldstein, Judith, 102, 103
Goode, Richard, 289
Goodrich, Leland M., 406
Gorbachev, Mikhail, 301
Gordenker, Leon, 401
Graham, Edward M., 68, 75
Gray J. Patrick, 208
Gray, Colin S., 146
Greenhouse, Steven, 144
Gross, Leo, 32
Grotius, Hugo, 16, 322
Gurr, Ted Robert, 199

Haas, Ernst B., 64, 409, 410
Haas, Michael, 234
Haggard, Stephen, 298
Haig, Alexander, 132
Hall, Edward T., 233
Halperin, Morton H., 128
Halperin, Sandra, 198
Halpern, Nina, 102
Haney, Patrick J., 137
Hanink, Dean H., 263
Hartung, William D., 174
Hartz, Louis, 97
Havas, M., 311
Hayes, Carlton J., 46
Head, Ivan L., 296
Hermann, Margaret G., 115, 117, 119, 232
Herz, John H., 5
Hey, Jeanne A. K., 137
Hobsbaum, Eric J., 37
Hobson, John A., 38, 267
Hocking, Brian, 136
Hoffmann, Stanley, 97, 264, 365
Hollist, W. Ladd, 195
Holsti, Kalevi J., 142, 189, 190, 197
Holsti, Ole R., 119, 134, 243
Holtfrerich, Ludwig, 246
Honan, William H., 157
Hook, Steven W., 269, 333

Hopkins, Raymond, 130
Howard, Michael, 4
Howard, Richard, 124
Hudson, Valerie M., 115
Hufbauer, Gary Clyde, 247, 258, 273
Hughes, Barry B., 134
Hume, David, 162
Hunt, Michael H., 97
Huntington, Samuel, 104, 196, 273
Husain, Mir Zohair, 99
Hutchinson, John, 46
Hutchinson, T. C., 311
Huth, Paul K., 169

Ikle, Fred, 224, 233
Inkeles, Alex, 284
Isaak, Robert A., 242, 256

Jackson, John H., 259
Jacobson, Harold K., 64
Jaffe, Mark, 319
Jaguaribe, H., 285
James, Alan, 32
Janis, Irving L., 115, 130
Janis, Mark, W., 35
Jarvis, Anthony P., 61
Jenkins, Rhys, 292, 304
Jensen, Jane S., 122
Jensen, Lloyd, 68, 222, 229, 233
Jervis, Robert, 120, 122, 210
Johnston, Douglas, 102
Jowett, Garth S., 158

Kane, Hal, 253, 299, 311, 320, 315
Kant, Immanuel, 16
Kanter, Arnold, 128
Kaplan, Stephen S., 169
Kapstein, Ethan B., 275
Karatnycky, Adrian, 98, 201
Karlsson, Gail V., 326
Karns, Margaret, 407
Kauppi, Mark V., 286
Kegley, Charles W., Jr., 13, 21, 131, 163, 187, 192, 193, 194, 269
Kennan, George F., 18

Kennedy, Paul, 39, 151, 154, 322
Keohane, Robert O., 21, 102, 103, 243
Kinder, Donald R., 121, 123
King, Alexander, 60, 299, 305
Kissinger, Henry A., 128, 129, 233
Klineberg, Otto, 92
Knorr, Klaus, 8
Knutson, J. N., 117
Kolb, Deborah M., 235
Kolko, Gabriel, 20
Korb, Lawrence J., 205
Korbin, Stephen, 275
Korey, William, 377
Kotkin, Joel, 353
Krasner, Stephen D., 254, 295, 421
Kratochwil, Friedrich, 421
Kugler, Jacek, 156, 194
Kurth, James R., 204
Kurtz, David L., 259
Kuttner, Robert, 242

Lairson, Thomas D., 242, 283, 289, 305
Lake, David A., 242
Lall, Arthur, 224
Lambelet, John C., 196
Lane, Robert E., 114
Laquer, Walter, 176
Lasswell, Harold, 114
Lebow, Richard Ned, 120, 169, 210
Lenin, Vladimir, 38, 268
Lenssen, Nicholas, 299, 311, 315, 320
Lerner, Daniel, 284
Lerner, Max, 15
Levering, Ralph B., 134
Levi, Werner, 102
Lewis, Paul, 150
Lichtheim, George, 339
Licklider, Roy, 86
Lindsay, James M., 273
Lippmann, Walter, 124
Locke, John, 360, 416
Lockhart, Charles, 230
Lorenz, Konrad, 206
Lowenthal, Richard, 20

# Author Index 439

Luard, Evan, 188, 189, 200
Luttwak, Edward N., 153
Lutz, Wolfgang, 382

Machiavelli, Niccolo, 15
Mackinder, Sir Halford, 146
Maddison, Angus, 246
Mahan, Alfred T., 146
Mansfield, Edward D., 421
Manyon, L. A., 6
Maslow, Abraham, 114
McGowan, Patrick J., 134, 187
McKeown, Timothy, 115
McKibben, Bill, 315
McNamara, Robert S., 209
McNeill, William F., 41
Mendez, Ruben, 415
Michel, James H., 303
Mills, C. Wright, 204
Mingst, Karen A. 137
Mitchell, George H., Jr., 303
Mitchell, J. M., 157
Mitrany, David A., 334, 403
Montesquieu, Charles-Louis de Secondat, 95
Morgenthau, Hans J., 18, 155, 268
Morris, Desmond, 206
Morrow, James D., 196
Munoz, Heraldo, 75
Murphy, Cornelius J., 32
Muskie, Edmund, 131

Nardin, Terry, 390
Neack, Laura, 137
Neff, Stephen C., 254
Neidle, Alan, 229
Nicholson, Harold, 214, 217
Nincic, Miroslav, 97
Norchi, Charles H., 380
Nugent, Neill, 335
Nye, Joseph S., Jr., 21

O'Brien, James C., 392
O'Connell, Mary Ellen, 361, 366
O'Donnel, Victoria, 158
O'Neal, John R., 292
O'Prey, Kevin, 173
Olson, R. S., 273
Oppenheimer, Michael, 319

Oren, Ido, 195
Organski, A. F. K., 156, 159, 194
Ostrom, Charles W., 195
Oye, Kenneth A., 20

Pachter, Henry M., 365
Palan, Ronen, 198
Paolini, Albert J., 61
Paret, Peter, 4
Pearson, Frederic S., 173, 204
Perre, Jean, 189
Pestel, Eduard, 305
Pfaff, William, 37, 46
Plischke, Elmer, 212
Poggi, Gianfranco, 28
Polochak, Solomon William, 280
Pool, Ithiel de Sola, 135
Post, Jerrold M., 116
Postel, Sandra, 308, 310
Prebisch, Raul, 285
Prestowitz, Clyde V., 242
Putnam, Robert D., 226

Ray, James Lee, 159, 335
Raymond, Gregory A., 163, 193, 194, 195
Reed, Bruce, 339
Rejai, Mustafa, 90
Richardson, Lewis F., 195
Robins, Robert S., 116
Robinson, Neal, 93
Robock, Stefan H., 255
Rochester, J. Martin, 201, 204
Rohatyn, Felix, 75
Roodman, David Malin, 315
Rosati, Jerel A., 97
Rosecrance, Richard N., 39, 143, 244, 285, 298, 322
Rosenau, James N., 21, 135, 137, 350, 422
Rosenblum, Mort, 72
Ross, Marc Howard, 208
Rostow, W. W., 285
Rothstein, Robert L., 39
Rourke, John T., 262
Rousseau, Jean-Jacques, 95, 416
Rowen, Hobart, 257
Rowny, Edward L., 232
Royama, Michio, 145
Rubin, Barry, 176

Rubin, Trudy, 351
Rummel, Rudolph J., 191, 200, 202
Rusk, Dean, 220
Russett, Bruce M., 200, 280, 369

Sabine, George H., 32
Sabrosky, Alan N., 195
Sachs, Jeffrey R., 263
Sampson, Cynthia, 102
Sanger, David E., 256
Santer, Jacques, 339
Sarkees, Meredith, 189
Sarkesian, Sam C., 204
Saunders, Harold, 232
Sawyer, Ralph D., 15
Sayrs, Lois W., 280
Scheinman, Lawrence, 365
Schneider, Bertrand, 60, 299, 305
Schott, Jeffrey J., 273
Schwartz, David C., 169
Schwartz, Morton, 116
Sciolino, Elaine, 156, 269
Scowroft, Brent, 131
Seabury, Paul, 191
Shaw, George Bernard, 204
Shaw, Lisa D., 180
Shotwell, James T., 49
Siann, Gerda, 206
Siegel, Sidney, 230
Silverberg, James, 208
Simai, Mihaly, 351
Singer, Eric, 115
Singer, J. David, 8, 159, 187, 188, 189, 194, 196
Singer, Max, 39
Sivard, Ruth, 147, 170
Siverson, Randolph M., 243
Skidmore, David, 242, 283, 289, 305
Small, Melvin, 187 188, 189, 196
Smith, Anthony, 46
Smith, Dale L., 335
Smith, David H., 284
Smith, Donald, E., 102
Smith, M. H.,, 126
Smith, Steve, 127
Snyder, Glenn H., 209
Spar, Debora L., 269
Spykman, Nicholas J., 147

## Author Index

Stavis, Benedict, 301
Stein, Janice Gross, 120, 169, 228
Steinbrunner, John, 170
Stewart, Abigail J., 119
Stoessinger, John, 120, 124, 208
Stoll, Richard J., 195
Sullivan, John D., 89
Sullivan, Michael, 115
Sun Tzu, 15, 17
Sun, Sun, Yan, 93
Susser, Bernard, 90
Sutterlin, James S., 415
Sylvester, Christine, 22

Tanter, Raymond, 202
Taylor, Paul, 65
Taylor, Telford, 390
Thompson, Kenneth W., 18, 155, 268
Thorson, Thomas Landon, 32
Thucydides, 15
Tickner, J. Ann, 22
Tilly, Charles, 28, 46
Tower, John, 131
Triandis, Harry C., 104, 232
Tuch, Hans N., 157
Tuchman, Barbara, 49, 208
Tucker, Robert W., 103
Turner, Stansfield, 178
Tyson, Laura, 242

Ulam, Adam, 300
Ullmann, Walter, 27
Urquhart, Brian, 415
Ury, William, 231

Valenzuela, Arturo, 285
Valenzuela, J. Samuel, 285
Van Boven, Theo, 393
Van Dyke, Vernon, 90, 377
Vattel, Emmerich de, 17
Verba, Sidney, 8
Vernadsky, V. I., 76
Vernon, Raymond, 269
Vertzberger, Yaacov Y. I., 95, 120
Veseth, Michael, 242
Viotti, Paul R., 180, 286

Wagar, Warren, 21
Walensteen, Peter, 188, 273
Walker, Andrew, 275
Walker, Rachel, 61
Walkling, Sarah, 174
Wallace, Michael D., 196
Wallerstein, Immanuel, 20, 286
Walters, F. P. 64, 401
Waltz, Kenneth N., 20, 144, 192
Watson, Adam, 215
Watson, R. T., 316
Webb, Walter, 72
WeWeigley, Russell F., 175
Weiner, Myron, 284

Weinstein, Franklin B., 274
Weiss, Janet A., 121, 123
Weiss, Thomas G., 67, 406, 410
Welch, Charles E., Jr., 381
Wesson, Robert G., 93, 269
White, Ralph K., 120, 210
Whiting Allen S., 210
Wicker, Tom, 170
Wildavsky, Aaron, 39
Wilkenfeld, Jonathan, 202
Wilkinson, David, 365
Wilkinson, Paul, 175
Williams, Philip, 126
Wilson, Marjorie Kerr, 206
Winham, Gilbert R., 230
Winter, David G., 119
Wittkopf, Eugene R., 131
Wren, Christopher S., 99
Wright, Quincy, 144, 200, 202, 268, 389, 390, 401, 409
Wright, Shelley, 382
WuDunn, Sheryl, 258

Yeager, Leland B., 262
Young, Oran R., 421

Zacher, Mark W., 410
Zartman, I. William., 63, 230
Zillman, Dolf, 207
Zysman, John, 242

# Subject Index

Aberrant personalities, 115
ABM. *See* anti-ballistic missile
Accidental war, 170
Acid rain, 311
Afghanistan War, 273
   Soviet involvement, 141, 146, 193, 362, 392, 418
   U.S. response, 119, 174, 272, 273
African Charter on Human and People's Rights, 381
Aggression
   as an innate response, 206
   as learned behavior, 207
Aimes, Aldrich, 270
Algeria, 387, 388
   threat of Islamic fundamentalism in, 387
Allende, Salvador, 121, 201, 420
Alliance(s), 49, 88, 113, 143, 162, 173, 193, 227, 269, 399, 417, 418, 420
   contemporary compared to earlier, 64
Alsace, 42, 241
Amalgamation, 332
Amazon basin, 305
Amazonian rain forest, 317
Ambassador
   latitude in decision making, 214
American Convention on Human Rights, 381
Amin, Hafizullah, 362
Amnesty International, 63, 66, 380
Anarchy, 6, 15, 19, 21, 42, 264
Angola, 158
   civil war, 68, 219, 236
Antarctica, 319, 324
Antarctica Treaty, 324, 325, 327
Antiballistic missile, 181
Anti-Semitism, 9
Antigua, 60
Apartheid, 102, 249, 360, 378, 387, 388, 394, 397
Appeasement policy, 120
Aquino, Corazon, 235
Arab League, 420

Arabs, 232, 233
Arafat, Yasir, 228, 233, 235
Arbitration, defined, 369
Argentina, 102, 124, 149, 150, 210, 298, 324, 366, 383
Aristide, Jean-Bertrand, 107, 225
Armenian-Azerbaijani war, 61
Armenians, 33
Arms control, 179–186
   effect of the end of the cold war upon, 183
   history of, 180
   list of agreements, 181
   on-site inspection, 223, 224
   open skies, 227
   relationship to deterrence, 179
Arms races, 195–197
Aryan race, 103
Asia-Pacific Economic Cooperation (APEC), 256
Association of South-East Asian Nations (ASEAN), 4, 63, 340, 420
Associations of governments, 4
Athens, 200
Australia, 38
   independence of, 39
Austria, 16, 30, 36, 37, 40, 42, 49, 144, 255, 335, 382, 393, 399
Austro-Hungarian empire, 4, 45, 399
Austro-Prussian war, 42, 268
Autarky, 149
Authoritarian decision-making structures, 125
Authoritarian economic planning
   failure of, 301
Authoritarian personality, 114
Axis powers, 12, 50, 103
Aztecs, 33, 120

Bacon, Sir Francis, 131
Baker, James, 13, 124, 219, 399
   and Mideast negotiations, 228
Balance of payments, 243, 260–262, 277, 279, 288, 299
   components of, 262

Balance of power, 12, 18, 162–163, 191, 196, 205, 212, 333, 399
   effect of preponderance in, 194
   in the Middle East, 162
   prerequisites of, 162
   role of Britain in, 162
Baldwin, Stanley, 167
Ball, George, 116
Baltic states, 61
*Banco Nacional de Cuba* v. *Sabbatino*, 367
Bandaranaike, Sirimavo, 235
Bangladesh, 312
Bangladesh war, 56, 235
Barbarians, 27
Barbary pirates, 178
Basques, 61
Bay of Pigs invasion, 134
Beagle Channel dispute, 102
Beggar-thy-neighbor, 247
Begin, Menachem, 116, 135
Beijing, 3, 312, 380, 398
Belarus, 171
Belgium, 50, 253, 274, 335
Belief systems, 83
   compared, 101
   functions of, 87
   future of, 104
   impact of, 102
   in foreign policy, 86–90
   rationalization of, 103
   religious, 98
Benelux, 253
Bengal, 56
Berlin, 103, 120, 219, 366
Berlin wall, 3, 148
Bhutto, Ali, 202
Biafra, 58
Biodiversity, 318
Biological Convention, 180
Biological weapons, 180–182
Biosphere, 76–79, 80, 315, 325
   threats to, 76
Bismarck, Otto von, 37, 42
Bohemia, 30
Bolshevik Revolution, 216
Bosch, Juan, 121

441

Bosnia, 87, 98, 120, 136, 152, 164, 165, 174, 219, 220, 225, 226, 270, 272, 273, 281, 350, 388, 411, 423
  arms embargo of, 205
  lessons of, 349
  NATO troops in, 236, 419
Bourgeoisie, 90, 244
Boutros-Ghali, Boutros, 304, 415
Boycott, 68, 85, 149, 201, 270, 277
  oil (1973–1974), 68
Brain drain, 296
Brazil, 105, 150, 173, 298, 299, 305, 317, 318
Bretton Woods, 260, 262, 287
Brezhnev, Leonid, 106, 173, 221, 234, 301
Briand, Aristide, 333
Britain, 9, 10, 28, 33, 35, 36, 38, 45, 50, 51, 56, 58, 60, 61, 84, 95, 104, 115, 117, 120, 121, 124, 129, 142, 144, 146–148, 150, 154, 162, 164, 172, 194, 196, 200, 208, 210, 218, 221, 236, 245, 254, 269, 277, 278, 298, 337, 366, 405, 406
British Channel, 146
Brundtland, Gro, 235
Buddhism, 98
Bureaucracies, 111, 125, 131, 132, 137, 217, 273, 365
  effect on decision making of, 127
  fragmentation of policy in, 129
  power of compared with chief executives, 127
  trends in, 128
Bush, George, 86, 113, 120, 158, 178, 184, 192, 219, 227, 228, 250, 252, 319, 320, 348, 366, 414
  attitude toward China of, 221
Byzantium, 41, 198

Cambodia, 191, 236, 375
Cambodian Empire, 33
Camp David Accord, 135, 219, 223, 228, 269
Canada, 39, 61, 105, 149, 150, 157, 199, 219, 221, 253, 255, 279, 290, 292, 332, 333, 340, 341, 348, 358, 368, 369, 408, 418
  independence of, 39
Capabilities
  compared to influence, 142
  tangible, 141
  military, 151-154
Capitalism, 20, 90–92, 101, 106, 110, 126, 203, 244, 246, 263, 264, 267, 268, 285, 286, 306, 307, 345, 363
  growth of, 263
Carbon dioxide emissions, 314–318, 327
Caribbean, 59, 199
Cartel, 304, 332
Carter, Jimmy, 84, 107, 119, 134, 135, 168, 175, 178, 214, 219, 221, 223, 225, 228, 273
  changes view of Soviet Union, 119
  negotiating SALT II, 135
  private diplomacy after presidency, 225
Cartoons
  Auth: "Ethnic Cleansing," 349
  Azar: Clinton, Arafat, Rabin, 133
  Batchelor: "Come on in. I'll treat you right," 11
  Darling, "Eventually, Why Not Now?" 371
  Gable: "While you're at it . . . could you get us a cup of coffee?" 22
  Herblock: "Watch Out or We'll Hurl Another Threat at You," 166
  Kahil: "Give Us Back the Cold War," 203
  Low: Rendezvous, 89
  Mario: "Try the IMF," 271
  Moir: "Mummy, I Wish We Were Whales," 386
  Oliphant: "Take it Off! Take it all off!" 248
  Renault: "You're acting like a bunch . . . of capitalists!" 297
  Wasserman: "I bomb to fight technology," 74
  Wright: "Turn up the Cloud Machine," 222
Caspian Sea, 61
Castlereagh, Viscount, 112, 115
Castro, Fidel, 121, 142, 165, 272, 367, 420
Catholicism, 26, 30, 40, 41, 47, 67, 87, 98, 102, 348
Ceausescu, Nicolae, 149
Cedras, General Raoul, 107
Central American Common Market, 340
Central Intelligence Agency (CIA), 184, 217, 270, 372
  involvement in assassination plots, 201
Chamberlain, Neville, 9
Chamorro, Violeta Barrios de, 219
Charlemagne, 28, 198
Chechnya, 184, 190, 270, 347, 348, 385
Chemical arms treaty, 182
Chemical weapons, 172, 180, 182
Chernobyl, 76, 309
Chiang Kai-shek, 91,
Chile, 102, 121, 201, 256, 323, 324, 383, 420
China, Ancient, 26, 33, 46, 213, 269
China, Peoples Republic of, 89–91, 93–96, 102, 114, 126, 132, 142, 143, 144, 148, 150, 153, 156, 157, 159, 171, 172, 173, 182, 191, 210, 216, 221, 252, 263, 298, 313, 403
  and the Korean war, 405–406
  and nuclear testing, 182
  copyright and patent infringements in, 132
  economic development of, 300–303
  Great Cultural Revolution in, 145, 300
  ideology of, 93–96
  policy on Tibet of, 345, 348
  population policy of, 312, 321
  power of, 159

## Subject Index

recognition of, 216
response to World Conference on Women, 398
Chlorofluorocarbons (CFCs), 79, 320
Christianity, 18, 41, 44, 58, 87, 98, 99, 110
Churchill, Winston, 9, 72, 116, 226
Ciller, Tansu, 156
Civil defense, 168
Clinton, Bill, 107, 112, 116, 117, 132, 136, 156, 164, 182, 219, 221, 226, 256, 258, 280, 319
  Bosnian policy of, 125, 134
  effect of Vietnam on views of, 120
  policy of on North Korean nuclear program, 112
  trade policy of, 249
Cognitive consistency, 121, 122
Cognitive dissonance, coping with, 122
Collective security, 398
  and the League of Nations, 398
  bipolarity and, 405
  in Ethiopian case, 400
  failure of in the League, 400
  in the Persian Gulf War, 413
  lessons from the Persian Gulf War, 414
  requirements of, 399
  under the U.N. Security Council, 403
Colombia, 249
Colonialism, 13, 33, 38, 41, 44, 58, 72, 85, 149, 293, 295, 326, 366, 405
  ending of, 368
  positive and negative effects of, 72
Columbus, Christopher, 76
Command and control functions, 171
Commission on Human Rights, 377
Committee on the Peaceful Uses of Outer Space, 422
Commodity prices
  instability of, 299

Commons, global, 324
Common heritage of humankind, 321, 324
Common market, 253
Commonwealth of Independent States, 61, 93, 153
Communication capabilities, 71, 72, 73, 147, 157–158
Communism, 3, 11, 13, 72, 85, 87, 91, 92, 97, 101, 103, 104, 110, 122, 141, 257, 289, 300, 302, 340, 346, 386, 420
  economic reform of communist systems, 300
Communist party, 92, 93, 126, 183, 300, 301, 369
Comprehensive nuclear test ban, 182
Concert of Europe, 361
Conflict resolution, 64, 224, 344
Confucius, 9
Congress of Vienna, 42
Constantinople, 44
Contadora Group, 219
Convention Against the Taking of Hostages, 179
Convention for the Prevention and Suppression of Terrorism, 179
Convention on Climate Change, 319
Convention to Combat Desertification, 319
Conventional arms spread, 173–174
Conventional Forces in Europe Agreement (CFE), 183
  Russian desire to renegotiate, 184
Conventional weapons
  measurement of, 153
  world trade in, 173
Copernicus, 14
Council of Europe, 8, 381
Crimean War, 202, 268
Crimes Against Humanity, 391
Crimes Against the Peace, 389
Croatia, 87, 219
Cruise missile, 152, 153
  use of in Gulf War and in Bosnia, 152

Crusades, 98, 103
Cuban Missile Crisis, 208, 365, 366
  effect on Kennedy's popularity of, 134
Currency, 128, 243, 254, 260, 262, 263, 276, 277, 281, 288, 337
  convertibility, 260
  devaluation, 260, 263, 277
  hard and soft, 260
  effect of values on trade and investment, 262
Customs Union, 253, 254, 256
Cyprus, 59, 408
Czech Republic, 347, 380
Czechoslovakia, 61, 106, 121, 346–348, 369, 417
  1968 invasion of, 106, 417
  split in 1992 of, 61

Damage limiting strategies, 168
Danube River Commission, 397
Dayton Peace Accord, 220, 226, 411
De Gaulle, Charles, 88
de Klerk, F. W., 388
Debt burden, 299, 304
Decision makers, motivation of, 116
Decolonization, 47, 50, 56, 59, 214, 346
Defense budgets, 153
Deforestation, 316
Delivery Systems, 151
  continued relevance of bombers as, 153
  U.S.-Soviet compared, 153
Democracy, 3, 56, 84, 88, 90, 95–97, 102–104, 107, 109, 110, 124, 125, 198, 200, 201, 209, 225, 252, 270, 321, 334, 338, 346, 354, 380, 387
  growth of, 384
Demographic profile, 148
Deng Xiaoping, 300
Denmark, 37, 245, 254, 255, 290, 335, 393
Dependency, 70

Dependency theory, 75, 285–287
  LDC disadvantages in terms of trade, 286
Deterrence, 163–171
  communicating credible threats through, 164
  destabilizing factors in, 169
  effectiveness of, 168
  military, 163–164
  stabilizing, 171
  strategies, 166–168
Developed countries (DCs), 75, 85, 190, 203, 224, 248, 249, 255, 260, 279, 281, 282, 284, 292, 308, 309, 341
Developing countries, 148, 149, 203, 253, 286, 288, 292, 296, 297, 299, 303, 312, 314, 316, 326
Development, sustainable, 326
Dictators, 85, 113, 122, 126
Diplomacy, 96, 157, 161, 165, 212–215, 218, 220–222, 225, 228, 234, 236, 237, 293, 338, 398, 417
  changes in, 218
  pre-Westphalian, 212
  strains in modern, 214
Diplomatic immunity, 216, 359
Diplomatic recognition
  de jure and de facto, 215
Diplomats
  in consular service, 217
  ethical standards of, 214
  functions of, 216
  involvement in negotiation of, 217
  persona non grata, 216
Disarmament. *See* arms control
Disarmament Commission, 222
Disintegration of states, 345
Dollar diplomacy, 96
Domestic jurisdiction, 360
Dominica, 60
Dominican Republic, 96, 121, 419
  U.S. intervention in, 419
Domino theory, 120
Dove's dilemma, 205
Drake, Sir Francis, 84
Drug trafficking, 71, 366

Dubcek, Alexander, 417
Dulles, John Foster, 119
Dumping, 128, 251, 252

Earth Summit (Rio), 136, 218, 219, 225
East Germany, 148, 347, 417, 418
Eastern European 1989 revolutions, 92
Economic Community of West African States (ECOWAS), 340, 420
Economic determinism, 90
Economic development, 283, 305
  and the problem of scarcity, 308
  as a threat to the environment, 315
Economic imperialism, 90, 266–268, 281, 339
  role of capitalism in, 267
Economic interdependence, 278
  effect upon conflict of, 280
Economic liberalism
  defined, 243
  successes of, 302
Economic power
  disparity in, 147
Economic sanctions, 107, 160, 236, 249, 266, 270, 272–274, 281, 321, 397, 400, 401, 413, 414, 423
  against Ethiopia, 272
  against Haiti, 107
  against Iraq, 413
  blacklisting, 272
  effectiveness of, 272
  embargoes, 270
Economic self-sufficiency, 296
Economic wealth
  measurement of, 150
Ecuador, 323
Egypt, 26, 88, 99, 142, 219, 228, 269, 271, 291, 332, 406, 409
Eighteen Nation Disarmament Conference, 224
Eisenhower, Dwight D., 119, 127, 167, 204, 205, 214
El Salvador, 102, 176, 236
Embassy, 123, 134, 179, 216, 217, 272

Energy, 21, 75–77, 79, 85, 92, 123, 124, 129, 144, 158, 159, 186, 217, 220, 243, 269, 284, 286, 307–311, 321, 327, 34
Environment, 19, 68, 71, 85, 136, 161, 212, 219, 233, 236, 243, 251, 254, 259, 264, 308, 314, 315, 317, 319, 320, 321, 324–326, 344, 345, 421
Esquipulus Group, 219
Estonia, 332
Ethiopia, 50, 103, 272, 301, 400, 401, 422
Ethnic conflict, 45, 353, 387
Ethnicity, 33
Ethnocentrism, 198
Euro, 254
Europe
  between World War I and II, 50
  economic development of, 283
  medieval, 28
  new economic motors of, 351
European Coal and Steel Community, 253, 334
European Commission, 337
European Community, 63, 219, 248, 254, 255, 319, 320, 340, 344, 347, 354
European Convention for the Protection of Human Rights and Fundamental Freedoms, 380
European Defense Community, 334
European Economic Community, 253, 335
European Free Trade Association, 255, 340
European Parliament, 337
European Union, 4, 64, 65, 75, 79, 89, 105, 144, 156, 179, 213, 250, 253–256, 258, 275, 280, 298, 333, 335, 337, 338–340, 344, 351, 353, 368
  Council of Ministers of, 337
  evaluation of, 338
  historical background of, 333

membership of, 335
organization of, 335
Exclusive Economic Zone, 323
Extended deterrence, 168
Exxon Valdez oil spill, 76

F-scale (fascism scale), 114
Falklands [Malvinas] crisis, 202, 366
  misperception in, 209
Fascism, 9
Feminism, 19, 21, 24, 106
Ferdinand, King of Austria, 30
Ferdinand, Archduke, 210
Feudalism, 42, 91, 101, 244
Fiji, 59
Finite deterrence, 168
Finland, 255, 335, 400
Fishing rights
  conflict between Spain and Canada over, 368
Fixed exchange rates, 260
Flexible response, 167
Floating exchange rate, 260
Ford, Gerald R., 134, 158
Foreign aid, 114, 122, 135, 160, 236, 262, 266, 268–270, 274, 281, 288, 289, 291
  defined, 268
  Marxist view of, 268
  political purposes of, 269
  U.S. restrictions on use of, 269
  World Bank aid in, 288
Foreign direct investment (FDI), 74, 84, 93, 147, 154, 216, 249, 255, 256, 262, 263, 267, 268, 270, 275, 277, 279, 285–288, 291, 293, 300, 303, 319, 340, 344, 387, 420
  amounts involved in, 292
Foreign offices, 128
  growth of, 215
Foreign policy, 5, 7, 363
  of democratic contrasted with authoritarian regimes, 124
  effect of domestic structures on, 124
  goals of, 83
  miscalculation in, 121
  morality of, 106

Fort Sumter, 202
Fractionation, 228
France, 9, 10, 28, 30, 33, 36, 37, 42, 50, 60, 61, 71, 75, 88, 98, 103, 104, 142, 144, 149, 150, 154, 155, 157, 162, 164, 173, 196, 198, 200, 202, 208, 215, 218, 219, 221, 224, 241, 250, 253, 262, 269, 272, 278, 290, 334, 335, 337, 350, 358, 403, 405, 406, 413
  and nuclear testing, 182
  colonial map of, in 1900, 51
  terrorism in French revolution, 175
Franco, Francisco, 200
Franco-German War, 268
Franco-Prussian War, 42, 335
Frederick the Great, 42, 161, 189
Free trade area, 253
Free trade theory
  theory of comparative advantage, 285
Freedom House, 125, 201
Freedom of the seas, 322
  reasons for, 363
French Revolution, 103, 175
Frustration-aggression hypothesis, 206
Functionalism, 334, 337, 403
  spillover effect in, 337

Gadhafi, Mu'ammar Muhammad al-, 98
  U.S. aerial attack upon, 178
Galileo, 14
Galtieri, General Leopoldo, 124, 210
Gambia, 59
Gandhi, Indira, 123, 235
Gandhi, Mohandas K., 17
Gandhi, Rajiv, 56
GDP. *See* gross domestic product
General Agreement on Tariffs and Trade (GATT), 241, 251–253, 256, 258, 259, 287
General Motors, 68
Generalized system of preferences, 303

Geneva Protocol, 181, 182
Genghis Khan, 375
Geopolitics, 145–146, 351–354
German Empire, 42
Germany, 33, 49, 105, 128, 191, 253, 389
  aid and investment in, 292
  economy of, 150, 245, 280, 299
  and European unity, 333–335
  history of, 37, 42, 189
  power of, 143, 144, 149
  reunification of, 418
Germany (Nazi), 8–10, 50, 88, 120–123, 390
  world war II conquests by, 52, 121, 122
Ghana, 75
Glasnost, 92, 104, 143
Global Change Boxes
  belief systems, 104
  character of states and EU, 338
  decision making structures, 125
  diplomacy and negotiation, 218
  economic development, 299
  economic patterns, 254
  environment, 314
  human rights, 382
  intergovernmental organizations, 402
  international law, 364
  investment and finance, 277
  military force, 167
  power, 147
  20th century, 60
  world system: 1648–1919, 36
Global commons, 322
Global warming, 79, 315
Globalism, 19, 20–21, 14, 70
Goa, 368
Goals
  hierarchy of, 85
Gold standard, 244, 262, 317
Gorbachev, Mikhail, 13, 93, 104, 112, 116, 119, 155, 184, 220, 221, 227, 233, 234, 301
  and the collapse of the Soviet bloc, 369
  and the disintegration of the USSR, 346

Gorbachev, Mikhail (*cont.*)
  departure from Marxist thought of, 92
  diplomatic skills of, 156
  position on a united Germany of, 418
  unilateral initiatives of, 126, 183
Goring, Hermann, 122
Government, 4, 5, 6, 35
  world, 357
Grachev, Pavel, 184
Gramsci, Antonio, 9
Great Depression, 11
Great man theory, 9
Great Powers
  intervention by, 362
  rise and fall of as result of military spending, 154
  role in the global order of, 361
  since 1700, 144
Greek city-states, 162
Greenhouse effect, 314, 315
Greenpeace, 4, 272
Grenada, 60, 165
  1983 invasion of, 96
Gross domestic product, 159, 270, 273, 290, 321
  compared to production of MNCs, 69
  defined, 150
  in 1994 and 2020 (projection), 150
Gross world product
growth of, 310
Grotius, Hugo, 16, 322
Group of Seven, 150, 219, 341
Group of 77, 295, 297, 304
Groupthink, 130
Guantanamo, 229
Gulf Oil Co., 277
Guyana, 59

Hague Peace Conferences, 390
Haig, Alexander, 132
Hamilton, Alexander, 242
Han dynasty, 46
Hapsburg empire, 37, 40, 44, 49, 61, 215

Haushofer, Karl, 147
Heartland, 146, 147
Hegemonic stability theory, 243, 417
Helsinki process, 418
Helsinki Watch Committee, 419
High Commissioner for Human Rights, 380
Hijacked aircraft, 178
Hinduism, 98, 345
Hiroshima, 9, 151
Historical determinism, 91
Historical experiences
  effect on foreign policy beliefs, 119
Hitler, Adolph, 8–10, 50, 88, 97, 103, 117, 120–122, 145–147, 155, 159, 166, 182, 192, 198, 272, 277, 375, 390, 400–401
  attack on Poland and Czechoslovakia by, 121
Holy Roman Empire, 28, 30, 33, 37
Honduras, 165
Hong Kong, 144, 298, 303
Hot line, 171
Human rights, 66, 84, 136, 156, 269, 281, 366, 372, 374–383, 385–389, 391–395
  creating international standards of, 375
  enforcement of, 377
  regional efforts for, 380
  World Conference on (1993), 380
Human Rights Committee, 377
Human rights treaties
  list of, 378
Human Rights Watch, 380
Hungary, 347
Hussein, Saddam, 25, 99, 101, 120, 124, 165, 172, 182, 191, 192, 205, 225, 249, 268, 348, 349, 362, 363, 366, 368, 387, 413, 414
  miscalculation of U.S. resolve, 158
Hutu, 393
Hypothesis, 14

I-FOR, 411, 413
ICBM. *See* intercontinental ballistic missile
Iceland, 145
Idealism, 16, 17, 24, 106, 385
Ideology, 49, 87, 89–91, 93, 94, 96, 102, 105, 106, 108–110, 154, 163, 213, 215, 237, 264, 304, 306, 387
Idiosyncratic variables, 112–124
  when they make a difference, 113
Imperialism, 44, 47–50, 55, 56, 58, 60, 61, 86, 90, 197, 211, 266–268, 281, 339
  and decolonization after World War II, 50
  and demise of European colonialism, 41
  European, 38, 58
  Soviet, 105
Inca, 33
India, 26, 38, 47, 67, 88, 123, 142, 148, 150, 171, 173, 182, 190, 199, 202, 298, 314, 345, 368, 369
  map of—1900 and today, 57
  population policy of, 321
  profile box of, 56
Indonesia, 116, 150, 219, 298, 305
Industrial revolution, 35, 76, 245, 283, 284, 286, 308, 311, 314, 315
Infopolitics, 353
Integration, 68
  defined, 332
Intelligence capabilities, 156–157
Intercontinental ballistic missiles, 151, 152, 180
  Minuteman III, 220
  MX, 152
Interdependence, 136, 271–281, 238–230, 221, 344
  defined, 70, 332
  of security, 70
Intergovernmental organizations (IGOs), 7, 20, 63–66, 293, 342–344, 357, 396, 402
  challenge to sovereignty of, 396
  defined, 63
  effect on foreign policy of, 65

## Subject Index

Intermediate Nuclear Force Treaty (INF), 180
International Bank for Reconstruction and Development (IBRD), 260, 287, 288
International Civil Aviation Organization (ICAO), 421
International Commission of Jurists, 380
International Court of Justice (ICJ), 357, 370, 372
International Covenant on Civil and Political Rights, 376
International Covenant on Economic, Social and Cultural Rights, 376
International criminal court, 393
International Development Association (IDA), 288
International economic system
  effect of the private sector upon, 275
International Finance Corporation (IFC), 287, 288
International Labor Organization (ILO), 4
International law, 33, 355
  and foreign policy, 363
  changes in subjects of, 364
  codified and customary, 359
  compared with domestic law, 356
  compliance due to threat of retaliation, 368
  importance of reciprocity in, 359
  in conflict with foreign policy, 367
  monist and dualist views of, 355
  natural, 356, 391
  positive, 355, 391
  purposes of, 357
  rights and duties of states in, 358
  uses in foreign policy of, 364
International Monetary Fund (IMF), 254, 260, 265, 275, 287, 288
International monetary system, 259–261

International organizations. *See* intergovernmental organizations
International Physicians for the Prevention of Nuclear War, 66
International political economy
  perspectives on, 241
International regime, 20
  building, 421
International Seabed Authority (ISA), 324
International Telecommunications Satellite Organization (INTELSAT), 402
International Telecommunications Union (ITU), 63, 402
International Telegraph Union (ITU), 402
International terrorism
  frequency, 176
  how to deal with, 177
  state-sponsored, 176
International Trade Organization (ITO,) 251
Internet, 351
INTERPOL, 402
Intervention
  humanitarian, 107
Investment. *See* foreign direct investment
Invulnerable retaliatory capability, 171
Iran, 89, 98, 99, 103, 110, 112, 122, 131, 162, 173–176, 178, 182, 185, 190, 191, 223, 272, 333, 346
Iran-contra affair, 131, 178, 223
Iran-Iraq war, 89, 346
Iranian hostage situation, 99, 123, 141, 175, 179, 223
  effect on President Carter's popularity of, 134
Iraq, 25, 89, 90, 99, 124, 162, 172–174, 176, 182, 186, 188, 191, 192, 201, 249, 272, 273, 333, 345, 346, 348, 362, 366, 408, 413, 414
Ireland, 61, 98, 236, 254, 335, 345, 348, 380, 388

Irish Republican Army, 236
Iroquois, 33
Islam, 44, 99–101
  disunity in, 99
Islamic beliefs, 98–101
Islamic fundamentalism, 174
Isolationism, 95–96
Israel, 114, 132, 144, 165, 172, 178, 182, 190, 191, 224, 228, 232, 236, 269, 271, 272, 291, 346, 366, 388, 406, 409
  and terrorism, 178
  bombing of Osirak reactor by, 172
Italian city-states, 213
Italy, 12, 33, 37, 47, 50, 105, 142, 144, 150, 156, 219, 253, 272, 290, 335, 400, 401, 408
  and the Ethiopian crisis, 400

Jackson, Jesse, 225
Japan, 9, 10, 44, 50, 53, 67, 105, 142, 144, 145, 147, 148, 150, 156, 199, 200, 205, 219, 241, 245, 250, 257, 258, 263, 270, 277, 278, 280, 283, 290, 298, 299, 303, 312, 341, 382, 389, 400, 401
  decline in militarism of, 200
  opening to the West of, 44
*Jihad*, 98
John Paul II, Pope, 102
Johnson, Lyndon B., 106, 121, 178, 220, 419
  appeasement concerns of, 120
  and Dominican Republic invasion, 121
Joint Chiefs of Staff, 220
Jordan, 99, 178, 269, 291

Kahn, Ayub, 116
Kautilya, 213
Kazakhstan, 171
Keiretsu system, 258
Kellogg-Briand Treaty, 390
Kenya, 33
KGB, 217

Khomeini, Ayatollah, 98, 99, 123
Khrushchev, Nikita, 91, 92, 116, 126, 156, 164, 165, 173, 234
  peaceful coexistence policy of, 126
Kim Il Sung, 112
Kim Jong Il, 112
Kissinger, Henry A., 17, 128, 129
  view of U.S.-Soviet negotiating behavior, 233
Koran, 98
Korean War, 114, 196, 210, 405
  U.S. misperception in, 210
Kumaranatunga, Chandrika, 235
Kurds, 47, 333, 345, 348, 392
Kuwait, 25, 99, 120, 124, 149, 158, 162, 180, 190–192, 201, 249, 268, 272, 273, 291, 295, 346, 362, 366, 413, 414, 423

Laissez-faire, 95, 243, 307, 322
Landmines, 174
Latin America, 85, 96, 181, 199, 201, 202, 214, 338, 381, 419, 420
Latin American Free Trade Association, 340
Latin American Integration Association (LAIA), 340
Latvia, 332
Law of the sea, 322
  changes in, 364
  expansion of territorial limits in, 323
  mining operations in, 324
Law of the Sea Treaty (1982), 323, 327
League of Nations, 12, 60, 64, 96, 179, 272, 333, 370, 389, 390, 398–401, 405, 422
  Ethiopia case, 400–401
  origin and development of, 64
  successes of, 401
Lebanon, 61, 132, 176, 178, 3 46
  invasion of, 116
*Lebensraum*, 103, 147
Lenin, Vladimir, 38, 91–93, 267, 268, 300–302,
  adaptation of Marxism by, 91
  theory of imperialism of, 267
Less developed countries (LDCs), 85, 190, 203, 224, 248, 260, 274, 279, 281, 282, 286
  power of, 278
Levels of analysis, 8, 13, 23, 192
Liberal democracies, 84, 109
  rise in numbers of, 201
Liberal-democratic beliefs, 95–98
Liberal economic order,
  dominant position of, 287, 302–303
Liberia, 420
Libya, 99, 173, 175, 176, 178, 185
Limited Test Ban Treaty, 182, 229
Lincoln, Abraham, 202
Li Peng, 398
List, Friedrich, 242
Lithuania, 332
Locarno Treaties, 389
Locke, John, 95, 110, 360, 416
London Charter, 390, 391
Long postwar peace, 192
Lorraine, 42, 335
Louis XIII, 157
Louis XIV, 84, 157
Ludwig, King of Bavaria, 116
Luther, Martin, 157
Luxembourg, 253, 335

Maastricht Treaty, 255, 337, 338
Machiavelli, Niccolo, 15, 17
Macmillan, Harold, 120
Madrid Protocol, 325
Maginot Line, 10
Mahan, Admiral Alfred T., 146
Managed trade, 257
Manchurian affair, 400
Mandela, Nelson, 272, 389, 397
Manifest destiny, 84
Mao Zedong, 93, 116, 300
  digression from Marxist thought by, 93
Marcos, Ferdinand, 85
Marshall Plan, 289, 290
Martyrdom, 99
Marx, Karl, 9, 90–93, 104, 244, 267, 302

Marxism, 90–95, 109
  critique of economic liberalism in, 244
Marxist-Leninist beliefs, 90–95
Masai, 33
Mass media, 73, 133, 135
Massive retaliation doctrine, 167
McFarlane, Robert, 132
McNamara, Robert S.
  *mea culpa* on Vietnam of, 209
Mediator, 223
Meir, Golda, 114, 235
Mercantilism, 84, 242–244, 264
  defined, 242
Mesopotamia, 25
Metternich, Klemens von, 42, 112
Mexico, 26, 33, 46, 67, 75, 105, 132, 150, 253, 255, 298, 299, 312, 314, 332, 340, 341
  economic bailout of, 132
Michelangelo virus, 71
Middle Ages, 28, 41
Migration
  right of, 382
Militarism, 9
Military aid, 270
Military power, 151–154
  functions of, 161
  problems in measurement, 153
Military weapons
  overseas sales of, 205
Military-industrial complex, 90, 198, 203–206, 211
Mill, John Stuart, 243, 278
Milosevic, Slobodan, 273
Mineral resources, 149
Minorities
  numbers of threatened by violence, 199
MIRV. *See* multiple-independently targeted reentry vehicle
Misperception, 103, 119, 123, 124, 192, 206, 208–210, 212
  effect on foreign policy of, 123
  reasons for, 208
Missile Technology Control Regime, 173, 402
Mitsubishi Corporation, 63, 68
Mitterrand, Francois, 221
Modernization theory, 284

## Subject Index

Moghul empire, 56
Monetary system, 240, 243, 254, 259, 264
  and currency valuation, 260
Mongella, Gertrude, 398
Monroe Doctrine, 96
Montreal Protocol on Substances that Deplete the Ozone Layer, 320
Moody's, 147
Moralism, 96–8
Morality in foreign policy, 106–109, 205
Most favored nation (MFN), 252
Mozambique, 102, 236, 301
Mugabe, Robert, 121
Muhammad, 98
Multiculturalism, 47
  compared with nationalism, 105
Multinational corporations, 67–70, 75, 276, 279, 281, 296, 341, 344
  effect upon international integration of, 344
  exploitation of LDCs by, 279
  growth of, 67
  power compared to states of, 68, 277
  undermining of sanctions by, 272
Multiple-independently targeted reentry vehicle (MIRV), 152
Multipolarity, 192–195
Munich Conference, 120
Muslims, 56, 85, 87, 99, 110, 116, 226, 348, 362, 411
  map showing location of, 100
Mussolini, Benito, 9, 88
  attitude on population growth of, 149
Mutual assured destruction (MAD), 167
Myanmar (Burma), 72

NAFTA. *See* North American Free Trade Agreement
Nagasaki, 9, 151
Napoleon, 37, 42, 97, 167, 197, 198, 202, 228
Napoleon III, 42

Napoleon, Louis, 202
Narco-terrorism, 71
Nasser, Abdul, 88, 142, 406, 409
Nation-state(s)
  contrasted with other actors, 32
  evolution of, 37
  foundations of, 30
  micro states as, 59
  new since 1945, 54
National interests, 84–6
National morale, 158
National security
  continuing threats to, 171–179
Nationalism, 11, 30, 33, 37, 42, 44, 45, 48–50, 56, 60, 61, 79, 84, 88, 99, 105, 106, 109, 110, 198, 209, 211, 245, 280, 334, 339, 345, 346, 350
  as a cause of war, 198
  as a response to imperialism, 61
  compared with multiculturism, 105
  defined, 45
  positive aspects of, 198
Nationalization of property, 367
NATO. *See* North Atlantic Treaty Organization
Natural law, 356, 391
Nauru, 45, 60
Need for affiliation, 117
Need for power, 117
Neoliberalism, 19, 21, 70
Neorealism, 19–20, 24
Negotiating style
  Asian, 234
  high and low context, 233
  Japanese, 233
  of women, 234
Negotiation
  approach-avoidance bargaining in, 230
  bilateral and multilateral, 223
  effect of negotiating structures on, 217
  impact of domestic politics on, 225
  mediator role in, 223
  Middle East, 228

non-agreement objectives of, 227
  open and closed, 221
  reciprocity of concessions in, 228
  role of non-state actors in, 225
  shuttle diplomacy in, 228
  summit, 218
  used for propaganda, 222
Nehru, Jawaharlal, 88, 142
Neomercantilism, 242
Neorealism, 19
Netherlands, 30, 33, 36, 40, 50, 144, 148, 150, 253, 269, 290, 292, 335, 337, 361, 370, 408
New International Economic Order (NIEO), 295, 296, 306
New Zealand, 38
  independence of, 39
Newly industrialized countries (NICs), 297–299
Nicaragua, 96
Nigeria, 58, 59, 158, 291, 314
  ethnic map of, 59
  profile of, 58
Nixon, Richard M., 135, 205
  and arms control, 220
  summit negotiations of, 221
  suspension of dollar convertibility by, 262
  view of bureaucracy of, 129
  visit to China by, 93, 126, 216
Nobel Prize for Peace, 66
Nobel, Alfred, 167
Nongovernmental organizations (NGOs), 66–68, 79, 310, 325, 357, 380, 398
  defined, 66
Nonstate actors, 4, 23, 63, 79, 147, 218, 225, 226
  growth of, 63
Nonaligned Bloc, 219
Nontariff barriers, 253
Noriega, Manuel, 366
Normative Dilemmas Boxes
  challenge of the information society, 345
  communications revolution, 72

Normative Dilemmas Boxes *(cont.)*
  environmental threats compete with other values, 321
  impact of European colonialism, 72
  in negotiation, 226
  in the use of violence, 205
  in U.S. decision to intervene in Haiti, 107
  in using economic leverage, 274
  making law conform to social norms, 368
  normative challenges to realpolitik, 155
  regarding human rights, 388
  self-interested judges, collective goods and police power, 416
  social problems with development, 305
  the use of force, 174
  What foreign policy choice to make? 132
  winners and losers in trade agreements, 259
Normative values
  defined, 86
Norms, 86, 107, 108, 130, 215, 236, 272, 280, 350, 351, 357, 359, 363, 365, 368, 372, 392, 421
North American Free Trade Agreement (NAFTA), 105, 135, 255–258, 340, 341
North Atlantic Treaty Organization (NATO), 4, 64, 89, 128, 136, 152, 164, 168, 174, 181, 183, 219, 227, 236, 281, 350, 381, 411, 413, 415–419, 423
North Korea, 95, 158, 173, 175, 176, 182, 406, 408
  nuclear agreement with U.S., 269
North, Oliver, 112, 131
North-south relations, 282
  economic gap in, 283
Northern Ireland, 345, 380
Nuclear Non-Proliferation Treaty (NPT), 182
Nuclear proliferation, 172–173

Nuclear stability, 168–171
  effect of the disintegration of the USSR on, 171
Nuclear weapons
  measurement of explosive power of, 151
  stockpiles of, 151
Nunn, Sam, 107
Nuremberg, 375, 382, 389–392, 395
Nye Commission, 204

Official development assistance (ODA), 289, 291, 293
  amounts provided in, 290
Oil consumption, 308
OPEC (Organization of Petroleum Exporting Countries), 75, 128, 144, 149, 290, 304, 332
  oil boycott by (1970s), 75, 149, 304
Open Door policy, 96
Organization for Economic Cooperation and Development (OECD), 295
Organization for Security and Cooperation in Europe (OSCE), 381, 418, 419
Organization of African Unity (OAU), 64, 379, 381, 420
Organization of American States (OAS), 63, 179, 379, 381, 419, 420
Ortega, Daniel, 219
Ottoman empire, 29, 33, 44, 45, 49, 61, 70, 99, 162, 197, 346
  collapse of, 45
Outer space
  changes in law of, 364
  Outer Space Treaty, 422
Oxfam, 4
Ozone layer, 76, 79, 85, 315, 319, 320, 325, 327
  destruction of, 319
  hole in, 76

Palestine, 33
Palestinians, 47, 178, 232, 235
Panama, 96, 135, 158, 229, 366
Panama Canal Treaties, 135, 229

Paris Four-Power summit (1960), 220
Park, Chung Hee, 85
Parliamentary diplomacy, 221
Partnership for Peace, 418
Peacekeeping, 174, 402, 406, 408–412, 415, 420, 421, 423
  after Cold War, 410–413
  compared with collective security, 408
  growth of in U.N., 402
  map of U.N. peacekeeping operations, 411
  purposes of, 408
  regional efforts at, 415
Pearl Harbor, 9, 50, 123, 130, 210
Perception, 119
Perestroika, 92, 301
Permanent Court of International Justice, 370
Perot, Ross, 135
Persian Gulf War, 86, 97, 158, 162, 165, 182, 186, 188, 191, 205, 268, 348, 362, 366, 387
  casualties in, 188
  collective action against Iraq in, 362
  impact upon Iraqi people of, 387
Persona non grata, 216, 359
Personal expectations
  effect on foreign policy views of, 121
Personalities
  compatibility between, 116
  effect on foreign policy of, 113
Peru, 26, 323
  war with Ecuador, 190
Pétain, Philippe, 9
Peter I (the Great), 44
Philippines, 85, 123, 146, 383
Plato, 9, 267
PLO (Palestinian Liberation Organization), 178, 228, 366
Pluralism, 19, 21
Pol Pot, 375
Poland, 13, 33, 50, 120, 121, 146, 216, 347, 389, 390, 417

Political economy, 7, 238–240
Pollution, 21, 76, 77, 155, 309, 322, 324, 369
Pope, 28
Population, 9, 12, 35, 40, 47, 49, 50, 56, 58, 61, 74, 88, 109, 122, 133, 135, 141–143, 145, 151, 155, 158–160, 167, 168, 189, 198, 199, 201, 210, 211, 216, 263, 273, 299, 300, 306–314, 317, 321, 323, 326, 327, 339, 344, 364, 382, 387, 390, 391
  growth, 76, 77–78
  growth by region, 313
  growth since 1 A.D., 312
  impact upon capabilities of, 148–149
  problem of overpopulation, 149
  problems of excess growth of, 311–314
Portugal, 28, 33, 36, 40, 44, 50, 61, 148, 254, 255, 335, 368
Positive law, 355, 391
Post-industrial economies, 105
Poverty, 287
Powell, General Colin, 107
Power
  effect of domestic events on, 145
  geopolitical sources of, 145
  inequality of states with regard to, 36
  intangible sources of, 154
  meaning of, 141
  measurement of, 159
  politics, 18, 19
Power transition model, 194
Preventive strike, 166
Primary products, 75
Princip, Gavrilo, 4
Proletariat, 90, 244
Propaganda, 88, 157
Protectionism, 11, 247–249, 259
Protestant Reformation, 30
Protestantism, 40, 47, 98, 348
Prussia, 37
Public diplomacy, 157
Public opinion
  impact on foreign policy of, 133
  manipulation of, 134

Quebec, 33
*Quebecois*, 333
Quotas, import and export, 250

Rabin, Yitzak, 228, 235
Racism, 9
Radioactive fallout, 79
Rainbow Warrior, 272
Rann of Kutch, 369
Rational actor model, 111–112
  advantages and disadvantages of, 112
  defined, 111
Raw materials
  scarcity, 309
Reagan, Ronald, 13, 72, 113, 118, 123, 143, 156, 165, 167, 170, 178, 183, 205, 221, 227, 269, 273, 280, 370, 420
  calls Soviet Union an evil empire, 218
  defense policy of, 154
  image of Soviet Union, 119
  impact of bureaucracy on, 132
  Reykjavik summit meeting of, 220
Realism, 14–16, 18, 19, 24, 106, 108, 164, 242, 334
Reasons of state, 108
Reciprocal Trade Agreements Act, 251
Reciprocity, 322, 359, 360, 363
Red Cross, 63, 67
Regimes, 20, 66, 98, 121, 124–127, 137, 198, 200–202, 301, 302, 325, 345, 383, 417, 421–423
Regional organizations
  membership in, 342
Reification, 10
Religion
  positive functions of, 101
  proselytizing efforts of, 84
  schisms in, 99
Reprisals, 216, 368
Revisionism, 11
Reykjavik summit, 220

Ricardo, David, 136, 246
Rio Earth Summit. *See* Earth Summit (Rio)
Risk-averse behavior, 117, 194
Roman empire, 26, 213
Romania, 61, 92, 149, 347
Roosevelt, Franklin D., 9, 10, 60, 116, 126, 127, 135
Roosevelt, Theodore, 146
Root, Elihu, 219
Rusk, Dean, 220
Russia (*see also* Soviet Union), 13, 33, 36, 41, 44, 61, 71, 97, 129, 142, 144, 146, 149, 150, 156, 168, 170–173, 180, 184–186, 190, 197, 202, 208, 267, 270, 300, 301, 340, 347, 348, 400, 403, 413, 418
Russo-Japanese war, 202, 268
Rwanda, 45, 392

Sadat, Anwar, 99, 135, 219
SALT I, 181, 220
SALT II, 119, 135, 181, 183, 214
San Marino, 59
Sanctions. *See* economic sanctions
Sandinistas, 143, 370
Saudi Arabia, 290, 295
Schlesinger, James, 126
Schmidt, Helmut, 221
Scud missiles, 174
Security-community
  pluralistic and amalgamated, 339
Security dilemma, 174
Security studies, 7, 138–139
Selective perception, 122
Self-determination, 50, 84
Self-actualizer, 114
Self-esteem, effect on foreign policy of, 114
Separatist movements, 345
  in the Soviet Union, 346
Serbia, 87, 208, 210, 219, 272, 273
Seven Years War (1756–1763), 189
Seward, William, 202
Seychelles, 60

## Subject Index

Sherman, General William Tecumseh, 175
Shevardnadze, Eduard, 219, 234, 362
Shiites, 98
Short-range attack missiles (SRAMs), 153
Sikhs, 345
Singapore, 143, 298
Sinhalese, 345
Sinn Fein, 236
Six-Day war, 409
SLBM. *See* submarine-launched ballistic missile
Slovakia, 347
Small powers
  contrasted with great powers, 39
Smith, Adam, 95, 307
Social Darwinism, 147
Somalia, 61, 165, 188
South Africa, 102, 149, 219, 249, 272, 274, 281, 299, 360, 387, 388, 389, 397
  human rights treatment in, 387
South Korea, 85, 134, 269, 298, 383, 405
Sovereignty, 20, 32, 35, 36, 38, 39, 47, 48, 68, 70, 140, 147, 163, 166, 229, 236, 242, 253, 255, 275, 276, 326, 327, 333, 355–358, 364, 372, 385, 395–397
  and state equality, 33
  requirements of, 35
Soviet Bloc, 12, 347
  map of successor states of, 62
Soviet Union (*see also* Russia), 3, 9, 10, 12, 13, 45, 49, 50, 66, 88, 91, 92, 93, 97, 116, 119, 126, 135, 141, 143–147, 150, 153, 154, 155, 157, 159, 160, 166–168, 171–173, 180, 183, 184, 190, 191, 193, 194, 198, 199, 205, 214, 216, 217–221, 223, 224, 230, 232–234, 252, 260, 270, 272, 273, 300, 301, 332, 338, 345, 346, 347, 365, 369, 381, 400, 403, 405, 406, 416–418, 422, 423

collapse of, 61
effect of World War II upon, 148
relations with China, 89
Spain, 33, 36, 44, 45, 50, 61, 98, 104, 144, 150, 154, 200, 202, 254, 290, 333, 335, 345, 368
Spanish-American War, 268
Sparta, 200
Special Drawing Rights (SDRs), 262
Sputnik, 422
Sri Lanka, 98, 214, 235, 345
St. Kitts, 60
Stalin, Joseph, 9, 91, 116, 135, 173, 226, 234, 300, 375, 405
  retreat from Marxist dogma of, 91
Standard Oil Company, 63
START I, 181, 184
START II, 181, 184, 185
  nuclear forces allowed by, 185
Stereotypes, 114, 130
Stinger missiles, 174
Strategic Arms Limitation Talks. *See* SALT I and II
Strategic Arms Reduction Talks. *See* START I and II
Strategic Defense Initiative (SDI), 113, 167, 170, 220, 227
Strauss, Robert, 225
Subcommission on Prevention of Discrimination and Protection of Minorities, 379
Submarine-launched ballistic missiles, 152, 181, 183, 185
Subsidies, 257
Subsystem, 12
Suez crisis, 406
Suez invasion, 164, 225
Sukarno, 116
Summit negotiations, 218–221
  advantages and disadvantages of, 220
  U.S.-Soviet, 119, 221, 227
Sunni, 99
Super 301, 258
Supranational organizations, 64
Sustainable development, 326

Sweden, 30, 33, 144, 198, 245, 255, 290, 291, 298, 335, 377
Switzerland, 30, 40, 67, 98, 145, 201, 255, 292
  history of, 40
Syria, 70, 99, 162, 175, 176, 225, 228, 236, 269, 332

Taft, William Howard, 96
Taiwan, 144, 150, 173, 263, 298, 303
Talleyrand, Charles de, 157
Tamils, 345
Tanzania, 33
Tariffs, 247, 249–255, 257, 258, 264, 295, 335
Territory, 145
  advantages of large, 145
  effect of, 146
  effect of location of, 146
Terrorism
  history of, 175
  misconceptions about, 176
Thailand, 65, 298
Thatcher, Margaret, 116–118, 129, 156, 221, 235, 280, 366
  criticism of U.S. budget deficits by, 280
  view of foreign policy bureaucracy of, 129
  view of Gorbachev of, 116
Third U.N. Conference on the Law of the Sea (UNCLOS III), 323
Third world, 72, 95, 180, 211, 299, 305, 312, 324, 339, 345, 354
Thirty Years War (1618–1648), 30, 34–36, 40, 42, 44, 47, 49, 71, 98, 189, 226, 333
Thothmes II, 25
Threats, 164–5
Tiananmen, 3, 94, 126, 252, 301
Tibet, 33, 345
Tito, Josip Broz, 142, 200
Toughness dilemma, 226
Tower Committee, 131
Trade, 245–259
  debate over free, 245
  increase in, 244

## Subject Index

preferences, 295
regional organizations, 253
retaliation, 258
subsidies, 257
the case for protectionism in, 248
U.S.-Japanese, 257
Trade negotiations, 224
  Kennedy Round, 230, 252
  Tokyo Round, 225, 252
  Uruguay Round, 252, 256
Trade protectionism
  import quotas, 250
  nontariff barriers, 250
  tariff, 249
  voluntary export restraints (VERs), 257
Transfer of technology problem, 296
Travel, speed of, 70
Treaties, 359
  growth in numbers of, 215
  pacta sunt servanda in, 360
  rebus sic stantibus in, 360
Treaty of Rome, 339
Treaty of Westphalia, 30, 37, 42, 47, 70, 98, 139, 338, 355, 358
Tripoli, 70
Truman, Harry S., 9, 120, 127, 134, 135, 251, 323
Tunis, 70
Turkey, 33, 36, 44, 49, 103, 127, 144, 156, 171, 291, 333
  joins the European system, 44
Tutsi, 393
Tutu, Bishop Desmond, 102
Twentieth Party Congress, 126

U Thant, 409
U.N. Charter, 375
U.N. Conference on the Environment and Development (UNCED), 319 (*see also* Earth Summit)
U.N. Conference on Trade and Development (UNCTAD), 295
U.N. Development Program (UNDP), 293, 421
U.N. Economic and Social Council (ECOSOC), 377, 379, 402
U.N. Emergency Force (UNEF), 409
U.N. Environmental Program (UNEP), 318
U.N. Force in Cyprus (UNFICYP), 408
U.N. General Assembly, 377
U.N. peacekeeping operations, 406–412
U.N. Protection Force in Bosnia (UNPROFOR), 411
U.N. Secretary-General, 398, 409, 415
U.N. Security Council, 361, 401
  economic boycott of Iraq, 201
U.N. Women's Conference (Beijing), 380, 398
Uganda, 33, 249
Ukraine, 171, 347
UNESCO (United Nations Educational, Scientific, and Cultural Organization), 379
United Arab Emirates, 291, 295
United Arab Republic, 332
United Fruit Company, 96, 278, 344
United Nations
  effectiveness of, 409
  organizational chart of, 404
  origin of, 64
  peacekeeping by, 406–412
  permanent members of, 361, 403
United Nations Charter
  Article 33, 224, 403
  Military Staff Committee, 414
United States
  assistance to multinational corporations, 278
  effect of 1973 oil boycott on, 149
  efforts to affect domestic economies of others by, 280
  interventionism of, 95
  isolationism of, 95
  joins the Westphalian system, 44
  moralism in foreign policy of, 96
  use of foreign aid for political purposes by, 269
Uniting for Peace resolution, 405
Universal Postal Union (UPU), 63, 65, 402
Universal Declaration of Human Rights, 375, 394
Ural Mountains, 61
USSR. *See* Soviet Union
U.S. Department of Defense, 205, 223
U.S. Department of State, 115, 128, 223
U.S. Trade Representatives Office, 226

Vance-Owen plan, 226
Vasco da Gama, 56
Venezuela, 75Il:
Versailles Treaty, 42, 49, 398
Vienna Convention for the Protection of the Ozone Layer, 320
Viet Cong, 226
Vietnam, 17, 33, 90, 95, 97, 112, 120, 134, 141, 142, 158, 188, 193, 209, 274, 362, 363, 392, 418
Vietnam syndrome, 158
Vietnam war, 112, 120, 134, 418
  U.S. bombing, 112
Voluntary export restraints, 257

War, 187–211
  bipolarity and, 192
  by type and region, 187, 190
  casualties since 1815 in, 189
  civil, 190–191
  civilians killed in, 187
  effect of democracy upon, 200
  multipolarity and, 193
  statistics of, 187
  systemic explanations of, 192
  war crimes, 390
War crimes trials, 382, 389–394
  former Yugoslavia, 350, 392, 395
  future of, 394
  Nuremberg, 389, 395
  Rwanda, 392, 395

War crimes trials (*cont.*)
  Tokyo, 389, 395
War, causes of, 191–210
  arms races and, 195–197
  democracy and, 200–201
  domestic instability and, 201–201
  human aggression and, 206–208
  military-industrial complex and, 203–205
  miscalculation in, 208–210
  nationalism and, 197–199
  power polarity and, 192–195
Warsaw Pact, 106, 128, 153, 181, 183, 270, 381, 415, 417, 418, 423
Washington Conference on Naval Limitation, 157
Water resources, 314
Watergate, 135
Wealth, 13, 72, 74, 101, 149, 150, 211, 240, 242–245, 264, 275, 282, 291, 307, 310, 324, 326, 353, 367, 387
Weighted voting, 256, 289
Wessin y Wessin, General, 121
West Africa, 26
Westphalia, 25, 32, 34, 35, 36, 45, 50, 237, 350, 385
Westphalia, Peace of, 30, 37, 47, 70, 139, 355
Western Europe, 25, 34, 65, 70, 72, 76, 91, 146, 199, 249, 283, 311, 334, 337, 338, 341, 345, 347, 354, 382

Wilhelm I, Kaiser, 42
Wilson, Woodrow, 17, 84, 104, 115, 198, 218, 399
  frailties of, 117
  and the League of Nations, 96, 221
  view on national self-determination, 198
Win-win outcome, 230
Women, 21, 148, 149, 159, 234, 235, 287, 377–380, 398
  negotiating style of, 234–236
  rights of, 364, 380, 382
World Bank. *See* International Bank for Reconstruction and Development
World Court. *See* International Court of Justice
World Health Organization (WHO), 299
World Intellectual Property Organization (WIPO), 402
World Meteorological Organization (WMO), 319, 402
World order, 19
World Trade Organization (WTO), 252, 259
World War I, 4, 10, 18, 36, 37, 41, 42, 44, 45, 50, 63, 64, 66, 84, 87, 96, 97, 162, 170, 182, 188, 204, 210, 218, 221, 245, 268, 333, 335, 346, 391, 398, 399
  role of misperception in, 208

  U.S. entry into, 49
World War II, 5, 8–12, 16, 18, 24, 44, 50, 58, 63, 64, 66, 85, 88, 90, 91, 97, 98, 110, 120, 126, 128, 134, 146, 148, 155, 156, 163, 165, 186–188, 196, 215, 229, 243, 245, 251, 252, 254, 257, 272, 274, 280, 283, 287, 288, 300, 302, 312, 333–335, 337, 338, 346, 358, 366, 375, 380, 387, 389–391, 393–395, 398, 401, 403, 409, 417, 419, 420, 423
Worldwatch Institute, 310

Yalta, 226
Yalta Conference, 226
Yalu River, 406
Yeltsin, Boris, 93, 116, 184, 221, 234, 385
Yen, 254, 260, 263
Yom Kippur war, 235, 269, 271
Yugoslavia, 49, 61, 87, 102, 142, 188, 190, 198–200, 345, 346, 349, 382, 387, 392, 393, 395, 414

Zaire, 249
Zambia, 249
Zero-sum, 243
Zhou Enlai, 221
Zia ul-Haq, General Mohammad, 98